Prealgebra

SECOND EDITION

Prealgebra

Richard N. Aufmann
Palomar College, California

Vernon C. Barker
Palomar College, California

Joanne S. Lockwood
Plymouth State College, New Hampshire

HOUGHTON MIFFLIN COMPANY
Boston New York

Senior Sponsoring Editor: Maureen O'Connor
Associate Editor: Dawn Nuttall
Associate Project Editor: Tamela Ambush
Editorial Assistant: Jodi O'Rourke
Senior Production/Design Coordinator: Carol Merrigan
Senior Manufacturing Coordinator: Florence Cadran
Marketing Manager: Sara Whittern

Cover design: Deborah Azerrad Savona
Cover image: France, Burgundy, rustic ivy-clad house front with blue shutters. Michael Busselle for Tony Stone Worldwide Images.

Printed in the U.S.A.
Library of Congress Catalog Card Number: 97-72435

ISBN Numbers:
Student Text: 0-395-87093-3
Instructor's Annotated Edition: 0-395-87094-1

3456789-CK-01 00 99

Contents

CHAPTER 1
APPLICATIONS

The Arts, 16; Astronomy, 16, 17; Aviation, 18, 65; Business, 39, 65, 81, 82; Computer Science, 64, 65; Construction, 82; Consumerism, 16, 39; Cost of Living, 63; Demography, 37; The Economy, 82; Education, 18, 33, 72; Finances, 38, 39, 65, 72; Geography, 16, 17, 39, 72; Geology, 36; Health, 17, 38, 59; History, 16, 38; Investments, 66; Mathematics, 37; Nutrition, 16, 38, 64; Physics, 18, 40; Sports, 16, 38, 64, 82; Statistics, 40; Temperature, 72; Travel, 17, 66, 72, 82

4 *Decimals and Real Numbers* *223*

CHAPTER 4
APPLICATIONS

Accounting, 260, 261, 266; Astronautics, 276; Business, 233, 252, 256, 260, 265, 266, 285, 292, 294; Chemistry, 292; Community Service, 285; Computers, 286; Consumerism, 234, 258, 259, 261, 266, 292, 294; Cost of Living, 265, 266; Demography, 260; Earth Science, 275; Economics, 233, 234; Education, 285, 286, 292; Environmental Science, 260; Exchange Rates, 254; Finances, 258, 259, 261, 285; Health, 233, 285; History, 292; Investments, 258, 260; Labor, 291, 293; Measurement, 233; Physics, 261, 266, 275, 276, 292, 294; Sports, 233, 286; Temperature, 258, 294; Travel, 258; Utilities, 261

10 *Statistics and Probability* *549*

Preface

The second edition of *Prealgebra* is designed to be a transition from the concrete aspects of arithmetic to the symbolic world of algebra. The text has been created to meet the needs of both the traditional college student and returning students whose mathematical proficiency may have declined during the years away from formal education.

In *Prealgebra,* careful attention has been given to implementing the standards suggested by AMATYC. There is an abundance of real sourced data graphs and tables integrated throughout the text. Each chapter begins with a Focus on Problem Solving that introduces students to various problem-solving strategies. At the end of each section there are Critical Thinking Exercises and Writing Exercises. The chapter ends with Projects in Mathematics that can be used for cooperative learning activities.

Besides the Index of Applications on the inside front cover, a chapter-by-chapter index of the variety of application problems can be found in the Table of Contents. This additional index highlights the importance and scope of the applications of mathematics in an easily accessible location.

Instructional Features

Early Introduction to Variables

One way to provide the transition from arithmetic to algebra is through an early introduction to variables. In *Prealgebra* variables are introduced in a natural way, through applications and variable expressions, in Chapter 1. This early introduction of variables and their integrated use throughout the text provide the student with the practice required to make the passage from concrete arithmetic to symbolic algebra.

Interactive Approach

Prealgebra uses an interactive style that provides a student with an opportunity to try a skill as it is presented. Each section is divided into objectives, and every objective contains one or more sets of matched-pair examples. The first example in each set is worked out; the second example, called "You Try It," is for the student to work. By solving this problem, the student practices concepts as they are presented in the text. There are complete worked-out solutions to these examples in an appendix at the end of the text. By comparing their solution to the solution in the appendix, students are able to obtain immediate feedback on and reinforcement of the skill being learned.

Emphasis on Problem-Solving Strategies

Besides a presentation of the fundamental mathematics concepts, *Prealgebra* contains a variety of contemporary application problems. The solution of each application problem features a carefully developed approach to problem solving that emphasizes developing strategies to solve problems. Students are encouraged to develop their own strategies and to write these strategies as part of the solution to a problem.

Completely Integrated Learning System Organized by Objectives

Each chapter begins with a list of learning objectives included within that chapter. Each of the objectives is then restated in the chapter to remind the student of the current topic of discussion. The same objectives that organize the text are used as the structure for exercises, testing programs, and the Computer Tutor.

An Interactive Approach

Instructors have long realized the need for a text that requires the student to use a skill as it is being taught. *Prealgebra* uses an interactive technique that meets this need. Every objective, including the one shown below, contains at least one pair of examples in which one example is worked. The second example of the pair is not worked so that the student may "interact" with the text by solving it. To provide immediate feedback, a complete worked-out solution to the You Try It example is provided in the Solutions Section. The benefit of this interactive strategy is that the student can immediately determine whether a new skill has been learned before attempting a homework assignment or moving on to the next skill.

An explanatory passage begins each objective.

Paired examples follow the explanatory passage.

The interactive key is the You Try It in each pair. It has not been worked so that the student may practice the skill, referring to the worked example at the left if necessary.

Reference to the Solutions Section allows the student to check full solutions immediately.

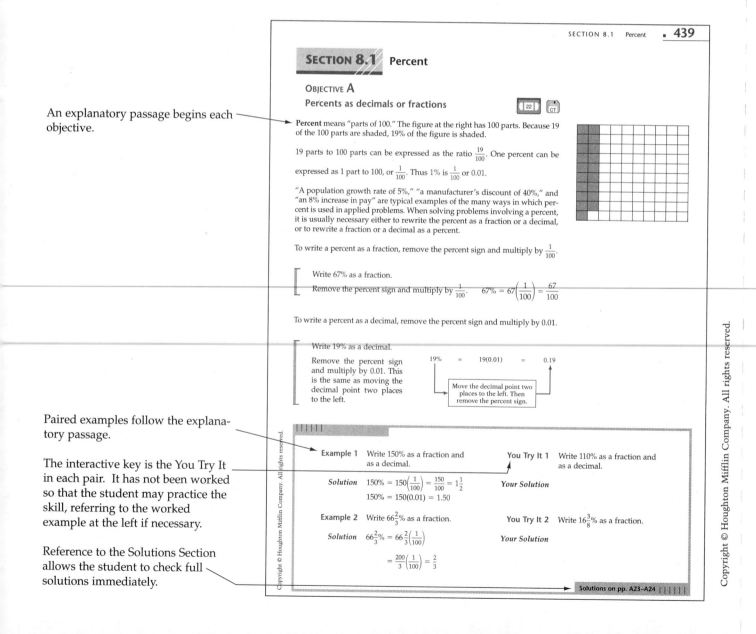

SECTION 8.1   Percent   . 439

SECTION 8.1   Percent

OBJECTIVE A
Percents as decimals or fractions

Percent means "parts of 100." The figure at the right has 100 parts. Because 19 of the 100 parts are shaded, 19% of the figure is shaded.

19 parts to 100 parts can be expressed as the ratio $\frac{19}{100}$. One percent can be expressed as 1 part to 100, or $\frac{1}{100}$. Thus 1% is $\frac{1}{100}$ or 0.01.

"A population growth rate of 5%," "a manufacturer's discount of 40%," and "an 8% increase in pay" are typical examples of the many ways in which percent is used in applied problems. When solving problems involving a percent, it is usually necessary either to rewrite the percent as a fraction or a decimal, or to rewrite a fraction or a decimal as a percent.

To write a percent as a fraction, remove the percent sign and multiply by $\frac{1}{100}$.

Write 67% as a fraction.

Remove the percent sign and multiply by $\frac{1}{100}$. $67\% = 67\left(\frac{1}{100}\right) = \frac{67}{100}$

To write a percent as a decimal, remove the percent sign and multiply by 0.01.

Write 19% as a decimal.

Remove the percent sign and multiply by 0.01. This is the same as moving the decimal point two places to the left.

$19\% \qquad = \qquad 19(0.01) \qquad = \qquad 0.19$

Move the decimal point two places to the left. Then remove the percent sign.

Example 1   Write 150% as a fraction and as a decimal.

Solution $150\% = 150\left(\frac{1}{100}\right) = \frac{150}{100} = 1\frac{1}{2}$
$150\% = 150(0.01) = 1.50$

You Try It 1   Write 110% as a fraction and as a decimal.

Your Solution

Example 2   Write $66\frac{2}{3}\%$ as a fraction.

Solution $66\frac{2}{3}\% = 66\frac{2}{3}\left(\frac{1}{100}\right)$

$= \frac{200}{3}\left(\frac{1}{100}\right) = \frac{2}{3}$

You Try It 2   Write $16\frac{3}{8}\%$ as a fraction.

Your Solution

Solutions on pp. A23–A24

An Applied Approach

The traditional approach to teaching an introduction to algebra neglects the difficulties that students have in making the transition from arithmetic to algebra. One of the most troublesome and uncomfortable transitions for the student is the one from concrete arithmetic to symbolic algebra. *Prealgebra* recognizes the formidable task the student faces by introducing variables in a very natural way—through applications of mathematics. A secondary benefit of this approach is that the student becomes aware of the value of algebra as a real-life tool.

The solution of an application problem in *Prealgebra* is always accompanied by two parts: Strategy and Solution. The strategy is a written description of the steps that are necessary to solve the problem; the solution is the implementation of the strategy. Using this format provides students with a structure for problem solving. It also encourages students to write strategies for solving problems which, in turn, fosters organizing problem-solving strategies in a logical way.

Early introduction of variables through application problems helps the student make the transition from the concrete to the abstract.

A strategy which the student may use in solving an application problem is stated.

The strategy is used in the solution of the worked example.

Students are encouraged to write a strategy for the application problem they solve.

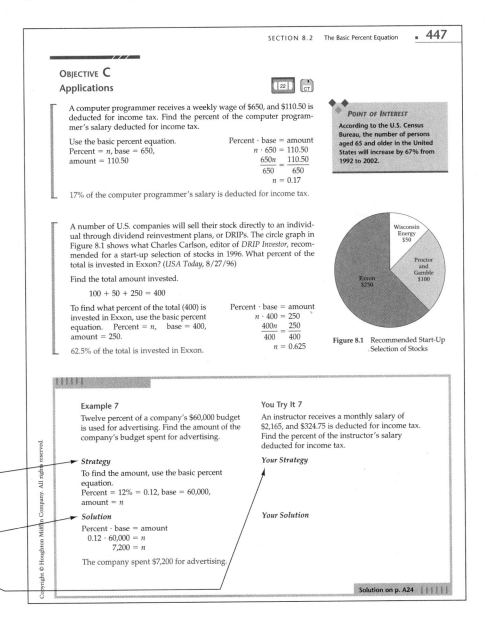

OBJECTIVE C
Applications

A computer programmer receives a weekly wage of $650, and $110.50 is deducted for income tax. Find the percent of the computer programmer's salary deducted for income tax.

Use the basic percent equation.
Percent = n, base = 650,
amount = 110.50

Percent · base = amount
$n \cdot 650 = 110.50$
$\dfrac{650n}{650} = \dfrac{110.50}{650}$
$n = 0.17$

17% of the computer programmer's salary is deducted for income tax.

POINT OF INTEREST
According to the U.S. Census Bureau, the number of persons aged 65 and older in the United States will increase by 67% from 1992 to 2002.

A number of U.S. companies will sell their stock directly to an individual through dividend reinvestment plans, or DRIPs. The circle graph in Figure 8.1 shows what Charles Carlson, editor of *DRIP Investor*, recommended for a start-up selection of stocks in 1996. What percent of the total is invested in Exxon? (*USA Today*, 8/27/96)

Find the total amount invested.

$100 + 50 + 250 = 400$

To find what percent of the total (400) is invested in Exxon, use the basic percent equation. Percent = n, base = 400, amount = 250.

Percent · base = amount
$n \cdot 400 = 250$
$\dfrac{400n}{400} = \dfrac{250}{400}$
$n = 0.625$

62.5% of the total is invested in Exxon.

Wisconsin Energy $50
Proctor and Gamble $100
Exxon $250

Figure 8.1 Recommended Start-Up Selection of Stocks

Example 7

Twelve percent of a company's $60,000 budget is used for advertising. Find the amount of the company's budget spent for advertising.

Strategy
To find the amount, use the basic percent equation.
Percent = 12% = 0.12, base = 60,000, amount = n

Solution
Percent · base = amount
$0.12 \cdot 60,000 = n$
$7,200 = n$

The company spent $7,200 for advertising.

You Try It 7

An instructor receives a monthly salary of $2,165, and $324.75 is deducted for income tax. Find the percent of the instructor's salary deducted for income tax.

Your Strategy

Your Solution

Solution on p. A24

Objective-Specific Approach

Many texts in mathematics are not organized in a manner that facilitates management of learning. Typically, students are left to wander through a maze of apparently unrelated lessons, exercise sets, and tests. *Prealgebra* solves this problem by organizing all lessons, exercise sets, and tests around a carefully constructed hierarchy of objectives. The advantage of this objective-by-objective organization is that it enables the student who is uncertain at any step in the learning process to refer easily to the original presentation and review that material.

The Objective-Specific Approach also allows the instructor greater control over the management of student progress. The Computerized Testing Program and the Printed Testing Program are organized by the same objectives as the text. The answers to the test items contain a reference to the objective from which the question was taken. This allows the instructor to quickly determine those objectives for which a student may need additional instruction.

The Computer Tutor is also organized around the objectives of the text. As a result, supplemental instruction is available for the specific objectives that are troublesome for a student. Also, objective-specific reports are available to the instructor using the new classroom management system along with the tutor. The management system also offers many other types of reports showing the progress of an entire class or a particular student.

A numbered objective statement names the topic of each lesson.

SECTION 8.1 **Percent**

OBJECTIVE A
Percents as decimals or fractions

The exercise sets correspond to the objectives in the text.

8.1 EXERCISES

OBJECTIVE A

Write as a fraction and as a decimal.

1. 5%	**2.** 60%	**3.** 30%	**4.** 90%
5. 250%	**6.** 140%	**7.** 28%	**8.** 66%
9. 35%	**10.** 85%	**11.** 6%	**12.** 8%
13. 122%	**14.** 166%	**15.** 29%	**16.** 83%

The answers to the odd-numbered exercises are provided in the Answer Section.

8.1 EXERCISES (*pages 441–442*)

1. $\frac{1}{20}$, 0.05   **3.** $\frac{3}{10}$, 0.30   **5.** $\frac{5}{2}$, 2.50   **7.** $\frac{7}{25}$, 0.28   **9.** $\frac{7}{20}$, 0.35   **11.** $\frac{3}{50}$, 0.06   **13.** $\frac{61}{50}$, 1.22   **15.** $\frac{29}{100}$, 0.29
17. $\frac{1}{9}$ **19.** $\frac{3}{8}$ **21.** $\frac{1}{32}$ **23.** $\frac{5}{11}$ **25.** $\frac{3}{800}$ **27.** $\frac{1}{200}$ **29.** $\frac{5}{6}$ **31.** $\frac{7}{8}$ **33.** 0.073   **35.** 0.158

The answers to the Chapter Review Exercises show the objective to study if the student incorrectly answers the exercise.

CHAPTER REVIEW EXERCISES (*pages 471–472*)

1. $\frac{8}{25}$ (Objective 8.1A)   **2.** 0.22 (Objective 8.1A)   **3.** $\frac{1}{4}$, 0.25 (Objective 8.1A)   **4.** $\frac{17}{500}$ (Objective 8.1A)   **5.** 17.5%
(Objective 8.1B)   **6.** 128.6% (Objective 8.1B)   **7.** 280% (Objective 8.1B)   **8.** 21 (Objective 8.2A/8.2B)   **9.** 500%

The answers to the Cumulative Review Exercises also show the objective that relates to the exercise.

CUMULATIVE REVIEW EXERCISES (*pages 473–474*)

1. 24.954 (Objective 4.2A)   **2.** 625 (Objective 1.3B)   **3.** 14.04269 (Objective 4.2B)   **4.** $4x^2 - 16x + 15$ (Objective 5.5B)
5. $1\frac{25}{62}$ (Objective 3.4B)   **6.** $6a^3b^3 - 8a^4b^4 + 2a^3b^4$ (Objective 5.5A)   **7.** 42 (Objective 8.2A/8.2B)   **8.** -3

Additional Learning Aids

Focus on Problem Solving

Each chapter begins with a Focus on Problem Solving, the purpose of which is to introduce the student to various successful problem-solving strategies. Each Focus consists of a problem and an appropriate strategy to solve the problem. Strategies such as guessing, trying to solve a simpler but similar problem, drawing a diagram, and looking for patterns are some of the techniques that are demonstrated.

Projects in Mathematics

The Projects in Mathematics feature occurs at the end of each chapter. These projects can be used as extra credit or cooperative learning activities. Each project is an extended application. Frequently, the problem-solving strategy presented in the Focus on Problem Solving can be used to solve these problems.

Chapter Summaries

At the end of each chapter there is a Chapter Summary which includes the Key Words and Essential Rules that we covered in the chapter. These Chapter Summaries provide one focus for the student when preparing for a test.

Study Tips

Interspersed throughout Chapter 1 are suggestions for how to use this text and approaches to creating good study habits.

Computer Tutor

The Computer Tutor is a networkable, interactive, algorithmically-driven software package. This powerful ancillary features full-color graphics, a glossary, extensive hints, animated solution steps, and a comprehensive class management system.

Exercises

End-of-Section Exercises

There are a wide variety of exercise sets in *Prealgebra*. At the end of each section there are exercise sets that are keyed to the corresponding learning objective. The exercises are carefully developed to ensure that students can apply the concepts in the section to a variety of problem situations.

Critical Thinking Exercises

The End-of-Section Exercises are followed by Critical Thinking Exercises. These exercises require that a student investigate a certain concept in more depth or detail.

Writing Exercises

At the end of each Critical Thinking exercise set are Writing Exercises denoted by ✐ . These exercises ask students to write a few paragraphs about a topic in the section or to research and report on a related topic.

Chapter Review Exercises

Chapter Review Exercises are found at the end of each chapter. These exercises are selected to help the student integrate all of the topics presented in the chapter. The answers to all review exercises are given in the answer section. Along with the answer, there is a reference to the objective that pertains to the exercise.

Cumulative Review Exercises

Cumulative Review Exercises, which appear at the end of each chapter (beginning with Chapter 2), help the student maintain skills learned in previous chapters. The answers to all cumulative review exercises are given in the answer section. Along with the answer, there is a reference to the objective that pertains to the exercise.

New to This Edition

Topical Coverage

The study of pictographs, circle graphs, bar graphs, and line graphs has been moved to Chapter 1. This placement allows us to integrate statistical information throughout the remainder of the text, providing students with constant reinforcement of the skill of reading graphs and tables, as well as providing them with applications that involve real-life data. Exercises that incorporate these features are marked with a ✍.

Solving equations is now integrated throughout the text, beginning in Chapter 1. However, those instructors who prefer to postpone equation solving until later in the course can easily omit the material on equations in the earlier chapters. Notes to the instructor in the Instructor's Annotated Edition clearly indicate references to equations which may be omitted.

Chapter 6 still contains complete coverage of solving linear equations. However, this edition presents the rectangular coordinate system and graphing linear equations in the last section of this chapter.

Computer Tutor

The Computer Tutor has been completely revised. It is now an algorithmically based tutor that includes color and animation. The algorithms have been carefully crafted to present a variety of problem types from easy to difficult. A complete solution to each problem is provided.

The Computer Tutor is an interactive tutorial that requires students to respond to questions about the topic in the current lesson. A Glossary can be accessed at any time so that students can look up the words whose definitions they may have forgotten.

A printed report of the progress of the student is available. This optional report gives the student's name, the objective studied, the number of problems attempted, the number of problems correct, and the percent correct. The instructor also has the option of creating a cumulative report via the Tutor's new class management system.

Margin Notes

Point of Interest notes are interspersed throughout the text. These notes are interesting sidelights of the topic being discussed.

Take Note alerts students that a procedure may be particularly involved or reminds students that certain checks of their work should be performed.

Calculator Notes describe some of the functions of a calculator. These notes demonstrate for students the ways in which calculators can be used in problem solving.

Instructor Notes are printed only in the Instructor's Annotated Edition. These notes provide suggestions for presenting a lesson or related material that can be used in class.

Supplements for Instructor Use

Instructor's Annotated Edition

The Instructor's Annotated Edition is an exact replica of the student text except that answers to all exercises are given in the text. Also, there are Instructor Notes in the margins that offer suggestions for presenting the material in a lesson.

The Instructor's Annotated Edition also depicts the calculator logo alongside exercises for which students may need to use a calculator. These logos were placed only in the Instructor's Annotated Edition so that students can learn, without the visual clue of an icon, when a calculator is an appropriate problem-solving tool.

Instructor's Resource Manual with Solutions Manual

The Instructor's Resource Manual includes suggestions for course sequencing and outlines for the answers to the writing exercises. The Solutions Manual contains worked-out solutions for all the exercises in the text.

Computerized Test Generator

The Computerized Test Generator is the first source of testing material. The database contains more than 2,000 test items. The Test Generator is designed to provide an unlimited number of tests for each chapter, cumulative chapter tests, and a final exam. It is available for the IBM PC and compatible computers and the Macintosh. The Windows version also provides **on-line testing** and **gradebook** functions.

Printed Test Bank with Chapter Tests

The Printed Test Bank, the second component of the testing material, is a printout of all items in the Computerized Test Generator. Instructors who do not have access to a computer can use the test bank to select items to be included on a test being prepared by hand. The Chapter Tests comprise the third source of testing material. Four printed tests, two free response and two multiple choice, are provided for each chapter. In addition, there are cumulative tests after Chapters 4, 7, and 10, and a final exam.

Supplements for Student Use

Student Solutions Manual

The Student Solutions Manual contains the complete solutions to all odd-numbered exercises in the text.

Computer Tutor

The Computer Tutor is an interactive instructional computer program with algorithmically generated exercises for student use. Each objective of the text is supported by a tutorial. These tutorials contain an interactive lesson that covers the material in the objective. The randomly generated exercises following the lesson are created by algorithms and have been carefully designed to provide the student with a variety of appropriate practice problems.

The Computer Tutor can be used in several ways: (1) to cover material the student missed because of an absence; (2) to reinforce instruction on a concept that the student has not yet mastered; (3) to review material in preparation for exams. This tutorial is available for the IBM PC and compatible computers and the Macintosh.

Within each section of the book, a computer icon appears next to each objective. The icon serves as a reminder that there is a Computer Tutor lesson corresponding to that objective.

Videotapes

Within each section of the text, a videotape icon appears next to an objective for which there is a corresponding video. The icon contains the reference number of the appropriate video. Each video topic is motivated through an application and the necessary mathematics to solve that problem are then presented.

Acknowledgments

We sincerely wish to thank the following reviewers, who reviewed the manuscript in various stages of development, for their valuable contributions.

Annette Burden, *Youngstown State University, OH*

Irene Doo, *Austin Community College, TX*

Richard Eells, *Roxbury Community College, MA*

Michael A. Jones

JoAnne Kennedy, *City University of New York, La Guardia Community College*

Frank Mauz, *Honolulu Community College, HI*

Kenneth McClain, *The University of Memphis, TN*

Deana J. Richmond

Helen Salzburg, *Rhode Island College*

Sharon Verholek, *Pennsylvania State University—Shenango Campus*

Prealgebra

CHAPTER 1

Whole Numbers

FOCUS On Problem Solving

You encounter problem-solving situations every day. Some problems are easy to solve, and you may mentally solve these problems without considering the steps you are taking in order to draw a conclusion. Others may be more challenging and require more thought and consideration.

Suppose a friend suggests that you both take a trip over spring break. You'd like to go. What questions go through your mind? You might ask yourself some of the following questions:

How much will the trip cost? What will be the cost for travel, hotel rooms, meals, etc.?

Are some costs going to be shared by both me and my friend?

Can I afford it?

How much money do I have in the bank?

How much more money than I have now do I need?

How much time is there to earn that much money?

How much can I earn in that amount of time?

How much money must I keep in the bank in order to pay the next tuition bill (or some other expense)?

These questions require different mathematical skills. Determining the cost of the trip requires *estimation*; for example, you must use your knowledge of air fares or the cost of gasoline to arrive at an estimate of these costs. If some of the costs are going to be shared, you need to *divide* those costs by 2 in order to determine your share of the expense. The question regarding how much more money you need requires *subtraction*: the amount needed minus the amount currently in the bank. To determine how much money you can earn in the given amount of time requires *multiplication*—for example, the amount you earn per week times the number of weeks to be worked. To determine whether the amount you can

earn in the given amount of time is sufficient, you need to use your knowledge of *order relations* to compare the amount you can earn with the amount needed.

Facing the problem-solving situation described above may not seem difficult to you. The reason may be that you have faced similar situations before and, therefore, know how to work through this one. You may feel better prepared to deal with a circumstance such as this one because you know what questions to ask. An important aspect of learning to solve problems is learning what questions to ask. As you work through application problems in this text, try to become more conscious of the mental process you are going through. You might begin the process by asking yourself the following questions whenever you are solving an application problem:

1. Have I read the problem enough times to be able to understand the situation being described?

2. Will restating the problem in different words help me to better understand the problem situation?

3. What facts are given? (You might make a list of the information contained in the problem.)

4. What information is being asked for?

5. What relationship exists between the given facts? What relationship exists between the given facts and the solution?

6. What mathematical operations are needed in order to solve the problem?

Try to focus on the problem-solving situation, not the computation or getting the answer quickly. And remember, the more problems you solve, the better able you are to solve other problems in the future, partly because you are learning what questions to ask.

 **Introduction to Whole Numbers**

Objective A
Order relations between whole numbers

The **natural numbers** are 1, 2, 3, 4, 5, 6, 7, 8, 9, 10, 11, . . .

The three dots mean that the list continues on and on and there is no largest natural number. The natural numbers are also called the **counting numbers.**

The **whole numbers** are 0, 1, 2, 3, 4, 5, 6, 7, 8, 9, 10, 11, . . . Note that the whole numbers include the natural numbers and zero.

Just as distances are associated with markings on the edge of a ruler, the whole numbers can be associated with points on a line. This line is called the **number line** and is shown below.

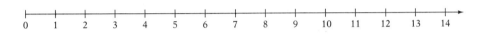

The arrowhead at the right indicates that the number line continues to the right.

The **graph** of a whole number is shown by placing a heavy dot on the number line directly above the number. Shown below is the graph of 6 on the number line.

On the number line, the numbers get larger as we move from left to right. The numbers get smaller as we move from right to left. Therefore, the number line can be used to visualize the order relation between two whole numbers.

A number that appears to the right of a given number is **greater than** the given number. The symbol for *is greater than* is $>$.

8 is to the right of 3.
8 is greater than 3.
$8 > 3$

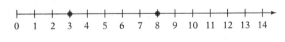

A number that appears to the left of a given number is **less than** the given number. The symbol for *is less than* is $<$.

5 is to the left of 12.
5 is less than 12.
$5 < 12$

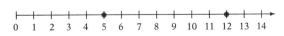

An **inequality** expresses the relative order of two mathematical expressions. $8 > 3$ and $5 < 12$ are inequalities.

Example 1 Graph 4 on the number line.

Solution

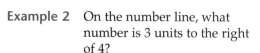

You Try It 1 Graph 9 on the number line.

Your Solution

Example 2 On the number line, what number is 3 units to the right of 4?

Solution

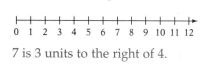

7 is 3 units to the right of 4.

You Try It 2 On the number line, what number is 4 units to the left of 11?

Your Solution

Example 3 Place the correct symbol, < or >, between the two numbers.

a. 38   23   **b.** 0   54

Solution   **a.** 38 > 23   **b.** 0 < 54

You Try It 3 Place the correct symbol, < or >, between the two numbers.

a. 47   19   **b.** 26   0

Your Solution

Example 4 Write the given numbers in order from smallest to largest.

16, 5, 47, 0, 83, 29

Solution   0, 5, 16, 29, 47, 83

You Try It 4 Write the given numbers in order from smallest to largest.

52, 17, 68, 0, 94, 3

Your Solution

Solutions on p. A5

OBJECTIVE B
Place value

When a whole number is written using the digits 0, 1, 2, 3, 4, 5, 6, 7, 8, and 9, it is said to be in **standard form.** The position of each digit in the number determines the digit's **place value.** The diagram below shows a **place-value chart** naming the first twelve place values. The number 64,273 is in standard form and has been entered in the chart.

In the number 64,273, the position of the digit 6 determines that its place value is ten-thousands.

When a number is written in standard form, each group of digits separated by a comma is called a **period.** The number 5,316,709,842 has four periods. The period names are shown in color in the place-value chart above.

To write a number in words, start from the left. Name the number in each period. Then write the period name in place of the comma.

5,316,709,842 is read "five billion three hundred sixteen million seven hundred nine thousand eight hundred forty-two."

To write a whole number in standard form, write the number named in each period, and replace each period name with a comma.

Six million fifty-one thousand eight hundred seventy-four is written 6,051,874. The zero is used as a place holder for the hundred-thousands' place.

The whole number 37,286 can be written in **expanded form** as

    30,000 + 7,000 + 200 + 80 + 6

The place-value chart can be used to find the expanded form of a number.

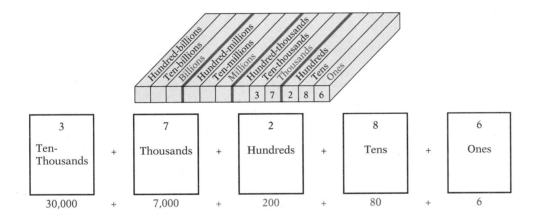

Write the number 510,409 in expanded form.

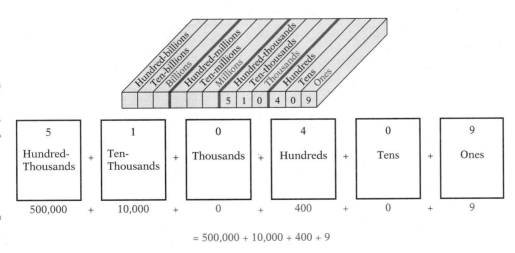

    = 500,000 + 10,000 + 400 + 9

Example 5

Write 82,593,071 in words.

Solution

eighty-two million five hundred ninety-three thousand seventy-one

You Try It 5

Write 46,032,715 in words.

Your Solution

Example 6

Write four hundred six thousand nine in standard form.

Solution

406,009

You Try It 6

Write nine hundred twenty thousand eight in standard form.

Your Solution

Example 7

Write 32,598 in expanded form.

Solution

30,000 + 2,000 + 500 + 90 + 8

You Try It 7

Write 76,245 in expanded form.

Your Solution

Solutions on p. A5

OBJECTIVE C
Rounding

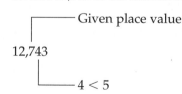

When the distance to the sun is given as 93,000,000 mi, the number represents an approximation to the true distance. Giving an approximate value for an exact number is called **rounding**. A number is rounded to a given place value.

48 is closer to 50 than it is to 40. 48 rounded to the nearest ten is 50.

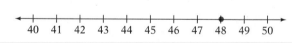

4,872 rounded to the nearest ten is 4,870.

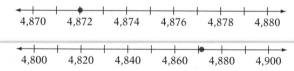

4,872 rounded to the nearest hundred is 4,900.

A number is rounded to a given place value without using the number line by looking at the first digit to the right of the given place value.

If the digit to the right of the given place value is less than 5, that digit and all digits to the right are replaced by zeros.

Round 12,743 to the nearest hundred.

┌──────── Given place value
12,743
└──── 4 < 5

12,743 rounded to the nearest hundred is 12,700.

If the digit to the right of the given place value is greater than or equal to 5, increase the digit in the given place value by 1, and replace all other digits to the right by zeros.

Round 46,738 to the nearest thousand.

Given place value

46,738

7 > 5

46,738 rounded to the nearest thousand is 47,000.

Round 29,873 to the nearest thousand.

Given place value

29,873

8 > 5   Round up by adding 1 to the 9 (9 + 1 = 10).
Carry the 1 to the ten-thousands' place (2 + 1 = 3).

29,873 rounded to the nearest thousand is 30,000.

Example 8

Round 435,278 to the nearest ten-thousand.

Solution

Given place value

435,278

5 = 5

435,278 rounded to the nearest ten-thousand is 440,000.

You Try It 8

Round 529,374 to the nearest ten-thousand.

Your Solution

Example 9

Round 1,967 to the nearest hundred.

Solution

Given place value

1,967

6 > 5

1,967 rounded to the nearest hundred is 2,000.

You Try It 9

Round 7,985 to the nearest hundred.

Your Solution

Solutions on p. A5

OBJECTIVE D
Applications and statistical graphs

Graphs are displays that provide a pictorial representation of data. The advantage of graphs is that they present information in a way that is easily read. The disadvantage of graphs is that they can be misleading. See the Projects in Mathematics at the end of this chapter.

A **pictograph** uses symbols to represent information. The symbol chosen usually has a connection to the data it represents.

Figure 1.1 represents the net worth of the world's richest billionaires. Each symbol represents one billion dollars.

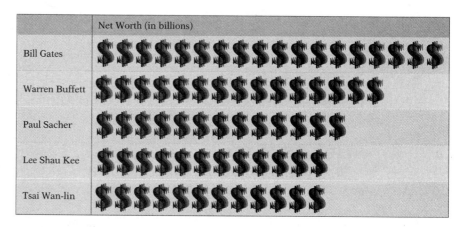

Figure 1.1 Net Worth of the World's Richest Billionaires
Source: Forbes

From the pictograph, we can see that Bill Gates has the greatest net worth. Warren Buffett's net worth is $2 billion more than Paul Sacher's net worth.

A typical household in the United States has an average after-tax income of $40,550. The **circle graph** in Figure 1.2 represents how this annual income is spent. The complete circle represents the total amount, $40,550. Each sector of the circle represents the amount spent on a particular expense.

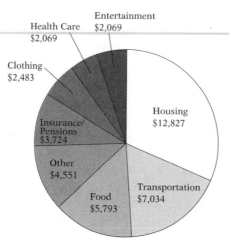

Figure 1.2 Average Annual Expenses in a U.S. Household
Source: American Demographics

From the circle graph, we can see that the largest amount is spent on housing. We can see that the amount spent on food ($5,793) is less than the amount spent on transportation ($7,034).

The **bar graph** in Figure 1.3 shows the expected U.S. population aged 100 and over.

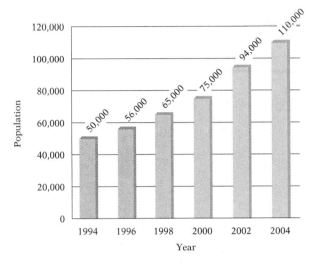

Figure 1.3   Expected U.S. Population Aged 100 and Over
Source: Census Bureau

In this bar graph, the horizontal axis is labeled with the years (1994, 1996, 1998, etc.) and the vertical axis is labeled with the numbers for the population. For each year, the height of the bar indicates the population for that year. For example, we can see that the expected population of those aged 100 and over in the year 2000 is 75,000. The graph indicates that the population of people aged 100 and over keeps increasing.

A **double-bar graph** is used to display data for the purposes of comparison. The double-bar graph in Figure 1.4 shows the number of seconds it takes computers with different operating speeds, measured in megahertz (mHz), to process a spreadsheet, a word processing document, a graphics program, and a database file.

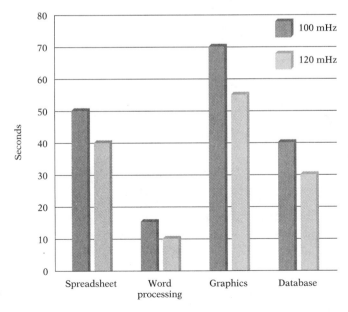

Figure 1.4

From the graph, we can see that the 100-mHz computer took more time to process the spreadsheet (50 s) than it took the 120-mHz computer to process the spreadsheet (40 s).

The **broken-line graph** in Figure 1.5 shows the effect of inflation on the value of a $100,000 life insurance policy. (An inflation rate of 5 percent is used here.)

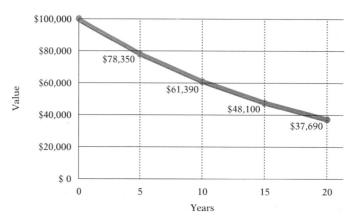

Figure 1.5   Effect of Inflation on the Value of a $100,000 Life Insurance Policy

According to the line graph, after five years the purchasing power of the $100,000 has decreased to $78,350. We can see that the value of the $100,000 keeps decreasing over the 20-year period.

Two broken-line graphs are used so that data can be compared. Figure 1.6 shows the population of California and of Texas. The figures are those of the U.S. Census for the years 1900, 1930, 1960, and 1990. The numbers are rounded to the nearest thousand.

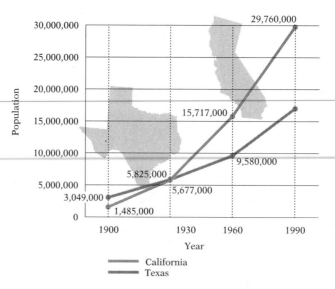

Figure 1.6   Populations of California and Texas

From the graph, we can see that the population was greater in Texas in 1900 and 1930, while the population was greater in California in 1960 and 1990.

To solve an application problem, first read the problem carefully. The **Strategy** involves identifying the quantity to be found and planning the steps that are necessary to find that quantity. The **Solution** involves performing each operation stated in the Strategy and writing the answer.

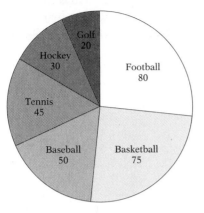

The circle graph in Figure 1.7 shows the result of a survey of 300 people who were asked to name their favorite sport. Use this graph for Example 10 and You Try It 10.

Figure 1.7   Distribution of Responses in a Survey

Example 10

According to Figure 1.7, which sport was named by the least number of people?

Strategy

To find the sport named by the least number of people, find the smallest number given in the circle graph.

Solution

The smallest number given in the graph is 20.

The sport named by the least number of people was golf.

You Try It 10

According to Figure 1.7, which sport was named by the greatest number of people?

Your Strategy

Your Solution

Example 11

The distance between St. Louis, Missouri, and Portland, Oregon, is 2,057 mi. The distance between St. Louis, Missouri, and Seattle, Washington, is 2,135 mi. Which distance is greater, St. Louis to Portland or St. Louis to Seattle?

Strategy

To find the greater distance, compare the numbers 2,057 and 2,135.

Solution

2,135 > 2,057

The greater distance is from St. Louis to Seattle.

You Try It 11

The distance between Los Angeles, California, and San Jose, California, is 347 mi. The distance between Los Angeles, California, and San Francisco, California, is 387 mi. Which distance is shorter, Los Angeles to San Jose or Los Angeles to San Francisco?

Your Strategy

Your Solution

Solutions on p. A5

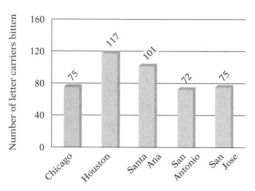

The bar graph in Figure 1.8 shows the number of letter carriers bitten in 1995 in five cities in the United States. It was in these cities that the most dog bites occurred. Use this graph for Example 12 and You Try It 12.

Figure 1.8 Number of Letter Carriers Bitten in 1995

Source: U.S. Postal Service, Humane Society of the U.S.

Example 12

According to Figure 1.8, in which city were the most letter carriers bitten?

Strategy

To determine the city in which the most letter carriers were bitten, locate the city that corresponds to the highest bar.

Solution

The highest bar corresponds to the city of Houston.

The most letter carriers were bitten in Houston.

You Try It 12

According to Figure 1.8, in which city were fewer letter carriers bitten, Santa Ana or San Jose?

Your Strategy

Your Solution

Example 13

The land area of the United States is 3,539,341 mi². What is the land area of the United States to the nearest ten-thousand square miles?

Strategy

To find the land area to the nearest ten-thousand square miles, round 3,539,341 to the nearest ten-thousand.

Solution

3,539,341 rounded to the nearest ten-thousand is 3,540,000.

To the nearest ten-thousand square miles, the land area of the United States is 3,540,000 mi².

You Try It 13

The land area of Canada is 3,851,809 mi². What is the land area of Canada to the nearest thousand square miles?

Your Strategy

Your Solution

Solutions on p. A5

1.1 EXERCISES

OBJECTIVE A

Graph the number on the number line.

1. 2

0 1 2 3 4 5 6 7 8 9 10 11 12

2. 7

0 1 2 3 4 5 6 7 8 9 10 11 12

3. 10

0 1 2 3 4 5 6 7 8 9 10 11 12

4. 1

0 1 2 3 4 5 6 7 8 9 10 11 12

5. 5

0 1 2 3 4 5 6 7 8 9 10 11 12

6. 11

0 1 2 3 4 5 6 7 8 9 10 11 12

On the number line, which number is:

7. 4 units to the left of 9

8. 5 units to the left of 8

9. 3 units to the right of 2

10. 4 units to the right of 6

11. 7 units to the left of 7

12. 8 units to the left of 11

Place the correct symbol, $<$ or $>$, between the two numbers.

13. 27   39

14. 68   41

15. 0   52

16. 61   0

17. 273   194

18. 419   502

19. 2,761   3,857

20. 3,827   6,915

21. 4,610   4,061

22. 5,600   56,000

23. 8,005   8,050

24. 92,010   92,001

Write the given numbers in order from smallest to largest.

25. 21, 14, 32, 16, 11

26. 18, 60, 35, 71, 27

27. 72, 48, 84, 93, 13

Write the given numbers in order from smallest to largest.

28. 54, 45, 63, 28, 109

29. 26, 49, 106, 90, 77

30. 505, 496, 155, 358, 271

31. 736, 662, 204, 981, 399

32. 440, 404, 400, 444, 4,000

33. 377, 370, 307, 3,700, 3,077

Objective B

Write the number in words.

34. 704

35. 508

36. 374

37. 635

38. 2,861

39. 4,790

40. 48,297

41. 53,614

42. 563,078

43. 246,053

44. 6,379,482

45. 3,842,905

Write the number in standard form.

46. seventy-five

47. four hundred ninety-six

48. two thousand eight hundred fifty-one

49. fifty-three thousand three hundred forty

50. one hundred thirty thousand two hundred twelve

51. five hundred two thousand one hundred forty

52. eight thousand seventy-three

53. nine thousand seven hundred six

Write the number in standard form.

54. six hundred three thousand one hundred thirty-two

55. five million twelve thousand nine hundred seven

56. three million four thousand eight

57. eight million five thousand ten

Write the number in expanded form.

58. 6,398

59. 7,245

60. 46,182

61. 532,791

62. 328,476

63. 5,064

64. 90,834

65. 20,397

66. 400,635

67. 402,708

68. 504,603

69. 8,000,316

OBJECTIVE C

Round the number to the given place value.

70. 3,049 Tens

71. 7,108 Tens

72. 1,638 Hundreds

73. 4,962 Hundreds

74. 17,639 Hundreds

75. 28,551 Hundreds

76. 5,326 Thousands

77. 6,809 Thousands

78. 84,608 Thousands

79. 93,825 Thousands

80. 389,702 Thousands

81. 629,513 Thousands

82. 746,898 Ten-thousands

83. 352,876 Ten-thousands

84. 36,702,599 Millions

85. 71,834,250 Millions

OBJECTIVE D

Solve.

86. *Sports* During his baseball career, Eddie Collins had a record of 743 stolen bases. Max Carey had a record of 738 stolen bases during his baseball career. Who had more stolen bases, Eddie Collins or Max Carey?

87. *Sports* During his baseball career, Ty Cobb had a record of 892 stolen bases. Billy Hamilton had a record of 937 stolen bases during his baseball career. Who had more stolen bases, Ty Cobb or Billy Hamilton?

88. *Nutrition* The figure at the right shows the annual per capita turkey consumption in different countries. (a) What is the annual per capita turkey consumption in the United States? (b) In which country is the annual per capita turkey consumption the highest?

89. *The Arts* The play *Hello Dolly* was performed 2,844 times on Broadway. The play *Fiddler on the Roof* was performed 3,242 times on Broadway. Which play had the greater number of performances, *Hello Dolly* or *Fiddler on the Roof*?

Britain	🦃 🦃 🦃 🦃
Canada	🦃 🦃 🦃 🦃 🦃
France	🦃 🦃 🦃 🦃 🦃 🦃
Ireland	🦃 🦃 🦃 🦃
Israel	🦃 🦃 🦃 🦃 🦃 🦃 🦃 🦃 🦃 🦃 🦃
Italy	🦃 🦃 🦃 🦃 🦃
U.S.	🦃 🦃 🦃 🦃 🦃 🦃 🦃 🦃 🦃

Each 🦃 represents 2 lb.

Per Capita Turkey Consumption
Source: National Turkey Federation

90. *The Arts* The play *Annie* was performed 2,377 times on Broadway. The play *My Fair Lady* was performed 2,717 times on Broadway. Which play had the greater number of performances, *Annie* or *My Fair Lady*?

91. *Nutrition* Two tablespoons of peanut butter contain 190 calories. Two tablespoons of grape jelly contain 114 calories. Which contains more calories, two tablespoons of peanut butter or two tablespoons of grape jelly?

92. *History* In 1892, the diesel engine was patented. In 1844, Samuel F. B. Morse patented the telegraph. Which was patented first, the diesel engine or the telegraph?

93. *Geography* The distance between St. Louis, Missouri, and Reno, Nevada, is 1,892 mi. The distance between St. Louis, Missouri, and San Diego, California, is 1,833 mi. Which is the shorter distance, St. Louis to Reno or St. Louis to San Diego?

94. *Consumerism* The circle graph at the right shows the result of a survey of 150 people who were asked, "What bothers you most about movie theaters?" (a) Among the respondents, what was the most often mentioned complaint? (b) What was the least often mentioned complaint?

95. *Astronomy* As measured at the equator, the diameter of the planet Uranus is 32,200 mi and the diameter of the planet Neptune is 30,800 mi. Which planet is smaller, Uranus or Neptune?

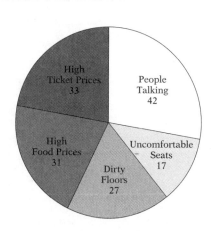

Distribution of Responses in a Survey

Solve.

96. *Astronomy*   The diameter of Callisto, one of the moons orbiting Jupiter, is 4,890 mi. The diameter of Ganymede, another of Jupiter's moons, is 5,216 mi. Which is the larger moon, Callisto or Ganymede?

97. *Travel*   The figure below shows the number of highway fatalities during each of the major holidays during a recent year. (a) What was the number of highway fatalities over Memorial Day? (b) During which holiday was the number of highway fatalities greatest?

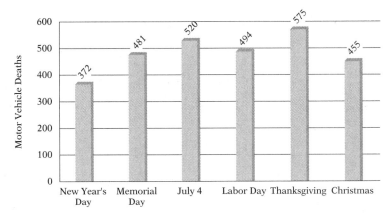

Holiday Highway Fatalities
Source: National Highway Traffic Safety Administration

98. *Geography*   The land area of Alaska is 570,833 mi². What is the land area of Alaska to the nearest thousand square miles?

99. *Geography*   The acreage of the Appalachian Trail is 161,546. What is the acreage of the Appalachian Trail to the nearest ten-thousand acres?

100. *Health*   The figure below shows the leading causes of death for males and females in the United States during a recent year. In which category is the number of female deaths greater than the number of male deaths?

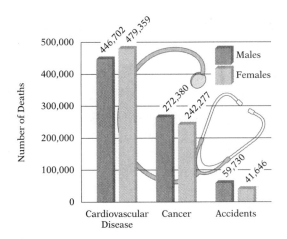

Leading Causes of Death for Males and Females in the U.S.
Source: National Center for Health Statistics and the American Heart Association

Solve.

101. *Education* The student enrollment in the Newfound Area School District in New Hampshire is shown at the right. The jagged line at the bottom of the vertical axis indicates that this scale is missing the hundreds between 0 and 1,300. (a) During which school year was enrollment the lowest? (b) During which school year were less than 1,400 students enrolled?

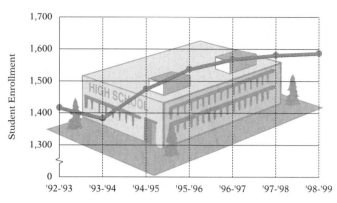

Enrollment in the Newfound Area School District
Source: Newfound Area School District Newsletter

102. *Aviation* The cruising speed of a Boeing 747 is 589 mph. What is the cruising speed of a Boeing 747 to the nearest ten miles per hour?

103. *Physics* Light travels at a speed of 299,800 km/s. What is the speed of light to the nearest thousand kilometers per second?

CRITICAL THINKING

104. Find the land area of the seven continents. List the continents in order from largest to smallest. List the oceans on Earth from largest to smallest.

105. What is the largest three-digit number? What is the smallest five-digit number?

106. What is the total enrollment of your school? To what place value would it be reasonable to round this number? Why? To what place value is the population of your town or city rounded? Why? To what place value is the population of your state rounded? To what place value is the population of the United States rounded?

107. Look at the visual illusion pictured at the right. Describe it in words. Find other examples of visual illusions.

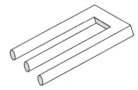

108. What is the national debt of the United States? What does this figure mean?

109. Prepare a report on the symbols used in the Egyptian numeration system.

SECTION 1.2   Addition and Subtraction of Whole Numbers

OBJECTIVE A
Addition of whole numbers

Addition is the process of finding the total of two or more numbers.

On Arbor Day, a community group planted 3 trees along one street and 5 trees along another street. By counting, we can see that there were a total of 8 trees planted.

$$3 \quad + \quad 5 \quad = \quad 8$$

The 3 and 5 are called **addends.** The **sum** is 8.

The basic addition facts for adding one digit to one digit should be memorized. Addition of larger numbers requires the repeated use of the basic addition facts.

To add large numbers, begin by arranging the numbers vertically, keeping the digits of the same place value in the same column.

Find the sum of 211, 45, 23, and 410.

Remember that a *sum* is the answer to an addition problem.
Arrange the numbers vertically, keeping digits of the same place value in the same column.
Add the numbers in each column.

$$\begin{array}{r} 211 \\ 45 \\ 23 \\ + 410 \\ \hline 689 \end{array}$$

The phrase *the sum of* was used in the example above to indicate the operation of addition. All of the phrases listed below indicate addition. An example of each is shown at the right of each phrase.

added to	6 added to 9	$9 + 6$
more than	3 more than 8	$8 + 3$
the sum of	the sum of 7 and 4	$7 + 4$
increased by	2 increased by 5	$2 + 5$
the total of	the total of 1 and 6	$1 + 6$
plus	8 plus 10	$8 + 10$

STUDY TIPS   *Know Your Instructor's Requirements*

To do your best in this course, you must know exactly what your instructor requires. Instructors ordinarily explain course requirements during the first few days of class. Course requirements may be stated in a *syllabus,* which is a printed outline of the main topics of the course, or they may be presented orally. When they are listed in a syllabus or on other printed pages, keep them in a safe place. When course requirements are presented orally, make sure to take complete notes. In either case, understand them completely and follow them exactly.

When the sum of the numbers in a column exceeds 9, addition involves "carrying."

Add: 359 + 478

Add the ones' column.
9 + 8 = 17 (1 ten + 7 ones).
Write the 7 in the ones' column and carry the 1 ten to the tens' column.

```
     Hundreds Tens Ones
          1
      3 | 5 | 9
      4 | 7 | 8
          |   | 7
```

Add the tens' column.
1 + 5 + 7 = 13 (1 hundred + 3 tens).
Write the 3 in the tens' column and carry the 1 hundred to the hundreds' column.

```
    1 1
    359
  + 478
     37
```

Add the hundreds' column.
1 + 3 + 4 = 8 (8 hundreds).
Write the 8 in the hundreds' column.

```
    1 1
    359
  + 478
    837
```

The bar graph in Figure 1.9 shows the seating capacity in 1996 of the five largest National Football League stadiums. What is the total seating capacity of these five stadiums? *Note:* The jagged line below 75,000 on the vertical axis indicates that this scale is missing the numbers less than 75,000.

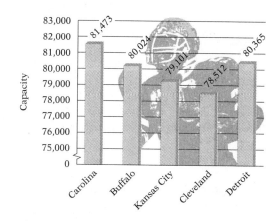

Figure 1.9 Seating Capacity of the Five Largest NFL Stadiums

81,473 + 80,024 + 79,101 + 78,512 + 80,365 = 399,475

The total capacity of the five stadiums is 399,475 people.

STUDY TIPS *Survey the Chapter*

Before you begin reading a chapter, take a few minutes to survey it. Glancing through the chapter will give you an overview of its content and help you see how the pieces fit together as you read.

Begin by reading the chapter title. The title summarizes what the chapter is about. Next read the section headings. The section headings summarize the major topics presented in the chapter. Then read the objectives under each section heading. The objective headings describe the learning goals for that section. Keep these headings in mind as you work through the material. They provide direction as you study.

An important skill in mathematics is the ability to determine whether an answer to a problem is reasonable. One method of determining whether an answer is reasonable is to use estimation. An **estimate** is an approximation.

Estimation is especially valuable when using a calculator. Suppose that you are adding 1,497 and 2,568 on a calculator. You enter the number 1,497 correctly, but you inadvertently enter 256 instead of 2,568 for the second addend. The sum reads 1,753. If you quickly make an estimate of the answer, you can determine that the sum 1,753 is not reasonable and that an error has been made.

$$
\begin{array}{r}
1{,}497 \\
+\ 2{,}568 \\
\hline
4{,}065
\end{array}
\qquad
\begin{array}{r}
1{,}497 \\
+\ \ \ 256 \\
\hline
1{,}753
\end{array}
$$

To estimate the answer to a calculation, round each number to the highest place value of the number; the first digit of each number will be nonzero and all other digits will be zero. Perform the calculation using the rounded numbers.

$$
\begin{array}{r}
1{,}497 \longrightarrow \ \ \ 1{,}000 \\
2{,}568 \longrightarrow +\ 3{,}000 \\
\hline
4{,}000
\end{array}
$$

As shown above, the sum 4,000 is an estimate of the sum of 1,497 and 2,568; it is very close to the actual sum, 4,065. 4,000 is not close to the incorrectly calculated sum, 1,753.

Just as the word *it* is used in language to stand for an object, a letter of the alphabet can be used in mathematics to stand for a number. Such a letter is called a **variable.**

A mathematical expression that contains one or more variables is a **variable expression.** Replacing the variables in a variable expression with numbers and then simplifying the numerical expression is called **evaluating the variable expression.**

Evaluate $a + b$ when $a = 678$ and $b = 294$.

Replace a with 678 and b with 294.

$$a + b$$
$$678 + 294$$

Arrange the numbers vertically.

$$
\begin{array}{r}
{\scriptstyle 1\ 1} \\
678 \\
+\ 294 \\
\hline
972
\end{array}
$$

Add.

Variables are often used in algebra to describe mathematical relationships. Variables are used below to describe three properties, or rules, of addition. An example of each property is shown at the right.

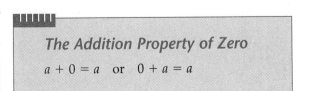

The Addition Property of Zero

$a + 0 = a$ or $0 + a = a$ $5 + 0 = 5$

The Addition Property of Zero states that the sum of a number and zero is the number. The variable a is used here to represent any whole number. It can even represent the number zero because $0 + 0 = 0$.

The Commutative Property of Addition

$a + b = b + a$

$5 + 7 = 7 + 5$
$12 = 12$

The Commutative Property of Addition states that two numbers can be added in either order; the sum will be the same. Here the variables a and b represent any whole numbers. Therefore, if you know that the sum of 5 and 7 is 12, then you also know that the sum of 7 and 5 is 12, because $5 + 7 = 7 + 5$.

The Associative Property of Addition

$(a + b) + c = a + (b + c)$

$(2 + 3) + 4 = 2 + (3 + 4)$
$5 + 4 = 2 + 7$
$9 = 9$

The Associative Property of Addition states that when adding three or more numbers, the numbers can be grouped in any order; the sum will be the same. Note in the example at the right above that we can add the sum of 2 and 3 to 4, or we can add 2 to the sum of 3 and 4. In either case, the sum of the three numbers is 9.

> Rewrite the expression by using the Associative Property of Addition.
>
> $(3 + x) + y$
>
> The Associative Property of Addition states that addends can be grouped in any order. $(3 + x) + y = 3 + (x + y)$

An **equation** expresses the equality of two numerical or variable expressions. In the example above, $(3 + x) + y$ is an expression; it does not contain an equal sign. $(3 + x) + y = 3 + (x + y)$ is an equation; it contains an equal sign.

POINT OF INTEREST

The equal sign (=) is generally credited to Robert Recorde. In his 1557 treatise on algebra, *The Whetstone of Whit*, he wrote, "No two things could be more equal (than two parallel lines)." His equal sign gained popular usage, even though Continental mathematicians preferred a dash.

Here is another example of an equation. The **left side** of the equation is the variable expression $n + 4$. The **right side** of the equation is the number 9. $n + 4 = 9$

Just as a statement in English can be true or false, an equation may be true or false. The equation shown above is *true* if the variable is replaced by 5.

$n + 4 = 9$
$5 + 4 = 9$ True

The equation is *false* if the variable is replaced by 8. $8 + 4 = 9$ False

A **solution** of an equation is a number that, when substituted for the variable, results in a true equation. The solution of the equation $n + 4 = 9$ is 5 because replacing n by 5 results in a true equation. When 8 is substituted for n, the result is a false equation; therefore, 8 is not a solution of the equation.

10 is a solution of $x + 5 = 15$ because $10 + 5 = 15$ is a true equation.

20 is not a solution of $x + 5 = 15$ because $20 + 5 = 15$ is a false equation.

Is 9 a solution of the equation $11 = 2 + x$?

Replace x by 9.

Simplify the right side of the equation. Compare the results. If the results are equal, the given number is a solution of the equation. If the results are not equal, the given number is not a solution.

$$11 = 2 + x$$
$$11 \mid 2 + 9$$
$$11 = 11$$

Yes, 9 is a solution of the equation.

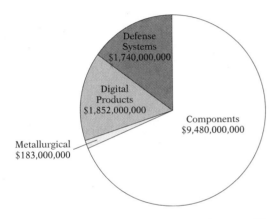

The circle graph in Figure 1.10 shows Texas Instruments' 1995 net revenues by segment. Use this graph for Example 1 and You Try It 1.

Figure 1.10   Texas Instruments 1995 Segment Net Revenues
Source: Texas Instruments 1995 Annual Report

Example 1   Use Figure 1.10 to determine Texas Instruments' total net revenue from its Digital Products and Metallurgical segments.

Solution   The net revenue from the Digital Products segment is $1,852,000,000. The net revenue from the Metallurgical segment is $183,000,000.

$$\begin{array}{r} 1{,}852{,}000{,}000 \\ +\quad 183{,}000{,}000 \\ \hline 2{,}035{,}000{,}000 \end{array}$$

The total net revenue from these two segments is $2,035,000,000.

You Try It 1   Use Figure 1.10 to determine the sum of the net revenues from Texas Instruments' Defense Systems and Components segments.

Your Solution

Solution on p. A5

Example 2 Estimate the sum of 379, 842, 693, and 518.

Solution

$$
\begin{array}{rcl}
379 & \longrightarrow & 400 \\
842 & \longrightarrow & 800 \\
693 & \longrightarrow & 700 \\
518 & \longrightarrow & +\ 500 \\
\hline
& & 2{,}400
\end{array}
$$

You Try It 2 Estimate the total of 6,285, 3,972, and 5,140.

Your Solution

Example 3 Evaluate $x + y + z$ when $x = 8{,}427$, $y = 3{,}659$, and $z = 6{,}281$.

Solution $x + y + z$
$8{,}427 + 3{,}659 + 6{,}281$

$$
\begin{array}{r}
{\scriptstyle 1\ 11} \\
8{,}427 \\
3{,}659 \\
+\ 6{,}281 \\
\hline
18{,}367
\end{array}
$$

You Try It 3 Evaluate $x + y + z$ when $x = 1{,}692$, $y = 4{,}783$, and $z = 5{,}046$.

Your Solution

Example 4 Identify the property that justifies the statement.

$7 + 2 = 2 + 7$

Solution The Commutative Property of Addition

You Try It 4 Identify the property that justifies the statement.

$33 + 0 = 33$

Your Solution

Example 5 Is 6 a solution of the equation $9 + y = 14$?

Solution

$$
\begin{array}{c|c}
9 + y = 14 \\
\hline
9 + 6 & 14 \\
15 \neq 14
\end{array}
$$
 The symbol $\neq$ is read "is not equal to."

No, 6 is not a solution of the equation $9 + y = 14$.

You Try It 5 Is 7 a solution of the equation $13 = b + 6$?

Your Solution

Solutions on p. A5

STUDY TIPS *Become an Active Participant*

Many students feel that they will never understand math, whereas others appear to do very well with little effort. Oftentimes what makes the difference is that successful students take an active role in the learning process.

Learning mathematics requires your *active* participation. Although doing your homework is one way you can actively participate, it is not the only way. First, you must attend class regularly and become an active participant. Second, you must become actively involved with the textbook. *Prealgebra* was written and designed with you in mind as a participant.

OBJECTIVE B
Subtraction of whole numbers

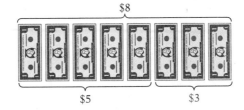

Subtraction is the process of finding the difference between two numbers.

By counting, we see that the difference between $8 and $5 is $3.

$$\$8 \quad - \quad \$5 \quad = \quad \$3$$

Minuend − Subtrahend = Difference

Note that addition and subtraction are related.

Subtrahend	5
+ Difference	+ 3
= Minuend	8

The fact that the sum of the subtrahend and the difference equals the minuend can be used to check subtraction.

To subtract large numbers, begin by arranging the numbers vertically, keeping the digits of the same place value in the same column. Then subtract the numbers in each column.

Find the difference between 8,955 and 2,432.

A *difference* is the answer to a subtraction problem.

$$\begin{array}{r} 8\,9\,5\,5 \\ -\,2\,4\,3\,2 \\ \hline 6\,5\,2\,3 \end{array}$$

Check:	Subtrahend	2,432
	+ Difference	+ 6,523
	= Minuend	8,955

In the subtraction example above, the lower digit in each place value is smaller than the upper digit. When the lower digit is larger than the upper digit, subtraction involves "borrowing."

Subtract: 692 − 378

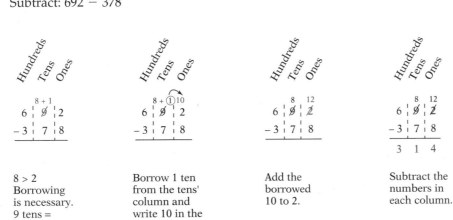

8 > 2
Borrowing
is necessary.
9 tens =
8 tens + 1 ten

Borrow 1 ten
from the tens'
column and
write 10 in the
ones' column.

Add the
borrowed
10 to 2.

Subtract the
numbers in
each column.

Subtraction may involve repeated borrowing.

Subtract: 7,325 − 4,698

$$
\begin{array}{r}
\overset{1}{}\overset{15}{} \\
7,3\ \cancel{2}\ \cancel{5} \\
-4,6\ 9\ 8 \\
\hline
7
\end{array}
$$

Borrow 1 ten
(10 ones) from the
tens' column and
add 10 to the 5 in
the ones' column.
Subtract 15 − 8.

$$
\begin{array}{r}
\overset{11}{} \\
\overset{2}{}\ \cancel{\ }\ \overset{15}{} \\
7,\ \cancel{3}\ \cancel{2}\ \cancel{5} \\
-4,\ 6\ 9\ 8 \\
\hline
2\ 7
\end{array}
$$

Borrow 1 hundred
(10 tens) from the
hundreds' column
and add 10 to the 1
in the tens' column.
Subtract 11 − 9.

$$
\begin{array}{r}
\overset{12}{}\ \overset{11}{} \\
6\ \cancel{2}\ \cancel{1}\ \mathbf{15} \\
\cancel{7},\ \cancel{3}\ \cancel{2}\ \cancel{5} \\
-4,\ 6\ 9\ 8 \\
\hline
2,\ 6\ 2\ 7
\end{array}
$$

Borrow 1 thousand
(10 hundreds) from
the thousands' column
and add 10 to the 2 in
the hundreds' column.
Subtract 12 − 6 and
6 − 4.

When there is a zero in the minuend, subtraction involves repeated borrowing.

Subtract: 3,904 − 1,775

$$
\begin{array}{r}
\overset{8}{}\ \overset{10}{} \\
3,\ 9\ 0\ 4 \\
-1,\ 7\ 7\ 5 \\
\hline
\end{array}
$$

There is a 0 in the
tens' column. Borrow
1 hundred (10 tens)
from the hundreds'
column and write 10
in the tens' column.

$$
\begin{array}{r}
\overset{9}{} \\
8\ \cancel{10}\ 14 \\
3,\ \cancel{9}\ \cancel{0}\ \cancel{4} \\
-1,\ 7\ 7\ 5 \\
\hline
\end{array}
$$

Borrow 1 ten from
the tens' column
and add 10 to the
4 in the ones'
column.

$$
\begin{array}{r}
\overset{9}{} \\
8\ \cancel{10}\ 14 \\
3,\ \cancel{9}\ \cancel{0}\ \cancel{4} \\
-1,\ 7\ 7\ 5 \\
\hline
2,\ 1\ 2\ 9
\end{array}
$$

Subtract the
numbers in
each column.

Note that, for the preceding example, the borrowing could be performed as shown below.

Borrow 1 from 90. (90 − 1 = 89. The 8 is in the hundreds' column. The 9 is in the tens' column.) Add 10 to
the 4 in the ones' column. Then subtract the numbers in
each column.

$$
\begin{array}{r}
8\ 914 \\
3,9\cancel{0}\cancel{4} \\
-1,775 \\
\hline
2,129
\end{array}
$$

Estimate the difference between 27,843 and 19,206.

Round each number to the nearest ten-
thousand.
Subtract the rounded numbers.

$$
\begin{array}{rcr}
27,843 & \longrightarrow & 30,000 \\
19,206 & \longrightarrow & -\ 20,000 \\
\hline
& & 10,000
\end{array}
$$

10,000 is an estimate of the difference between
27,843 and 19,206.

The total number of games won by each of the eight teams that qualified for a playoff spot for the Eastern Division of the 1996 NBA finals are shown in the bar graph in Figure 1.11. Find the difference between the number of games that Orlando won and the number of games that Miami won.

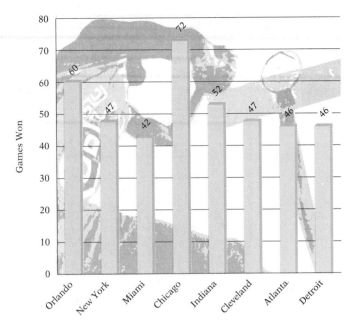

Figure 1.11   Games Won by Qualifying Teams

Orlando won 60 games.

Miami won 42 games. $60 - 42 = 18$

Subtract 42 from 60.   The difference is 18 games.

The phrase *the difference between* was used in the example above to indicate the operation of subtraction. All of the phrases listed below indicate subtraction. An example of each is shown at the right of each phrase.

minus	10 minus 3	$10 - 3$
less	8 less 4	$8 - 4$
less than	2 less than 9	$9 - 2$
the difference between	the difference between 6 and 1	$6 - 1$
decreased by	7 decreased by 5	$7 - 5$

TAKE NOTE

Note the order in which the numbers are subtracted when the phrase *less than* is used. Suppose that you have $10 and I have $6 *less than* you do; then I have $6 less than $10, or $10 - $6 = $4.

Evaluate $c - d$ when $c = 6,183$ and $d = 2,759$.

Replace c with 6,183 and d with 2,759.

$$c - d$$
$$6,183 - 2,759$$

Arrange the numbers vertically and then subtract.

$$\begin{array}{r} {}^{5\ \ 11\ 7\ 13} \\ 6,\cancel{1}\cancel{8}\cancel{3} \\ -2,759 \\ \hline 3,424 \end{array}$$

Is 23 a solution of the equation $41 - n = 17$?

Replace n by 23.
Simplify the left side of the equation.
The results are not equal.

$$41 - n = 17$$
$$\frac{41 - 23 \ \vert \ 17}{}$$
$$18 \neq 17$$

No, 23 is not a solution of the equation.

POINT OF INTEREST

Someone who is our equal is our peer. Two make a pair. Both of the words *peer* and *pair* come from the Latin *par, paris,* meaning "equal."

The graph in Figure 1.12 shows the number of Pier 1 Imports stores world-wide in the years 1992 through 1996. Use this graph for Example 6 and You Try It 6.

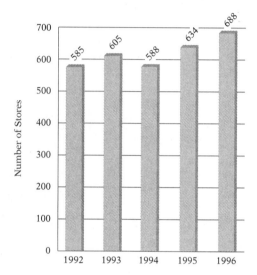

Figure 1.12 Number of Pier 1 Imports Stores Worldwide
Source: Pier 1 Imports 1996 Annual Report

Example 6 Use Figure 1.12 to find the difference between the number of Pier 1 Imports stores in 1996 and 1992.

Solution The number of stores in 1996 was 688. The number of stores in 1992 was 585.

$$\begin{array}{r} 688 \\ -585 \\ \hline 103 \end{array}$$

The difference is 103 stores.

You Try It 6 Use Figure 1.12 to find the difference between the number of Pier 1 Imports stores in 1995 and 1993.

Your Solution

Example 7 Subtract and check:
57,004 − 26,189

Solution

$$\begin{array}{r} {\scriptstyle 6\ \ 9\ 9\ 14} \\ 5\,\cancel{7},\cancel{0}\,\cancel{0}\,4 \\ -2\,6\,,1\,8\,9 \\ \hline 3\,0\,,8\,1\,5 \end{array}$$

$$\text{Check:} \quad \begin{array}{r} 26,189 \\ +30,815 \\ \hline 57,004 \end{array}$$

You Try It 7 Subtract and check:
49,002 − 31,865

Your Solution

Example 8 Estimate the difference between 7,261 and 4,315. Then find the exact answer.

Solution
$$7,261 \longrightarrow 7,000$$
$$4,315 \longrightarrow -4,000$$
$$3,000$$

$$7,261$$
$$-4,315$$
$$2,946$$

You Try It 8 Estimate the difference between 8,544 and 3,621. Then find the exact answer.

Your Solution

Example 9 Evaluate $x - y$ when $x = 3,506$ and $y = 2,477$

Solution $x - y$
$3,506 - 2,477$

$$\overset{4\ \ 9\ \ 16}{3,\cancel{506}}$$
$$-2,477$$
$$1,029$$

You Try It 9 Evaluate $x - y$ when $x = 7,061$ and $y = 3,229$.

Your Solution

Example 10 Is 39 a solution of the equation $24 = m - 15$?

Solution
$$24 = m - 15$$
$$24 \mid 39 - 15$$
$$24 = 24$$

Yes, 39 is a solution of the equation.

You Try It 10 Is 11 a solution of the equation $46 = 58 - p$?

Your Solution

Solutions on p. A6

STUDY TIPS   *Take Careful Notes in Class*

Attending class is vital if you are to succeed in this course. You need a notebook in which to keep class notes and records about assignments. Make sure to take complete and well-organized notes. Your instructor will explain text material that may be difficult for you to understand on your own, will provide practice in the skills you are learning, and may provide important information that is not provided in the textbook. Be sure to include in your notes everything that is written on the chalkboard.

Information recorded in your notes about assignments should explain exactly what they are, how they are to be done, and when they are due. Information about tests should include exactly what text materials and topics will be covered on each test and the dates on which the tests will be given.

Remember that you are responsible for *everything* that happens in class, even if you are absent. If you must be absent from a class session:

1. Deliver due assignments to the instructor as soon as possible.

2. Contact a classmate to learn about assignments or tests announced during your absence.

3. Hand-copy or photocopy notes taken by a classmate while you were absent.

OBJECTIVE C

Applications and formulas

In this section, some of the phrases used to indicate the operations of addition and subtraction were presented. In solving application problems, you might also look for the types of questions listed below.

Addition	*Subtraction*
How many . . . altogether?	How many more (or fewer) . . . ?
How many . . . in all?	How much is left?
How many . . . and . . . ?	How much larger (or smaller) . . . ?

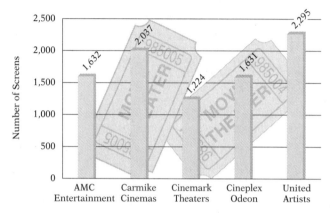

The bar graph in Figure 1.13 shows the number of screens owned by each of the top five movie theater chains in North America. Use this graph for Example 11 and You Try It 11.

Figure 1.13 Screens Owned by the Top Five Movie Theater Chains in North America

Source: National Association of Theatre Owners

	Example 11	According to Figure 1.13, of the top five chains, how many more screens are owned by the chain that owns the most screens than by the chain that owns the fewest screens?	You Try It 11	Use Figure 1.13 to find the total number of screens owned by the three theater chains that own the most screens.

Strategy To find how many more are owned by the chain that owns the most screens:

▶ Find the number of screens owned by the chain that owns the most screens and the number owned by the chain that owns the fewest.
▶ Subtract the smaller number from the larger.

Your Strategy

Solution The most: United Artists, 2,295
The fewest: Cinemark Theaters, 1,224

$$2,295 - 1,224 = 1,071$$

United Artists owns 1,071 more screens than Cinemark Theaters.

Your Solution

Solution on p. A6

Example 12

The height of the Carnegie Tower in New York City is 756 ft. The height of the Trump Tower in New York City is 664 ft. Find the difference in height between the Carnegie Tower and the Trump Tower.

Strategy

To find the difference, subtract the height of the Trump Tower (664) from the height of the Carnegie Tower (756).

Solution

$$\begin{array}{r} 756 \\ -\ 664 \\ \hline 92 \end{array}$$

The difference in height is 92 ft.

You Try It 12

The height of the World Trade Center in New York City is 1,368 ft. The height of the Empire State Building is 1,250 ft. How much taller is the World Trade Center than the Empire State Building?

Your Strategy

Your Solution

Solution on p. A6

Study Tips   Use the Textbook to Learn the Material

For each objective studied, read the objective statement carefully so you understand the learning goal that is being presented. Next, read very carefully all of the objective material. As you read, note carefully the words printed in **boldface** type. These words indicate important concepts that you should familiarize yourself with. Study each in-text example carefully, noting the techniques and strategies used to solve the example.

You will then come to the key learning feature of this text, the *boxed examples.* These examples have been designed to aid you in a very specific way. Notice that in each example box, the example on the left is completely worked out and the example on the right is not. The reason for this is that *you* are expected to work the right-hand example (in the space provided) in order to immediately test your understanding of the material you have just studied.

Carefully study the worked-out example on the left by working through each step presented. This allows you to focus on each step and reinforces the technique for solving that type of problem. Then you should solve the right-hand example using the problem-solving techniques that you have just studied. When you have completed your solution, check your work by turning to the page in the Appendix where the complete solution is provided. The page number on which the solution appears is printed at the bottom right-hand side of the example box. By checking your solution, you will know immediately whether you understand the skill just studied. If your answer is incorrect, check each step of your solution against the steps given in the answer section. It may be helpful to review the example on the left also. Determine where you made your mistakes. Do this for each pair of examples.

Next, do the problems in the exercise set that correspond to the objective just studied. An exercise set follows every section in the textbook. Exercise sets are identified by objective. For example, the exercises corresponding to Section 2.2, Objective A are labeled 2.2 Exercises, Objective A. The answers to all the odd-numbered exercises appear in the answer section in the back of the book. Check your answers to the exercises against these. If you have difficulty solving problems in the exercise set, review the material in the text. Many examples are solved within the text material. Review the solutions to these problems. Then restudy the boxed examples provided for the objective.

Example 13

What is the price of a pair of skates that cost a business $109 and has a markup of $49? Use the formula $P = C + M$, where P is the price of a product to the consumer, C is the cost paid by the store for the product, and M is the markup.

Strategy

To find the price, replace C by 109 and M by 49 in the given formula and solve for P.

Solution

$P = C + M$
$P = 109 + 49$
$P = 158$

The price of the skates is $158.

You Try It 13

What is the price of a leather jacket that cost a business $148 and has a markup of $74? Use the formula $P = C + M$, where P is the price of a product to the consumer, C is the cost paid by the store for the product, and M is the markup.

Your Strategy

Your Solution

Solution on p. A6

STUDY TIPS   *Keep Up to Date with Coursework*

College terms start out slowly. Then they gradually get busier and busier, reaching a peak of activity at final examination time. If you fall behind in the work for a course, you will find yourself trying to catch up at a time when you are very busy with all your other courses. Don't fall behind—keep up to date with course work.

Keeping up with course work is doubly important for a course in which information and skills learned early in the course are needed to learn information and skills later in the course. Prealgebra is such a course. Skills must be learned immediately and reviewed often.

Your instructor gives assignments to help you acquire a skill or understand a concept. Do each assignment as it is assigned or you may well fall behind and have great difficulty catching up. Keeping up with course work also makes it easier to prepare for each exam.

1.2 EXERCISES

OBJECTIVE A

Add.

1. 732,453
   + 651,206

2. 563,841
   + 726,053

3. 2,879
   + 3,164

4. 9,857
   + 1,264

5. 45,825
   + 66,327

6. 56,442
   + 71,289

7. 4,037
   3,342
   + 5,169

8. 5,242
   7,883
   + 4,165

9. 67,390
   42,761
   + 89,405

10. 34,801
   97,302
   + 68,945

11. 54,097
   33,432
   97,126
   64,508
   + 78,310

12. 23,086
   44,697
   67,302
   83,441
   + 19,843

Solve.

13. What is 88,123 increased by 80,451?

14. What is 44,765 more than 82,003?

15. What is 654 added to 7,293?

16. Find the sum of 658, 2,709, and 10,935.

17. Find the total of 216, 8,707, and 90,714.

18. Write the sum of x and y.

19. Education   Use the figure on the right to find the total number of undergraduates enrolled at the college in 1996.

20. Education   Use the figure on the right to find the total number of undergraduates enrolled at the college in 1997.

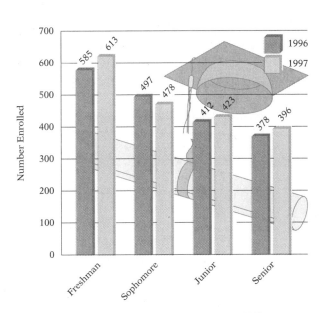

Undergraduates Enrolled in a Private College

Estimate by rounding. Then find the exact answer.

21. 6,742 + 8,298

22. 5,426 + 1,732

23. 972,085 + 416,832

24. 23,774 + 38,026

25.
```
   387
   295
   614
 + 702
```

26.
```
   528
   163
   947
 + 275
```

27.
```
  224,196
    7,074
 + 98,531
```

28.
```
    1,607
  873,925
 + 28,744
```

Evaluate the variable expression $x + y$ for the given values of x and y.

29. $x = 574; y = 698$

30. $x = 359; y = 884$

31. $x = 4,752; y = 7,398$

32. $x = 6,047; y = 9,283$

33. $x = 38,229; y = 51,671$

34. $x = 74,376; y = 19,528$

Evaluate the variable expression $a + b + c$ for the given values of a, b, and c.

35. $a = 693; b = 508; c = 371$

36. $a = 177; b = 892; c = 405$

37. $a = 4,938; b = 2,615; c = 7,038$

38. $a = 6,059; b = 3,774; c = 5,136$

39. $a = 12,897; b = 36,075; c = 48,441$

40. $a = 52,847; b = 49,036; c = 24,717$

Identify the property that justifies the statement.

41. $9 + 12 = 12 + 9$

42. $8 + 0 = 8$

43. $11 + (13 + 5) = (11 + 13) + 5$

44. $0 + 16 = 16 + 0$

45. $0 + 47 = 47$

46. $(7 + 8) + 10 = 7 + (8 + 10)$

Use the given property of addition to complete the statement.

47. The Addition Property of Zero
$28 + 0 = ?$

48. The Commutative Property of Addition
$16 + ? = 7 + 16$

49. The Associative Property of Addition
$9 + (? + 17) = (9 + 4) + 17$

50. The Addition Property of Zero
$0 + ? = 51$

51. The Commutative Property of Addition
$? + 34 = 34 + 15$

52. The Associative Property of Addition
$(6 + 18) + ? = 6 + (18 + 4)$

53. Is 38 a solution of the equation $42 = n + 4$?

54. Is 17 a solution of the equation $m + 6 = 13$?

55. Is 13 a solution of the equation $2 + h = 16$?

56. Is 41 a solution of the equation $n = 17 + 24$?

57. Is 30 a solution of the equation $32 = x + 2$?

58. Is 29 a solution of the equation $38 = 11 + z$?

OBJECTIVE B

Subtract.

59.
$$\begin{array}{r} 883 \\ -\ 467 \\ \hline \end{array}$$

60.
$$\begin{array}{r} 591 \\ -\ 238 \\ \hline \end{array}$$

61.
$$\begin{array}{r} 360 \\ -\ 172 \\ \hline \end{array}$$

62.
$$\begin{array}{r} 950 \\ -\ 483 \\ \hline \end{array}$$

63.
$$\begin{array}{r} 657 \\ -\ 193 \\ \hline \end{array}$$

64.
$$\begin{array}{r} 762 \\ -\ 659 \\ \hline \end{array}$$

65.
$$\begin{array}{r} 407 \\ -\ 199 \\ \hline \end{array}$$

66.
$$\begin{array}{r} 805 \\ -\ 147 \\ \hline \end{array}$$

Subtract.

67. 6,814
− 3,257

68. 7,361
− 4,575

69. 5,000
− 2,164

70. 4,000
− 1,873

71. 3,400
− 1,963

72. 7,300
− 2,562

73. 30,004
− 9,856

74. 70,003
− 8,246

Solve.

75. Find the difference between 2,536 and 918.

76. What is 1,623 minus 287?

77. What is 5,426 less than 12,804?

78. Find 14,801 less 3,522?

79. Find 85,423 decreased by 67,875.

80. Write the difference between x and y.

81. ✍ *Geology* Use the figure on the right to find the difference between the maximum height to which Great Fountain erupts and the maximum height to which Valentine erupts.

82. ✍ *Geology* According to the figure on the right, how much higher is the eruption of the Giant than that of Old Faithful?

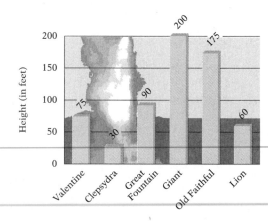

The Maximum Heights of the Eruptions of Six Geysers at Yellowstone National Park

Estimate by rounding. Then find the exact answer.

83. 7,355 − 5,219

84. 8,953 − 2,217

85. 59,126 − 20,843

86. 63,051 − 29,478

87. 36,287
− 5,092

88. 58,316
− 19,072

89. 224,196
− 98,531

90. 873,925
− 28,744

Evaluate the variable expression $x - y$ for the given values of x and y.

91. $x = 50; y = 37$

92. $x = 80; y = 33$

93. $x = 914; y = 271$

94. $x = 623; y = 197$

95. $x = 740; y = 385$

96. $x = 870; y = 243$

97. $x = 8,672; y = 3,461$

98. $x = 7,814; y = 3,512$

99. $x = 1,605; y = 839$

100. $x = 1,406; y = 968$

101. $x = 23,409; y = 5,178$

102. $x = 56,397; y = 8,249$

103. Is 24 a solution of the equation $29 = 53 - y$?

104. Is 31 a solution of the equation $48 - p = 17$?

105. Is 44 a solution of the equation $t - 16 = 60$?

106. Is 25 a solution of the equation $34 = x - 9$?

107. Is 27 a solution of the equation $82 - z = 55$

108. Is 28 a solution of the equation $72 = 100 - d$?

OBJECTIVE C

Solve.

109. *Mathematics* What is the sum of all the whole numbers less than 21?

110. *Mathematics* Find the sum of all the natural numbers greater than 89 and less than 101.

111. *Mathematics* Find the difference between the smallest four-digit number and the largest two-digit number.

112. *Demography* The figure on the right shows the expected U.S. population aged 100 and over every two years from 1994 to 2004. (a) During which two-year period is the smallest increase in the number of people aged 100 and over? (b) During which two-year period is the greatest increase?

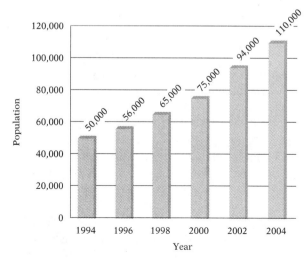

Expected U.S. Population Aged 100 and Over
Source: Census Bureau

Solve.

113. *Nutrition*   You eat an apple and one cup of cornflakes with one table-spoon of sugar and one cup of milk for breakfast. Find the total number of calories consumed if one apple contains 80 calories, one cup of corn-flakes has 95 calories, one tablespoon of sugar has 45 calories, and one cup of milk has 150 calories.

114. *Health*   You are on a diet to lose weight and are limited to 1,500 calo-ries per day. If your breakfast and lunch contained 950 calories, how many more calories can you consume during the rest of the day?

The figure on the right shows the top five professional women tennis play-ers with the most victories through 1994. Use this graph for Exercises 115 to 117.

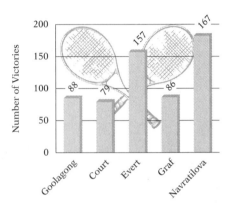

Number of Victories for the Top Five
Professional Women Tennis Players
Source: 1996 Information Please Sports Almanac

115. *Sports*   (a) What is the total number of wins for the women with the two highest number of wins? (b) What is the total number of wins for the women with the two lowest number of wins?

116. *Sports*   Is the total number of victories for the two women with the lowest number of wins greater than or less than the total number of wins for the woman with the highest number of wins?

117. *Sports*   Is the total number of victories for the three women with the lowest number of wins greater than or less than the total number of wins for the two women with the highest number of wins?

Solve.

118. *History*   The Gemini-Titan 7 space flight made 206 orbits of Earth. The Apollo-Saturn 7 space flight made 163 orbits of Earth. How many more orbits did the Gemini-Titan 7 flight make than the Apollo-Saturn 7 flight?

119. *Finances*   You had $1,054 in your checking account before making a deposit of $870. Find the amount in your checking account after you made the deposit.

120. *Sports*   The seating capacity of the Kingdome in Seattle is 57,748. The seating capacity of Fenway Park in Boston is 34,182. Find the difference between the seating capacity of the Kingdome and Fenway Park.

121. *Finances*   The repair bill on your car includes $179 for parts, $78 for labor, and a sales tax of $15. What is the total amount owed?

Auto Repair Bill	
Parts	$ 179
Labor	$ 78
Sales tax	$ 15
Total	
	01373

Solve.

122. *Finances*   The computer system you would like to purchase includes an operating system priced at $830, a monitor that costs $245, an extended keyboard priced at $175, and a printer that sells for $395. What is the total cost of the computer system?

123. *Geography*   The area of Lake Superior is 81,000 mi²; the area of Lake Michigan is 67,900 mi²; the area of Lake Huron is 74,000 mi²; the area of Lake Erie is 32,630 mi²; and the area of Lake Ontario is 34,850 mi². Estimate the total area of the five Great Lakes.

124. *Consumerism*   The odometer on your car read 58,376 this time last year. It now reads 77,912. Estimate the number of miles your car has been driven during the past year.

The figure on the right shows the number of cars sold by a dealership for the first four months of 1996 and 1997. Use this graph for Exercises 125 to 127.

125. *Business*   Between which two months did car sales decrease the most for the two years shown in the graph? What was the amount of decrease?

126. *Business*   Between which two months did car sales increase the most for the two years shown in the graph? What was the amount of increase?

127. *Business*   In which year were more cars sold during the four months shown?

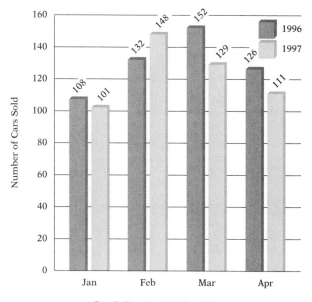

Car Sales at a Dealership

Solve.

128. *Finances*   Use the formula $A = P + I$, where A is the value of an investment, P is the original investment, and I is the interest earned, to find the value of an investment that earned $775 in interest on an original investment of $12,500.

129. *Finances*   Use the formula $A = P + I$, where A is the value of an investment, P is the original investment, and I is the interest earned, to find the value of an investment that earned $484 in interest on an original investment of $8,800.

130. *Finances*   What is the mortgage loan amount on a home that sells for $145,000 with a down payment of $14,500? Use the formula $M = S - D$, where M is the mortgage loan amount, S is the selling price, and D is the down payment.

131. *Finances*   What is the mortgage loan amount on a home that sells for $118,000 with a down payment of $23,600? Use the formula $M = S - D$, where M is the mortgage loan amount, S is the selling price, and D is the down payment.

Solve.

132. *Physics*   What is the ground speed of an airplane traveling into a 25 mph head wind with an air speed of 375 mph? Use the formula $g = a - h$, where g is the ground speed, a is the air speed, and h is the speed of the head wind.

133. *Physics*   Find the ground speed of an airplane traveling into a 15 mph head wind with an air speed of 425 mph? Use the formula $g = a - h$, where g is the ground speed, a is the air speed, and h is the speed of the head wind.

In some states, the speed limit on certain sections of highway is 70 mph. To test drivers' compliance with the speed limit, the highway patrol conducted a one-week study during which they recorded the speeds of motorists on one of these sections of highway. The results are recorded in the table at the right. Use this table for Exercises 134 to 137.

Speed	Number of Cars
> 80	1,708
76 – 80	2,503
71 – 75	3,651
66 – 70	3,717
61 – 65	2,984
< 61	2,870

134. *Statistics*   (a) How many drivers were traveling at 70 mph or less? (b) How many drivers were traveling at 76 mph or more?

135. *Statistics*   Looking at the data in the table, is it possible to tell how many motorists were driving at 70 mph? Explain your answer.

136. *Statistics*   Looking at the data in the table, is it possible to tell how many motorists were driving at less than 70 mph? Explain your answer.

137. *Statistics*   Are more people driving at or below the posted speed limit, or are more people driving above the posted speed limit?

CRITICAL THINKING

138. If you roll two ordinary six-sided dice and add the two numbers that appear on top, how many different sums are possible?

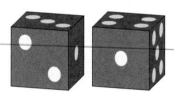

139. How many two-digit numbers are there? How many three-digit numbers are there?

140. Determine whether the statement is always true, sometimes true, or never true.
   a. If a is any whole number, then $a - 0 = a$.
   b. If a is any whole number, then $a - a = 0$.

141. Find the circulation of your local newspaper and the population of the area served by that paper. What is the difference between the area's population and the newspaper's circulation? Why would this figure be of concern to the owner of the newspaper?

142. What estimate is given for the size of the population in your state by the year 2000? What is the estimate of the size of the population in the United States by the year 2000? Estimates differ. On what basis was the estimate you recorded derived?

 **SECTION 1.3** **Multiplication and Division of Whole Numbers**

OBJECTIVE A
Multiplication of whole numbers

A store manager orders six boxes of telephone answering machines. Each box contains eight answering machines. How many answering machines are ordered?

The answer can be calculated by adding 6 eights.

$$8 + 8 + 8 + 8 + 8 + 8 = 48$$

This problem involves repeated addition of the same number. The answer can be calculated by a shorter process called multiplication. **Multiplication** is the repeated addition of the same number.

There is a total of 48 dots on the 6 dominoes.

The numbers that are multiplied are called **factors.** The answer is called the **product.**

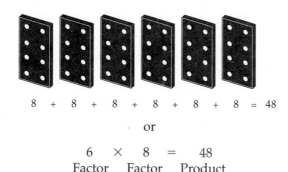

$$8 \;+\; 8 \;+\; 8 \;+\; 8 \;+\; 8 \;+\; 8 \;=\; 48$$

or

$$6 \quad \times \quad 8 \quad = \quad 48$$
Factor   Factor   Product

The time sign "×" is one symbol that is used to mean multiplication. Each of the expressions below also represents multiplication.

$$6 \cdot 8 \qquad 6(8) \qquad (6)(8) \qquad 6a \qquad 6(a) \qquad ab$$

The expression $6a$ means "6 times a." The expression ab means "a times b."

The basic facts for multiplying one-digit numbers should be memorized. Multiplication of larger numbers requires the repeated use of the basic multiplication facts.

Multiply: 37(4)

Multiply $4 \cdot 7$.

$4 \cdot 7 = 28$ (2 tens + 8 ones).

Write the 8 in the ones' column and carry the 2 to the tens' column.

$$\begin{array}{r} \overset{2}{3}\,7 \\ \times 4 \\ \hline 8 \end{array}$$

The 3 in 37 is 3 tens.

Multiply $4 \cdot 3$ tens.
Add the carry digit.

Write the 14.

$$4 \cdot 3 \text{ tens} = \begin{array}{r} 12 \text{ tens} \\ + \; 2 \text{ tens} \\ \hline 14 \text{ tens} \end{array}$$

$$\begin{array}{r} \overset{2}{3}\,7 \\ \times 4 \\ \hline 14\;8 \end{array}$$

In the preceding example, a number was multiplied by a one-digit number. The examples that follow illustrate multiplication by larger numbers.

Multiply: (47)(23)

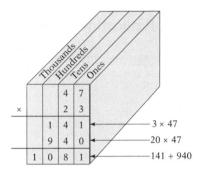

Multiply by the ones' digit.	Multiply by the tens' digit.	Add.
3 · 47 = 141.	2 · 47 = 94.	

```
       47              47              47
     × 23            × 23            × 23
      141             141             141
                       94              94
                                    1,081
```

— 3 × 47
— 20 × 47
— 141 + 940

The last digit is written in the ones' column.	The last digit is written in the tens' column.	The place-value chart illustrates the placement of the products.

Note the placement of the products when multiplying by a factor that contains a zero.

Multiply: 439(206)

```
     439
   × 206
   2 634
   0 00
  87 8
  90,434
```

When working the problem, usually only one zero is written, as shown at the right. Writing this zero ensures the proper placement of the products.

```
     439
   × 206
   2 634
  87 80
  90,434
```

Note the pattern when the following numbers are multiplied.

Multiply the nonzero part of the factors. Attach the same number of zeros in the product as the total number of zeros in the factors.

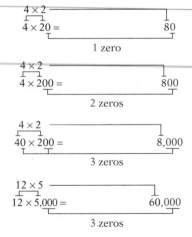

```
4 × 2
4 × 20 =              80
          1 zero

4 × 2
4 × 200 =            800
          2 zeros

4 × 2
40 × 200 =        8,000
          3 zeros

12 × 5
12 × 5,000 =     60,000
          3 zeros
```

Find the product of 600 and 70.

Remember that a *product* is the answer to a multiplication problem.

$$600 \cdot 70 = 42,000$$

Multiply: 3(20)(10)(4)

Multiply the first two numbers.	3(20)(10)4 = 60(10)(4)
Multiply the product by the third number.	= (600)(4)
Continue multiplying until all the numbers have been multiplied.	= 2,400

Figure 1.14 shows what leading commencement speakers earn for one graduation speech. Using these figures, what would be an athlete's earnings for speaking at four graduations next May?

Multiply the number of speeches (4) times the amount earned for one speech ($30,000).

4(30,000) = 120,000

An athlete would earn $120,000 for speaking at four graduations.

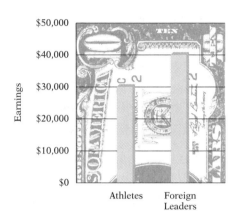

Figure 1.14 Leading Commencement Speakers' Earnings Per Speech
Source: Leading Authorities, Inc.

STUDY TIPS   *Practice Good Study Habits*

Find a place to study where you are comfortable and can concentrate well. Many students find the campus library to be a good place. You might select two or three places at the college library where you like to study. Or there may be a small, quiet lounge on the third floor of a building where you find you can study well. Take the time to find places that promote good study habits.

Spaced practice is generally superior to massed practice. For example, four half-hour study periods will produce more learning than one two-hour study session. The following suggestions may help you decide when you will study.

1. A free period immediately before class is the best time to study about the lecture topic for class.

2. A free period after class is the best time to review notes taken during a class and to begin your homework assignment. It allows you to immediately apply the concepts you have just learned in class.

3. A brief period of time is good for reciting or reviewing information.

4. A long period of time of an hour or more is good for doing challenging activities such as learning to solve a new type of problem.

5. Free periods just before you go to sleep are good times for learning information. (There is evidence that information learned just before sleep is remembered longer than information learned at other times.)

Study followed by reward is usually productive. Schedule something enjoyable to do following study sessions. If you know that you have only two hours to study because you have scheduled a pleasant activity for yourself, you may be inspired to make the best use of the two hours that you have set aside for studying.

Estimate the product of 345 and 92.

Round each number to its highest place value.

$$345 \longrightarrow 300$$
$$92 \longrightarrow 90$$

Multiply the rounded numbers.

$$300 \cdot 90 = 27{,}000$$

27,000 is an estimate of the product of 345 and 92.

The phrase *the product of* was used in the example above to indicate the operation of multiplication. All of the phrases below indicate multiplication. An example of each is shown at the right of each phrase.

times	8 times 4	$8 \cdot 4$
the product of	the product of 9 and 5	$9 \cdot 5$
multiplied by	7 multiplied by 3	$3 \cdot 7$
twice	twice 6	$2 \cdot 6$

Evaluate xyz when $x = 50$, $y = 2$, and $z = 7$.

xyz means $x \cdot y \cdot z$.

$$xyz$$

Replace each variable by its value.

$$50 \cdot 2 \cdot 7$$

Multiply the first two numbers.

$$= 100 \cdot 7$$

Multiply the product by the next number.

$$= 700$$

As for addition, there are properties of multiplication.

The Multiplication Property of Zero

$a \cdot 0 = 0$ or $0 \cdot a = 0$

$8 \cdot 0 = 0$

The Multiplication Property of Zero states that the product of a number and zero is zero. The variable a is used here to represent any whole number. It can even represent the number zero because $0 \cdot 0 = 0$.

The Multiplication Property of One

$a \cdot 1 = a$ or $1 \cdot a = a$

$1 \cdot 9 = 9$

The Multiplication Property of One states that the product of a number and one is the number. Multiplying a number by 1 does not change the number.

The Commutative Property of Multiplication

$a \cdot b = b \cdot a$

$$4 \cdot 9 = 9 \cdot 4$$
$$36 - 36$$

The Commutative Property of Multiplication states that two numbers can be multiplied in either order; the product will be the same. Here the variables a and b represent any whole numbers. Therefore, for example, if you know that the product of 4 and 9 is 36, then you also know that the product of 9 and 4 is 36 because $4 \cdot 9 = 9 \cdot 4$.

The Associative Property of Multiplication

$(a \cdot b) \cdot c = a \cdot (b \cdot c)$

$$(2 \cdot 3) \cdot 4 = 2 \cdot (3 \cdot 4)$$
$$6 \cdot 4 = 2 \cdot 12$$
$$24 = 24$$

The Associative Property of Multiplication states that when multiplying three numbers, the numbers can be grouped in any order; the product will be the same. Note in the example at the right above that we can multiply the product of 2 and 3 by 4, or we can multiply 2 by the product of 3 and 4. In either case, the product of the three numbers is 24.

What is the solution of the equation $5x = 5$?

By the Multiplication Property of One, the product of a number and 1 is the number.

$$5x = 5$$
$$5(1) \quad 5$$
$$5 = 5$$

The solution is 1.

The check is shown at the right.

Is 7 a solution of the equation $3m = 21$?

Replace m by 7.

$$3m = 21$$

Simplify the left side of the equation.

$$3(7) \quad 21$$
$$21 = 21$$

The results are equal.

Yes, 7 is a solution of the equation.

STUDY TIPS   *Get Adequate Practice*

Instructors often advise students to spend twice the amount of time outside of class studying as they spend in the classroom. For example, if a course meets for three hours each week, customarily instructors advise students to study for six hours each week outside of class. However, only you can decide how much time you should spend studying.

Your prealgebra course requires the learning of skills, which are abilities acquired through practice. It is often necessary to practice a skill more than a teacher requires. For example, this textbook may provide 50 practice problems on a specific objective and the instructor may assign only 25 of them. However, some students may need to do 30, 40, or all 50 of the problems.

If you are an accomplished athlete, musician, or dancer, you know that long hours of practice are necessary to acquire a skill. Do not cheat yourself of the practice you need to develop the abilities taught in this course.

Figure 1.15 shows the costs per day of business travel in five cities in the United States. Use this graph for Example 1 and You Try It 1.

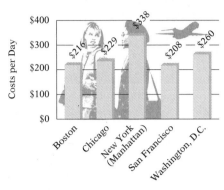

Costs per Day

$400
$300
$200
$100
$0

$216 $229 $338 $208 $260

Boston   Chicago   New York (Manhattan)   San Francisco   Washington, D.C.

Figure 1.15   Costs Per Day of Business Travel (average costs for breakfast, lunch, dinner, and business-class lodging)
Source: Runzheimer International Meal-Lodging Cost Index

Example 1   According to Figure 1.15, what would be the cost of 12 days of business travel in New York?

Solution   The cost of travel in New York is $338 per day.

$$\begin{array}{r} 338 \\ \times\ 12 \\ \hline 676 \\ 3\ 38 \\ \hline 4{,}056 \end{array}$$

The cost would be $4,056.

You Try It 1   According to Figure 1.15, what would be the cost of 14 days of business travel in Chicago?

Your Solution

Example 2   Estimate the product of 2,871 and 49.

Solution $2{,}871 \longrightarrow 3{,}000$
$49 \longrightarrow 50$

$3{,}000 \cdot 50 = 150{,}000$

You Try It 2   Estimate the product of 8,704 and 93.

Your Solution

Example 3   Evaluate $3ab$ when $a = 10$ and $b = 40$.

Solution $3ab$

$3(10)(40) = 30(40)$
$\qquad\qquad = 1{,}200$

You Try It 3   Evaluate $5xy$ when $x = 20$ and $y = 60$.

Your Solution

Solutions on p. A6

Example 4   What is 800 times 300?

Solution $800 \cdot 300 = 240{,}000$

You Try It 4   What is 90 multiplied by 7,000?

Your Solution

Example 5   Complete the statement by using the Associative Property of Multiplication.

$(7 \cdot 8) \cdot 5 = 7 \cdot (? \cdot 5)$

Solution $(7 \cdot 8) \cdot 5 = 7 \cdot (8 \cdot 5)$

You Try It 5   Complete the statement by using the Multiplication Property of Zero.

$? \cdot 10 = 0$

Your Solution

Example 6   Is 9 a solution of the equation $82 = 9q$?

Solution $82 = 9q$

$\dfrac{82 \mid 9(9)}{}$

$82 \neq 81$

No, 9 is not a solution of the equation.

You Try It 6   Is 11 a solution of the equation $7a = 77$?

Your Solution

Solutions on p. A6 |||||

Objective **B**
Exponents

Repeated multiplication of the same factor can be written in two ways:

$4 \cdot 4 \cdot 4 \cdot 4 \cdot 4$ or $4^5 \longleftarrow$ exponent
$ \uparrow \!\!\!\text{\_\_\_\_\_ base}$

The expression 4^5 is in **exponential form.** The **exponent,** 5, indicates how many times the **base,** 4, occurs as a factor in the multiplication.

It is important to be able to read numbers written in exponential form.

$2 = 2^1$ read "two to the first power" or just "two." Usually the 1 is not written.

$2 \cdot 2 = 2^2$ read "two squared" or "two to the second power."

$2 \cdot 2 \cdot 2 = 2^3$ read "two cubed" or "two to the third power."

$2 \cdot 2 \cdot 2 \cdot 2 = 2^4$ read "two to the fourth power."

$2 \cdot 2 \cdot 2 \cdot 2 \cdot 2 = 2^5$ read "two to the fifth power."

Variable expressions can contain exponents.

$x^1 = x$ x to the first power is usually written simply as x.

$x^2 = x \cdot x$ x^2 means x times x.

$x^3 = x \cdot x \cdot x$ x^3 means x occurs as a factor 3 times.

$x^4 = x \cdot x \cdot x \cdot x$ x^4 means x occurs as a factor 4 times.

> **POINT OF INTEREST**
>
> Lao-tzu, founder of Taoism, wrote: Counting gave birth to Addition, Addition gave birth to Multiplication, Multiplication gave birth to Exponentiation, Exponentiation gave birth to all the myriad operations.

Each place value in the place-value chart can be expressed as a power of 10.

$$\begin{aligned}
\text{Ten} &= 10 = 10 = 10^1 \\
\text{Hundred} &= 100 = 10 \cdot 10 = 10^2 \\
\text{Thousand} &= 1{,}000 = 10 \cdot 10 \cdot 10 = 10^3 \\
\text{Ten-thousand} &= 10{,}000 = 10 \cdot 10 \cdot 10 \cdot 10 = 10^4 \\
\text{Hundred-thousand} &= 100{,}000 = 10 \cdot 10 \cdot 10 \cdot 10 \cdot 10 = 10^5 \\
\text{Million} &= 1{,}000{,}000 = 10 \cdot 10 \cdot 10 \cdot 10 \cdot 10 \cdot 10 = 10^6
\end{aligned}$$

Note that the exponent on 10 when the number is written in exponential form is the same as the number of zeros in the number written in standard form. For example, $10^5 = 100{,}000$; the exponent on 10 is 5, and the number 100,000 has 5 zeros.

To evaluate a numerical expression containing exponents, write each factor as many times as indicated by the exponent and then multiply.

$$5^3 = 5 \cdot 5 \cdot 5 = 25 \cdot 5 = 125$$
$$2^3 \cdot 6^2 = (2 \cdot 2 \cdot 2) \cdot (6 \cdot 6) = 8 \cdot 36 = 288$$

Evaluate the variable expression c^3 when $c = 4$.

Replace c with 4 and then evaluate the exponential expression.

$$\begin{aligned}
c^3 \\
4^3 &= 4 \cdot 4 \cdot 4 \\
&= 16 \cdot 4 = 64
\end{aligned}$$

A calculator can be used to evaluate an exponential expression. The y^x key is used to enter the exponent.

POINT OF INTEREST

A billion is too large a number for most of us to comprehend. If a computer were to start counting from 1 to 1 billion, writing to the screen one number every second of every day, it would take over 31 years for the computer to complete the task.

And if a billion is a large number, consider a googol. A googol is 1 with 100 zeros after it, or 10^{100}. Edward Kasner is the mathematician credited with thinking up this number, and his nine-year-old nephew is said to have thought up the name. The two then coined the word googolplex, which is 10^{googol}.

Example 7 Write $7 \cdot 7 \cdot 7 \cdot 4 \cdot 4$ in exponential form.

Solution $7 \cdot 7 \cdot 7 \cdot 4 \cdot 4 = 7^3 \cdot 4^2$

You Try It 7 Write $2 \cdot 2 \cdot 2 \cdot 3 \cdot 3 \cdot 3 \cdot 3$ in exponential form.

Your Solution

Example 8 Evaluate 8^3.

Solution $8^3 = 8 \cdot 8 \cdot 8 = 64 \cdot 8 = 512$

You Try It 8 Evaluate 6^4.

Your Solution

Example 9 Evaluate 10^7.

Solution $10^7 = 10{,}000{,}000$

(The exponent on 10 is 7. There are 7 zeros in 10,000,000.)

You Try It 9 Evaluate 10^8.

Your Solution

Solutions on p. A6

Example 10   Evaluate $3^3 \cdot 5^2$.

Solution $3^3 \cdot 5^2 = (3 \cdot 3 \cdot 3) \cdot (5 \cdot 5)$
$= 27 \cdot 25 = 675$

You Try It 10   Evaluate $2^4 \cdot 3^2$.

Your Solution

Example 11   What is the square of 9?

Solution   The square of 9
$= 9^2 = 9 \cdot 9 = 81$

You Try It 11   Find the cube of 5.

Your Solution

Example 12   Evaluate $x^2 y^3$ when $x = 4$ and $y = 2$.

Solution $x^2 y^3$ ($x^2 y^3$ means x^2 times y^3.)

$4^2 \cdot 2^3 = (4 \cdot 4) \cdot (2 \cdot 2 \cdot 2)$
$= 16 \cdot 8$
$= 128$

You Try It 12   Evaluate $x^4 y^2$ when $x = 1$ and $y = 3$.

Your Solution

Solutions on p. A6 | | | | | |

OBJECTIVE C
Division of whole numbers

Division is used to separate objects into equal groups.

A grocer wants to equally distribute 24 new products on 4 shelves. From the diagram, we see that the grocer would place 6 products on each shelf.

The grocer's problem could be written:

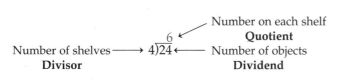

Number of shelves ⟶ 4)24 ← Number of objects
Divisor   **Dividend**

Number on each shelf
Quotient

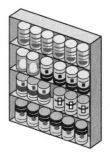

> **POINT OF INTEREST**
>
> The Chinese divided a day into 100 k'o, which was a unit equal to a little less than 15 min. Sundials were used to measure time during the daylight hours and, by A.D. 500, candles, water clocks, and incense sticks were used to measure time at night.

Notice that the quotient multiplied by the divisor equals the dividend.

$\overset{6}{4)24}$ because | 6 Quotient | × | 4 Divisor | = | 24 Dividend |

Division is also represented by the symbol ÷ or by a fraction bar. Both are read "divided by."

$\overset{6}{9)54}$ $54 \div 9 = 6$ $\dfrac{54}{9} = 6$

The fact that the quotient times the divisor equals the dividend can be used to illustrate properties of division.

$0 \div 4 = 0$ because $0 \cdot 4 = 0$.

$4 \div 4 = 1$ because $1 \cdot 4 = 4$.

$4 \div 1 = 4$ because $4 \cdot 1 = 4$.

$4 \div 0 = ?$ What number can be multiplied by 0 to get 4? There is no number whose product with 0 is 4 because the product of a number and zero is 0. **Division by zero is undefined.** $? \cdot 0 = 4$

The properties of division are stated below. In these statements, the symbol $\neq$ is read "is not equal to." Recall that the variable a represents any whole number. Therefore, for the first two properties, we must state that $a \neq 0$ in order to ensure that we are not dividing by zero.

Division Properties of Zero and One

If $a \neq 0, 0 \div a = 0$.	Zero divided by any number other than zero is zero.
If $a \neq 0, a \div a = 1$.	Any number other than zero divided by itself is one.
$a \div 1 = a$	A number divided by one is the number.
$a \div 0$ is undefined.	Division by zero is undefined.

The example below illustrates division of a larger whole number by a one-digit number.

Divide and check: $3{,}192 \div 4$

$$
\begin{array}{r}
7 \\
4)\overline{3{,}192} \\
-28 \\
\hline
39
\end{array}
$$
Think $31 \div 4$.
Subtract 7×4.
Bring down the 9.

$$
\begin{array}{r}
79 \\
4)\overline{3{,}192} \\
-28 \\
\hline
39 \\
-36 \\
\hline
32
\end{array}
$$
Think $39 \div 4$.
Subtract 9×4.
Bring down the 2.

$$
\begin{array}{r}
798 \\
4)\overline{3{,}192} \\
-28 \\
\hline
39 \\
-36 \\
\hline
32 \\
-32 \\
\hline
0
\end{array}
$$
Think $32 \div 4$.
Subtract 8×4.

Check:
$$
\begin{array}{r}
798 \\
\times 4 \\
\hline
3{,}192
\end{array}
$$

The place-value chart is used to show why
this method works.

$$
\begin{array}{r}
\text{Hundreds Tens Ones} \\
7\ 9\ 8 \\
4\overline{)3,1\ 9\ 2} \\
-2\ 8\ 0\ 0 \qquad \text{7 hundreds} \times 4\\
\hline
3\ 9\ 2 \\
-3\ 6\ 0 \qquad \text{9 tens} \times 4\\
\hline
3\ 2 \\
-3\ 2 \qquad \text{8 ones} \times 4\\
\hline
0
\end{array}
$$

Sometimes it is not possible to separate objects into a whole number of equal
groups.

A packer at a bakery has 14 muffins to pack
into 3 boxes. Each box will hold 4 muffins.
From the diagram, we see that after the
packer places 4 muffins in each box, there
are 2 muffins left over. The 2 is called the
remainder.

The packer's division problem could be written:

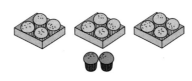

Number in each box
Quotient

Number of boxes ⟶ $3\overline{)14}$ ⟵ Total number of muffins   or $3\overline{)14}^{\,4\ \text{r}2}$
Divisor $\underline{-12}$ **Dividend**
                 2 ⟵ Number left over
                      Remainder

For any division problem, **(quotient · divisor) + remainder = dividend.** This
result can be used to check a division problem.

Find the quotient of 389 and 24.

$$
\begin{array}{r}
16\ \text{r}5 \\
24\overline{)389} \\
-24 \\
\hline
149 \\
-144 \\
\hline
5
\end{array}
$$
    Check: $(16 \cdot 24) + 5 = 384 + 5 = 389$

The phrase *the quotient of* was used in the example above to indicate the operation of division. The phrase *divided by* also indicates division.

the quotient of	the quotient of 8 and 4	$8 \div 4$
divided by	9 divided by 3	$9 \div 3$

Estimate the result when 56,497 is divided by 28.

Round each number to its highest place value.   56,497 $\longrightarrow$ 60,000

28 $\longrightarrow$ 30

Divide the rounded numbers. $60,000 \div 30 = 2,000$

2,000 is an estimate of $56,497 \div 28$.

Evaluate $\dfrac{x}{y}$ when $x = 4,284$ and $y = 18$. $\dfrac{x}{y}$

Replace x with 4,284 and y with 18.

$\dfrac{4,284}{18}$ means $4,284 \div 18$. $\dfrac{4,284}{18} = 238$

Is 42 a solution of the equation $\dfrac{x}{6} = 7$? $\dfrac{x}{6} = 7$

Replace x by 42.

Simplify the left side of the equation. $\dfrac{42}{6}$ | 7

The results are equal. $7 = 7$

42 is a solution of the equation.

Example 13   What is the quotient of 8,856 and 42?

You Try It 13   What is 7,694 divided by 24?

Solution

$$\begin{array}{r} 210\ \text{r}36 \\ 42\overline{)8,856} \\ \underline{8\ 4}\ \ \ \ \\ 45\ \ \ \\ \underline{-42}\ \ \ \\ 36 \\ \underline{-\ 0} \\ 36 \end{array}$$

36 Think $42\overline{)36}$.

$-\ 0$ Subtract $0 \cdot 42$.

Check: $(210 \cdot 42) + 36$
$= 8,820 + 36 = 8,856$

Your Solution

Solution on p. A6

Figure 1.16 shows a household's annual expenses of $44,000. Use this graph for Example 14 and You Try It 14.

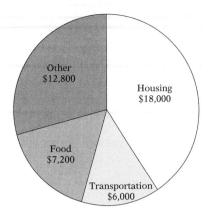

Figure 1.16   Annual Household Expenses

Example 14   Use Figure 1.16 to find the household's monthly expense for housing.

Solution   The annual expense for housing is $18,000.

$18,000 \div 12 = 1,500$

The monthly expense is $1,500.

You Try It 14   Use Figure 1.16 to find the household's monthly expense for food.

Your Solution

Example 15   Estimate the quotient of 55,272 and 392.

Solution $55,272 \longrightarrow 60,000$
$392 \longrightarrow 400$

$60,000 \div 400 = 150$

You Try It 15   Estimate the quotient of 216,936 and 207.

Your Solution

Example 16   Evaluate $\dfrac{x}{y}$ when $x = 342$ and $y = 9$.

Solution $\dfrac{x}{y}$

$\dfrac{342}{9} = 38$

You Try It 16   Evaluate $\dfrac{x}{y}$ when $x = 672$ and $y = 8$.

Your Solution

Example 17   Is 28 a solution of the equation $\dfrac{x}{7} = 4$?

Solution $\dfrac{x}{7} = 4$

$\dfrac{28}{7} \,\Big|\, 4$

$4 = 4$

Yes, 28 is a solution of the equation.

You Try It 17   Is 12 a solution of the equation $\dfrac{60}{y} = 2$?

Your Solution

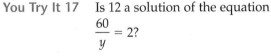

Solutions on p. A7

OBJECTIVE D
Factors and prime factorization

Natural number factors of a number divide that number evenly (there is no remainder).

1, 2, 3, and 6 are natural number factors of 6 because they divide 6 evenly.

$$\begin{array}{cccc} 6 & 3 & 2 & 1 \\ 1\overline{)6} & 2\overline{)6} & 3\overline{)6} & 6\overline{)6} \end{array}$$

Notice that both the divisor and the quotient are factors of the dividend.

To find the factors of a number, try dividing the number by 1, 2, 3, 4, 5, . . . Those numbers that divide the number evenly are its factors. Continue this process until the factors start to repeat.

Find all the factors of 42.

$42 \div 1 = 42$	1 and 42 are factors of 42.
$42 \div 2 = 21$	2 and 21 are factors of 42.
$42 \div 3 = 14$	3 and 14 are factors of 42.
$42 \div 4$	4 will not divide 42 evenly.
$42 \div 5$	5 will not divide 42 evenly.
$42 \div 6 = 7$	6 and 7 are factors of 42.
$42 \div 7 = 6$	7 and 6 are factors of 42.

The factors are repeating.
All the factors of 42 have been found.

The factors of 42 are 1, 2, 3, 6, 7, 14, 21, and 42.

The following rules are helpful in finding the factors of a number.

2 is a factor of a number if the digit in the ones' place of the number is 0, 2, 4, 6, or 8.	436 ends in 6. Therefore, 2 is a factor of 436. $(436 \div 2 = 218)$
3 is a factor of a number if the sum of the digits of the number is divisible by 3.	The sum of the digits of 489 is $4 + 8 + 9 = 21$. 21 is divisible by 3. Therefore, 3 is a factor of 489. $(489 \div 3 = 163)$
4 is a factor of a number if the last two digits of the number are divisible by 4.	556 ends in 56. 56 is divisible by 4. $(56 \div 4 = 14)$ Therefore, 4 is a factor of 556. $(556 \div 4 = 139)$
5 is a factor of a number if the ones' digit of the number is 0 or 5.	520 ends in 0. Therefore, 5 is a factor of 520. $(520 \div 5 = 104)$

A **prime number** is a natural number greater than 1 that has exactly two natural number factors, 1 and the number itself. 7 is prime because its only factors are 1 and 7. If a number is not prime, it is a **composite** number. Because 6 has factors of 2 and 3, 6 is a composite number. The prime numbers less than 50 are

2, 3, 5, 7, 11, 13, 17, 19, 23, 29, 31, 37, 41, 43, 47.

The **prime factorization** of a number is the expression of the number as a product of its prime factors. To find the prime factors of 90, begin with the smallest prime number as a trial divisor and continue with prime numbers as trial divisors until the final quotient is prime.

Find the prime factorization of 90.

$$
\begin{array}{r} 45 \\ 2)\overline{90} \end{array}
\qquad
\begin{array}{r} 15 \\ 3)\overline{45} \\ 2)\overline{90} \end{array}
\qquad
\begin{array}{r} 5 \\ 3)\overline{15} \\ 3)\overline{45} \\ 2)\overline{90} \end{array}
$$

Divide 90 by 2.   45 is not divisible by 2.   Divide 15 by 3.
                     Divide 45 by 3.   5 is prime.

The prime factorization of 90 is $2 \cdot 3 \cdot 3 \cdot 5$, or $2 \cdot 3^2 \cdot 5$.

Finding the prime factorization of larger numbers can be more difficult. Try each prime number as a trial divisor. Stop when the square of the trial divisor is greater than the number being factored.

Find the prime factorization of 201.

$$
\begin{array}{r} 67 \\ 3)\overline{201} \end{array}
$$

67 cannot be divided evenly by 2, 3, 5, 7, or 11. Prime numbers greater than 11 need not be tried because $11^2 = 121$ and $121 > 67$.

The prime factorization of 201 is $3 \cdot 67$.

Example 18	Find all the factors of 40.	**You Try It 18** Find all the factors of 30.

Solution

$40 \div 1 = 40$
$40 \div 2 = 20$
$40 \div 3$ Does not divide evenly.
$40 \div 4 = 10$
$40 \div 5 = 8$
$40 \div 6$ Does not divide evenly.
$40 \div 7$ Does not divide evenly.
$40 \div 8 = 5$ The factors are repeating.

The factors of 40 are 1, 2, 4, 5, 8, 10, 20, and 40.

Your Solution

Solution on p. A7

STUDY TIPS   *Get Help for Academic Difficulties*

If you have trouble in this course, a teacher, counselor, or advisor can help. They usually know of study groups, tutors, or other sources of help that are available. They may suggest visiting an office of academic skills, a learning center, a tutorial service, or some other department or service on campus. If your college has a math lab available, use it. If your school has tutoring available, make an appointment to receive tutoring *immediately*. Don't wait until this obstacle blocks your continued progress.

Example 19 Find the prime factorization of 84.

Solution

$$
\begin{array}{r}
7 \\
3\overline{)21} \\
2\overline{)42} \\
2\overline{)84}
\end{array}
$$

$84 = 2 \cdot 2 \cdot 3 \cdot 7 = 2^2 \cdot 3 \cdot 7$

You Try It 19 Find the prime factorization of 88.

Your Solution

Example 20 Find the prime factorization of 141.

Solution

$$
\begin{array}{r}
47 \\
3\overline{)141}
\end{array}
$$
Try only 2, 3, 5, and 7 because $7^2 = 49$ and $49 > 47$.

$141 = 3 \cdot 47$

You Try It 20 Find the prime factorization of 295.

Your Solution

Solutions on p. A7

STUDY TIPS **Use the Review Material at the End of Each Chapter**

To help you review the material presented within a chapter, a Chapter Summary appears at the end of each chapter. In the Chapter Summary, definitions of the important terms and concepts introduced in the chapter are provided under Key Words. Listed under Essential Rules are the formulas and procedures presented in the chapter. After completing a chapter, be sure to read the Chapter Summary. Use it to check your understanding of the material presented and to determine what concepts you need to review.

At the end of each chapter are Chapter Review Exercises. These problems summarize what you should have learned when you have finished the chapter. Do these exercises as you prepare for an examination. We suggest that you do these exercises a few days before your actual exam. Complete the Chapter Review Exercises in a quiet place and try to complete the exercises in the same amount of time as you will be allowed for your exam. When completing the Chapter Review Exercises, practice the strategies of successful test-takers:

1. Look over the entire test before you begin to solve any problem.

2. Write down any rules or formulas you may need so they are readily available.

3. Read the directions carefully.

4. Work the problems that are easiest for you first.

5. Check your work, looking particularly for careless errors.

After completing the Chapter Review Exercises, check your answers against those in the back of the book. Answers to every exercise in the Chapter Review are provided there. The objective being tested by any particular problem is written in parentheses following the answer. For example, in the answers to the Chapter Review Exercises for Chapter 2, the answer "8. 4 (Objective 2.2A)" indicates that Exercise 8 corresponds to Chapter 2, Section 2, Objective A. For any problem you answer incorrectly, review the material corresponding to that objective in the textbook. Determine *why* your answer was wrong.

Cumulative Review Exercises are included in every chapter after Chapter 1. Doing these exercises allows you to review the skills you learned in previous chapters. This is very important in mathematics. By consistently reviewing previous materials, you will retain the previous skills as you build new ones.

OBJECTIVE E

Applications and formulas

In this section, some of the phrases used to indicate the operations of multiplication and division were presented. In solving application problems, you might also look for the following types of questions.

Multiplication
per . . . How many altogether?
each . . . What is the total number of . . ?
every . . . Find the total . . .

Division
What is the hourly rate?
Find the amount per . . .
How many does each . . . ?

Figure 1.17 shows the cost of a first-class postage stamp from the 1950s to 1997. Use this graph for Example 21 and You Try It 21.

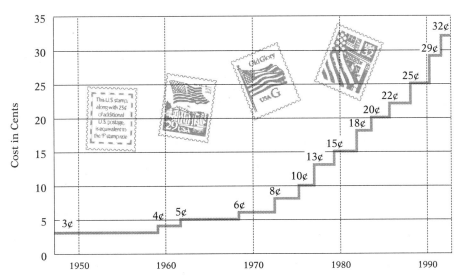

Figure 1.17   Cost of a First-Class Postage Stamp

> **POINT OF INTEREST**
> One calorie is the energy required to lift 1 kilogram a distance of 427 meters. (See You Try It 22.)

Example 21

How many times more expensive was a stamp in 1980 than in 1950? Use Figure 1.17.

Strategy

To find how many times more expensive a stamp was, divide the cost in 1980 (15) by the cost in 1950 (3).

Solution

15 ÷ 3 = 5

A stamp was 5 times more expensive in 1980.

You Try It 21

How many times more expensive was a stamp in 1997 than in 1960? Use Figure 1.17.

Your Strategy

Your Solution

Solution on p. A7

Example 22

An auto mechanic receives a salary of $525 each week. How much does the auto mechanic earn in 6 weeks?

Strategy

To find the mechanic's earnings for 6 weeks, multiply the weekly salary (525) by the number of weeks (6).

Solution

$$\begin{array}{r} 525 \\ \times\ \ 6 \\ \hline 3{,}150 \end{array}$$

The mechanic earns $3,150 in 6 weeks.

You Try It 22

A person playing tennis burns 480 calories per hour. Find the number of calories burned by a person playing 4 h of tennis.

Your Strategy

Your Solution

Example 23

At what rate of speed would you need to travel in order to drive a distance of 294 mi in 6 h? Use the formula $r = \dfrac{d}{t}$, where r is the average rate of speed, d is the distance, and t is the time.

Strategy

To find the rate of speed, replace d by 294 and t by 6 in the given formula and solve for r.

Solution

$$r = \frac{d}{t}$$

$$r = \frac{294}{6} = 49$$

You would need to travel at a speed of 49 mph.

You Try It 23

At what rate of speed would you need to travel in order to drive a distance of 486 mi in 9 h? Use the formula $r = \dfrac{d}{t}$, where r is the average rate of speed, d is the distance, and t is the time.

Your Strategy

Your Solution

Solutions on p. A7

STUDY TIPS   *Review Material Often*

Reviewing material is the repetition that is essential for learning. Much of what we learn is soon forgotten unless we review it. If you find that you do not remember a skill that you studied previously, you probably have not reviewed it sufficiently. *You will remember best what you review most.*

One method of reviewing material is to begin a study session by reviewing a concept you have studied previously. For example, before trying to solve a new type of problem, spend a few minutes solving a kind of problem you already know how to solve. Not only will you provide yourself with the review practice you need, but you are also likely to put yourself in the right frame of mind for learning how to solve the new type of problem.

1.3 EXERCISES

OBJECTIVE A

Multiply.

1. (9)(127)

2. (4)(623)

3. (6,709)(7)

4. (3,608)(5)

5. 8 · 58,769

6. 7 · 60,047

7. 683
× 71

8. 591
× 92

9. 7,053
× 46

10. 6,704
× 58

11. 3,285
× 976

12. 5,327
× 624

Solve.

13. Find the product of 500 and 3.

14. Find 30 multiplied by 80.

15. What is 40 times 50?

16. What is twice 700?

17. What is the product of 400, 3, 20, and 0?

18. Write the product of f and g.

19. Write the product of q, r, and s.

20. *Health* The figure to the right shows the number of calories burned on three different exercise machines during one hour of a light, moderate, or vigorous workout. How many calories would you burn by (a) working out vigorously on a stair climber for a total of 6 hours? (b) working out moderately on a treadmill for a total of 12 hours?

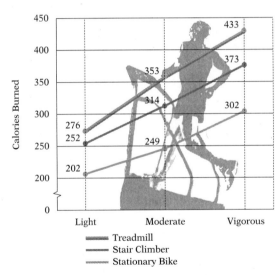

Calories Burned on Exercise Machines
Source: Journal of American Medical Association

Estimate by rounding. Then find the exact answer.

21. 3,467 · 359

22. 8,745(63)

23. (39,246)(29)

24. 64,409 · 67

Estimate by rounding. Then find the exact answer.

25. 745(63) **26.** 432 · 91 **27.** (8,941)(726) **28.** 2,837(216)

Evaluate the expression for the given values of the variables.

29. *ab*, when *a* = 465 and *b* = 32

30. *cd*, when *c* = 381 and *d* = 25

31. 7*a*, when *a* = 465

32. 6*n*, when *n* = 382

33. *xyz*, when *x* = 5, *y* = 12, and *z* = 30

34. *abc*, when *a* = 4, *b* = 20, and *c* = 50

35. 2*xy*, when *x* = 67 and *y* = 23

36. 4*ab*, when *a* = 95 and *b* = 33

Identify the property that justifies the statement.

37. 1 · 29 = 29

38. (10 · 5) · 8 = 10 · (5 · 8)

39. 43 · 1 = 1 · 43

40. 0(76) = 0

Use the given property of multiplication to complete the statement.

41. The Commutative Property of Multiplication
19 · ? = 30 · 19

42. The Associative Property of Multiplication
(? · 6)100 = 5(6 · 100)

43. The Multiplication Property of Zero
45 · 0 = ?

44. The Multiplication Property of One
? · 77 = 77

45. Is 6 a solution of the equation $4x = 24$?

46. Is 0 a solution of the equation $4 = 4n$?

47. Is 23 a solution of the equation $96 = 3z$?

48. Is 14 a solution of the equation $56 = 4c$?

49. Is 19 a solution of the equation $2y = 38$?

50. Is 11 a solution of the equation $44 = 3a$?

OBJECTIVE B

Write in exponential form.

51. $2 \cdot 2 \cdot 2 \cdot 7 \cdot 7 \cdot 7 \cdot 7 \cdot 7$

52. $3 \cdot 3 \cdot 3 \cdot 3 \cdot 3 \cdot 3 \cdot 5 \cdot 5 \cdot 5$

53. $2 \cdot 2 \cdot 3 \cdot 3 \cdot 3 \cdot 5 \cdot 5 \cdot 5 \cdot 5$

54. $7 \cdot 7 \cdot 11 \cdot 11 \cdot 11 \cdot 19 \cdot 19 \cdot 19 \cdot 19$

55. $c \cdot c$

56. $d \cdot d \cdot d$

57. $x \cdot x \cdot x \cdot y \cdot y \cdot y$

58. $a \cdot a \cdot b \cdot b \cdot b \cdot b$

Evaluate.

59. 2^5

60. 2^6

61. 10^6

62. 10^9

63. $2^3 \cdot 5^2$

64. $2^4 \cdot 3^2$

65. $3^2 \cdot 10^3$

66. $2^4 \cdot 10^2$

67. $0^2 \cdot 6^2$

68. $4^3 \cdot 0^3$

69. $2^2 \cdot 5 \cdot 3^3$

70. $5^2 \cdot 2 \cdot 3^4$

Solve.

71. Find the square of 12.

72. What is the cube of 6?

Solve.

73. Find the cube of 8.

74. What is the square of 11?

75. Write the fourth power of a.

76. Write the fifth power of t.

Evaluate the expression for the given values of the variables.

77. x^3y, when $x = 2$ and $y = 3$

78. x^2y, when $x = 3$ and $y = 4$

79. ab^6, when $a = 5$ and $b = 2$

80. ab^3, when $a = 7$ and $b = 4$

81. c^2d^2, when $c = 3$ and $d = 5$

82. m^3n^3, when $m = 5$ and $n = 10$

OBJECTIVE C

Divide.

83. $9\overline{)2{,}763}$

84. $4\overline{)2{,}160}$

85. $5\overline{)1{,}549}$

86. $8\overline{)1{,}636}$

87. $15{,}300 \div 6$

88. $43{,}500 \div 5$

89. $681 \div 32$

90. $879 \div 41$

91. $9{,}152 \div 62$

92. $4{,}161 \div 23$

93. $7{,}408 \div 37$

94. $5{,}207 \div 26$

95. $31{,}546 \div 78$

96. $38{,}976 \div 64$

97. $7{,}713 \div 476$

98. $8{,}947 \div 223$

Solve.

99. Find the quotient of 7,256 and 8.

100. What is the quotient of 8,172 and 9?

Solve.

101. What is 6,168 divided by 7?

102. Find 4,153 divided by 9.

Average Annual Rental Costs in Ten Inexpensive Locations Nationwide	
Location	Annual Rental Costs
Corbin, KY	$2,400
Newport, TN	$3,060
Scottsboro, AL	$3,060
Hennessey, OK	$3,180
Casper, WY	$3,180
Roanoke Rapids, NC	$3,540
Midland, TX	$3,600
Austin, MN	$3,660
Hobbs, NM	$3,660
Lafayette, LA	$3,720

103. Write the quotient of c and d.

104. *Cost of Living*   The table at the right shows the average annual cost of renting a three-room apartment in ten different cities in the United States (*Source:* Runzheimer International, 1994). (a) What would be the average monthly rent for an apartment in Scottsboro, Alabama? (b) What would be the average monthly rent for an apartment in Hobbs, New Mexico?

Estimate by rounding. Then find the exact answer.

105. $36{,}472 \div 47$

106. $62{,}176 \div 58$

107. $389{,}804 \div 76$

108. $637{,}072 \div 29$

109. $79\overline{)38{,}984}$

110. $53\overline{)11{,}792}$

111. $219\overline{)332{,}004}$

112. $324\overline{)632{,}124}$

Evaluate the variable expression $\dfrac{x}{y}$ for the given values of x and y.

113. $x = 48; y = 1$

114. $x = 56; y = 56$

115. $x = 79; y = 0$

116. $x = 0; y = 23$

117. $x = 39{,}200; y = 4$

118. $x = 16{,}200; y = 3$

119. Is 9 a solution of the equation $\dfrac{36}{z} = 4$?

120. Is 60 a solution of the equation $\dfrac{n}{12} = 5$?

121. Is 49 a solution of the equation $56 = \dfrac{x}{7}$?

122. Is 16 a solution of the equation $6 = \dfrac{48}{y}$?

OBJECTIVE D

Find all the factors of the number.

123. 10   **124.** 20   **125.** 12   **126.** 9   **127.** 8

128. 16   **129.** 13   **130.** 17   **131.** 18   **132.** 24

133. 25   **134.** 36   **135.** 56   **136.** 45   **137.** 28

138. 32   **139.** 48   **140.** 64   **141.** 54   **142.** 75

Find the prime factorization of the number.

143. 16   **144.** 24   **145.** 12   **146.** 27   **147.** 15

148. 36   **149.** 40   **150.** 50   **151.** 37   **152.** 83

153. 65   **154.** 80   **155.** 28   **156.** 49   **157.** 42

158. 81   **159.** 51   **160.** 89   **161.** 46   **162.** 120

OBJECTIVE E

Solve.

163. *Nutrition*   One ounce of cheddar cheese contains 115 calories. Find the number of calories in 4 oz of cheddar cheese.

Nutrition Facts	Amount/Serving	% DV*	Amount/Serving	% DV*
	Total Fat 9g	**14%**	**Total Carb.** 1g	**0%**
Serv. Size 1 oz.	Sat Fat 5g	**25%**	Fiber 0g	**0%**
Servings Per Package 12				
Calories 115	**Cholest.** 30mg	**10%**	Sugars 0g	
Fat Cal. 80	**Sodium** 170mg	**7%**	**Protein** 7g	
*Percent Daily Values (DV) are based on a 2,000 calorie diet	Vitamin A 6% • Vitamin C 0% • Calcium 20% • Iron 0%			

164. *Sports*   During his football career, John Riggins ran the ball 2,916 times. He averaged about 4 yd per carry. About how many total yards did he gain during his career?

165. *Computer Science*   A computer can store 2,211,840 bytes of information on 6 disks. How many bytes of information can be stored on one disk?

Solve.

166. *Aviation*   A plane flying from Los Angeles to Boston uses 865 gal of jet fuel each hour. How many gallons of jet fuel are used on a 5-hour flight?

167. *Business*   The figure on the right shows the cost of residential construction in four states. (a) According to the graph, how much would it cost to construct six homes in Georgia? (b) Provide an estimate of the cost of constructing eight homes in Texas. (c) How much more expensive would it be to construct 20 single-family homes in a development in Florida than in Arizona?

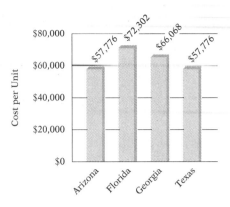

Per-Unit Cost of Residential Construction
Source: Gables Residential Trust Investment Co. survey for Urban Land Institute

168. *Computer Science*   A computer graphics screen has 640 rows of pixels and there are 480 pixels per row. Find the total number of pixels on the screen.

169. *Finances*   A computer analyst doing consulting work received $5,376 for working 168 h on a project. Find the hourly rate the consultant charged.

170. *Business*   A buyer for a department store purchased 215 suits at $83 each. Estimate the total cost of the order.

171. *Finances*   Financial advisors may predict how much money we should have saved for retirement by the ages of 35, 45, 55, and 65. One such prediction is included in the table below. (a) A couple has earnings of $100,000 per year. According to the table, by how much should their savings grow per year from age 45 to 55? (b) A couple has earnings of $50,000 per year. According to the table, by how much should their savings grow per year from age 55 to 65?

Minimum Levels of Savings Required for Married Couples to Be Prepared for Retirement				
	Savings Accumulation by Age			
Earnings	35	45	55	65
$50,000	8,000	23,000	90,000	170,000
$75,000	17,000	60,000	170,000	310,000
$100,000	34,000	110,000	280,000	480,000
$150,000	67,000	210,000	490,000	840,000

172. *Finances*   Find the total amount paid on a loan when the monthly payment is $285 and the loan is paid off in 24 months. Use the formula $A = MN$, where A is the total amount paid, M is the monthly payment, and N is the number of payments.

173. *Finances*   Find the total amount paid on a loan when the monthly payment is $187 and the loan is paid off in 36 months. Use the formula $A = MN$, where A is the total amount paid, M is the monthly payment, and N is the number of payments.

Solve.

174. *Travel* Use the formula $t = \dfrac{d}{r}$, where t is the time, d is the distance, and r is the average rate of speed, to find the time it would take to drive 513 mi at an average speed of 57 mph.

175. *Travel* Use the formula $t = \dfrac{d}{r}$, where t is the time, d is the distance, and r is the average rate of speed, to find the time it would take to drive 432 mi at an average speed of 54 mph.

176. *Investments* The current value of the stocks in a mutual fund is $10,500,000. The number of shares outstanding is 500,000. Find the value per share of the fund. Use the formula $V = \dfrac{C}{S}$, where V is the value per share, C is the current value of the stocks in the fund, and S is the number of shares outstanding.

177. *Investments* The current value of the stocks in a mutual fund is $4,500,000. The number of shares outstanding is 250,000. Find the value per share of the fund. Use the formula $V = \dfrac{C}{S}$, where V is the value per share, C is the current value of the stocks in the fund, and S is the number of shares outstanding.

CRITICAL THINKING

178. There are 52 weeks in a year. Is this an exact figure or an approximation?

179. 13,827 is not divisible by 4. By rearranging the digits, find the largest possible number that is divisible by 4.

180. A palindromic number is a whole number that remains unchanged when its digits are written in reverse order. For example, 818 is a palindromic number. Find the smallest three-digit multiple of 6 that is a palindromic number.

181. Determine whether the statement is always true, sometimes true, or never true.
  a. Let a be any whole number. Then $a \cdot 0 = a$.
  b. Let a be any whole number. Then $a \cdot 1 = 1$.

182. According to the National Safety Council, in a recent year, a death resulting from an accident occurred at the rate of one every 5 min. At this rate, how many accidental deaths occurred each hour? each day? throughout the year? Explain how you arrived at your answers.

183. Prepare a monthly budget for a family of four. Explain how you arrived at the cost of each item. Annualize the budget you prepared.

Monthly Budget	
Rent	$675
Electricity	
Telephone	
Gas	
Food	

SECTION 1.4  Solving Equations with Whole Numbers

OBJECTIVE **A**

Solving equations

Recall that a **solution** of an equation is a number that, when substituted for the variable, results in a true equation.

The solution of the equation $x + 5 = 11$ is 6 because when 6 is substituted for x, the result is a true equation.

$$x + 5 = 11$$
$$6 + 5 = 11$$

If 2 is subtracted from each side of the equation $x + 5 = 11$, the resulting equation is $x + 3 = 9$. Note that the solution of this equation is also 6.

$$x + 5 = 11$$
$$x + 5 - 2 = 11 - 2$$
$$x + 3 = 9 \qquad 6 + 3 = 9$$

This illustrates the subtraction property of equations.

> **The same number can be subtracted from each side of an equation without changing the solution of the equation.**

The subtraction property is used to *solve* an equation. To **solve an equation** means to find a solution of the equation. That is, to solve an equation you must find a number that, when substituted for the variable, results in a true equation.

An equation such as $x = 8$ is easy to solve. The solution is 8, the number that when substituted for the variable results in the true equation $8 = 8$. In solving an equation, the goal is to get the variable alone on one side of the equation; the number on the other side of the equation is the solution.

To solve an equation in which a number is added to a variable, use the subtraction property of equations: Subtract that number from each side of the equation.

Solve: $x + 5 = 11$

Note the effect of subtracting 5 from each side of the equation and then simplifying. The variable, x, is on one side of the equation; a number, 6, is on the other side.

$$x + 5 = 11$$
$$x + 5 - 5 = 11 - 5$$
$$x + 0 = 6$$
$$x = 6$$

The solution is 6.

Check:
$$\frac{x + 5 = 11}{6 + 5 \,\big|\, 11}$$
$$11 = 11$$

Note that we have checked the solution. You should always check the solution of an equation.

Solve: $19 = 11 + m$

11 is added to m. Subtract 11 from each side of the equation.

$$19 = 11 + m$$
$$19 - 11 = 11 - 11 + m$$
$$8 = 0 + m$$
$$8 = m$$

The solution is 8.

Check:
$$\frac{19 = 11 + m}{19 \,\big|\, 11 + 8}$$
$$19 = 19$$

TAKE NOTE

An *equation* always has an equal sign (=). An *expression* does not have an equal sign.

$x + 5 = 11$ is an equation.

$x + 5$ is an expression.

TAKE NOTE

For this equation, the variable is on the right side. The goal is to get the variable alone on the right side.

The solution of the equation $4y = 12$ is 3 because when 3 is substituted for y, the result is a true equation.

$$4y = 12$$
$$4(3) = 12$$
$$12 = 12$$

If each side of the equation $4y = 12$ is divided by 2, the resulting equation is $2y = 6$. Note that the solution of this equation is also 3.

$$4y = 12$$
$$\frac{4y}{2} = \frac{12}{2}$$
$$2y = 6 \qquad 2(3) = 6$$

This illustrates the division property of equations.

Each side of an equation can be divided by the same number (except zero) without changing the solution of the equation.

Solve: $30 = 5a$

a is multiplied by 5. To get a alone on the right side, divide each side of the equation by 5.

$$30 = 5a$$
$$\frac{30}{5} = \frac{5a}{5}$$
$$6 = 1a$$
$$6 = a$$

Check:
$$30 = 5a$$
$$30 \mid 5(6)$$
$$30 = 30$$

The solution is 6.

Example 1 Solve: $9 + n = 28$

Solution
$$9 + n = 28$$
$$9 - 9 + n = 28 - 9$$
$$0 + n = 19$$
$$n = 19$$

Check: $9 + n = 28$
$$9 + 19 \mid 28$$
$$28 = 28$$

The solution is 19.

You Try It 1 Solve: $37 = a + 12$

Your Solution

Example 2 Solve: $20 = 5c$

Solution
$$20 = 5c$$
$$\frac{20}{5} = \frac{5c}{5}$$
$$4 = 1c$$
$$4 = c$$

Check: $20 = 5c$
$$20 \mid 5(4)$$
$$20 = 20$$

The solution is 4.

You Try It 2 Solve: $3z = 36$

Your Solution

Solutions on p. A7

OBJECTIVE B
Applications and formulas

Recall that an equation states that two mathematical expressions are equal. To translate a sentence into an equation requires recognition of the words or phrases that mean "equals." Some of these phrases are

equals	is	was
is equal to	represents	is the same as

The number of scientific calculators sold by Evergreen Electronics last month is three times the number of graphing calculators the company sold last month. If it sold 225 scientific calculators last month, how many graphing calculators were sold last month?

Strategy   To find the number of graphing calculators sold, write and solve an equation using x to represent the number of graphing calculators sold.

Strategy

The number of scientific calculators sold	is	three times the number of graphing calculators sold

$$225 \quad = \quad 3x$$

$$\frac{225}{3} = \frac{3x}{3}$$

$$75 = x$$

Evergreen Electronics sold 75 graphing calculators last month.

> **TAKE NOTE**
>
> Sentences or phrases that begin "how many . . .," "how much . . .," "find . . .," and "what is . . ." are followed by a phrase that indicates what you are looking for. (In the problem at the left, the phrase is *graphing calculators:* "How many *graphing calculators . . .*") Look for these phrases to determine the unknown.

Example 3

The product of seven and a number equals twenty-eight. Find the number.

Solution

The unknown number: n

The product of seven and a number	equals	twenty-eight

$$7n = 28$$

$$\frac{7n}{7} = \frac{28}{7}$$

$$n = 4$$

The number is 4.

You Try It 3

A number increased by four is seventeen. Find the number.

Your Solution

Solution on p. A7

Example 4

The number of basketball games won by UNLV in 1990 was four more than the number of basketball games won by UCLA in 1995. UNLV won 35 basketball games in 1990. How many basketball games did UCLA win in 1995?

Strategy

To find the number of games won by UCLA, write and solve an equation using x to represent the number of basketball games won by UCLA.

Solution

The number of games won by UNLV	was	four more than the number of games won by UCLA

$$35 = x + 4$$
$$35 - 4 = x + 4 - 4$$
$$31 = x$$

UCLA won 31 games in 1995.

Example 5

Use the formula $A = P + I$, where A is the value of an investment, P is the original investment, and I is the interest earned, to find the interest earned on an original investment of $12,000 that now has a value of $14,280.

Strategy

To find the interest earned, replace A by 14,280 and P by 12,000 in the given formula and solve for I.

Solution

$$A = P + I$$
$$14,280 = 12,000 + I$$
$$14,280 - 12,000 = 12,000 - 12,000 + I$$
$$2,280 = I$$

The interest earned on the investment is $2,280.

You Try It 4

The price of a used Ford Windstar is $900 more than the price of a used Mercury Villager. The Ford Windstar is priced at $16,495. What is the price of the Mercury Villager?

Your Strategy

Your Solution

You Try It 5

Use the formula $A = P + I$, where A is the value of an investment, P is the original investment, and I is the interest earned, to find the interest earned on an original investment of $18,000 that now has a value of $21,060.

Your Strategy

Your Solution

1.4 EXERCISES

OBJECTIVE A

Solve.

1. $x + 9 = 23$ **2.** $y + 17 = 42$ **3.** $8 + b = 33$ **4.** $15 + n = 54$

5. $3m = 15$ **6.** $8z = 32$ **7.** $52 = 4c$ **8.** $60 = 5d$

9. $16 = w + 9$ **10.** $72 = t + 44$ **11.** $28 = 19 + p$ **12.** $33 = 18 + x$

13. $10y = 80$ **14.** $12n = 60$ **15.** $41 = 41d$ **16.** $93 = 93m$

17. $b + 7 = 7$ **18.** $q + 23 = 23$ **19.** $15 + t = 91$ **20.** $79 + w = 88$

OBJECTIVE B

Solve.

21. Sixteen added to a number is equal to forty. Find the number.

22. The sum of eleven and a number equals fifty-two. Find the number.

23. Five times a number is thirty. Find the number.

24. The product of ten and a number is equal to two hundred. Find the number.

25. Fifteen is three more than a number. Find the number.

$$15 = x + 3$$
$$\underline{-3 \quad -3}$$
$$12 = x$$

26. One thousand represents three hundred fifty plus a number. Find the number.

27. A number increased by fourteen equals seventy-two. Find the number.

28. A number multiplied by twenty equals four hundred. Find the number.

Solve.

29. *Education* The tuition for a student this semester is $75 more than the student's tuition last semester. The tuition for the student this semester is $835. What was the student's tuition last semester?

30. *Temperature* The average daily low temperature in Duluth, Minnesota, in June is eight times the average daily low temperature in Duluth in December. The average daily low temperature in Duluth in June is 48°. Find the average daily low temperature in Duluth in December.

31. *Geography* The table at the right lists the distances from four cities in Texas to Austin, Texas. The distance from Galveston to Austin is twenty-two miles more than the distance from Fort Worth, Texas, to Austin. Find the distance from Fort Worth to Austin.

32. *Geography* The table at the right lists the distances from four cities in Texas to Austin, Texas. The distance from Houston to Austin is twice the distance from San Antonio, Texas, to Austin. Find the distance from San Antonio to Austin.

City in Texas	Number of Miles to Austin, Texas
Corpus Christi	215
Dallas	195
Galveston	212
Houston	160

33. *Finances* Use the formula $A = MN$, where A is the total amount paid, M is the monthly payment, and N is the number of payments, to find the number of payments made on a loan for which the total amount paid is $13,968 and the monthly payment is $582.

34. *Finances* Use the formula $A = MN$, where A is the total amount paid, M is the monthly payment, and N is the number of payments, to find the number of payments made on a loan for which the total amount paid is $17,460 and the monthly payment is $485.

35. *Travel* Use the formula $d = rt$, where d is distance, r is rate of speed, and t is time, to find how long it would take to travel a distance of 1,120 mi at a speed of 140 mph.

36. *Travel* Use the formula $d = rt$, where d is distance, r is rate of speed, and t is time, to find how long it would take to travel a distance of 825 mi at a speed of 165 mph.

CRITICAL THINKING

37. Write an equation of the form $ax = b$, where a and b are numbers and x is a variable, that (a) has 0 as the solution of the equation and (b) has 1 as the solution of the equation.

38. Write two word problems for a classmate to solve, one that is a number problem (like Exercises 21–28 above) and another that involves using a formula (like Exercises 33–36 above).

SECTION 1.5   The Order of Operations Agreement

OBJECTIVE A

The Order of Operations Agreement

More than one operation may occur in a numerical expression. For example, the expression

$4 + 3(5)$

includes two arithmetic operations, addition and multiplication. The operations could be performed in different orders.

If we multiply first and then add, we have:

$4 + 3(5)$
$4 + 15$
19

If we add first and then multiply, we have:

$4 + 3(5)$
$7(5)$
35

To prevent more than one answer to the same problem, an Order of Operations Agreement is followed. By this agreement, 19 is the only correct answer.

The Order of Operations Agreement

Step 1 Do all operations inside parentheses.

Step 2 Simplify any numerical expressions containing exponents.

Step 3 Do multiplication and division as they occur from left to right.

Step 4 Do addition and subtraction as they occur from left to right.

CALCULATOR NOTE

Many calculators use the Order of Operations Agreement shown at the left.

Enter 4 [+] 3 [×] 5 [=] into your calculator. If the answer is 19, your calculator uses the Order of Operations Agreement.

Simplify: $2(4 + 1) - 2^3 + 6 \div 2$

	$2(4 + 1) - 2^3 + 6 \div 2$
Perform operations in parentheses.	$= 2(5) - 2^3 + 6 \div 2$
Simplify expressions with exponents.	$= 2(5) - 8 + 6 \div 2$
Do multiplication and division as they occur from left to right.	$= 10 - 8 + 6 \div 2$ $= 10 - 8 + 3$
Do addition and subtraction as they occur from left to right.	$= 2 + 3$ $= 5$

CALCULATOR NOTE

Here is an example of using the parentheses keys on a calculator. To evaluate 28(103 − 78), enter:

28 [×] [(] 103 [−] 78 [)] [=].

Note that [×] is required on most calculators.

One or more of the above steps may not be needed to simplify an expression. In that case, proceed to the next step in the Order of Operations Agreement.

Simplify: $8 + 9 \div 3$

There are no parentheses (Step 1). There are no exponents (Step 2).	$8 + 9 \div 3$
Do the division (Step 3).	$= 8 + 3$
Do the addition (Step 4).	$= 11$

◆◆

POINT OF INTEREST

Try this: Use the same one-digit number three times to write an expression that is equal to 30.

Evaluate $5a - (b + c)^2$ when $a = 6$, $b = 1$, and $c = 3$.

$$5a - (b + c)^2$$

Replace a with 6, b with 1, and c with 3.

$$5(6) - (1 + 3)^2$$

Use the Order of Operations Agreement to simplify the resulting numerical expression. Perform operations inside parentheses.

$$= 5(6) - (4)^2$$

Simplify expressions with exponents.

$$= 5(6) - 16$$

Do the multiplication.

$$= 30 - 16$$

Do the subtraction.

$$= 14$$

Example 1

Simplify: $18 \div (6 + 3) \cdot 9 - 4^2$

Solution

$$\begin{aligned}
18 \div (6 + 3) \cdot 9 - 4^2 &= 18 \div 9 \cdot 9 - 4^2 \\
&= 18 \div 9 \cdot 9 - 16 \\
&= 2 \cdot 9 - 16 \\
&= 18 - 16 \\
&= 2
\end{aligned}$$

You Try It 1

Simplify: $4 \cdot (8 - 3) \div 5 - 2$

Your Solution

Example 2

Simplify: $20 + 24(8 - 5) \div 2^2$

Solution

$$\begin{aligned}
20 + 24(8 - 5) \div 2^2 &= 20 + 24(3) \div 2^2 \\
&= 20 + 24(3) \div 4 \\
&= 20 + 72 \div 4 \\
&= 20 + 18 \\
&= 38
\end{aligned}$$

You Try It 2

Simplify: $16 + 3(6 - 1)^2 \div 5$

Your Solution

Example 3

Evaluate $(a - b)^2 + 3c$ when $a = 6$, $b = 4$, and $c = 1$.

Solution

$$(a - b)^2 + 3c$$
$$\begin{aligned}
(6 - 4)^2 + 3(1) &= (2)^2 + 3(1) \\
&= 4 + 3(1) \\
&= 4 + 3 \\
&= 7
\end{aligned}$$

You Try It 3

Evaluate $(a - b)^2 + 5c$ when $a = 7$, $b = 2$, and $c = 4$.

Your Solution

Solutions on p. A8

STUDY TIPS **Remember . . .**

To be successful, attend class regularly; read the textbook carefully; actively participate in class; work with your textbook using the boxed examples for immediate feedback and reinforcement of each skill; do all the homework assignments; review constantly; and work carefully.

1.5 Exercises

Objective A

Simplify.

1. $8 \div 4 + 2$

2. $12 - 9 \div 3$

3. $6 \cdot 4 + 5$

4. $5 \cdot 7 + 3$

5. $4^2 - 3$

6. $6^2 - 14$

7. $5 \cdot (6 - 3) + 4$

8. $8 + (6 + 2) \div 4$

9. $9 + (7 + 5) \div 6$

10. $14 \cdot (3 + 2) \div 10$

11. $13 \cdot (1 + 5) \div 13$

12. $14 - 2^3 + 9$

13. $6 \cdot 3^2 + 7$

14. $18 + 5 \cdot 3^2$

15. $14 + 5 \cdot 2^3$

16. $20 + (9 - 4) \cdot 2$

17. $10 + (8 - 5) \cdot 3$

18. $3^2 + 5 \cdot (6 - 2)$

19. $2^3 + 4(10 - 6)$

20. $3^2 \cdot 2^2 + 3 \cdot 2$

21. $6(7) + 4^2 \cdot 3^2$

22. $14 - 2(6)$

23. $18 + 3(7)$

24. $2(9 - 2) + 5$

25. $6(8 - 3) - 12$

26. $15 - (7 - 1) \div 3$

27. $16 - (13 - 5) \div 4$

Simplify.

28. $11 + 2 - 3 \cdot 4 \div 3$

29. $17 + 1 - 8 \cdot 2 \div 4$

30. $3(5 + 3) \div 8$

Evaluate the expression for the given values of the variables.

31. $x - 2y$, where $x = 8$ and $y = 3$

32. $x + 6y$, where $x = 5$ and $y = 4$

33. $x^2 + 3y$, where $x = 6$ and $y = 7$

34. $3x^2 + y$, where $x = 2$ and $y = 9$

35. $x^2 + y \div x$, where $x = 2$ and $y = 8$

36. $x + y^2 \div x$, where $x = 4$ and $y = 8$

37. $4x + (x - y)^2$, where $x = 8$ and $y = 2$

38. $(x + y)^2 - 2y$, where $x = 3$ and $y = 6$

39. $x^2 + 3(x - y) + z^2$, where $x = 2$, $y = 1$, and $z = 3$

40. $x^2 + 4(x - y) \div z^2$, where $x = 8$, $y = 6$, and $z = 2$

━━━━━━━━━━━━━━━━━

CRITICAL THINKING

41. What is the smallest prime number greater than $15 + (8 - 3)(2^4)$?

42. Write an expression that can be used to determine whether your calculator uses the Order of Operations Agreement. Use the expression to determine whether your calculator uses the Order of Operations Agreement.

43. ✎ What was the trade deficit of the United States last year? (*Hint:* Find the value of imports to the United States and the value of exports from the United States last year.)

44. ✎ Simplify $(47 + 48 + 49 + 51 + 52 + 53) \div 100$. What do you notice that will allow you to mentally calculate the answer?

PROJECTS IN MATHEMATICS

Garbology   Save your garbage for one week. Weigh it. Estimate how much trash you throw away per year.

How much of the garbage you collected is recyclable? If these materials were recycled, how much refuse would this save over a year's time?

Multiply the amount of garbage you collected by the number of people living in your community. Determine the amount of trash discarded by your community during a year.

Investigate a recycling program at your school or in your community.

> How many tons of garbage are recycled each year?
>
> What materials are accepted as recyclable in the program?
> Classify the products as glass, aluminum, paper, cardboard, plastic, yard waste, etc.
>
> What is the cost of the recycling program? Include the cost of labor in your figure.
>
> How much money does your school or community receive in payment for the recyclables?
>
> What is the cost of trash removal? What would be the cost of discarding the recyclables rather than recycling them?
>
> Does your school or community realize a profit as a result of recycling materials rather than discarding them? If so, how large a profit?
>
> Can the benefits of recycling be measured only by the profit or loss associated with a recycling program? Why or why not?

Applications of Patterns in Mathematics   For the circle shown at the left below, use a straight line to connect each dot on the circle with every other dot on the circle. How many different straight lines are there?

Follow the same procedure described above for each of the other circles shown below. How many different straight lines are there in each?

Find a pattern to describe the number of dots on a circle and the corresponding number of different lines drawn. Use the pattern to determine the number of different lines that would be drawn in a circle with 7 dots and in a circle with 8 dots.

You are arranging a tennis tournament with nine players. How many singles matches will be played among the nine players if each player plays each of the other players once?

Graphical Representation of Data

A graphical representation of data can sometimes be misleading. Consider the graphs shown below. An investment firm's financial advisor claims that an investment with the firm will grow as shown in the graph on the left, whereas an investment with a competitor will grow as shown in the graph on the right. Apparently, you would accumulate more money by choosing the investment on the left.

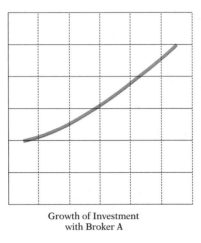

Growth of Investment
with Broker A

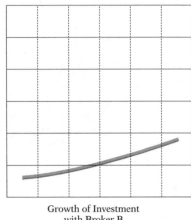

Growth of Investment
with Broker B

However, these graphs have a serious flaw. There are no labels on the horizontal and vertical axes. Therefore, it is impossible to tell which investment increased more or over what time interval. When labels are not placed on the axes of a graph, the data that graph represents are meaningless. It is one way advertisers use a visual impact to distort the true meaning of data.

The graphs below are the same as those drawn above except that scales have been drawn along each axis. Now it is possible to tell how each investment has performed. Note that each one has exactly the same performance.

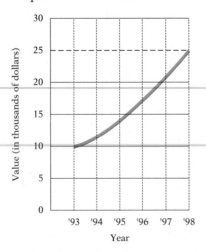

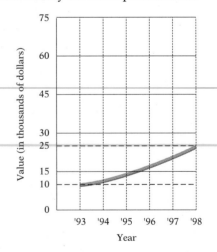

Drawing a circle graph as an oval is another way of distorting data. This is especially true if a three-dimensional representation is given. From the appearance of the circle graph at the left, region *A* is larger than region *B*. However, that isn't true. Measure each sector to see this for yourself.

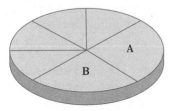

As you read newspapers and magazines, find examples of graphs that may distort the actual data. Discuss how these graphs should be drawn to be more accurate.

Surveys   On page 16 is a circle graph showing the results of a survey of 150 people who were asked, "What bothers you most about movie theaters?" Note that the responses included (1) people talking in the theater, (2) high ticket prices, (3) high prices for food purchased in the theater, (4) dirty floors, (5) and uncomfortable seats.

Conduct a similar survey in your class. Ask each classmate which of the five conditions stated above is most irritating. Record the number of students who answered each one of the five possible responses. Prepare a bar graph to display the results of the survey. A model is provided below to help you get started.

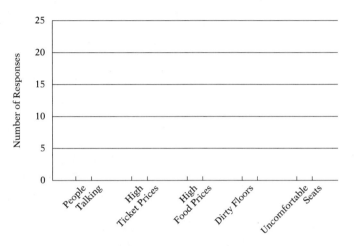

Responses to Theater-Goers Survey

CHAPTER SUMMARY

Key Words   The *natural numbers* or *counting numbers* are 1, 2, 3, 4, 5, 6, 7, 8, 9, 10, . . .

The *whole numbers* are 0, 1, 2, 3, 4, 5, 6, 7, 8, 9, 10, . . .

The symbol for *is less than* is <. The symbol for *is greater than* is >. A statement that uses the symbol < or > is an *inequality*.

When a whole number is written using the digits 0, 1, 2, 3, 4, 5, 6, 7, 8 and 9, it is said to be in *standard form*. The position of each digit in the number determines the digit's *place value*.

A *pictograph* represents data by using a symbol that is characteristic of the data. A *circle graph* represents data by the size of the sectors. A *bar graph* represents data by the height of the bars. A *broken-line graph* represents data by the position of the lines and shows trends or comparisons.

Addition is the process of finding the total of two or more numbers. The two numbers being added are called *addends.* The answer is the *sum.*

Subtraction is the process of finding the difference between two numbers. The *minuend* minus the *subtrahend* equals the *difference.*

Multiplication is the repeated addition of the same number. The numbers that are multiplied are called *factors.* The answer is the *product.*

Division is used to separate objects into equal groups. The *dividend* divided by the *divisor* equals the *quotient.*

The expression 3^5 is in *exponential form.* The *exponent,* 5, indicates how many times the *base,* 3, occurs as a factor in the multiplication.

Natural number *factors* of a number divide that number evenly (there is no remainder).

A number greater than 1 is a *prime* number if its only whole number factors are 1 and itself. If a number is not prime, it is a *composite* number.

The *prime factorization* of a number is the expression of the number as a product of its prime factors.

A *variable* is a letter that is used to stand for a number. A mathematical expression that contains one or more variables is a *variable expression.* Replacing the variables in a variable expression with numbers and then simplifying the numerical expression is called *evaluating the variable expression.*

An *equation* expresses the equality of two numerical or variable expressions. An equation contains an equals sign. A *solution* of an equation is a number that, when substituted for the variable, results in a true equation.

Essential Rules

To round a number to a given place value: If the digit to the right of the given place value is less than 5, that digit and all digits to the right are replaced by zeros. If the digit to the right of the given place value is greater than or equal to 5, increase the digit in the given place value by 1, and replace all other digits to the right by zeros.

To estimate the answer to a calculation: Round each number to the highest place value of that number. Perform the calculation using the rounded numbers.

Addition Property of Zero	$a + 0 = a$ or $0 + a = a$
Commutative Property of Addition	$a + b = b + a$
Associative Property of Addition	$(a + b) + c = a + (b + c)$
Multiplication Property of Zero	$a \cdot 0 = 0$ or $0 \cdot a = 0$
Multiplication Property of One	$a \cdot 1 = a$ or $1 \cdot a = a$
Commutative Property of Multiplication	$a \cdot b = b \cdot a$
Associative Property of Multiplication	$(a \cdot b) \cdot c = a \cdot (b \cdot c)$
Division Properties of Zero and One	If $a \neq 0$, $0 \div a = 0$.
	If $a \neq 0$, $a \div a = 1$.
	$a \div 1 = a$.
	$a \div 0$ is undefined.

Subtraction Property of Equations

The same number can be subtracted from each side of an equation without changing the solution of the equation.

Division Property of Equations

Each side of an equation can be divided by the same number (except zero) without changing the solution of the equation.

The Order of Operations Agreement

Step 1 Do all operations inside parentheses.

Step 2 Simplify any numerical expressions containing exponents.

Step 3 Do multiplication and division as they occur from left to right.

Step 4 Do addition and subtraction as they occur from left to right.

CHAPTER REVIEW EXERCISES

1. Graph 8 on the number line.

2. Evaluate 10^4.

3. Find the difference between 4,207 and 1,624.

4. Write $3 \cdot 3 \cdot 5 \cdot 5 \cdot 5 \cdot 5$ in exponential notation.

5. Add: $319 + 358 + 712$

6. Round 38,729 to the nearest hundred.

7. Place the correct symbol, $<$ or $>$ between the two numbers.

   247   163

8. Write thirty-two thousand five hundred nine in standard form.

9. Evaluate $2xy$ when $x = 50$ and $y = 7$.

10. Find the quotient of 15,642 and 6.

11. Subtract: $6,407 - 2,359$

12. Estimate the sum of 482, 319, 570, and 146.

13. Find all the factors of 50.

14. Is 7 a solution of the equation $24 - y = 17$?

15. Simplify: $16 + 4(7 - 5)^2 \div 8$

16. Identify the property that justifies the statement.
    $10 + 33 = 33 + 10$

17. Write 4,927,036 in words.

18. Evaluate $x^3 y^2$ when $x = 3$ and $y = 5$.

19. *Business* The figure on the right shows the uses of cash by Virco Mfg. Corporation, the largest producer of educational furniture in the United States. How many times greater was Virco's use of cash to purchase property, plant and equipment than for all other purposes?

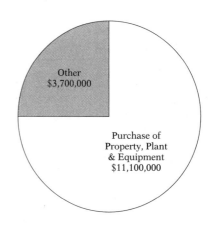

Uses of Cash by Virco Mfg. Corporation
Source: Virco Mfg. Corporation 1995 Annual Report

20. Divide: $6{,}234 \div 92$

21. Find the product of 4 and 659.

22. Evaluate $x - y$ when $x = 270$ and $y = 133$.

23. Find the prime factorization of 90.

24. Evaluate $\dfrac{x}{y}$ when $x = 480$ and $y = 6$.

25. Complete the statement by using the Multiplication Property of One.

$? \cdot 82 = 82$

26. Solve: $36 = 4x$

27. Evaluate $x + y$ when $x = 683$ and $y = 249$.

28. Multiply: $18 \cdot 24$

29. Evaluate $(a + b)^2 - 2c$ when $a = 5$, $b = 3$, and $c = 4$.

Solve.

30. *Sports* During his professional basketball career, Kareem Abdul-Jabbar had 17,440 rebounds. Elvin Hayes had 16,279 rebounds during his professional basketball career. Who had more rebounds, Abdul-Jabbar or Hayes?

31. *Construction* A contractor quotes the cost of work on a new house, which is to have 2,800 ft² of floor space, at $65 per square foot. Find the total cost of the contractor's work on the house.

32. *The Economy* The following figure shows the number of people unemployed in the United States during the last three months of 1995 and the first three months of 1996. Between which two months did the number of unemployed people increase the most? What was the amount of increase?

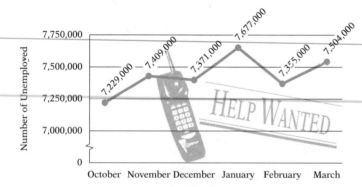

Number of Unemployed Persons October 1995 to March 1996
Source: Labor Department

33. *Travel* Use the formula $d = rt$, where d is distance, r is rate of speed, and t is time, to find the distance traveled in 3 h by a cyclist traveling at a speed of 14 mph.

34. *Business* Find the markup on a word processor which cost a business $1,775 and which sold for $2,224. Use the formula $M = S - C$, where M is the markup on a product, S is the selling price of the product, and C is the cost of the product to the business.

Integers

FOCUS On Problem Solving

INSTRUCTOR NOTE

The feature entitled Focus on Problem Solving appears at the beginning of every chapter of the text. It provides optional material to enhance your students' problem-solving skills.

How do you best remember something? Do you remember best what you hear? The word "aural" means *pertaining to the ear*; people with a strong aural memory remember best those things that they hear. The word "visual" means *pertaining to the sense of sight*; people with a strong visual memory remember best that which they see written down. Some people claim their memory is in their writing hand—they remember something only if they write it down! The method by which you best remember something is probably also the method by which you can best learn something new.

In problem-solving situations, try to capitalize on your strengths. If you tend to understand material better when you hear it spoken, read application problems aloud or have someone else read them to you. If writing helps you to organize ideas, rewrite application problems in your own words.

No matter what your main strength, visualizing a problem can be a valuable aid in problem solving. A drawing, sketch, diagram, or chart can be a useful tool in problem solving, just as calculators and computers are tools. A diagram can be helpful in gaining an understanding of the relationships inherent in a problem-solving situation. A sketch will help you to organize the given information and can lead to your being able to focus on the method by which the solution can be determined.

A tour bus drives 5 mi south, then 4 mi west, then 3 mi north, then 4 mi east. How far is the tour bus from the starting point?

Draw a diagram of the given information.

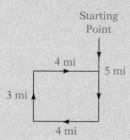

From the diagram, we can see that the solution can be determined by subtracting 3 from 5: $5 - 3 = 2$.

The bus is 2 mi from the starting point.

If you roll two ordinary six-sided dice and multiply the two numbers that appear on top, how many different products are there?

Make a chart of the possible products. In the chart below, repeated products are marked with an asterisk.

$1 \cdot 1 = 1$	$2 \cdot 1 = 2 \, (*)$	$3 \cdot 1 = 3 \, (*)$
$1 \cdot 2 = 2$	$2 \cdot 2 = 4 \, (*)$	$3 \cdot 2 = 6 \, (*)$
$1 \cdot 3 = 3$	$2 \cdot 3 = 6 \, (*)$	$3 \cdot 3 = 9$
$1 \cdot 4 = 4$	$2 \cdot 4 = 8$	$3 \cdot 4 = 12 \, (*)$
$1 \cdot 5 = 5$	$2 \cdot 5 = 10$	$3 \cdot 5 = 15$
$1 \cdot 6 = 6$	$2 \cdot 6 = 12$	$3 \cdot 6 = 18$
$4 \cdot 1 = 4 \, (*)$	$5 \cdot 1 = 5 \, (*)$	$6 \cdot 1 = 6 \, (*)$
$4 \cdot 2 = 8 \, (*)$	$5 \cdot 2 = 10 \, (*)$	$6 \cdot 2 = 12 \, (*)$
$4 \cdot 3 = 12 \, (*)$	$5 \cdot 3 = 15 \, (*)$	$6 \cdot 3 = 18 \, (*)$
$4 \cdot 4 = 16$	$5 \cdot 4 = 20 \, (*)$	$6 \cdot 4 = 24 \, (*)$
$4 \cdot 5 = 20$	$5 \cdot 5 = 25$	$6 \cdot 5 = 30 \, (*)$
$4 \cdot 6 = 24$	$5 \cdot 6 = 30$	$6 \cdot 6 = 36$

By counting the products that are not repeats, we can see that there are 18 different products.

In this chapter, you will notice that frequently a number line is used: to help you to visualize the integers, as an aid in ordering integers, to help you to understand the concepts of opposite and absolute value, to illustrate addition of integers. As you begin your work with integers, you may find that sketching a number line will prove helpful in your understanding of a problem or in working through a calculation involving integers.

SECTION 2.1   Introduction to Integers

OBJECTIVE **A**

Integers and the number line

In Chapter 1, only zero and numbers greater than zero were discussed. In this chapter, numbers less than zero are introduced. Phrases such as "7 degrees below zero," "$50 in debt," and "20 feet below sea level" refer to numbers less than zero.

Numbers greater than zero are called **positive numbers.** Numbers less than zero are called **negative numbers.**

Positive and Negative Numbers

A number n is positive if $n > 0$.
A number n is negative if $n < 0$.

A positive number can be indicated by placing the sign + in front of the number. For example, we can write +4 instead of 4. Both +4 and 4 represent "positive 4." Usually, however, the plus sign is omitted and it is understood that the number is a positive number.

A negative number is indicated by placing a negative sign (−) in front of the number. The number −1 is read "negative one," −2 is read "negative two," and so on.

The number line can be extended to the left of zero to show negative numbers.

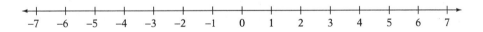

The **integers** are . . . −4, −3, −2, −1, 0, 1, 2, 3, 4, . . . The integers to the right of zero are the **positive integers.** The integers to the left of zero are the **negative integers.** Zero is an integer, but it is neither positive nor negative. The point corresponding to 0 on the number line is called the **origin.**

On a number line, the numbers get larger as we move from left to right. The numbers get smaller as we move from right to left. Therefore, a number line can be used to visualize the order relation between two integers.

A number that appears to the right of a given number is greater than (>) the given number. A number that appears to the left of a given number is less than (<) the given number.

2 is to the right of −3 on the number line.
2 is greater than −3.
$2 > -3$

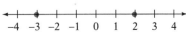

−4 is to the left of 1 on the number line.
−4 is less than 1.
$-4 < 1$

POINT OF INTEREST

Chinese manuscripts dating from about 250 B.C. contain the first recorded use of negative numbers. However, it was not until late in the fourteenth century that most mathematicians generally accepted these numbers.

> ### Order Relations
> $a > b$ if a is to the right of b on the number line.
> $a < b$ if a is to the left of b on the number line.

Example 1 On the number line, what number is 5 units to the right of −2?

Solution

3 is 5 units to the right of −2.

You Try It 1 On the number line, what number is 4 units to the left of 1?

Your Solution

Example 2 If G is 2 and I is 4, what numbers are B and D?

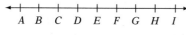

Solution

```
←+---+---+---+---+---+---+---+---+→
 −4  −3  −2  −1   0   1   2   3   4
```

B is −3, and D is −1.

You Try It 2 If G is 1 and H is 2, what numbers are A and C?

```
←+---+---+---+---+---+---+---+---+→
 A   B   C   D   E   F   G   H   I
```

Your Solution

Example 3 Place the correct symbol, < or >, between the two numbers.

   a. −3   −1   **b.** 1   −2

Solution **a.** −3 is to the left of −1 on the number line.

$$-3 < -1$$

   b. 1 is to the right of −2 on the number line.

$$1 > -2$$

You Try It 3 Place the correct symbol, < or >, between the two numbers.

   a. 2   −5   **b.** −4   3

Your Solution

Example 4 Write the given numbers in order from smallest to largest.

5, −2, 3, 0, −6

Solution −6, −2, 0, 3, 5

You Try It 4 Write the given numbers in order from smallest to largest.

−7, 4, −1, 0, 8

Your Solution

Solutions on p. A8

OBJECTIVE B
Opposites

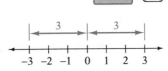

The distance from 0 to 3 on the number line is 3 units. The distance from 0 to −3 on the number line is 3 units. 3 and −3 are the same distance from 0 on the number line, but 3 is to the right of 0 and −3 is to the left of 0.

Two numbers that are the same distance from zero on the number line but on opposite sides of zero are called **opposites.**

    −3 is the opposite of 3   and   3 is the opposite of −3.

For any number n, the opposite of n is $-n$ and the opposite of $-n$ is n.

We can now define the **integers** as the whole numbers and their opposites.

A negative sign can be read as "the opposite of."

    −(3) = −3   The opposite of positive 3 is negative 3.

    −(−3) = 3   The opposite of negative 3 is positive 3.

Therefore, $-(a) = -a$ and $-(-a) = a$.

Note that with the introduction of negative integers and opposites, the symbols + and − can be read in different ways.

6 + 2	"six plus two"	+ is read "plus"
+2	"positive two"	+ is read "positive"
6 − 2	"six minus two"	− is read "minus"
−2	"negative two"	− is read "negative"
−(−6)	"the opposite of negative six"	− is read first as "the opposite of" and then as "negative"

When the symbols + and − indicate the operations of addition and subtraction, spaces are inserted before and after the symbol. When the symbols + and − indicate the sign of a number (positive or negative), there is no space between the symbol and the number.

Example 5   Find the opposite number.

  a. −8   **b.** 15   **c.** a

Solution   **a.** 8   **b.** −15   **c.** $-a$

You Try It 5   Find the opposite number.

  a. 24   **b.** −13   **c.** $-b$

Your Solution

Solution on p. A8

Example 6 Write the expression in words.

 a. $7 - (-9)$ **b.** $-4 + 10$

 Solution **a.** seven minus negative nine

 b. negative four plus ten

You Try It 6 Write the expression in words.

 a. $-3 - 12$ **b.** $8 + (-5)$

Your Solution

Example 7 Simplify.

 a. $-(-27)$ **b.** $-(-c)$

 Solution **a.** $-(-27) = 27$

 b. $-(-c) = c$

You Try It 7 Simplify.

 a. $-(-59)$ **b.** $-(y)$

Your Solution

Solutions on p. A8

OBJECTIVE C

Absolute value

The **absolute value** of a number is the distance from zero to the number on the number line. Distance is never a negative number. Therefore, the absolute value of a number is a positive number or zero. The symbol for absolute value is "| |."

The distance from 0 to 3 is 3 units. Thus $|3| = 3$ (the absolute value of 3 is 3).

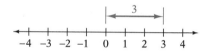

The distance from 0 to -3 is 3 units. Thus $|-3| = 3$ (the absolute value of -3 is 3).

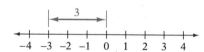

Because the distance from 0 to 3 and the distance from 0 to -3 are the same,

 $|3| = |-3| = 3$.

Absolute Value

The absolute value of a positive number is positive. $|5| = 5$

The absolute value of a negative number is positive. $|-5| = 5$

The absolute value of zero is zero. $|0| = 0$

TAKE NOTE

It is important to be aware that the negative sign is *in front of the absolute value.* Compare $-|7| = -7$ but $|-7| = 7$.

Evaluate $-|7|$.

The negative sign is *in front of* the absolute value symbol.

Recall that a negative sign can be read as "the opposite of."

Therefore, $-|7|$ can be read "the opposite of the absolute value of 7."

 $-|7| = -7$

Example 8   Find the absolute value of
 a. 6 and **b.** −9.

Solution   **a.** |6| = 6

b. |−9| = 9

You Try It 8   Find the absolute value of
 a. −8 and **b.** 12.

Your Solution

Example 9   Evaluate **a.** |−27| and **b.** −|−14|.

Solution   **a.** |−27| = 27

b. −|−14| = −14

You Try It 9   Evaluate **a.** |0| and **b.** −|35|.

Your Solution

Example 10   Evaluate |−x|, where x = −4.

Solution |−x| = |−(−4)| = |4| = 4

You Try It 10   Evaluate |−y|, where y = 2.

Your Solution

Example 11   Write the given numbers in order from smallest to largest.

|−7|, −5, |0|, −(−4), −|−3|

Solution |−7| = 7, |0| = 0,
−(−4) = 4, −|−3| = −3

−5, −|−3|, |0|, −(−4), |−7|

You Try It 11   Write the given numbers in order from smallest to largest.

|6|, |−2|, −(−1), −4, −|−8|

Your Solution

Solutions on p. A8

Objective D
Applications

Data that are represented by negative numbers on a bar graph are shown below the horizontal axis. For instance, Figure 2.1 shows the lowest recorded temperatures, in Fahrenheit, for selected states in the United States. Hawaii's lowest recorded temperature is 14°F, which is a positive number, so the bar that represents that temperature is above the horizontal axis. The bars for the other states are below the horizontal axis and therefore represent negative numbers.

We can see from the graph that the state with the lowest recorded temperature is New York, with a temperature of −52°F.

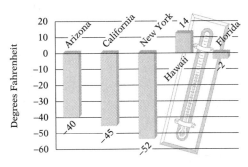

Figure 2.1   Lowest Recorded Temperatures

In a golf tournament, scores below par are recorded as negative numbers; scores above par are recorded as positive numbers. The winner of the tournament is the player who has the lowest score.

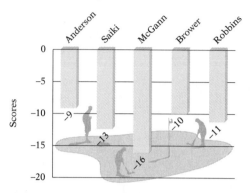

Figure 2.2 shows the number of strokes under par for the five best finishers in the 1996 Youngstown-Warren LPGA Classic held in Warren, Ohio. Use this graph for Example 12 and You Try It 12.

Figure 2.2 The Top Finishers in the 1996 Youngstown-Warren LPGA Classic

Example 12

Use Figure 2.2 to name the player who won the tournament.

Strategy

From the bar graph, find the player with the lowest number for a score.

Solution

$-16 < -13 < -11 < -10 < -9$

The lowest number among the scores is -16.

McGann won the tournament.

Example 13

Which is the colder temperature, $-18°F$ or $-15°F$?

Strategy

To determine which is the colder temperature, compare the numbers -18 and -15. The lower number corresponds to the colder temperature.

Solution

$-18 < -15$

The colder temperature is $-18°F$.

You Try It 12

Use Figure 2.2 to name the player who came in third in the tournament.

Your Strategy

Your Solution

You Try It 13

Which is closer to blast off, -9 s and counting or -7 s and counting?

Your Strategy

Your Solution

2.1 EXERCISES

OBJECTIVE A

Graph the number on the number line.

1. −5

$$\xleftarrow{\hspace{0.3em}}\overset{}{\underset{\substack{-6\ -5\ -4\ -3\ -2\ -1\ \ 0\ \ 1\ \ 2\ \ 3\ \ 4\ \ 5\ \ 6}}{|\ \ |\ \ |\ \ |\ \ |\ \ |\ \ |\ \ |\ \ |\ \ |\ \ |\ \ |\ \ |}}\xrightarrow{\hspace{0.3em}}$$

2. −1

$$\xleftarrow{\hspace{0.3em}}\overset{}{\underset{\substack{-6\ -5\ -4\ -3\ -2\ -1\ \ 0\ \ 1\ \ 2\ \ 3\ \ 4\ \ 5\ \ 6}}{|\ \ |\ \ |\ \ |\ \ |\ \ |\ \ |\ \ |\ \ |\ \ |\ \ |\ \ |\ \ |}}\xrightarrow{\hspace{0.3em}}$$

3. −6

$$\xleftarrow{\hspace{0.3em}}\overset{}{\underset{\substack{-6\ -5\ -4\ -3\ -2\ -1\ \ 0\ \ 1\ \ 2\ \ 3\ \ 4\ \ 5\ \ 6}}{|\ \ |\ \ |\ \ |\ \ |\ \ |\ \ |\ \ |\ \ |\ \ |\ \ |\ \ |\ \ |}}\xrightarrow{\hspace{0.3em}}$$

4. −2

$$\xleftarrow{\hspace{0.3em}}\overset{}{\underset{\substack{-6\ -5\ -4\ -3\ -2\ -1\ \ 0\ \ 1\ \ 2\ \ 3\ \ 4\ \ 5\ \ 6}}{|\ \ |\ \ |\ \ |\ \ |\ \ |\ \ |\ \ |\ \ |\ \ |\ \ |\ \ |\ \ |}}\xrightarrow{\hspace{0.3em}}$$

5. x, where $x = 5$

$$\xleftarrow{\hspace{0.3em}}\overset{}{\underset{\substack{-6\ -5\ -4\ -3\ -2\ -1\ \ 0\ \ 1\ \ 2\ \ 3\ \ 4\ \ 5\ \ 6}}{|\ \ |\ \ |\ \ |\ \ |\ \ |\ \ |\ \ |\ \ |\ \ |\ \ |\ \ |\ \ |}}\xrightarrow{\hspace{0.3em}}$$

6. x, where $x = 0$

$$\xleftarrow{\hspace{0.3em}}\overset{}{\underset{\substack{-6\ -5\ -4\ -3\ -2\ -1\ \ 0\ \ 1\ \ 2\ \ 3\ \ 4\ \ 5\ \ 6}}{|\ \ |\ \ |\ \ |\ \ |\ \ |\ \ |\ \ |\ \ |\ \ |\ \ |\ \ |\ \ |}}\xrightarrow{\hspace{0.3em}}$$

7. x, where $x = -4$

$$\xleftarrow{\hspace{0.3em}}\overset{}{\underset{\substack{-6\ -5\ -4\ -3\ -2\ -1\ \ 0\ \ 1\ \ 2\ \ 3\ \ 4\ \ 5\ \ 6}}{|\ \ |\ \ |\ \ |\ \ |\ \ |\ \ |\ \ |\ \ |\ \ |\ \ |\ \ |\ \ |}}\xrightarrow{\hspace{0.3em}}$$

8. x, where $x = -3$

$$\xleftarrow{\hspace{0.3em}}\overset{}{\underset{\substack{-6\ -5\ -4\ -3\ -2\ -1\ \ 0\ \ 1\ \ 2\ \ 3\ \ 4\ \ 5\ \ 6}}{|\ \ |\ \ |\ \ |\ \ |\ \ |\ \ |\ \ |\ \ |\ \ |\ \ |\ \ |\ \ |}}\xrightarrow{\hspace{0.3em}}$$

On the number line, which number is:

9. 3 units to the right of −2?

10. 5 units to the right of −3?

11. 4 units to the left of 3?

12. 2 units to the left of −1?

13. 6 units to the right of −3?

14. 4 units to the right of −4?

For Exercises 15–18, use the following number line.

$$\xleftarrow{\hspace{0.3em}}\overset{}{\underset{\substack{A\ \ \ B\ \ \ C\ \ \ D\ \ \ E\ \ \ F\ \ \ G\ \ \ H\ \ \ I}}{|\ \ \ |\ \ \ |\ \ \ |\ \ \ |\ \ \ |\ \ \ |\ \ \ |\ \ \ |}}\xrightarrow{\hspace{0.3em}}$$

15. If F is 1 and G is 2, what numbers are A and C?

16. If G is 1 and H is 2, what numbers are B and D?

17. If H is 0 and I is 1, what numbers are A and D?

18. If G is 2 and I is 4, what numbers are B and E?

Place the correct symbol, $<$ or $>$, between the two numbers.

19. $-2 \quad -5$

20. $-6 \quad -1$

21. $3 \quad -7$

22. $-11 \quad -8$

23. $-42 \quad 27$

24. $21 \quad -34$

25. $53 \quad -46$

26. $-27 \quad -39$

27. $-51 \quad -20$

28. $-136 \quad 0$

29. $-131 \quad 101$

30. $127 \quad -150$

Write the given numbers in order from smallest to largest.

31. $3, -7, 0, -2$

32. $-4, 8, 6, -1$

33. $-3, 1, -5, 4$

34. $-6, 2, -8, 7$

35. $9, -4, 5, 0$

36. $6, -9, -12, 8$

37. $-10, 4, 12, -5, -7$

38. $11, -8, -1, 7, -6$

39. $10, -11, -2, 5, -7$

OBJECTIVE B

Find the opposite of the number.

40. 22

41. 45

42. -31

43. -88

44. c

45. n

46. $-w$

47. $-d$

Write the expression in words.

48. $-(-11)$

49. $-(-13)$

50. $-(-d)$

51. $-(-p)$

52. $-2 + (-5)$

53. $5 + (-10)$

54. $6 - (-7)$

55. $-14 - (-3)$

56. $9 - 12$

57. $-13 - 8$

58. $-a - b$

59. $m + (-n)$

Simplify.

60. $-(-5)$ **61.** $-(-7)$ **62.** $-(-38)$ **63.** $-(-61)$

64. $-(29)$ **65.** $-(46)$ **66.** $-(-52)$ **67.** $-(-73)$

68. $-(-m)$ **69.** $-(-z)$ **70.** $-(b)$ **71.** $-(p)$

OBJECTIVE C

Find the absolute value of the number.

72. 4 **73.** -4 **74.** -7 **75.** 9

76. -1 **77.** -11 **78.** 10 **79.** -12

Evaluate.

80. $|-15|$ **81.** $|-23|$ **82.** $-|33|$ **83.** $-|27|$

84. $|32|$ **85.** $|25|$ **86.** $-|-36|$ **87.** $-|-41|$

88. $-|-81|$ **89.** $-|-93|$ **90.** $|x|$, where $x = 7$ **91.** $|x|$, where $x = -10$

92. $|-x|$, where $x = 2$ **93.** $|-x|$, where $x = 8$ **94.** $|-y|$, where $y = -3$ **95.** $|-y|$, where $y = -6$

Place the correct symbol, $<$, $=$, or $>$, between the two numbers.

96. $|7|$ $|-9|$ **97.** $|-12|$ $|8|$ **98.** $|-5|$ $|-2|$ **99.** $|6|$ $|13|$

100. $|-8|$ $|3|$ **101.** $|-1|$ $|-17|$ **102.** $|-14|$ $|14|$ **103.** $|x|$ $|-x|$

Write the given numbers in order from smallest to largest.

104. $|-8|, -(-3), |2|, -|-5|$ **105.** $-|6|, -(4), |-7|, -(-9)$ **106.** $-(-1), |-6|, |0|, -|3|$

107. $-|-7|, -9, -(5), |4|$ **108.** $-|2|, -(-8), 6, |1|, -7$ **109.** $-(-3), -|-8|, |5|, -|10|, -(-2)$

Solve.

110. Find the values of a for which $|a| = 7$.

111. Find the values of y for which $|y| = 11$.

112. Given that x is an integer, find all values of x for which $|x| < 5$.

113. Given that c is an integer, find all values of c for which $|c| < 7$.

OBJECTIVE D

The table below gives equivalent temperatures for combinations of temperature and wind speed. For example, the combination of a temperature of 15°F and a wind blowing at 10 mph has a cooling power equal to −3°F. Use this table for Exercises 114–119.

Wind Chill Factors																
Wind Speed (mph)	Thermometer Reading (degrees Fahrenheit)															
	35	30	25	20	15	10	5	0	−5	−10	−15	−20	−25	−30	−35	−40	−45
5	33	27	21	19	12	7	0	−5	−10	−15	−21	−26	−31	−36	−42	−47	−52
10	22	16	10	3	−3	−9	−15	−22	−27	−34	−40	−46	−52	−58	−64	−71	−77
15	16	9	2	−5	−11	−18	−25	−31	−38	−45	−51	−58	−65	−72	−78	−85	−92
20	12	4	−3	−10	−17	−24	−31	−39	−46	−53	−60	−67	−74	−81	−88	−95	−103
25	8	1	−7	−15	−22	−29	−36	−44	−51	−59	−66	−74	−81	−88	−96	−103	−110
30	6	−2	−10	−18	−25	−33	−41	−49	−56	−64	−71	−79	−86	−93	−101	−109	−116
35	4	−4	−12	−20	−27	−35	−43	−52	−58	−67	−74	−82	−89	−97	−105	−113	−120
40	3	−5	−13	−21	−29	−37	−45	−53	−60	−69	−76	−84	−92	−100	−107	−115	−123
45	2	−6	−14	−22	−30	−38	−46	−54	−62	−70	−78	−85	−93	−102	−109	−117	−125

114. *Environmental Science* Find the wind chill factor when the temperature is 5°F and the wind speed is 15 mph.

115.   *Environmental Science*   Find the wind chill factor when the temperature is 10°F and the wind speed is 20 mph.

116.   *Environmental Science*   Find the cooling power of a temperature of −10°F and a 5 mph wind.

117.   *Environmental Science*   Find the cooling power of a temperature of −15°F and a 10 mph wind.

118.   *Environmental Science*   Which feels colder, a temperature of 0°F with a 15 mph wind or a temperature of 10°F with a 25 mph wind?

119.   *Environmental Science*   Which would feel colder, a temperature of −30°F with a 5 mph wind or a temperature of −20°F with a 10 mph wind?

120.   *Rocketry*   Which is closer to blast-off, −12 min and counting or −17 min and counting?

One of the measures used by a financial analyst to evaluate the financial strength of a company is *earnings per share*. This number is found by taking the total profit of the company and dividing by the number of shares of stock that the company has sold to investors. If the company has a loss instead of a profit, the earnings per share is a negative number. In a bar graph, a profit is shown by a bar extending above the horizontal axis, and a loss is shown by a bar extending below the horizontal axis. The figure on the right shows the earnings per share for Mycogen for the years 1990 through 1995. Use this graph for Exercises 121 to 124.

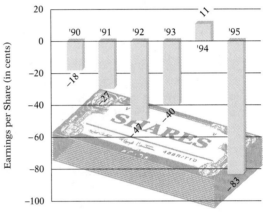

Mycogen Earnings per Share (in cents)

121.   *Business*   (a) What were the earnings per share for Mycogen in 1991? (b) What were the earnings per share for Mycogen in 1993?

122.   *Business*   For the years shown, in which year did Mycogen have the greatest loss?

123.   *Business*   For the years shown, did Mycogen ever have a profit? If so, in what year?

124.   *Business*   In which year was the Mycogen earnings per share lower, 1990 or 1992?

Solve.

125.   *Investments*   In the stock market, the net change in the price of a share of stock is recorded as a positive or a negative number. If the price rises, the net change is positive. If the price falls, the net change is negative. If the net change for a share of Stock A is −2 and the net change for a share of Stock B is −1, which stock showed the least net change?

Solve.

126. *Business* Some businesses show a profit as a positive number and a loss as a negative number. During the first quarter of this year, the loss experienced by a company was recorded as −12,575. During the second quarter of this year, the loss experienced by the company was −11,350. During which quarter was the loss greater?

127. *Business* Some businesses show a profit as a positive number and a loss as a negative number. During the third quarter of last year, the loss experienced by a company was recorded as −26,800. During the fourth quarter of last year, the loss experienced by the company was −24,900. During which quarter was the loss greater?

CRITICAL THINKING

128. *A* is a point on the number line halfway between −9 and 3. *B* is a point halfway between *A* and the graph of 1 on the number line. *B* is the graph of what number?

129. **a.** Name two numbers that are 4 units from 2 on the number line.
b. Name two numbers that are 5 units from 3 on the number line.

130. Determine whether the statement is always true, sometimes true, or never true.
a. The number $-n$ is a negative number.
b. A number and its opposite are different numbers.
c. $|x| > x$
d. $|x| > -x$
e. If n is a negative number, $-n$ is a positive number.
f. If n is a positive number, $-n$ is a negative number.

131. The $\boxed{+/-}$ key on a calculator changes the sign of the number in the calculator's display. In other words, it changes the number in the display to its opposite. Use the $\boxed{+/-}$ key on your calculator to display each of the following numbers.
a. −9　　**b.** −20　　**c.** −148　　**d.** −573

132. Given that z is an integer and $|z| < 15$, find all values of z for which $|z| > 10$.

133. Given that x is an integer and $|x| < 10$, find all values of x for which $|x| > 6$.

134. In your own words, describe (a) the opposite of a number, (b) the absolute value of a number, and (c) the difference between the words *negative* and *minus*.

135. Student A, Student B, Student C, and Student D were being questioned by their teacher. The teacher knew that one of the students had left an apple on the teacher's desk but did not know which one. Student A said it was either Student B or Student D. Student D said it was neither Student B nor Student C. If both statements were false, who left the apple on the teacher's desk? Explain how you arrived at your solution.

SECTION 2.2   Addition and Subtraction of Integers

OBJECTIVE A
Addition of integers

Not only can an integer be graphed on a number line, an integer can be represented anywhere along a number line by an arrow. A positive number is represented by an arrow pointing to the right. A negative number is represented by an arrow pointing to the left. The absolute value of the number is represented by the length of the arrow. The integers 5 and -4 are shown on the number line in the figure below.

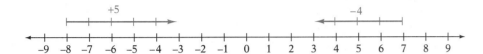

The sum of two integers can be shown on a number line. To add two integers, find the point on the number line corresponding to the first addend. At that point, draw an arrow representing the second addend. The sum is the number directly below the tip of the arrow.

$4 + 2 = 6$

$-4 + (-2) = -6$

$-4 + 2 = -2$

$4 + (-2) = 2$

Ths sums shown above can be categorized by the signs of the addends.

The addends have the same sign.

$\quad 4 + 2 \qquad$ positive 4 plus positive 2
$\quad -4 + (-2) \qquad$ negative 4 plus negative 2

The addends have different signs.

$\quad -4 + 2 \qquad$ negative 4 plus positive 2
$\quad 4 + (-2) \qquad$ positive 4 plus negative 2

The rule for adding two integers depends on whether the signs of the addends are the same or different.

Rule for Adding Two Integers

TO ADD INTEGERS WITH THE SAME SIGN, add the absolute values of the numbers. Then attach the sign of the addends.

TO ADD INTEGERS WITH DIFFERENT SIGNS, find the absolute values of the numbers. Subtract the smaller absolute value from the larger absolute value. Then attach the sign of the addend with the larger absolute value.

Add: $(-4) + (-9)$

The signs of the addends are the same.
Add the absolute values of the numbers.
$\quad |-4| = 4, |-9| = 9, 4 + 9 = 13$
Attach the sign of the addends.
(Both addends are negative.
The sum is negative.) $(-4) + (-9) = -13$

CALCULATOR NOTE
To add $-14 + (-47)$ with your calculator, enter the following:
$14\;\boxed{+/-}\;\boxed{+}\;47\;\boxed{+/-}\;\boxed{=}$

Add: $-14 + (-47)$

The signs are the same.
Add the absolute values of the numbers.
Attach the sign of the addends. $-14 + (-47) = -61$

Add: $6 + (-13)$

The signs of the addends are different.
Find the absolute values of the numbers.
$\quad |6| = 6, |-13| = 13$
Subtract the smaller absolute value from the larger absolute value.
$\quad 13 - 6 = 7$
Attach the sign of the number with the larger absolute value.
$\quad |-13| > |6|.$ Attach the negative sign. $6 + (-13) = -7$

Add: $162 + (-247)$

The signs are different. Find the difference between the absolute values of the numbers.
$\quad 247 - 162 = 85$
Attach the sign of the number with the larger absolute value. $162 + (-247) = -85$

Add: $-8 + 8$

The signs are different. Find the difference between the absolute values of the numbers.
$\quad 8 - 8 = 0$ $-8 + 8 = 0$

Note in this last example that we are adding a number and its opposite (-8 and 8), and the sum is 0. The opposite of a number is called its **additive inverse.** The opposite or additive inverse of -8 is 8, and the opposite or additive inverse of 8 is -8. The sum of a number and its additive inverse is always zero. This is known as the Inverse Property of Addition.

The properties of addition presented in Chapter 1 hold true for integers as well as whole numbers. These properties are repeated below, along with the Inverse Property of Addition.

The Addition Property of Zero	$a + 0 = a$ or $0 + a = a$
The Commutative Property of Addition	$a + b = b + a$
The Associative Property of Addition	$(a + b) + c = a + (b + c)$
The Inverse Property of Addition	$a + (-a) = 0$ or $-a + a = 0$

Add: $(-4) + (-6) + (-8) + 9$ $(-4) + (-6) + (-8) + 9$

Add the first two numbers. $= (-10) + (-8) + 9$

Add the sum to the third number. $= (-18) + 9$

Continue until all the numbers have been added. $= -9$

The price of Byplex Corporation's stock fell each trading day of the first week of June 1997. Use Figure 2.3 to find the change in the price of Byplex stock over the week's time.

Add the five changes in price.

$-2 + (-3) + (-1) + (-2) + (-1)$
$= (-5) + (-1) + (-2) + (-1)$
$= -6 + (-2) + (-1)$
$= -8 + (-1) = -9$

The change in the price was -9.

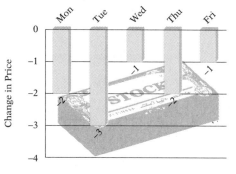

Figure 2.3   Change in Price of Byplex Corporation Stock

This means that the price of the stock fell $9 per share.

Evaluate $-x + y$ when $x = -15$ and $y = -5$.

Replace x with -15 and y with -5. $\begin{aligned} -x + y \\ -(-15) + (-5) \end{aligned}$

Simplify $-(-15)$. $= 15 + (-5)$

Add. $= 10$

Is -7 a solution of the equation $x + 4 = -3$?

	$x + 4 = -3$
Replace x by -7 and then simplify.	$-7 + 4 \mid -3$
The results are equal.	$-3 = -3$

-7 is a solution of the equation.

Example 1 Add: $42 + (-12) + (-30)$

Solution $42 + (-12) + (-30)$
$= 30 + (-30)$
$= 0$

You Try It 1 Add: $-36 + 17 + (-21)$

Your Solution

Example 2 What is -162 increased by 98?

Solution $-162 + 98 = -64$

You Try It 2 Find the sum of -154 and -37.

Your Solution

Example 3 Evaluate $-x + y$ when $x = -11$ and $y = -2$.

Solution $-x + y$
$-(-11) + (-2) = 11 + (-2)$
$= 9$

You Try It 3 Evaluate $-x + y$ when $x = -3$ and $y = -10$.

Your Solution

Example 4 Is -6 a solution of the equation $3 + y = -2$?

Solution $\dfrac{3 + y = -2}{3 + (-6) \mid -2}$
$-3 \neq -2$

No, -6 is not a solution of the equation.

You Try It.4 Is -9 a solution of the equation $2 = 11 + a$?

Your Solution

Solutions on p. A8

OBJECTIVE **B**

Subtraction of integers

Recall that the sign $-$ can indicate the sign of a number, as in -3 (negative 3), or can indicate the operation of subtraction, as in $9 - 3$ (nine minus three).

Look at each of the four subtraction expressions below and state whether the second number in each expression is a positive number or a negative number.

1. $(-10) - 8$

2. $(-10) - (-8)$

3. $10 - (-8)$

4. $10 - 8$

In (1) and (4), the second number is positive 8. In (2) and (3), the second number is negative 8.

Opposites are used to rewrite subtraction problems as related addition problems. Notice below that the subtraction of whole numbers is the same as the addition of the opposite number.

Subtraction		Addition of the Opposite	
$8 - 4$	$=$	$8 + (-4)$	$= 4$
$7 - 5$	$=$	$7 + (-5)$	$= 2$
$9 - 2$	$=$	$9 + (-2)$	$= 7$

Subtraction of integers can be written as the addition of the opposite number. To subtract two integers, rewrite the subtraction expression as the first number plus the opposite of the second number. Some examples are shown below.

First number	−	second number	=	First number	+	opposite of the second number
8	−	15	=	8	+	$(-15) = -7$
8	−	(-15)	=	8	+	$15 = 23$
-8	−	15	=	-8	+	$(-15) = -23$
-8	−	(-15)	=	-8	+	$15 = 7$

Rule for Subtracting Two Integers

To subtract two integers, add the opposite of the second integer to the first integer.

Subtract: $(-15) - 75$

Rewrite the subtraction operation as the sum of the first number and the opposite of the second number. The opposite of 75 is -75.

$(-15) - 75$

$= (-15) + (-75)$

Add. $= -90$

When subtraction occurs several times in an expression, rewrite each subtraction as addition of the opposite and then add.

Subtract: $-13 - 5 - (-8)$

Rewrite each subtraction as addition of the opposite.

Add.

$-13 - 5 - (-8)$
$= -13 + (-5) + 8$
$= -18 + 8$
$= -10$

Simplify: $-14 + 6 - (-7)$

This problem involves both addition and subtraction. Rewrite the subtraction as addition of the opposite.

Add.

$-14 + 6 - (-7)$
$= -14 + 6 + 7$
$= -8 + 7$
$= -1$

Evaluate $a - b$ when $a = -2$ and $b = -9$.

Replace a with -2 and b with -9.

Rewrite the subtraction as addition of the opposite.

Add.

$a - b$
$-2 - (-9)$
$= -2 + 9$
$= 7$

Is -4 a solution of the equation $3 - a = 11 + a$?

Replace a by -4 and then simplify.

$$\begin{array}{c|c} 3 - a & = 11 + a \\ \hline 3 - (-4) & 11 + (-4) \\ 3 + 4 & 7 \end{array}$$

The results are equal.

$$7 = 7$$

Yes, -4 is a solution of the equation.

The table below shows the boiling point and the melting point in degrees Celsius of three chemical elements. Use this table for Example 5 and You Try It 5.

Chemical Element	Boiling Point	Melting Point
Carbon	4,827	−3,550
Radon	−62	−71
Xenon	−107	−112

Example 5 Use the table above to find the difference between the boiling point and the melting point of carbon.

Solution The boiling point of carbon is 4,827.

The melting point of carbon is −3,550.

$$4,827 - (-3,550) = 4,827 + 3,550$$
$$= 8,377$$

The difference is 8,377°C.

You Try It 5 Use the table above to find the difference between the boiling point and the melting point of xenon.

Your Solution

Example 6 What is −12 minus 8?

Solution $-12 - 8 = -12 + (-8)$
$= -20$

You Try It 6 What is 14 less than −8?

Your Solution

Example 7 Simplify:
$-8 - 30 - (-12) - 7 - (-14)$

Solution $-8 - 30 - (-12) - 7 - (-14)$
$= -8 + (-30) + 12 + (-7) + 14$
$= -38 + 12 + (-7) + 14$
$= -26 + (-7) + 14$
$= -33 + 14$
$= -19$

You Try It 7 Simplify:
$-4 - (-3) + 12 - (-7) - 20$

Your Solution

Solutions on pp. A8–A9

Example 8 Evaluate $-x - y$ when $x = -4$ and $y = -3$.

Solution
$$-x - y$$
$$-(-4) - (-3) = 4 - (-3)$$
$$= 4 + 3$$
$$= 7$$

You Try It 8 Evaluate $x - y$ when $x = -9$ and $y = 7$.

Your Solution

Example 9 Is 8 a solution of the equation $-2 = 6 - x$?

Solution
$$-2 = 6 - x$$

-2	$6 - 8$
-2	$6 + (-8)$
$-2 = -2$	

Yes, 8 is a solution of the equation.

You Try It 9 Is -3 a solution of the equation $a - 5 = -8$?

Your Solution

Solutions on p. A9

OBJECTIVE C

Applications and formulas

Figure 2.4 shows the melting points in degrees Celsius of six chemical elements. The abbreviations of the elements are:

F - Fluorine	H - Hydrogen
S - Sulfur	N - Nitrogen
O - Oxygen	Li - Lithium

Use this graph for Example 10 and You Try It 10.

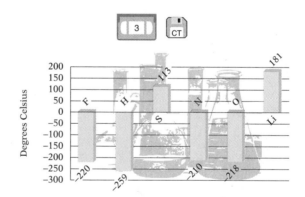

Figure 2.4   Melting Points of Chemical Elements

Example 10

Find the difference between the two lowest melting points shown in Figure 2.4.

Strategy

To find the difference, subtract the lowest melting point shown (-259) from the second lowest melting point shown (-220).

Solution

$$-220 - (-259) = -220 + 259 = 39$$

The difference is 39°C.

You Try It 10

Find the difference between the highest and lowest melting points shown in Figure 2.4.

Your Strategy

Your Solution

Solution on p. A9

Example 11

Find the temperature after an increase of 8°C from −5°C.

Strategy

To find the temperature, add the increase (8) to the previous temperature (−5).

Solution

−5 + 8 = 3

The temperature is 3°C.

You Try It 11

Find the temperature after an increase of 10°C from −3°C.

Your Strategy

Your Solution

Example 12

The average temperature on the sunlit side of the moon is approximately 215°F. The average temperature on the dark side is approximately −250°F. Find the difference between these average temperatures.

Strategy

To find the difference, subtract the average temperature on the dark side of the moon (−250) from the average temperature on the sunlit side (215).

Solution

$215 − (−250) = 215 + 250$
$= 465$

The difference is 465°F.

You Try It 12

The average temperature on the earth's surface is 57°F. The average temperature throughout the earth's stratosphere is −70°F. Find the difference between these average temperatures.

Your Strategy

Your Solution

Example 13

The distance, d, between point a and point b on the number line is given by the formula $d = |a − b|$. Use the formula to find d when $a = 7$ and $b = −8$.

Strategy

To find d, replace a by 7 and b by −8 in the given formula and solve for d.

Solution

$d = |a − b|$
$d = |7 − (−8)|$
$d = |7 + 8|$
$d = |15|$
$d = 15$

The distance between the two points is 15 units.

You Try It 13

The distance, d, between point a and point b on the number line is given by the formula $d = |a − b|$. Use the formula to find d when $a = −6$ and $b = 5$.

Your Strategy

Your Solution

Solutions on p. A9

2.2 EXERCISES

OBJECTIVE A

Add.

1. $3 + (-5)$

2. $6 + (-7)$

3. $-4 + (-5)$

4. $-12 + (-12)$

5. $-6 + 7$

6. $-9 + 8$

7. $(-5) + (-10)$

8. $(-3) + (-17)$

9. $-7 + 7$

10. $-11 + 11$

11. $(-15) + (-6)$

12. $(-18) + (-3)$

13. $0 + (-14)$

14. $-19 + 0$

15. $73 + (-54)$

16. $-89 + 62$

17. $2 + (-3) + (-4)$

18. $7 + (-2) + (-8)$

19. $-3 + (-12) + (-15)$

20. $9 + (-6) + (-16)$

21. $-17 + (-3) + 29$

22. $13 + 62 + (-38)$

23. $11 + (-22) + 4 + (-5)$

24. $-14 + (-3) + 7 + (-6)$

25. $-22 + 10 + 2 + (-18)$

26. $-6 + (-8) + 13 + (-4)$

27. $-25 + (-31) + 24 + 19$

28. $10 + (-14) + (-21) + 8$

Solve.

29. What is 3 increased by -21?

30. Find 12 plus -9.

31. What is 16 more than -5?

32. What is 17 added to -7?

33. Find the total of -3, -8, and 12.

34. Find the sum of 5, -16, and -13.

35. Write the sum of x and -7.

36. Write the total of $-a$ and b.

37. ✍ *Economics* A nation's balance of trade is the difference between its exports and imports. If the exports are greater than the imports, the result is a positive number and a *favorable balance of trade*. If the exports are less than the imports, the result is a negative number and an *unfavorable balance of trade*. The table at the right shows the unfavorable balance of trade in a recent year for the United States with four other countries. Find the total of the U.S. balance of trade with (a) Japan and China, (b) Canada and Germany, and (c) Japan and Germany.

U.S. Balance of Trade with Foreign Countries	
Japan	−65,888,000,000
China	−29,504,900,000
Canada	−13,967,400,000
Germany	−12,515,300,000

Source: Office of Trade and Economic Analysis, U.S. Department of Commerce

Evaluate the expression for the given values of the variables.

38. $x + y$, where $x = -5$ and $y = -7$

39. $-a + b$, where $a = -8$ and $b = -3$

40. $a + b$, where $a = -8$ and $b = -3$

41. $-x + y$, where $x = -5$ and $y = -7$

42. $a + b + c$, where $a = -4$, $b = 6$, and $c = -9$

43. $a + b + c$, where $a = -10$, $b = -6$, and $c = 5$

44. $x + y + (-z)$, where $x = -3$, $y = 6$, and $z = -17$

45. $-x + (-y) + z$, where $x = -2$, $y = 8$, and $z = -11$

Identify the property that justifies the statement.

46. $-12 + 5 = 5 + (-12)$

47. $-33 + 0 = -33$

48. $-46 + 46 = 0$

49. $-7 + (3 + 2) = (-7 + 3) + 2$

Use the given property of addition to complete the statement.

50. The Associative Property of Addition
$-11 + (6 + 9) = (? + 6) + 9$

51. The Addition Property of Zero
$-13 + ? = -13$

52. The Commutative Property of Addition
$-2 + ? = -4 + (-2)$

53. The Inverse Property of Addition
$? + (-18) = 0$

54. Is -3 a solution of the equation $x + 4 = 1$?

55. Is -8 a solution of the equation $6 = -3 + z$?

56. Is -6 a solution of the equation $6 = 12 + n$?

57. Is -8 a solution of the equation $-7 + m = -15$?

58. Is -2 a solution of the equation $3 + y = y + 3$?

59. Is -4 a solution of the equation $1 + z = z + 2$?

OBJECTIVE B

Subtract.

60. $7 - 14$

61. $6 - 9$

62. $-7 - 2$

63. $-9 - 4$

64. $7 - (-2)$

65. $3 - (-4)$

66. $-6 - (-6)$

67. $-4 - (-4)$

68. $-12 - 16$

69. $-10 - 7$

70. $(-9) - (-3)$

71. $(-7) - (-4)$

72. $4 - (-14)$

73. $-4 - (-16)$

74. $(-14) - (-7)$

75. $3 - (-24)$

76. $9 - (-9)$

77. $(-41) - 65$

78. $57 - 86$

79. $-95 - (-28)$

Solve.

80. How much larger is 5 than -11?

81. What is -10 decreased by -4?

82. Find -13 minus -8?

83. What is 6 less than -9?

84. Write the difference between $-y$ and 5.

85. Write $-t$ decreased by r.

The figure at the right shows the highest and lowest temperatures ever recorded for selected regions of the world. Use this graph for Exercises 86 to 88.

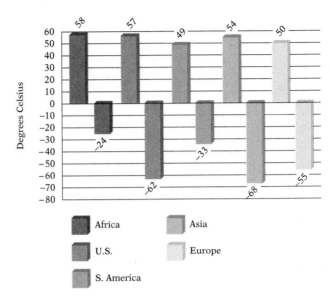

Highest and Lowest Temperatures Recorded (in Degrees Celsius)

86. ✍ *Temperature* What is the difference between the highest and lowest temperatures ever recorded in Africa?

87. ✍ *Temperature* What is the difference between the highest and lowest temperatures ever recorded in South America?

88. ✍ *Temperature* What is the difference between the lowest temperature recorded in Europe and the lowest temperature recorded in Asia?

Simplify.

89. $-4 - 3 - 2$

90. $4 - 5 - 12$

91. $12 - (-7) - 8$

92. $-12 - (-3) - (-15)$

93. $4 - 12 - (-8)$

94. $-30 - (-65) - 29 - 4$

95. $-16 - 47 - 63 - 12$

96. $42 - (-30) - 65 - (-11)$

97. $12 - (-6) + 8$

98. $-7 + 9 - (-3)$

99. $-8 - (-14) + 7$

100. $-4 + 6 - 8 - 2$

101. $9 - 12 + 0 - 5$

102. $11 - (-2) - 6 + 10$

103. $5 + 4 - (-3) - 7$

104. $-1 - 8 + 6 - (-2)$

105. $-13 + 9 - (-10) - 4$

106. $6 - (-13) - 14 + 7$

Evaluate the expression for the given values of the variables.

107. $-x - y$, where $x = -3$ and $y = 9$

108. $x - (-y)$, where $x = -3$ and $y = 9$

109. $-x - (-y)$, where $x = -3$ and $y = 9$

110. $a - (-b)$, where $a = -6$ and $b = 10$

Evaluate the expression for the given values of the variables.

111. $a - b - c$, where $a = 4$, $b = -2$, and $c = 9$

112. $a - b - c$, where $a = -1$, $b = 7$, and $c = -15$

113. $x - y - (-z)$, where $x = -9$, $y = 3$, and $z = 30$

114. $-x - (-y) - z$, where $x = 8$, $y = 1$, and $z = -14$

115. Is -3 a solution of the equation $x - 7 = -10$?

116. Is -4 a solution of the equation $1 = 3 - y$?

117. Is -2 a solution of the equation $-5 - w = 7$?

118. Is -8 a solution of the equation $-12 = m - 4$?

119. Is -6 a solution of the equation $-t - 5 = 7 + t$?

120. Is -7 a solution of the equation $5 + a = -9 - a$?

OBJECTIVE C

The elevation, or height, of places on the earth is measured in relation to sea level, or the average level of the ocean's surface. The table below shows height above sea level as a positive number and depth below sea level as a negative number. Use the table below for Exercises 121–123.

Continent	Highest Elevation (in meters)		Lowest Elevation (in meters)	
Africa	Mt. Kilimanjaro	5,895	Qattara Depression	-133
Asia	Mt. Everest	8,848	Dead Sea	-400
Europe	Mt. Elbrus	5,634	Caspian Sea	-28
America	Mt. Aconcagua	6,960	Death Valley	-86

121. *Geography*   What is the difference in elevation (a) between Mt. Aconcagua and Death Valley and (b) between Mt. Kilimanjaro and the Qattara Depression?

122. *Geography*   For which continent shown is the difference between the highest and lowest elevations greatest?

123. *Geography*   For which continent shown is the difference between the highest and lowest elevations smallest?

Solve.

124. *Temperature*   Find the temperature after a rise of 9°C from −6°C.

The table at the right shows the second quarter profits for 1996 and 1995 for selected companies. (*Note:* Negative profits indicate a loss.) Use the table at the right for Exercises 125 to 127.

Company Profits		
	2nd Quarter 1996	2nd Quarter 1995
Kmart	−99,000,000	−28,000,000
Polaroid	−26,000,000	−22,900,000
L A Gear	−7,480,000	−5,911,000
Pennzoil	24,543,000	−4,790,000

Source: Wall Street Journal, August 5, 1996

125. *Business*   What is the difference between (a) Kmart's second quarter profits for 1996 and 1995, and (b) Pennzoil's second quarter profits for 1996 and 1995?

126. *Business*   Of the companies shown in the table, which had the smallest difference between its second quarter profits for 1996 and 1995?

127. *Business*   Of the companies shown in the table, which had the largest difference between its second quarter profits for 1996 and 1995?

Solve.

128. *Sports*   Use the equation $S = N - P$, where S is a golfer's score relative to par in a tournament, N is the number of strokes made by the golfer, and P is par, to find a golfer's score relative to par when the golfer made 196 strokes and par is 208.

129. *Sports*   Use the equation $S = N - P$, where S is a golfer's score relative to par in a tournament, N is the number of strokes made by the golfer, and P is par, to find a golfer's score relative to par when the golfer made 49 strokes and par is 52.

130. *Mathematics*   The distance, d, between point a and point b on the number line is given by the formula $d = |a - b|$. Find d when $a = 6$ and $b = -15$.

131. *Mathematics*   The distance, d, between point a and point b on the number line is given by the formula $d = |a - b|$. Find d when $a = 7$ and $b = -12$.

CRITICAL THINKING

132. Given the list of numbers at the right, find the largest difference that can be obtained by subtracting one number in the list from a different number in the list?

   $5, -2, -9, 11, 14$

133. Determine whether the statement is always true, sometimes true, or never true.
   a. The difference between a number and its additive inverse is zero.
   b. The sum of a negative number and a negative number is a negative number.

134. Replace the question mark by $\leq$, $=$, or $\geq$ so that the resulting statement is always true for any two integers a and b.
   a. $|a + b|$? $|a| + |b|$ b. $|a - b|$? $||a| - |b||$

135. Describe the steps involved in using a calculator to simplify $-17 - (-8) + (-5)$.

SECTION 2.3 Multiplication and Division of Integers

OBJECTIVE A
Multiplication of integers

When 5 is multiplied by a sequence of decreasing integers, each product decreases by 5.

$$5(3) = 15$$
$$5(2) = 10$$
$$5(1) = 5$$
$$5(0) = 0$$

The pattern developed can be continued so that 5 is multiplied by a sequence of negative numbers. To maintain the pattern of decreasing by 5, the resulting products must be negative.

$$5(-1) = -5$$
$$5(-2) = -10$$
$$5(-3) = -15$$
$$5(-4) = -20$$

This example illustrates that the product of a positive number and a negative number is negative.

When -5 is multiplied by a sequence of decreasing integers, each product increases by 5.

$$-5(3) = -15$$
$$-5(2) = -10$$
$$-5(1) = -5$$
$$-5(0) = 0$$

The pattern developed can be continued so that -5 is multiplied by a sequence of negative numbers. To maintain the pattern of increasing by 5, the resulting products must be positive.

$$-5(-1) = 5$$
$$-5(-2) = 10$$
$$-5(-3) = 15$$
$$-5(-4) = 20$$

This example illustrates that the product of two negative numbers is positive.

The pattern for multiplication shown above is summarized in the following rule for multiplying integers.

POINT OF INTEREST

Operations with negative numbers were not accepted until the late thirteenth century. One of the first attempts to prove that the product of two negative numbers is positive was done in the book *Ars Magna* by Girolamo Cardan in 1545.

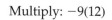

Rule for Multiplying Two Integers

TO MULTIPLY INTEGERS WITH THE SAME SIGN, multiply the absolute values of the factors. The product is **positive.**

TO MULTIPLY INTEGERS WITH DIFFERENT SIGNS, multiply the absolute values of the factors. The product is **negative.**

Multiply: $-9(12)$

The signs are different. The product is negative. $-9(12) = -108$

Multiply: $(-6)(-15)$

The signs are the same. The product is positive. $(-6)(-15) = 90$

CALCULATOR NOTE

To multiply $(-6)(-15)$ with your calculator, enter the following:

Figure 2.5 shows the melting point of bromine and mercury. The melting point of helium is 7 times the melting point of mercury. Find the melting point of helium.

Multiply the melting point of mercury (−39°C) by 7.

$$-39(7) = -273$$

The melting point of helium is −273°C.

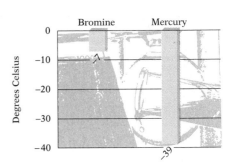

Figure 2.5 Melting Point of Chemical Elements (in Degrees Celsius)

The properties of multiplication presented in Chapter 1 hold true for integers as well as whole numbers. These properties are repeated below.

The Multiplication Property of Zero	$a \cdot 0 = 0$ or $0 \cdot a = 0$
The Multiplication Property of One	$a \cdot 1 = a$ or $1 \cdot a = a$
The Commutative Property of Multiplication	$a \cdot b = b \cdot a$
The Associative Property of Multiplication	$(a \cdot b) \cdot c = a \cdot (b \cdot c)$

TAKE NOTE

For the example at the right, the product is the same if the numbers are multiplied in a different order. For instance,

2(−3)(−5)(−7)=
2(−3)(35)=
2(−105)=
−210

Multiply: 2(−3)(−5)(−7)

Multiply the first two numbers.

Then multiply the product by the third number.

Continue until all the numbers have been multiplied.

$$2(-3)(-5)(-7)$$
$$= -6(-5)(-7)$$
$$= 30(-7)$$
$$= -210$$

By the Multiplication Property of One, $1 \cdot 6 = 6$ and $\mathbf{1} \cdot x = x$. Applying the rules for multiplication, we can extend this to $-1 \cdot 6 = -6$ and $\mathbf{-1} \cdot x = -x$.

TAKE NOTE

When variables are placed next to each other, it is understood that the operation is multiplication. −*ab* means "the opposite of *a* times *b*."

Evaluate $-ab$ when $a = -2$ and $b = -9$.

Replace a with −2 and b with 9.

Simplify $-(-2)$.

Multiply.

$$-ab$$
$$-(-2)(-9)$$
$$= 2(-9)$$
$$= -18$$

Is −4 a solution of the equation $5x = -20$?

Replace x by −4 and then simplify.

The results are equal.

$$\begin{array}{c|c} 5x = 20 \\ \hline 5(-4) & -20 \\ -20 = -20 \end{array}$$

Yes, −4 is a solution of the equation.

Example 1   Find −42 times 62.

Solution   −42 · 62 = −2,604

You Try It 1   What is −38 multiplied by 51?

Your Solution

Example 2   Multiply: −5(−4)(6)(−3)

Solution   −5(−4)(6)(−3) = 20(6)(−3)
                         = 120(−3)
                         = −360

You Try It 2   Multiply: −7(−8)(9)(−2)

Your Solution

Example 3   Evaluate −5x when x = −11.

Solution   −5x
                  −5(−11) = 55

You Try It 3   Evaluate −9y when y = 20.

Your Solution

Example 4   Is 5 a solution of the equation
                 30 = −6z?

Solution $\dfrac{30 = -6z}{30 \;\big|\; -6(5)}$
                  30 ≠ −30

No, 5 is not a solution of the
equation.

You Try It 4   Is −3 a solution of the
                    equation 12 = −4a?

Your Solution

Solutions on p. A9

Objective B
Division of integers

For every division problem, there is a related multiplication problem.

Division: $\dfrac{8}{2} = 4$ Related multiplication: $4(2) = 8$

This fact can be used to illustrate a rule for dividing integers.

$\dfrac{12}{3} = 4$ because $4(3) = 12$ and $\dfrac{-12}{-3} = 4$ because $4(-3) = -12.$

These two division examples suggest that the quotient of two numbers with the same sign is positive. Now consider these two examples.

$\dfrac{12}{-3} = -4$ because $-4(-3) = 12$ and

$\dfrac{-12}{3} = -4$ because $-4(3) = -12.$

These two division examples suggest that the quotient of two numbers with different signs is negative. This property is summarized next.

Rule for Dividing Two Integers

TO DIVIDE TWO NUMBERS WITH THE SAME SIGN, divide the absolute values of the numbers. The quotient is **positive.**

TO DIVIDE TWO NUMBERS WITH DIFFERENT SIGNS, divide the absolute values of the numbers. The quotient is **negative.**

Note from this rule that $\dfrac{12}{-3}$, $\dfrac{-12}{3}$, and $-\dfrac{12}{3}$ are all equal to -4.

If a and b are integers ($b \neq 0$), then $\dfrac{a}{-b} = \dfrac{-a}{b} = -\dfrac{a}{b}$.

Divide: $-36 \div 9$

The signs are different. The quotient is negative. $-36 \div 9 = -4$

CALCULATOR NOTE

To divide $(-105) \div (-5)$ with your calculator, enter the following:

105 $\boxed{+/-}$ $\boxed{\div}$ 5 $\boxed{+/-}$ $\boxed{=}$

Divide: $(-105) \div (-5)$

The signs are the same. The quotient is positive. $(-105) \div (-5) = 21$

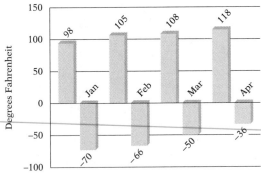

Figure 2.6   Record High and Low Temperatures, in Degrees Fahrenheit, in the United States for January, February, March, and April

Source: National Climatic Data Center, Asheville, NC, and Storm Phillips, STORMFAX, Inc.

Figure 2.6 shows record high and low temperatures in the United States for the first four months of the year. The record low temperature for April is four times the record low temperature for September. What is the record low temperature for September?

Divide the record low for April (-36) by 4. $-36 \div 4 = -9$

The record low temperature for September in the U.S. is $-9°$F.

The division properties of zero and one, which were presented in Chapter 1, hold true for integers as well as whole numbers. These properties are repeated here.

Division Properties of Zero and One

If $a \neq 0$, $\dfrac{0}{a} = 0$.   If $a \neq 0$, $\dfrac{a}{a} = 1$.

$\dfrac{a}{1} = a$ $\dfrac{a}{0}$ is undefined.

Evaluate $a \div (-b)$ when $a = -28$ and $b = -4$.

$a \div (-b)$

Replace a with -28 and b with -4. $-28 \div (-(-4))$

Simplify $-(-4)$. $= -28 \div (4)$

Divide. $= -7$

POINT OF INTEREST

Historical manuscripts indicate that mathematics is at least 4000 years old. Yet it was only 400 years ago that mathematicians started using variables to stand for numbers. Before that time, mathematics was written in words.

Is -4 a solution of the equation $\dfrac{-20}{x} = 5$?

$$\dfrac{-20}{x} = 5$$

Replace x by -4 and then simplify. $\dfrac{-20}{-4} \Big| 5$

The results are equal. $5 = 5$

Yes, -4 is a solution of the equation.

Example 5   Find the quotient of -23 and -23.

Solution $-23 \div (-23) = 1$

You Try It 5   What is 0 divided by -17?

Your Solution

Example 6   Divide: $\dfrac{95}{-5}$

Solution $\dfrac{95}{-5} = -19$

You Try It 6   Divide: $\dfrac{84}{-6}$

Your Solution

Example 7   Divide: $x \div 0$

Solution   Division by zero is not defined. $x \div 0$ is undefined.

You Try It 7   Divide: $x \div 1$

Your Solution

Solutions on p. A9

Example 8   Evaluate $\dfrac{-a}{b}$ when $a = -6$ and $b = -3$.

     Solution $\dfrac{-a}{b}$

$$\dfrac{-(-6)}{-3} = \dfrac{6}{-3} = -2$$

You Try It 8   Evaluate $\dfrac{a}{-b}$ when $a = -14$ and $b = -7$.

     Your Solution

Example 9   Is -9 a solution of the equation $-3 = \dfrac{x}{3}$?

     Solution $-3 = \dfrac{x}{3}$

$$-3 \,\bigg|\, \dfrac{-9}{3}$$

$$-3 = -3$$

Yes, -9 is a solution of the equation.

You Try It 9   Is -3 a solution of the equation $\dfrac{-6}{y} = -2$?

     Your Solution

Solutions on p. A9

OBJECTIVE C
Applications

Example 10

The daily low temperatures during one week were recorded as follows: $-10°, 2°, -1°, -9°, 1°, 0°, 3°$. Find the average daily low temperature for the week.

Strategy

To find the average daily low temperature:

▸ Add the seven temperature readings.
▸ Divide by 7.

Solution

$-10 + 2 + (-1) + (-9) + 1 + 0 + 3 = -14$

$-14 \div 7 = -2$

The average daily low temperature was $-2°$.

You Try It 10

The daily high temperatures during one week were recorded as follows: $-7°, -8°, 0°, -1°, -6°, -11°, -2°$. Find the average daily high temperature for the week.

Your Strategy

Your Solution

Solution on p. A9

2.3 EXERCISES

OBJECTIVE A

Multiply.

1. $-4 \cdot 6$
-24

2. $-7 \cdot 3$

3. $-2(-3)$
6

4. $-5(-1)$

5. $(9)(2)$
18

6. $(3)(8)$

7. $5(-4)$
-20

8. $4(-7)$

9. $-8(2)$
-16

10. $-9(3)$

11. $(-5)(-5)$
-25

12. $(-3)(-6)$

13. $(-7)(0)$
0

14. $-11(1)$

15. $14(3)$
42

16. $62(9)$

17. $-32(4)$
-128

18. $-24(3)$

19. $(-8)(-26)$
208

20. $(-4)(-35)$

21. $9(-27)$
-243

22. $8(-40)$

23. $-5 \cdot (23)$
-115

24. $-6 \cdot (38)$

25. $-7(-34)$
238

26. $-4(-51)$

27. $4 \cdot (-8) \cdot 3$
$32 \cdot 3 / 96$

28. $5 \cdot 7 \cdot (-2)$

29. $(-6)(5)(7)$
-210

30. $(-9)(-9)(2)$

31. $-8(-7)(-4)$
$8 \times 7 \cdot 56 \times 4$
-224

32. $-1(4)(-9)$

Solve.

33. What is twice -20? -40

34. Find the product of 100 and -7. -700

35. What is -30 multiplied by -6? 180

36. What is -9 times -40? 360

37. Write the product of $-q$ and r. $-qr$

38. Write the product of $-f$, g, and h. $-fgh$

The figure at the right shows earnings for the industries of entertainment, toy makers, and airlines for the first quarter of 1996. *Note:* Negative earnings indicate a loss.

39. *Business* If earnings were to continue throughout the year at the same level, what would be the annual earnings for (a) the entertainment industry, (b) the toy manufacturing industry, and (c) the airline industry? (*Note:* There are four quarters in one year.)

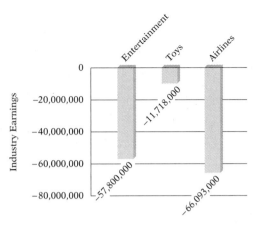

Industry Earnings (net on continuing operations), First Quarter 1996
Source: Wall Street Journal, May 6, 1996

Identify the property that justifies the statement.

40. $0(-7) = 0$

41. $1p = p$

42. $-8(-5) = -5(-8)$

43. $-3(9 \cdot 4) = (-3 \cdot 9)4$

Use the given property of multiplication to complete the statement.

44. The Commutative Property of Multiplication
$-3(-9) = -9(?)$

45. The Associative Property of Multiplication
$?(5 \cdot 10) = (-6 \cdot 5)10$

46. The Multiplication Property of Zero
$-81 \cdot ? = 0$

47. The Multiplication Property of One
$?(-14) = -14$

Evaluate the expression for the given values of the variables.

48. xy, when $x = -3$ and $y = -8$

49. $-xy$, when $x = -3$ and $y = -8$

50. $x(-y)$, when $x = -3$ and $y = -8$

51. $-xyz$, when $x = -6$, $y = 2$, and $z = -5$

52. $-8a$, when $a = -24$

53. $-7n$, when $n = -51$

Evaluate the expression for the given values of the variables.

54. $5xy$, when $x = -9$ and $y = -2$

55. $8ab$, when $a = 7$ and $b = -1$

56. $-4cd$, when $c = 25$ and $d = -8$

57. $-5st$, when $s = -40$ and $t = -8$

58. Is -4 a solution of the equation $6m = -24$?

59. Is -3 a solution of the equation $-5x = -15$?

60. Is -6 a solution of the equation $48 = -8y$?

61. Is 0 a solution of the equation $-8 = -8a$?

62. Is 7 a solution of the equation $-3c = 21$?

63. Is 9 a solution of the equation $-27 = -3c$?

OBJECTIVE B

Divide.

64. $12 \div (-6)$

65. $18 \div (-3)$

66. $(-72) \div (-9)$

67. $(-64) \div (-8)$

68. $0 \div (-6)$

69. $-49 \div 1$

70. $81 \div (-9)$

71. $-40 \div (-5)$

72. $\dfrac{72}{-3}$

73. $\dfrac{44}{-4}$

74. $\dfrac{-93}{-3}$

75. $\dfrac{-98}{-7}$

76. $-114 \div (-6)$

77. $-91 \div (-7)$

78. $-53 \div 0$

79. $(-162) \div (-162)$

80. $-128 \div 4$

81. $-130 \div (-5)$

82. $(-200) \div 8$

83. $(-92) \div (-4)$

Solve.

84. Find the quotient of -700 and 70.

85. Find 550 divided by -5.

86. What is -670 divided by -10?

87. What is the quotient of -333 and -3?

88. Write the quotient of $-a$ and b.

89. Write -9 divided by x.

The figure at the right shows the second quarter profits for 1996 for selected companies. (*Note:* Negative profits indicate a loss. One quarter of the year is three months.) Use this figure for Exercises 90 and 91.

Profit for the Second Quarter of 1996
Source: Wall Street Journal, August 5, 1996

90. *Business* For the quarter shown, what was the average monthly profit for Boise Cascade?

91. *Business* For the quarter shown, what was the average monthly profit for Sun Co.?

Evaluate the expression for the given values of the variables.

92. $a \div b$, where $a = -36$ and $b = -4$

93. $-a \div b$, where $a = -36$ and $b = -4$

94. $a \div (-b)$, where $a = -36$ and $b = -4$

95. $(-a) \div (-b)$, where $a = -36$ and $b = -4$

96. $\dfrac{x}{y}$, where $x = -42$ and $y = -7$

97. $\dfrac{-x}{y}$, where $x = -42$ and $y = -7$

98. $\dfrac{x}{-y}$, where $x = -42$ and $y = -7$

99. $\dfrac{-x}{-y}$, where $x = -42$ and $y = -7$

100. Is 20 a solution of the equation $\dfrac{m}{-2} = -10$?

101. Is 18 a solution of the equation $6 = \dfrac{-c}{-3}$?

102. Is 0 a solution of the equation $0 = \dfrac{a}{-4}$?

103. Is −3 a solution of the equation $\dfrac{21}{n} = 7$?

104. Is −6 a solution of the equation $\dfrac{x}{2} = \dfrac{-18}{x}$?

105. Is 8 a solution of the equation $\dfrac{m}{-4} = \dfrac{-16}{m}$?

OBJECTIVE C

Solve.

106. *Sports*   The combined scores of the top five golfers in a tournament equaled −10 (10 under par). What was the average score of the five golfers?

107. *Sports*   The combined scores of the top four golfers in a tournament equaled −12 (12 under par). What was the average score of the four golfers?

The following figure shows the record low temperatures, in degrees Fahrenheit, in the United States for each month. Use this figure for Exercises 108 to 110.

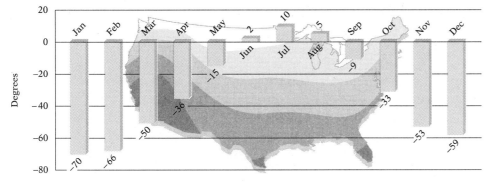

Record Low Temperatures, in Degrees Fahrenheit, in the United States
Source: National Climatic Data Center, Asheville, NC, and Storm Phillips, STORMFAX, Inc.

108. *Temperature*   What is the average record low temperature for July, August, and September?

109. *Temperature*   What is the average record low temperature for the first three months of the year?

110. *Temperature*   What is the average record low temperature for the four months with the lowest record low temperatures?

111. *Temperature* The daily low temperatures during one week were recorded as follows: $4°, -5°, 8°, -1°, -12°, -14°, -8°$. Find the average daily low temperature for the week.

112. *Temperature* The daily high temperatures during one week were recorded as follows: $-6°, -11°, 1°, 5°, -3°, -9°, -5°$. Find the average daily high temperature for the week.

113. *Environmental Science* The wind chill factor when the temperature is $-15°F$ and the wind is blowing at 20 mph is five times the wind chill factor when the temperature is $25°F$ and the wind is blowing at 35 mph. If the wind chill factor at $25°F$ with a 35 mph wind is $-12°$, what is the wind chill factor at $-15°F$ with a 20 mph wind?

A **geometric sequence** is a list of numbers in which each number after the first is found by multiplying the preceding number in the list by the same number. For example, in the sequence 1, 3, 9, 27, 81, . . . , each number after the first is found by multiplying the preceding number in the list by 3. To find the multiplier in a geometric sequence, divide the second number in the sequence by the first number; for the example above, $3 \div 1 = 3$.

114. *Mathematics* Find the next three numbers in the geometric sequence $-5, 15, -45, \ldots$

115. *Mathematics* Find the next three numbers in the geometric sequence 2, $-4, 8, \ldots$

116. *Mathematics* Find the next three numbers in the geometric sequence $-3, -12, -48, \ldots$

117. *Mathematics* Find the next three numbers in the geometric sequence $-1, -5, -25, \ldots$

CRITICAL THINKING.

118. Use repeated addition to show that the product of two integers with different signs is a negative number.

119. (a) Find the largest possible product of two negative integers whose sum is -18. (b) Find the smallest possible sum of two negative integers whose product is 16.

120. Determine whether the statement is always true, sometimes true, or never true. (a) The product of a number and its additive inverse is a negative number. (b) The product of an odd number of negative numbers is a negative number. (c) The square of a negative number is a positive number.

121. Find all negative integers x such that $1 - 3x < 12$.

122. Describe the steps involved in using a calculator to simplify $(-2491) \div (-47)$.

123. In your own words, describe the rules for multiplying and dividing integers.

SECTION 2.4 Solving Equations with Integers

OBJECTIVE A
Solving equations

Recall that an **equation** states that two expressions are equal. Two examples of equations are shown below.

$$3x = 36 \qquad -17 = y + 9$$

In Section 1.4, we solved equations using only whole numbers. In this section, we will extend the solutions of equations to include integers.

Solving an equation requires finding a number that when substituted for the variable results in a true equation. Two important properties that are used to solve equations were discussed earlier.

The same number can be subtracted from each side of an equation without changing the solution of the equation.

Each side of an equation can be divided by the same nonzero number without changing the solution of the equation.

A third property of equations involves *adding* the same number to each side of an equation.

As shown at the right, the solution of the equation $x + 6 = 13$ is 7.

$$x + 6 = 13$$
$$7 + 6 = 13$$
$$13 = 13$$

If 4 is added to each side of the equation $x + 6 = 13$, the resulting equation is $x + 10 = 17$. The solution of this equation is also 7.

$$x + 6 = 13$$
$$x + 6 + 4 = 13 + 4$$
$$x + 10 = 17 \qquad 7 + 10 = 17$$

This illustrates the addition property of equations.

The same number can be added to each side of an equation without changing the solution of the equation.

Solve: $x - 7 = 2$

7 is subtracted from the variable x. Add 7 to each side of the equation.

$$x - 7 = 2$$
$$x - 7 + 7 = 2 + 7$$

x is alone on the left side of the equation. The number on the right side is the solution.

$$x = 9$$

Check the solution.

Check:
$$\begin{array}{c|c} x - 7 = 2 \\ \hline 9 - 7 & 2 \\ \hline 2 = 2 \end{array}$$

The solution checks.

The solution is 9.

Solve: $-15 = t + 13$

13 is added to the variable t. Subtract 13 from each side of the equation.

$$-15 = t + 13$$
$$-15 - 13 = t + 13 - 13$$

t is alone on the right side of the equation. The number on the left side is the solution.

$$-28 = t$$

The solution is -28.

The division property of equations is also used with integers.

Solve: $5y = -30$

The variable y is multiplied by 5. Divide each side of the equation by 5.

$$5y = -30$$
$$\frac{5y}{5} = \frac{-30}{5}$$

y is alone on the left side of the equation. The number on the right side is the solution.

$$y = -6$$

Check the solution.

Check:
$$\frac{5y = -30}{5(-6) \mid -30}$$
$$-30 = -30$$

The solution checks.

The solution is -6.

Solve: $42 = -7a$

The variable a is multiplied by -7. Divide each side of the equation by -7.

$$42 = -7a$$
$$\frac{42}{-7} = \frac{-7a}{-7}$$

a is alone on the right side of the equation. The number on the left side is the solution. Remember to check the solution.

$$-6 = a$$

The solution is -6.

Example 1   Solve: $27 = v - 13$

Solution
$$27 = v - 13$$
$$27 + 13 = v - 13 + 13$$ 13 is sub-
$$40 = v$$ tracted from v. Add 13 to each side.

The solution is 40.

You Try It 1   Solve: $-12 = x + 12$

Your Solution

Example 2   Solve: $-24 = -4z$

Solution
$$-24 = -4z$$
$$\frac{-24}{-4} = \frac{-4z}{-4}$$
$$6 = z$$

The solution is 6.

You Try It 2   Solve: $14a = -28$

Your Solution

Solutions on p. A9

OBJECTIVE **B**
Applications

Recall that an equation states that two mathematical expressions are equal. To translate a sentence into an equation requires that you recognize the words or phrases that mean "equals." Some of these phrases are reviewed below.

equals	is	was
is equal to	represents	is the same as

Negative fifty-six equals negative eight times a number. Find the number.

Choose a variable to represent the unknown number.

The unknown number: m

Find two verbal expressions for the same value.

Negative fifty-six	equals	negative eight times a number

Translate the expressions and then write an equation. Solve the equation.

$$-56 = -8m$$
$$\frac{-56}{-8} = \frac{-8m}{-8}$$
$$7 = m$$

The number is 7.

The high temperature today is 7° lower than the high temperature yesterday. The high temperature today is −13°C. What was the high temperature yesterday?

Strategy   To find the high temperature yesterday, write and solve an equation using t to represent the high temperature yesterday.

Solution

The high temperature today	is	7° lower than the high temperature yesterday

$$-13 = t - 7$$
$$-13 + 7 = t - 7 + 7$$
$$-6 = t$$

The high temperature yesterday was −6°C.

A jeweler wants to make a profit of $250 on the sale of a gold bracelet that cost the jeweler $700. Use the formula $P = S - C$, where P is the profit on an item, S is the selling price, and C is the cost, to find the selling price of the bracelet.

Strategy   To find the selling price, replace P by 250 and C by 700 in the given formula and solve for S.

Solution

$$P = S - C$$
$$250 = S - 700$$
$$250 + 700 = S - 700 + 700$$
$$950 = S$$

The selling price for the gold bracelet should be $950.

Example 3

The average number of Federal Express employees in 1995 was 5,800 less than the average number of Federal Express employees in 1996. The average number of Federal Express employees in 1995 was 94,200. Find the average number of Federal Express employees in 1996.

Strategy

To find the average number of Federal Express employees in 1996, write and solve an equation using E to represent the average number of Federal Express employees in 1996.

Solution

The average number of Federal Express employees in 1995	was	5,800 less than the average number of Federal Express employees in 1996

$$94,200 = E - 5,800$$
$$94,200 + 5,800 = E - 5,800 + 5,800$$
$$100,000 = E$$

The average number of Federal Express employees in 1996 was 100,000.

You Try It 3

The number of people employed by Ryder System, Inc. in 1994 was 1,408 less than the number the company employed in 1995. The number of Ryder employees in 1994 was 43,095. Find the number of people employed by Ryder in 1995.

Your Strategy

Your Solution

Example 4

The ground speed of an airplane flying into a wind is given by the formula $g = a - h$, where g is the ground speed, a is the airspeed of the plane, and h is the speed of the head wind. Use this formula to find the airspeed of a plane whose ground speed is 624 mph and for which the head wind speed is 98 mph.

Strategy

To find the airspeed, replace g by 624 and h by 98 in the given formula and solve for a.

Solution

$$g = a - h$$
$$624 = a - 98$$
$$624 + 98 = a - 98 + 98$$
$$722 = a$$

The airspeed of the plane is 722 mph.

You Try It 4

The ground speed of an airplane flying into a wind is given by the formula $g = a - h$, where g is the ground speed, a is the airspeed of the plane, and h is the speed of the head wind. Use this formula to find the airspeed of a plane whose ground speed is 250 mph and for which the head wind speed is 50 mph.

Your Strategy

Your Solution

Solutions on pp. A9–A10

2.4 EXERCISES

OBJECTIVE A

Solve.

1. $x - 6 = 9$

2. $m - 4 = 6$

3. $8 = y - 3$

4. $12 = t - 4$

5. $x - 5 = -12$

6. $n - 7 = -21$

7. $-10 = z + 6$

8. $-21 = c + 4$

9. $x + 12 = 4$

10. $y + 7 = 2$

11. $-12 = c - 12$

12. $n - 9 = -9$

13. $3m = -15$

14. $6p = -54$

15. $-10 = 5v$

16. $-20 = 2z$

17. $-8x = -40$

18. $-4y = -28$

19. $-60 = -6v$

20. $3x = -39$

21. $5x = -100$

22. $-4n = 0$

23. $4x = 0$

24. $-15 = -15z$

OBJECTIVE B

Solve.

25. Ten less than a number is fifteen. Find the number.

26. The difference between a number and five is twenty-two. Find the number.

27. Zero is equal to fifteen more than some number. Find the number.

28. Twenty equals the sum of a number and thirty-one. Find the number.

29. Sixteen equals negative two times a number. Find the number.

30. The product of negative six and a number is negative forty-two. Find the number.

31. Zero is equal to the product of negative six and a number. Find the number.

32. Eight times some number is negative ninety-six. Find the number.

Solve. Use the table at the right for Exercises 33 and 34.

Year	U.S. Balance of Trade (in millions of dollars)
1950	1,043
1960	4,583
1970	2,325
1980	−24,245
1990	−101,012

Source: Office of Trade and Economic Analysis, U.S. Depart. of Commerce

33. *Economics* The U.S. balance of trade in 1950 was $116,611 million more than the U.S. balance of trade in 1993. What was the U.S. balance of trade in 1993?

34. *Economics* The U.S. balance of trade in 1980 was $126,384 million more than the U.S. balance of trade in 1994. What was the U.S. balance of trade in 1994?

35. *Temperature* The temperature now is 5° higher than it was this morning. The temperature now is 8°C. What was the temperature this morning?

36. *Temperature* The temperature now is 9° lower than it was yesterday at this time. The temperature now is −16°C. What was the temperature yesterday at this time?

37. *Business* A car dealer wants to make a profit of $925 on the sale of a car that cost the dealer $12,600. Use the equation $P = S - C$, where P is the profit on an item, S is the selling price, and C is the cost, to find the selling price of the car.

38. *Business* An office supplier wants to make a profit of $95 on the sale of a software package that cost the supplier $385. Use the equation $P = S - C$, where P is the profit on an item, S is the selling price, and C is the cost, to find the selling price of the software.

39. *Business* The net worth of a business is given by the formula $N = A - L$, where N is the net worth, A is the assets of the business (or the amount owned), and L is the liabilities of the business (or the amount owed). Use this formula to find the assets of a business that has a net worth of $11 million and liabilities of $4 million.

40. *Business* The net worth of ABL Electronics is $43 million and it has liabilities of $14 million. Use the net worth formula $N = A - L$, where N is the net worth, A is the assets of the business (or the amount owned), and L is the liabilities of the business (or the amount owed), to find the assets of ABL Electronics.

CRITICAL THINKING

41. For each part below, state whether the sentence is true or false. Give an example that supports your answer. (a) Zero cannot be the solution of an equation. (b) A negative number cannot be the solution of an equation. (c) If an equation contains a negative number, then the solution of the equation must be a negative number.

42. **a.** Find the value of $3y - 8$ given that $-3y = -36$.
    b. Find the value of $2x^2 - 18$ given that $x - 6 = -9$.

43. In your own words, explain the addition, subtraction, and division properties of equations.

Copyright © Houghton Mifflin Company. All rights reserved.

SECTION 2.5 The Order of Operations Agreement

OBJECTIVE A
The Order of Operations Agreement

The Order of Operations Agreement, used in Chapter 1, is repeated here for your reference.

> ### The Order of Operations Agreement
>
> **Step 1** Do all operations inside parentheses.
>
> **Step 2** Simplify any numerical expressions containing exponents.
>
> **Step 3** Do multiplication and division as they occur from left to right.
>
> **Step 4** Do addition and subtraction as they occur from left to right.

Note how the following expressions containing exponents are simplified.

$(-3)^2 = (-3)(-3) = 9$ The (-3) is squared. Multiply -3 by -3.

$-(3)^2 = -(3 \cdot 3) = -9$ Read $-(3^2)$ as "the opposite of three squared." 3^2 is 9. The opposite of 9 is -9.

$-3^2 = -(3^2) = -9$ The expression -3^2 is the same as $-(3^2)$.

TAKE NOTE

The -3 is squared only when the negative sign is *inside* the parentheses. In $(-3)^2$, we are squaring -3; in -3^2, we are finding the opposite of 3^2.

Simplify: $8 - 4 \div (-2)$

There are no operations inside parentheses (Step 1).

There are no exponents (Step 2).

Do the division (Step 3). $8 - 4 \div (-2) = 8 - (-2)$

Do the subtraction (Step 4). $= 8 + 2 = 10$

CALCULATOR NOTE

As shown above and at the left, the value of -3^2 is different from the value of $(-3)^2$. The keystrokes to evaluate each of these on your calculator are different.

To evaluate -3^2, enter:

3 $\boxed{x^2}$ $\boxed{+/-}$

To evaluate $(-3)^2$, enter:

3 $\boxed{+/-}$ $\boxed{x^2}$

Simplify: $(-3)^2 - 2(8 - 3) + (-5)$

Perform operations inside parentheses. $(-3)^2 - 2(8 - 3) + (-5)$
$= (-3)^2 - 2(5) + (-5)$

Simplify expressions with exponents. $= 9 - 2(5) + (-5)$

Do multiplication and division as they occur from left to right. $= 9 - 10 + (-5)$

Do addition and subtraction as they occur from left to right. $= 9 + (-10) + (-5)$
$= -1 + (-5)$
$= -6$

Evaluate $ab - b^2$ when $a = 2$ and $b = -6$.

	$ab - b^2$
Replace a with 2 and each b with -6.	$2(-6) - (-6)^2$
Use the Order of Operations Agreement to simplify the resulting numerical expression. Simplify the exponential expression.	$= 2(-6) - 36$
Do the multiplication.	$= -12 - 36$
Do the subtraction.	$= -12 + (-36)$
	$= -48$

Example 1   Simplify $(-4)^2$ and -4^2.

Solution $(-4)^2 = (-4)(-4) = 16$
$-4^2 = -(4 \cdot 4) = -16$

You Try It 1   Simplify $(-5)^2$ and -5^2.

Your Solution

Example 2   Simplify: $12 \div (-2)^2 - 5$

Solution $12 \div (-2)^2 - 5 = 12 \div 4 - 5$
$\qquad\qquad\qquad = 3 - 5$
$\qquad\qquad\qquad = 3 + (-5)$
$\qquad\qquad\qquad = -2$

You Try It 2   Simplify: $8 \div 4 \cdot 4 - (-2)^2$

Your Solution

Example 3   Simplify:
$(-3)^2(5 - 7)^2 - (-9) \div 3$

Solution $(-3)^2(5 - 7)^2 - (-9) \div 3$
$\quad = (-3)^2(-2)^2 - (-9) \div 3$
$\quad = (9)(4) - (-9) \div 3$
$\quad = 36 - (-9) \div 3$
$\quad = 36 - (-3)$
$\quad = 36 + 3$
$\quad = 39$

You Try It 3   Simplify:
$(-2)^2(3 - 7)^2 - (-16) \div (-4)$

Your Solution

Example 4   Evaluate $6a \div (-b)$ when $a = -2$ and $b = -3$.

Solution $6a \div (-b)$
$6(-2) \div (-(-3))$
$\quad = 6(-2) \div (3)$
$\quad = -12 \div 3$
$\quad = -4$

You Try It 4   Evaluate $3a - 4b$ when $a = -2$ and $b = 5$.

Your Solution

Solutions on p. A10

2.5 EXERCISES

OBJECTIVE A

Simplify.

1. $3 - 12 \div 2$

2. $-16 \div 2 + 8$

3. $2(3 - 5) - 2$

4. $2 - (8 - 10) \div 2$

5. $4 - (-3)^2$

6. $(-2)^2 - 6$

7. $4 \cdot (2 - 4) - 4$

8. $6 - 2 \cdot (1 - 3)$

9. $4 - (-2)^2 + (-3)$

10. $-3 + (-6)^2 - 1$

11. $3^3 - 4(2)$

12. $9 \div 3 - (-3)^2$

13. $3 \cdot (6 - 2) \div 6$

14. $4 \cdot (2 - 7) \div 5$

15. $2^3 - (-3)^2 + 2$

16. $6(8 - 2) \div 4$

17. $6 - 2(1 - 5)$

18. $(-2)^2 - (-3)^2 + 1$

19. $6 - (-4)(-3)^2$

20. $4 - (-5)(-2)^2$

21. $4 \cdot 2 - 3 \cdot 7$

22. $16 \div 2 - 9 \div 3$

23. $(-2)^2 - 5(3) - 1$

24. $4 - 2 \cdot 7 - 3^2$

25. $3 \cdot 2^3 + 5 \cdot (3 + 2) - 17$

26. $3 \cdot 4^2 - 16 - 4 + 3 - (1 - 2)^2$

27. $-12(6 - 8) + 1^3 \cdot 3^2 \cdot 2 - 6(2)$

28. $-3 \cdot (-2)^2 \cdot 4 \div 8 - (-12)$

29. $-27 - (-3)^2 - 2 - 7 + 6 \cdot 3$

30. $(-1) \cdot (4 - 7)^2 \div 9 + 6 - 3 - 4(2)$

31. $16 - 4 \cdot 8 + 4^2 - (-18) - (-9)$

32. $(-3)^2 \cdot (5 - 7)^2 - (-9) \div 3$

Evaluate the variable expression given $a = -2$, $b = 4$, $c = -1$, and $d = 3$.

33. $3a + 2b$

34. $a - 2c$

35. $16 \div (ac)$

36. $6b \div (-a)$

37. $bc \div (2a)$

38. $a^2 - b^2$

39. $b^2 - c^2$

40. $2a - (c + a)^2$

41. $(b - a)^2 + 4c$

42. $\dfrac{b + c}{d}$

43. $\dfrac{d - b}{c}$

44. $\dfrac{2d + b}{-a}$

45. $\dfrac{b - d}{c - a}$

46. $\dfrac{bd}{a} \div c$

47. $(d - a)^2 \div 5$

48. $(b + c)^2 + (a + d)^2$

49. $(d - a)^2 - 3c$

50. $(b + d)^2 - 4a$

CRITICAL THINKING

51. What is the smallest integer greater than $-2^2 - (-3)^2 + 5(4) \div 10 - (-6)$?

52. Evaluate.
   a. $1^3 + 2^3 + 3^3 + 4^3$
   b. $(-1)^3 + (-2)^3 + (-3)^3 + (-4)^3$
   c. $1^3 + 2^3 + 3^3 + 4^3 + 5^3$
   d. Based on your answers to parts (a), (b), and (c), evaluate
      $(-1)^3 + (-2)^3 + (-3)^3 + (-4)^3 + (-5)^3$.

53. **a.** Is -4 a solution of the equation $x^2 - 2x - 8 = 0$?
   b. Is -3 a solution of the equation $x^3 + 3x^2 - 5x - 15 = 0$?

54. Using the Order of Operations Agreement, explain how to evaluate Exercise 33.

55. Evaluate $a \div bc$ and $a \div (bc)$ when $a = 16$, $b = 2$, and $c = -4$. Explain why the answers are not the same.

PROJECTS IN MATHEMATICS

Addition and Multiplication Problems

The chart below is an addition table. Use it to answer Exercises 1 to 7 below.

+	Δ	‡	◊
Δ	‡	◊	Δ
‡	◊	Δ	‡
◊	Δ	‡	◊

1. Find the sum of Δ and ‡.

2. What is ◊ plus ◊?

3. In our number system, 0 can be added to any number without changing that number; 0 is called the **additive identity.** What is the additive identity for the system in the chart above? Explain your answer.

4. Does the Commutative Property of Addition apply to this system? Explain your answer.

5. What is −Δ (the opposite of Δ) equal to? Explain your answer.

6. What is −‡ (the opposite of ‡) equal to? Explain your answer.

7. Simplify −Δ + ‡ − ◊. Explain how you arrived at your answer.

The chart below is a multiplication table. Use it to answer Exercises 8 to 14 below.

×	£	¿	&
£	£	¿	&
¿	¿	¿	¿
&	&	¿	*

8. Find the product of £ and &.

9. What is ¿ times £?

10. Find the square of &.

11. Does the Commutative Property of Multiplication apply to this system? Explain your answer.

12. In our number system, the product of a number and 0 is 0. Is there an element in this system that corresponds to 0 in our system? Explain your answer.

13. In our number system, 1 can be multiplied by any number without changing that number; 1 is called the **multiplicative identity.** What is the multiplicative identity for the system in the chart above? Explain your answer.

14. Simplify & ÷ £ × ¿. Explain how you arrived at your answer.

Morse Code/Binary Code

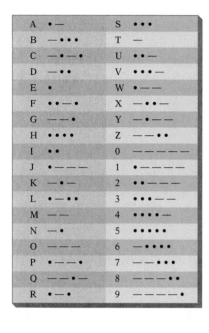

In this chapter, the symbol "−" was described as meaning *minus*, *negative*, or *opposite*. You have also seen this symbol used as a hyphen or a dash.

Morse Code is a system of communication in which letters and numbers are represented using dots and dashes. At the left is the representation of the letters of the alphabet and the numbers 0 through 9 in Morse Code.

1. Describe the pattern of dots and dashes used to represent the numbers 0 to 9.

2. Why might the letters E and T have been chosen to have the shortest representations (only 1 symbol)?

Note that Morse Code uses only two symbols. In our number system, the **decimal system,** we use 10 digits to represent any number: 0, 1, 2, 3, 4, 5, 6, 7, 8, and 9. Because 10 digits are used, the decimal system is also referred to as **base ten.**

Another number system, the **binary number system,** uses only two digits to represent any number: 0 and 1. Because two digits are used, the binary number system is also referred to as **base two.**

The first five place values of the digits in the decimal system are:

Ten-thousands   Thousands   Hundreds   Tens   Ones

Each place value is a power of 10 ($10 = 10^1$, $100 = 10^2$, $1,000 = 10^3$, $10,000 = 10^4$, and so on).

The first five place values of the digits in the binary number system are:

Sixteens   Eights   Fours   Twos   Ones

Each place value is a power of 2 ($2 = 2^1$, $4 = 2^2$, $8 = 2^3$, $16 = 2^4$, and so on).

The whole numbers are represented in base two as:

0
1
10 (1 two, zero ones)
11 (1 two, 1 one)
100 (1 four)
101 (1 four, 1 one)
110 (1 four, 1 two)
111 (1 four, 1 two, 1 one)
1000 (1 eight)
1001 (1 eight, 1 one), and so on.

3. Write the next ten whole numbers using base two.

4. What would be the sixth and seventh place values in base two?

The first four place values of the digits in a base eight number system are:

Five hundred twelves   Sixty-fours   Eights   Ones

Each place value is a power of 8 ($8 = 8^1$, $64 = 8^2$, $512 = 8^3$, and so on).

5. Represent the first 64 whole numbers in base 8.

6. How many numbers are representable by one-digit numbers in (a) base two, (b) base ten, and (c) base eight?

One of the most important applications of the binary number system is in the area of computers. Just as information in Morse Code is relayed using only two symbols, computers operate using only two symbols, 0 and 1. Because all data, or information, in a computer is represented using only 0's and 1's, the data are said to be in **binary code.** A computer works by electricity, and an electric switch has only two positions, on and off. A "0" in binary code indicates that an electric switch is to be turned off; a "1" indicates that the switch is to be turned on.

A **bit** is the smallest unit of code that computers read; it is a binary digit, either a 0 or a 1. Usually bits are grouped into **bytes,** or eight bits. Each byte stands for a letter, number, or any other symbol we might use in communicating information. For example, the letter W can be represented 01010111.

7. Find the definition of each of the following words: gigabit, kilobyte, megabyte, terabit.

8. Find out how the speed of a computer is described.

9. Write a description of machine language.

CHAPTER SUMMARY

Key Words

Positive numbers are numbers greater than zero. *Negative numbers* are numbers less than zero.

The *integers* are . . . $-4, -3, -2, -1, 0, 1, 2, 3, 4, \ldots$ The integers can be defined as the whole numbers and their opposites. *Positive integers* are to the right of zero on the number line. *Negative integers* are to the left of zero on the number line.

Opposite numbers are two numbers that are the same distance from zero on the number line but on opposite sides of zero. The opposite of a number is called its *additive inverse.*

The *absolute value* of a number is the distance from zero to the number on the number line. The absolute value of a number is a positive number or zero. The symbol for absolute value is "| |."

Essential Rules

To add integers with the same sign, add the absolute values of the numbers. Then attach the sign of the addends.

To add integers with different signs, find the difference between the absolute values of the numbers. Then attach the sign of the addend with the larger absolute value.

To subtract two integers, add the opposite of the second integer to the first integer.

To multiply integers with the same sign, multiply the absolute values of the factors. The product is positive.

To multiply integers with different signs, multiply the absolute values of the factors. The product is negative.

To divide two numbers with the same sign, divide the absolute values of the numbers. The quotient is positive.

To divide two numbers with different signs, divide the absolute values of the numbers. The quotient is negative.

Order Relations $a > b$ if a is to the right of b on the number line.
$a < b$ if a is to the left of b on the number line.

Addition Property of Zero	$a + 0 = a$ or $0 + a = a$
Commutative Property of Addition	$a + b = b + a$
Associative Property of Addition	$(a + b) + c = a + (b + c)$
Inverse Property of Addition	$a + (-a) = 0$ or $-a + a = 0$
Multiplication Property of Zero	$a \cdot 0 = 0$ or $0 \cdot a = 0$
Multiplication Property of One	$a \cdot 1 = a$ or $1 \cdot a = a$
Commutative Property of Multiplication	$a \cdot b = b \cdot a$
Associative Property of Multiplication	$(a \cdot b) \cdot c = a \cdot (b \cdot c)$
Division Properties of Zero and One	If $a \neq 0$, $0 \div a = 0$.
	If $a \neq 0$, $a \div a = 1$.
	$a \div 1 = a$
	$a \div 0$ is undefined.

Addition Property of Equations
The same number can be added to each side of an equation without changing the solution of the equation.

The Order of Operations Agreement

Step 1   Do all operations inside parentheses.

Step 2   Simplify any numerical expressions containing exponents.

Step 3   Do multiplication and division as they occur from left to right.

Step 4   Do addition and subtraction as they occur from left to right.

CHAPTER REVIEW EXERCISES

TAKE NOTE

On the Computer Tutor is a Chapter Review corresponding to each chapter of the text.

1. Write the expression $8 - (-1)$ in words.

2. Evaluate $-|-36|$.

3. Find the product of -40 and -5.

4. Evaluate $-a \div b$ when $a = -27$ and $b = -3$.

5. Add: $-28 + 14$

6. Simplify: $-(-13)$

7. Graph -2 on the number line.

$$\xleftarrow{\hspace{0.5em}\overset{\displaystyle +\ +\ +\ +\ +\ +\ +\ +\ +\ +\ +\ +\ +}{-6\ -5\ -4\ -3\ -2\ -1\ \ 0\ \ 1\ \ 2\ \ 3\ \ 4\ \ 5\ \ 6}}\hspace{0.5em}\rightarrow}$$

8. Solve: $-24 = -6y$

9. Divide: $-51 \div (-3)$

10. Find the quotient of 840 and -4.

11. Subtract: $-6 - (-7) - 15 - (-12)$

12. Evaluate $-ab$ when $a = -2$ and $b = -9$.

13. Find the sum of 18, -13, and -6.

14. Multiply: $-18(4)$

15. Simplify: $(-2)^2 - (-3)^2 \div (1 - 4)^2 \cdot 2 - 6$

16. Evaluate $-x - y$ when $x = -1$ and $y = 3$.

17. *Sports* The scores of four golfers after the third round of the August 1996 PGA golf tournament are shown in the figure at the right. What is the difference between the number of strokes Cochran had made after the third round and the number Crenshaw had made?

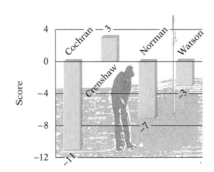

Golfers' Scores after Third Round of 1996 PGA Golf Tournament

18. Find the difference between −15 and −28.

19. Identify the property that justifies the statement
−11(−50) = −50(−11)

20. Is −9 a solution of −6 − t = 3?

21. Simplify: −9 + 16 − (−7)

22. Divide: $\dfrac{0}{-17}$

23. Multiply: −5(2)(−6)(−1)

24. Add: 3 + (−9) + 4 + (−10)

25. Evaluate $(a − b)^2 − 2a$ when $a = −2$ and $b = −3$.

26. Place the correct symbol, < or >, between the
two numbers.

−8 −10

27. Complete the statement by using the Inverse
Property of Addition.

−21 + ? = 0

28. Find the absolute value of −27.

Solve.

29. Forty-eight is the product of negative six and some number. Find the
number.

30. *Temperature* Which is colder, a temperature of −4°C or
−12°C?

31. ✍ *Chemisty* The figure at the right shows the boiling
point in degrees Celsius of three chemical elements.
The boiling point of neon is seven times the highest boiling
point shown in the table. What is the boiling point of neon?

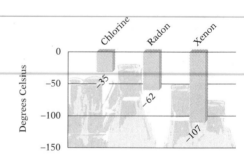
Boiling Points of Chemical Elements

32. *Temperature* Find the temperature after an increase of 5°C
from −8°C.

33. *Mathematics* The distance, *d*, between point *a* and point *b* on the num-
ber line is given by the formula $d = |a − b|$. Find *d* when $a = 7$ and
$b = −5$.

CUMULATIVE REVIEW EXERCISES

TAKE NOTE

On the Computer Tutor are Cumulative Reviews corresponding to Chapters 1–3, Chapters 4–6, and Chapters 7–10 of the text.

1. Find the difference between -27 and -32.

2. Estimate the product of 439 and 28.

3. Divide: $19,254 \div 6$

4. Simplify: $16 \div (3 + 5) \cdot 9 - 2^4$

5. Evaluate $-|-82|$.

6. Write three hundred nine thousand four hundred eighty in standard form.

7. Evaluate $5xy$ when $x = 80$ and $y = 6$.

8. What is -294 divided by -14?

9. Subtract: $-28 - (-17)$

10. Find the sum of -24, 16, and -32.

11. Find all the factors of 44.

12. Evaluate $x^4 y^2$ when $x = 2$ and $y = 11$.

13. Round 629,874 to the nearest thousand.

14. Estimate the sum of 356, 481, 294, and 117.

15. Evaluate $-a - b$ when $a = -4$ and $b = -5$.

16. Find the product of -100 and 25.

17. Find the prime factorization of 69.

18. Solve: $3x = -48$

19. Simplify: $(1 - 5)^2 \div (-6 + 4) + 8(-3)$

20. Evaluate $-c \div d$ when $c = -32$ and $d = -8$.

21. *Business* The table at the right shows United Parcel Service Air Pricing for packages of different weights. What would a business be charged for sending fifteen 9-lb packages via UPS next day air delivery?

UPS Air Pricing					
Weight (lb)	1	3	5	7	9
Next Day Air	$16	$18	$21	$25	$29

Note: All rates are as of February 1995.
The costs above do not reflect pickup charges.

22. Evaluate $\frac{a}{b}$ when $a = 39$ and $b = -13$.

23. Place the correct symbol, $<$ or $>$, between the two numbers.

$-62 \quad 26$

24. What is -18 multiplied by -7?

25. Solve: $12 + p = 3$

26. Write $2 \cdot 2 \cdot 2 \cdot 2 \cdot 2 \cdot 7 \cdot 7$ in exponential notation.

27. Evaluate $4a + (a - b)^3$ when $a = 5$ and $b = 2$.

28. Add: $5{,}971 + 482 + 3{,}609$

29. What is 5 less than -21?

30. Estimate the difference between 7,352 and 1,986.

31. Evaluate $3^4 \cdot 5^2$.

Solve.

32. *History*   The land area of the United States prior to the Louisiana Purchase was 891,364 mi². The land area of the Louisiana Purchase, which was purchased from France in 1803, was 831,321 mi². What was the land area of the United States immediately after the Louisiana Purchase?

33. *History*   Albert Einstein was born on March 14, 1879. He died on April 18, 1955. How old was Albert Einstein when he died?

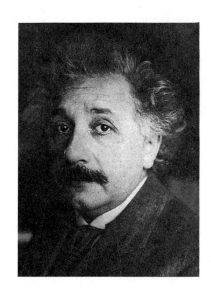

34. *Finances*   A customer makes a down payment of $3,550 on a car costing $17,750. Find the amount that remains to be paid.

35. *Real Estate*   A construction company is considering purchasing a 25-acre tract of land on which to build single-family homes. If the price is $3,690 per acre, what is the total cost of the land?

36. *Temperature*   Find the temperature after an increase of 7°C from −12°C.

37. *Business*   As a sales representative, your goal is to sell $120,000 in merchandise during the year. You sold $28,550 in merchandise during the first quarter of the year, $34,850 during the second quarter, and $31,700 during the third quarter. What must your sales for the fourth quarter be if you are to meet your goal for the year?

38. *Sports*   Use the equation $S = N - P$, where S is a golfer's score relative to par in a tournament, N is the number of strokes made by the golfer, and P is par, to find a golfer's score relative to par when the golfer made 198 strokes and par is 206.

CHAPTER 3

Fractions

FOCUS On
Problem Solving

An application problem may not provide all the information that is necessary to solve the problem. Sometimes, however, the necessary information is common knowledge.

You are traveling by bus from Boston to New York. The trip is 4 h long. If the bus leaves Boston at 10 A.M., what time should you arrive in New York?

What information do you need to know in order to solve this problem?

You need to know that, using a 12-hour clock, the hours run:

10 A.M.
11 A.M.
12 P.M.
1 P.M.
2 P.M.

Four hours after 10 A.M. is 2 P.M.

You should arrive in New York at 2 P.M.

You purchase a 32¢ stamp at the post office and hand the clerk a one-dollar bill. How much change do you receive?

What information do you need to know in order to solve this problem?

You need to know that there are 100¢ in one dollar.

Your change is 100¢ − 32¢.

$$100 - 32 = 68$$

You receive 68¢ in change.

What information do you need to know in order to solve each of the following problems?

1. You sell a dozen tickets to a fundraiser. Each ticket costs $10. How much money do you collect?

2. The weekly lab period for your science course is one hour and 20 minutes long. Find the length of the science lab period in minutes.

3. An employee's monthly salary is $1,750. Find the employee's annual salary.

4. A survey revealed that eighth graders spend an average of 3 h each day watching television. Find the total time an eighth grader spends watching TV each week.

Answers:

1. You need to know that there are 12 in one dozen.

2. You need to know that there are 60 minutes in one hour.

3. You need to know that there are 12 months in one year.

4. You need to know that there are 7 days in one week.

SECTION 3.1   Least Common Multiple and Greatest Common Factor

OBJECTIVE A

Least common multiple (LCM)

The **multiples** of a number are the products of that number and the numbers 1, 2, 3, 4, 5, . . .

$4 \cdot 1 = 4$
$4 \cdot 2 = 8$
$4 \cdot 3 = 12$
$4 \cdot 4 = 16$
$4 \cdot 5 = 20$ The multiples of 4 are 4, 8, 12, 16, 20, . . .
    .
    .
    .

A number that is a multiple of two or more numbers is a **common multiple** of those numbers.

The multiples of 6 are 6, 12, 18, 24, 30, 36, 42, 48, 54, 60, 66, 72, . . .
The multiples of 8 are 8, 16, 24, 32, 40, 48, 56, 64, 72, 80, 88, 96, . . .
Some common multiples of 6 and 8 are 24, 48, and 72.

The **least common multiple (LCM)** is the smallest common multiple of two or more numbers.

The least common multiple of 6 and 8 is 24. ·

Listing the multiples of each number is one way to find the LCM. Another way to find the LCM uses the prime factorization of each number.

To find the LCM of 6 and 8 using prime factorization:

Write the prime factorization of each number and circle the highest power of each prime factor.

$6 = 2 \cdot ③$
$8 = ②^3$

The LCM is the product of the circled factors.

$2^3 \cdot 3 = 8 \cdot 3 = 24$

The LCM of 6 and 8 is 24.

> Find the LCM of 32 and 36.
>
> Write the prime factorization of each number and circle the highest power of each prime factor.
>
> $32 = ②^5$
> $36 = 2^2 \cdot ③^2$
>
> The LCM is the product of the circled factors. $2^5 \cdot 3^2 = 32 \cdot 9 = 288$
>
> The LCM of 32 and 36 is 288.

Example 1 Find the LCM of 12, 18, and 40.

Solution

$12 = 2^2 \cdot 3$

$18 = 2 \cdot \boxed{3^2}$

$40 = \boxed{2^3} \cdot \boxed{5}$

The LCM $= 2^3 \cdot 3^2 \cdot 5$

$= 8 \cdot 9 \cdot 5 = 360.$

You Try It 1 Find the LCM of 16, 24, and 28.

Your Solution

Solution on p. A10

OBJECTIVE B
Greatest common factor (GCF)

Recall that a number that divides another number evenly is a **factor** of the number.

> 18 can be evenly divided by 1, 2, 3, 6, 9, and 18.
> 1, 2, 3, 6, 9, and 18 are factors of 18.

A number that is a factor of two or more numbers is a **common factor** of those numbers.

> The factors of 24 are 1, 2, 3, 4, 6, 8, 12, and 24.
> The factors of 36 are 1, 2, 3, 4, 6, 9, 12, 18, and 36.
> The common factors of 24 and 36 are 1, 2, 3, 4, 6, and 12.

The **greatest common factor (GCF)** is the largest common factor of two or more numbers.

> The greatest common factor of 24 and 36 is 12.

TAKE NOTE

12 is the GCF of 24 and 36 because 12 is the largest integer that divides evenly into both 24 and 36.

Listing the factors of each number is one way to find the GCF. Another way to find the GCF uses the prime factorization of each number.

To find the GCF of 24 and 36 using prime factorization:

Write the prime factorization of each number and circle the lowest power of each prime factor that occurs in *both* factorizations.

$24 = 2^3 \cdot \boxed{3}$

$36 = \boxed{2^2} \cdot 3^2$

The GCF is the product of the circled factors.

$2^2 \cdot 3 = 4 \cdot 3 = 12$

Find the GCF of 12 and 30.

Write the prime factorization of each number and circle the lowest power of each prime factor that occurs in both factorizations. The prime factor 5 occurs in the prime factorization of 30 but not in the prime factorization of 12. Since 5 is not a factor in both factorizations, do not circle 5.

$12 = 2^2 \cdot \boxed{3}$

$30 = \boxed{2} \cdot 3 \cdot 5$

The GCF is the product of the circled factors.

$2 \cdot 3 = 6$

The GCF of 12 and 30 is 6.

Example 2   Find the GCF of 14 and 27.

Solution $14 = 2 \cdot 7$
$27 = 3^3$

No common prime factor occurs in the factorizations.

The GCF is 1.

You Try It 2   Find the GCF of 25 and 52.

Your Solution

Example 3   Find the GCF of 16, 20, and 28.

Solution $16 = 2^4$
$20 = \boxed{2^2} \cdot 5$
$28 = 2^2 \cdot 7$

The GCF $= 2^2 = 4$.

You Try It 3   Find the GCF of 32, 40, and 56.

Your Solution

Solutions on p. A10

OBJECTIVE C
Applications

Example 4

Each month, copies of a national magazine are delivered to three different stores that have ordered 50, 75, and 125 copies, respectively. How many copies should be packaged together so that no package needs to be opened during delivery?

Strategy

To find the numbers of copies to be packaged together, find the GCF of 50, 75, and 125.

Solution

$50 = 2 \cdot \boxed{5^2}$
$75 = 3 \cdot 5^2$
$125 = 5^3$
The GCF $= 5^2 = 25$.

Each package should contain 25 copies of the magazine.

You Try It 4

A discount catalog offers blank diskettes at reduced prices. The customer must order 20, 50, or 100 diskettes. How many diskettes should be packaged together so that no package needs to be opened when a clerk is filling an order?

Your Strategy

Your Solution

Solution on p. A10

Example 5

To accommodate several activity periods and science labs after the lunch period and before the closing homeroom period, a high school wants to have both 25-minute class periods and 40-minute class periods running simultaneously in the afternoon class schedule. There is a 5-minute passing time between each class. How long a period of time must be scheduled if all students are to be in the closing homeroom period at the same time? How many 25-minute classes and 40-minute classes will be scheduled in that amount of time?

Strategy

To find the amount of time to be scheduled:

▶ Add the passing time (5 min) to the 25-minute class period and to the 40-minute class period to find the length of each period including the passing time.
▶ Find the LCM of the two time periods found in Step 1.

To find the number of 25-minute and 40-minute classes:

▶ Divide the LCM by each time period found in Step 1.

Solution

$25 + 5 = 30$
$40 + 5 = 45$

$30 = ② \cdot 3 \cdot ⑤$

$45 = ③^2 \cdot 5$

The LCM $= 2 \cdot 3^2 \cdot 5 = 90$

A 90-minute time period must be scheduled.

$90 \div 30 = 3$
$90 \div 45 = 2$

There will be three 25-minute class periods and two 40-minute class periods in the 90-minute period.

You Try It 5

You and a friend are running laps at the track. You run one lap every 3 min. Your friend runs one lap every 4 min. If you start at the same time from the same place on the track, in how many minutes will both of you be at the starting point again? Will you have passed each other at some other point on the track prior to that time?

Your Strategy

Your Solution

Solution on p. A10

3.1 EXERCISES

OBJECTIVE A

Find the LCM of the numbers.

1. 4 and 8	**2.** 3 and 9	**3.** 2 and 7	**4.** 5 and 11
5. 6 and 10	**6.** 8 and 12	**7.** 9 and 15	**8.** 14 and 21
9. 12 and 16	**10.** 8 and 14	**11.** 4 and 10	**12.** 9 and 30
13. 14 and 42	**14.** 16 and 48	**15.** 24 and 36	**16.** 16 and 28
17. 30 and 40	**18.** 45 and 60	**19.** 3, 5, and 10	**20.** 5, 10, and 20
21. 4, 8, and 12	**22.** 3, 12, and 18	**23.** 9, 36, and 45	**24.** 9, 36, and 72
25. 6, 9, and 15	**26.** 30, 40, and 60	**27.** 13, 26, and 39	**28.** 12, 48, and 72

OBJECTIVE B

Find the GCF of the numbers.

29. 9 and 12	**30.** 6 and 15	**31.** 18 and 30	**32.** 15 and 35
33. 14 and 42	**34.** 25 and 50	**35.** 16 and 80	**36.** 17 and 51
37. 21 and 55	**38.** 32 and 35	**39.** 8 and 36	**40.** 12 and 80
41. 12 and 76	**42.** 16 and 60	**43.** 24 and 30	**44.** 16 and 28
45. 24 and 36	**46.** 30 and 40	**47.** 45 and 75	**48.** 12 and 54

Find the GCF of the numbers.

49. 6, 10, and 12

50. 8, 12, and 20

51. 6, 15, and 36

52. 15, 20, and 30

53. 21, 63, and 84

54. 12, 28, and 48

55. 24, 36, and 60

56. 32, 56, and 72

OBJECTIVE C

Solve.

57. *Business*   Two machines are filling cereal boxes. One machine, which is filling 12-ounce boxes, fills one box every 2 min. The second machine, which is filling 18-ounce boxes, fills one box every 3 min. How often are the two machines starting to fill a box at the same time?

58. *Business*   A discount catalog offers stockings at reduced prices. The customer must order 3 pairs, 6 pairs, or 12 pairs of stockings. How many pairs should be packaged together so that no package needs to be opened when a clerk is filling an order?

59. *Business*   Each week, copies of a national magazine are delivered to three different stores that have ordered 75 copies, 100 copies, and 150 copies, respectively. How many copies should be packaged together so that no package needs to be opened during delivery?

60. *Sports*   You and a friend are swimming laps at a pool. You swim one lap every 4 min. Your friend swims one lap every 5 min. If you start at the same time from the same end of the pool, in how many minutes will both of you be at the starting point again? How many times will you have passed each other in the pool prior to that time?

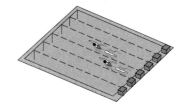

61. *Scheduling*   A mathematics conference is scheduling 30-minute sessions and 40-minute sessions. There will be a 10-minute break between each session. The sessions at the conference start at 9 A.M. At what time will all sessions begin at the same time once again? At what time should lunch be scheduled if all participants are to eat at the same time?

CRITICAL THINKING

62. If x is a prime number and y is a prime number, find the LCM of x and y. Find the GCF of x and y.

63. Find the LCM of x and $2x$. Find the GCF of x and $2x$.

64. In your own words, define the least common multiple of two numbers and the greatest common factor of two numbers.

65. Explain the meaning of relatively prime factors. List three pairs of relatively prime factors.

SECTION 3.2   Introduction to Fractions

OBJECTIVE **A**
Proper fractions, improper fractions, and mixed numbers

A recipe calls for $\frac{1}{2}$ cup of butter; a carpenter uses a $\frac{3}{8}$-inch screw; and a stock broker might say that Sears closed down $\frac{3}{4}$. The numbers $\frac{1}{2}$, $\frac{3}{8}$, and $\frac{3}{4}$ are fractions.

A **fraction** can represent the number of equal parts of a whole. The circle at the right is divided into 8 equal parts. 3 of the 8 parts are shaded. The shaded portion of the circle is represented by the fraction $\frac{3}{8}$.

Each part of a fraction has a name.

$$\text{Fraction bar} \longrightarrow \frac{3}{8} \begin{array}{l} \longleftarrow \textbf{Numerator} \\ \longleftarrow \textbf{Denominator} \end{array}$$

> **POINT OF INTEREST**
> The fraction bar was first used in 1050 by al-Hassar. It is also called a vinculum.

In a **proper fraction,** the numerator is smaller than the denominator. A proper fraction is less than one.

$$\frac{1}{2} \qquad \frac{3}{8} \qquad \frac{3}{4}$$
Proper fractions

In an **improper fraction,** the numerator is greater than or equal to the denominator. An improper fraction is a number greater than or equal to 1.

$$\frac{7}{3} \qquad \frac{4}{4}$$
Improper fractions

The shaded portion of the circles at the right is represented by the improper fraction $\frac{7}{3}$.

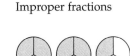

The shaded portion of the square at the right is represented by the improper fraction $\frac{4}{4}$.

A fraction bar can be read "divided by." Therefore, the fraction $\frac{4}{4}$ can be read "$4 \div 4$." Because a number divided by itself is equal to 1, $4 \div 4 = 1$ and $\frac{4}{4} = 1$.

The shaded portion of the square above can be represented as $\frac{4}{4}$ or 1.

Since the fraction bar can be read as "divided by" and any number divided by one is the number, any whole number can be represented as an improper fraction. For example, $5 = \frac{5}{1}$ and $7 = \frac{7}{1}$.

Because zero divided by any number other than zero is zero, **the numerator of a fraction can be zero.**

For example, $\frac{0}{6} = 0$ because $0 \div 6 = 0$.

Recall that division by zero is not defined. Therefore, **the denominator of a fraction cannot be zero.**

For example, $\frac{9}{0}$ is not defined because $\frac{9}{0} = 9 \div 0$, and division by zero is not defined.

A **mixed number** is a number greater than 1 with a whole number part and a fractional part.

The shaded portion of the circles at the right is represented by the mixed number $2\frac{1}{2}$.

Note from the diagram at the right that the improper fraction $\frac{5}{2}$ is equal to the mixed number $2\frac{1}{2}$.

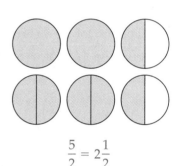

$$\frac{5}{2} = 2\frac{1}{2}$$

An improper fraction can be written as a mixed number.

To write $\frac{5}{2}$ as a mixed number, read the fraction bar as "divided by."

$\frac{5}{2}$ means $5 \div 2$.

Divide the numerator by the denominator.	To write the fractional part of the mixed number, write the remainder over the divisor.	Write the answer.
$\begin{array}{r} 2 \\ 2\overline{)5} \\ -4 \\ \hline 1 \end{array}$	$\begin{array}{r} 2\frac{1}{2} \\ 2\overline{)5} \\ -4 \\ \hline 1 \end{array}$	$\frac{5}{2} = 2\frac{1}{2}$

To write a mixed number as an improper fraction, multiply the denominator of the fractional part of the mixed number by the whole number part. The sum of this product and the numerator of the fractional part is the numerator of the improper fraction. The denominator remains the same.

Write $4\frac{5}{6}$ as an improper fraction.

$$4\frac{5}{6} = \frac{(6 \cdot 4) + 5}{6} = \frac{24 + 5}{6} = \frac{29}{6}$$

Example 1   Express the shaded portion of the circles as an improper fraction and as a mixed number.

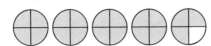

Solution $\frac{19}{4}$; $4\frac{3}{4}$

You Try It 1   Express the shaded portion of the circles as an improper fraction and as a mixed number.

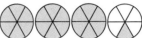

Your Solution

Example 2   Write $\frac{14}{5}$ as a mixed number.

Solution
$$\begin{array}{r} 2 \\ 5\overline{)14} \\ -10 \\ \hline 4 \end{array} \qquad \frac{14}{5} = 2\frac{4}{5}$$

You Try It 2   Write $\frac{26}{3}$ as a mixed number.

Your Solution

Example 3   Write $\frac{35}{7}$ as a whole number.

Solution
$$\begin{array}{r} 5 \\ 7\overline{)35} \\ -35 \\ \hline 0 \end{array} \qquad \frac{35}{7} = 5$$
Note: The remainder is zero.

You Try It 3   Write $\frac{36}{4}$ as a whole number.

Your Solution

Example 4   Write $12\frac{5}{8}$ as an improper fraction.

Solution $12\frac{5}{8} = \dfrac{(8 \cdot 12) + 5}{8} = \dfrac{96 + 5}{8}$

$\qquad\qquad = \dfrac{101}{8}$

You Try It 4   Write $9\frac{4}{7}$ as an improper fraction.

Your Solution

Example 5   Write 9 as an improper fraction.

Solution $9 = \dfrac{9}{1}$

You Try It 5   Write 3 as an improper fraction.

Your Solution

Solutions on p. A10

OBJECTIVE B
Equivalent fractions

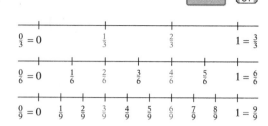

Fractions can be graphed as points on a number line. The number lines at the right show thirds, sixths, and ninths graphed from 0 to 1.

A particular point on the number line may be represented by different fractions, all of which are equal.

For example, $\frac{0}{3} = \frac{0}{6} = \frac{0}{9}$, $\frac{1}{3} = \frac{2}{6} = \frac{3}{9}$, $\frac{2}{3} = \frac{4}{6} = \frac{6}{9}$, and $\frac{3}{3} = \frac{6}{6} = \frac{9}{9}$.

Equal fractions with different denominators are called **equivalent fractions.**

$\frac{1}{3}, \frac{2}{6}$, and $\frac{3}{9}$ are equivalent fractions. $\frac{2}{3}, \frac{4}{6}$, and $\frac{6}{9}$ are equivalent fractions.

Note that we can rewrite $\frac{2}{3}$ as $\frac{4}{6}$ by multiplying both the numerator and denominator of $\frac{2}{3}$ by 2.

$$\frac{2}{3} = \frac{2 \cdot 2}{3 \cdot 2} = \frac{4}{6}$$

Also, we can rewrite $\frac{4}{6}$ as $\frac{2}{3}$ by dividing both the numerator and denominator of $\frac{4}{6}$ by 2.

$$\frac{4}{6} = \frac{4 \div 2}{6 \div 2} = \frac{2}{3}$$

This suggests the following property of fractions.

Equivalent Fractions

The numerator and denominator of a fraction can be multiplied by or divided by the same nonzero number. The resulting fraction is equivalent to the original fraction.

$$\frac{a}{b} = \frac{a \cdot c}{b \cdot c}, \quad \text{and} \quad \frac{a}{b} = \frac{a \div c}{b \div c}, \quad \text{where} \quad b \neq 0 \quad \text{and} \quad c \neq 0$$

Write an equivalent fraction with the given denominator: $\frac{3}{8} = \frac{}{40}$.

Divide the larger denominator by the smaller one. $40 \div 8 = 5$

Multiply the numerator and denominator of the given fraction by the quotient (5). $\frac{3}{8} = \frac{3 \cdot 5}{8 \cdot 5} = \frac{15}{40}$

A fraction is in **simplest form** when the numerator and denominator have no common factors other than 1. The fraction $\frac{3}{8}$ is in simplest form because 3 and 8 have no common factors other than 1. The fraction $\frac{15}{40}$ is not in simplest form because the numerator and denominator have a common factor of 5.

To write a fraction in simplest form, divide the numerator and denominator of the fraction by their common factors.

Write $\frac{12}{15}$ in simplest form.

12 and 15 have a common factor of 3. Divide the numerator and denominator by 3.

$$\frac{12}{15} = \frac{12 \div 3}{15 \div 3} = \frac{4}{5}$$

Simplifying a fraction requires that you recognize the common factors of the numerator and denominator. One way to do this is to write the prime factorization of the numerator and denominator and then divide by the common prime factors.

Write $\frac{30}{42}$ in simplest form.

Write the prime factorization of the numerator and denominator. Divide by the common factors.

$$\frac{30}{42} = \frac{\overset{1}{\cancel{2}} \cdot \overset{1}{\cancel{3}} \cdot 5}{\underset{1}{\cancel{2}} \cdot \underset{1}{\cancel{3}} \cdot 7} = \frac{5}{7}$$

Write $\frac{2x}{6}$ in simplest form.

Factor the numerator and denominator. Then divide by the common factors.

$$\frac{2x}{6} = \frac{\overset{1}{\cancel{2}} \cdot x}{\underset{1}{\cancel{2}} \cdot 3} = \frac{x}{3}$$

Example 6   Write an equivalent fraction with the given denominator:

$\frac{2}{5} = \frac{}{30}$.

Solution $30 \div 5 = 6$

$$\frac{2}{5} = \frac{2 \cdot 6}{5 \cdot 6} = \frac{12}{30}$$

$\frac{12}{30}$ is equivalent to $\frac{2}{5}$.

You Try It 6   Write an equivalent fraction with the given denominator:

$\frac{5}{8} = \frac{}{48}$.

Your Solution

Example 7   Write an equivalent fraction with the given denominator:

$3 = \frac{}{15}$.

Solution $3 = \frac{3}{1}$ $15 \div 1 = 15$

$$3 = \frac{3}{1} = \frac{3 \cdot 15}{1 \cdot 15} = \frac{45}{15}$$

$\frac{45}{15}$ is equivalent to 3.

You Try It 7   Write an equivalent fraction with the given denominator:

$8 = \frac{}{12}$.

Your Solution

Solutions on pp. A10–A11

Example 8 Write $\dfrac{18}{54}$ in simplest form.

Solution $\dfrac{18}{54} = \dfrac{\overset{1}{\cancel{2}} \cdot \overset{1}{\cancel{3}} \cdot \overset{1}{\cancel{3}}}{\underset{1}{\cancel{2}} \cdot \underset{1}{\cancel{3}} \cdot \underset{1}{\cancel{3}} \cdot 3} = \dfrac{1}{3}$

You Try It 8 Write $\dfrac{21}{84}$ in simplest form.

Your Solution

Example 9 Write $\dfrac{36}{20}$ in simplest form.

Solution $\dfrac{36}{20} = \dfrac{\overset{1}{\cancel{2}} \cdot \overset{1}{\cancel{2}} \cdot 3 \cdot 3}{\underset{1}{\cancel{2}} \cdot \underset{1}{\cancel{2}} \cdot 5} = \dfrac{9}{5}$

You Try It 9 Write $\dfrac{32}{12}$ in simplest form.

Your Solution

Example 10 Write $\dfrac{10m}{12}$ in simplest form.

Solution $\dfrac{10m}{12} = \dfrac{\overset{1}{\cancel{2}} \cdot 5 \cdot m}{\underset{1}{\cancel{2}} \cdot 2 \cdot 3} = \dfrac{5m}{6}$

You Try It 10 Write $\dfrac{11t}{11}$ in simplest form.

Your Solution

Solutions on p. A11 | | | | |

OBJECTIVE C

Order relations between two fractions

The number line can be used to determine the order relation between two fractions.

A fraction that appears to the left of a given fraction is less than the given fraction.

$\dfrac{3}{8}$ is to the left of $\dfrac{5}{8}$.

$\dfrac{3}{8} < \dfrac{5}{8}$

A fraction that appears to the right of a given fraction is greater than the given fraction.

$\dfrac{7}{8}$ is to the right of $\dfrac{3}{8}$.

$\dfrac{7}{8} > \dfrac{3}{8}$

To find the order relation between two fractions with the *same* denominator, compare the numerators. The fraction with the smaller numerator is the smaller fraction. The larger fraction is the fraction with the larger numerator.

$\dfrac{3}{8}$ and $\dfrac{5}{8}$ have the same denominator. $\dfrac{3}{8} < \dfrac{5}{8}$ because $3 < 5$.

$\dfrac{7}{8}$ and $\dfrac{3}{8}$ have the same denominator. $\dfrac{7}{8} > \dfrac{3}{8}$ because $7 > 3$.

Before comparing two fractions with *different* denominators, rewrite the fractions with a common denominator. The common denominator is the least common multiple (LCM) of the denominators of the fractions. The LCM of the denominators is sometimes called the lowest common denominator or LCD.

Find the order relation between $\frac{5}{12}$ and $\frac{7}{18}$.

Find the LCM of the denominators.

The LCM of 12 and 18 is 36.

Write each fraction as an equivalent fraction with the LCM as the denominator.

$$\frac{5}{12} = \frac{5 \cdot 3}{12 \cdot 3} = \frac{15}{36} \longleftarrow \text{Larger numerator}$$

$$\frac{7}{18} = \frac{7 \cdot 2}{18 \cdot 2} = \frac{14}{36} \longleftarrow \text{Smaller numerator}$$

Compare the fractions.

$$\frac{15}{36} > \frac{14}{36}$$

$$\frac{5}{12} > \frac{7}{18}$$

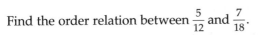

Example 11   Place the correct symbol, < or >, between the two numbers.

$$\frac{2}{3} \qquad \frac{4}{7}$$

Solution   The LCM of 3 and 7 is 21.

$$\frac{2}{3} = \frac{14}{21} \qquad \frac{4}{7} = \frac{12}{21}$$

$$\frac{14}{21} > \frac{12}{21}$$

$$\frac{2}{3} > \frac{4}{7}$$

You Try It 11   Place the correct symbol, < or >, between the two numbers.

$$\frac{4}{9} \qquad \frac{8}{21}$$

Your Solution

Example 12   Place the correct symbol, < or >, between the two numbers.

$$\frac{7}{12} \qquad \frac{11}{18}$$

Solution   The LCM of 12 and 18 is 36.

$$\frac{7}{12} = \frac{21}{36} \qquad \frac{11}{18} = \frac{22}{36}$$

$$\frac{21}{36} < \frac{22}{36}$$

$$\frac{7}{12} < \frac{11}{18}$$

You Try It 12   Place the correct symbol, < or >, between the two numbers.

$$\frac{17}{24} \qquad \frac{8}{21}$$

Your Solution

Solutions on p. A11 |||||

Company	Pounds Sold (in Millions)
Kellogg	895
General Mills	599
Post Cereal	411
Quaker Oats	199
Others	353

Source: North County Times, 8/15/96

OBJECTIVE D
Applications

The table at the left shows the number of pounds of cold cereal sold by U.S. companies for a 52-week period between 1995 and 1996. Use this table for Example 13 and You Try It 13.

Example 13

What fraction of the total sales did Kellogg have?

Strategy

To find the fraction:

▶ Add to find the total amount sold by the companies (895 + 599 + 411 + 199 + 353).
▶ Write a fraction with Kellogg's sales in the numerator and the total sales in the denominator.

Solution

$895 + 599 + 411 + 199 + 353 = 2,457$

$$\frac{895}{2,457}$$

Kellogg's fraction of the total sales was $\frac{895}{2,457}$.

You Try It 13

What fraction of the sales by the first four companies listed did Quaker Oats have?

Your Strategy

Your Solution

Example 14

$60 million was raised to create Atlanta's 21-acre Centennial Olympic Park. Of this amount, $10 million was donated by the Atlanta business community. What fraction of the amount raised was donated by the Atlanta business community?

Strategy

To find the fraction, write a fraction with the amount donated by the business community in the numerator and the total amount donated in the denominator.

Solution

$$\frac{10}{60} = \frac{1}{6}$$

$\frac{1}{6}$ was donated by the business community.

You Try It 14

For every dollar's worth of product sold by Coca-Cola outside the United States, Coke's profit is 30¢. What fraction of every dollar's worth of product sold outside the United States is profit for Coca-Cola?

Your Strategy

Your Solution

Solutions on p. A11

3.2 EXERCISES

OBJECTIVE A

Express the shaded portion of the circle as a fraction.

1.

2.

3.

4.

Express the shaded portion of the circles as an improper fraction and as a mixed number.

5.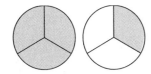

6.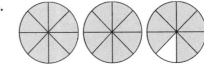

7.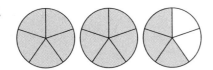

8.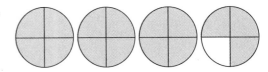

Write the improper fraction as a mixed number or a whole number.

9. $\dfrac{13}{4}$

10. $\dfrac{14}{3}$

11. $\dfrac{20}{5}$

12. $\dfrac{18}{6}$

13. $\dfrac{27}{10}$

14. $\dfrac{31}{3}$

15. $\dfrac{56}{8}$

16. $\dfrac{27}{9}$

17. $\dfrac{17}{9}$

18. $\dfrac{8}{3}$

19. $\dfrac{12}{5}$

20. $\dfrac{19}{8}$

21. $\dfrac{18}{1}$

22. $\dfrac{21}{1}$

23. $\dfrac{32}{15}$

24. $\dfrac{39}{14}$

25. $\dfrac{8}{8}$

26. $\dfrac{12}{12}$

27. $\dfrac{28}{3}$

28. $\dfrac{43}{5}$

Write the mixed number or whole number as an improper fraction.

29. $2\frac{1}{4}$

30. $4\frac{2}{5}$

31. $5\frac{1}{2}$

32. $3\frac{2}{3}$

33. $2\frac{4}{5}$

34. $6\frac{3}{8}$

35. $7\frac{5}{6}$

36. $9\frac{1}{5}$

37. 7

38. 4

39. $8\frac{1}{4}$

40. $1\frac{7}{9}$

41. $10\frac{1}{3}$

42. $6\frac{3}{7}$

43. $4\frac{7}{12}$

44. $5\frac{4}{9}$

45. 8

46. 6

47. $12\frac{4}{5}$

48. $11\frac{5}{8}$

OBJECTIVE B

Write an equivalent fraction with the given denominator.

49. $\frac{1}{2} = \frac{}{12}$

50. $\frac{1}{4} = \frac{}{20}$

51. $\frac{3}{8} = \frac{}{24}$

52. $\frac{9}{11} = \frac{}{44}$

53. $\frac{2}{17} = \frac{}{51}$

54. $\frac{9}{10} = \frac{}{80}$

55. $\frac{3}{4} = \frac{}{32}$

56. $\frac{5}{8} = \frac{}{32}$

57. $6 = \frac{}{18}$

58. $5 = \frac{}{35}$

59. $\frac{1}{3} = \frac{}{90}$

60. $\frac{3}{16} = \frac{}{48}$

61. $\frac{2}{3} = \frac{}{21}$

62. $\frac{4}{9} = \frac{}{36}$

63. $\frac{6}{7} = \frac{}{49}$

64. $\frac{7}{8} = \frac{}{40}$

65. $\frac{4}{9} = \frac{}{18}$

66. $\frac{11}{12} = \frac{}{48}$

67. $7 = \frac{}{4}$

68. $9 = \frac{}{6}$

Write the fraction in simplest form.

69. $\frac{3}{12}$

70. $\frac{10}{22}$

71. $\frac{33}{44}$

72. $\frac{6}{14}$

73. $\frac{4}{24}$

Write the fraction in simplest form.

74. $\dfrac{25}{75}$ **75.** $\dfrac{8}{33}$ **76.** $\dfrac{9}{25}$ **77.** $\dfrac{0}{8}$ **78.** $\dfrac{0}{11}$

79. $\dfrac{42}{36}$ **80.** $\dfrac{30}{18}$ **81.** $\dfrac{16}{16}$ **82.** $\dfrac{24}{24}$ **83.** $\dfrac{21}{35}$

84. $\dfrac{11}{55}$ **85.** $\dfrac{16}{60}$ **86.** $\dfrac{8}{84}$ **87.** $\dfrac{12}{20}$ **88.** $\dfrac{24}{36}$

89. $\dfrac{12m}{18}$ **90.** $\dfrac{20x}{25}$ **91.** $\dfrac{4y}{8}$ **92.** $\dfrac{14z}{28}$ **93.** $\dfrac{24a}{36}$

94. $\dfrac{28z}{21}$ **95.** $\dfrac{8c}{8}$ **96.** $\dfrac{9w}{9}$ **97.** $\dfrac{18k}{3}$ **98.** $\dfrac{24t}{4}$

OBJECTIVE C

Place the correct symbol, $<$ or $>$, between the two numbers.

99. $\dfrac{3}{8}$ $\dfrac{2}{5}$ **100.** $\dfrac{5}{7}$ $\dfrac{2}{3}$ **101.** $\dfrac{3}{4}$ $\dfrac{7}{9}$ **102.** $\dfrac{7}{12}$ $\dfrac{5}{8}$

103. $\dfrac{2}{3}$ $\dfrac{7}{11}$ **104.** $\dfrac{11}{14}$ $\dfrac{3}{4}$ **105.** $\dfrac{17}{24}$ $\dfrac{11}{16}$ **106.** $\dfrac{11}{12}$ $\dfrac{7}{9}$

107. $\dfrac{7}{15}$ $\dfrac{5}{12}$ **108.** $\dfrac{5}{8}$ $\dfrac{4}{7}$ **109.** $\dfrac{5}{9}$ $\dfrac{11}{21}$ **110.** $\dfrac{11}{30}$ $\dfrac{7}{24}$

111. $\dfrac{7}{12}$ $\dfrac{13}{18}$ **112.** $\dfrac{9}{11}$ $\dfrac{7}{8}$ **113.** $\dfrac{4}{5}$ $\dfrac{7}{9}$ **114.** $\dfrac{3}{4}$ $\dfrac{11}{13}$

115. $\dfrac{9}{16}$ $\dfrac{5}{9}$ **116.** $\dfrac{2}{3}$ $\dfrac{7}{10}$ **117.** $\dfrac{5}{8}$ $\dfrac{13}{20}$ **118.** $\dfrac{3}{10}$ $\dfrac{7}{25}$

OBJECTIVE D

Solve.

119. *Measurement*   A ton is equal to 2,000 lb. What fractional part of a ton is 250 lb?

120. *Measurement*   A pound is equal to 16 oz. What fractional part of a pound is 6 oz?

121. *Measurement*   If a history class lasts 50 min, what fractional part of an hour is the history class?

122. *Measurement*   If you sleep for 8 h one night, what fractional part of one day did you spend sleeping?

123. *Jewelry*   Gold is designated by karats. Pure gold is 24 karats. What fractional part of an 18-karat gold bracelet is pure gold?

124. *Measurement*   A thermos holds 2 qt of liquid. What fractional part of one gallon does the thermos hold?

125. *Card Games*   A standard deck of playing cards consists of 52 cards. What fractional part of a standard deck is spades?

126. *Card Games*   A standard deck of playing cards consists of 52 cards. What fractional part of a standard deck is aces?

$$\frac{\cancel{4}\,3}{52} = \frac{1}{4}$$
$$13 \cdot 4$$

127. *Education*   You answer 42 questions correctly on an exam of 50 questions. Did you answer more or less than $\frac{8}{10}$ of the questions correctly?

128. *Education*   To pass a real estate examination, you must answer at least $\frac{7}{10}$ of the questions correctly. If the exam has 200 questions and you answer 150 correctly, will you pass the exam?

129. *Sports*   Wilt Chamberlain holds the record for the most field goals in a basketball game. He had 36 field goals in 63 attempts. What fraction of the number of attempts did he not have a field goal?

130. *Sports*   In the 1995 Super Bowl game, Steve Young completed 24 of 36 attempted passes. What fraction of the number of attempted passes did he not complete?

The figure at the right shows the fraction of total sales contributed by each type of vehicle sold by a car dealership. Use this graph for Exercise 131.

131. _Business_   What type of vehicle had (a) the greatest fraction of total sales? (b) the least fraction of total sales?

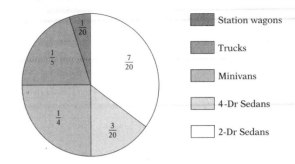

Distribution of Sales of 700 Vehicles

Solve.

132. _Sports_   In the 1996 Kickoff Classic, Penn State scored 14 points and USC scored 7 points. What fractional part of the total number of points scored were scored by Penn State?

133. _Sports_   In the 1994 Winter Olympics, United States skiers won 6 gold medals, 5 silver medals, and 2 bronze medals. Of the total medals won by the U.S. skiers, what fractional part were gold medals?

The figure at the right shows the number of women on active duty for four branches of the armed services. Use this graph for Exercise 134.

134. _The Military_   For the four branches shown, what fraction of the women were in (a) the Marines? (b) the Navy? (c) the Air Force? (d) the Army?

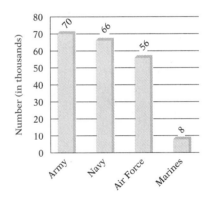

Number of Women on Active Duty, 1994
Source: 1996 Information Please Almanac

Solve.

135. _Sports_   In 1995, Steffi Graf won the U.S. Open Tennis Championship by beating Monica Seles. Graf won 7 games the first set, 0 games the second set, and 6 games the third set. Did Graf win more or less than one-half of the 28 games played between Graf and Seles?

136. _Geography_   What fraction of the states in the United States begin with the letter A?

The figure at the right shows how much money a telephone operator spends each month for various living expenses. Use this graph for Exercise 137.

137. _Finances_   What fraction of the total monthly expenses was spent on (a) entertainment? (b) taxes?

$$\frac{150}{2000} = \frac{15}{200} = \frac{3}{4}$$

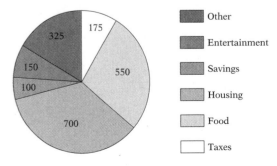

Monthly Expenses (in dollars)

CRITICAL THINKING

138. Is the expression $x < \frac{4}{9}$ true when $x = \frac{3}{8}$? Is it true when $x = \frac{5}{12}$? Is it true for any negative number?

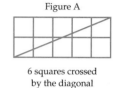

Figure A

6 squares crossed
by the diagonal

139. In Figure A, there are 2 rows of 5 squares, and in Figure B there are 3 rows of 7 squares. A diagonal line is drawn through each figure as shown, and the number of squares crossed by the diagonal is counted. Experiment with other arrangements of squares and develop a rule that will allow you to determine the number of squares crossed by the diagonal for m rows of n squares, where the GCF of m and n is 1.

Figure B

9 squares crossed
by the diagonal

140. $\frac{2}{3} < \frac{3}{4}$. Is $\frac{2+3}{3+4}$ less than $\frac{2}{3}$, greater than $\frac{3}{4}$, or between $\frac{2}{3}$ and $\frac{3}{4}$?

141. The following circle graphs represent the fractional part of a computer company's total annual sales for different areas of the United States. (a) According to the graphs, in which year were sales in the Northeast a greater fraction of total sales, 1996 or 1997? (b) In which year were sales in the Northwest a greater fraction of total sales, 1996 or 1997?

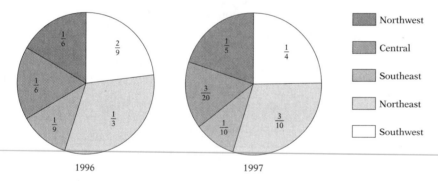

□ Northwest
□ Central
□ Southeast
□ Northeast
□ Southwest

1996   1997

Computer Sales, Contribution by Region

142. **(a)** On the number line, what fraction is halfway between $\frac{2}{a}$ and $\frac{4}{a}$?

(b) Find two fractions evenly spaced between $\frac{5}{b}$ and $\frac{8}{b}$.

143. Find the business section of your local newspaper. Choose a stock and record the fluctuations in the stock price for one week. Explain the part that fractions play in reporting the price and change in price of the stock.

144. It is now 1:15 P.M. Is it possible for you to arrive in a city for a 4:00 P.M. meeting if you drive 55 mph and the city is 140 mi away? Explain how you arrived at the answer.

SECTION 3.3   Addition and Subtraction of Fractions

OBJECTIVE A
Addition of fractions

Suppose you and a friend order a pizza. The pizza has been cut into 8 equal pieces. If you eat 3 pieces of the pizza and your friend eats 2 pieces, then together you have eaten $\frac{5}{8}$ of the pizza.

Note that in adding the fractions $\frac{3}{8}$ and $\frac{2}{8}$, the numerators are added and the denominator remains the same.

$$\frac{3}{8} + \frac{2}{8} = \frac{3+2}{8}$$

$$= \frac{5}{8}$$

Addition of Fractions

To add fractions with the same denominator, add the numerators and place the sum over the common denominator.

$$\frac{a}{b} + \frac{c}{b} = \frac{a+c}{b}, \text{ where } b \neq 0$$

Add: $\frac{5}{16} + \frac{7}{16}$

The denominators are the same. Add the numerators and place the sum over the common denominator.

$$\frac{5}{16} + \frac{7}{16} = \frac{5+7}{16}$$

Write the answer in simplest form.

$$= \frac{12}{16} = \frac{3}{4}$$

Add: $\frac{4}{x} + \frac{8}{x}$

The denominators are the same. Add the numerators and place the sum over the common denominator.

$$\frac{4}{x} + \frac{8}{x} = \frac{4+8}{x}$$

$$= \frac{12}{x}$$

Before two fractions can be added, the fractions must have the same denominator. To add fractions with different denominators, first rewrite the fractions as equivalent fractions with a common denominator. The common denominator is the least common multiple (LCM) of the denominators of the fractions. The LCM of denominators is sometimes called the least common denominator (LCD).

Find the sum of $\frac{5}{6}$ and $\frac{3}{8}$.

The common denominator is the LCM of 6 and 8.

The LCM of 6 and 8 is 24.

Write the fractions as equivalent fractions with the common denominator.

$$\frac{5}{6} + \frac{3}{8} = \frac{20}{24} + \frac{9}{24}$$

Add the fractions.

$$= \frac{20 + 9}{24}$$

$$= \frac{29}{24} = 1\frac{5}{24}$$

Figure 3.1 shows the approximate fractional amount of taxes the U.S. government collected from various sources in 1996. What fractional part of the total was collected from personal income taxes and corporate income taxes?

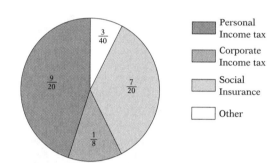

Figure 3.1 Taxes Collected by the U.S. Government in 1966

Add the fractional amount collected from personal income tax to the fractional amount collected from corporate income taxes.

$$\frac{9}{20} + \frac{1}{8} = \frac{18}{40} + \frac{5}{40} = \frac{23}{40}$$

$\frac{23}{40}$ of the total was collected from personal and corporate income taxes.

To add a fraction with a negative sign, rewrite the fraction with the negative sign in the numerator. Then add the numerators and place the sum over the common denominator.

Add: $-\frac{5}{6} + \frac{3}{4}$

The common denominator is the LCM of 4 and 6.

The LCM of 4 and 6 is 12.

Rewrite with the negative sign in the numerator.

$$-\frac{5}{6} + \frac{3}{4} = \frac{-5}{6} + \frac{3}{4}$$

Rewrite each fraction in terms of the common denominator.

$$= \frac{-10}{12} + \frac{9}{12}$$

Add the fractions.

$$= \frac{-10 + 9}{12}$$

Simplify the numerator and write the negative sign in front of the fraction.

$$= \frac{-1}{12} = -\frac{1}{12}$$

TAKE NOTE

Although the sum could have been left as $\frac{-1}{12}$, all answers in this text are written with the negative sign in front of the fraction.

Add: $-\dfrac{2}{3} + \left(-\dfrac{4}{5}\right)$

Rewrite each negative fraction with the negative sign in the numerator.

$$-\dfrac{2}{3} + \left(-\dfrac{4}{5}\right) = \dfrac{-2}{3} + \dfrac{-4}{5}$$

Rewrite each fraction as an equivalent fraction using the LCM as the denominator.

$$= \dfrac{-10}{15} + \dfrac{-12}{15}$$

Add the fractions.

$$= \dfrac{-10 + (-12)}{15}$$

$$= \dfrac{-22}{15} = -1\dfrac{7}{15}$$

Is $-\dfrac{2}{3}$ a solution of the equation $\dfrac{3}{4} + y = -\dfrac{1}{12}$?

$$\dfrac{3}{4} + y = -\dfrac{1}{12}$$

Replace y by $-\dfrac{2}{3}$. Then simplify.

$$\dfrac{3}{4} + \left(-\dfrac{2}{3}\right) \,\bigg|\, -\dfrac{1}{12}$$

The common denominator is 12.

$$\dfrac{9}{12} + \left(\dfrac{-8}{12}\right) \,\bigg|\, -\dfrac{1}{12}$$

$$\dfrac{9 + (-8)}{12} \,\bigg|\, -\dfrac{1}{12}$$

The results are not equal.

$$\dfrac{1}{12} \neq -\dfrac{1}{12}$$

No, $-\dfrac{2}{3}$ is not a solution of the equation.

The mixed number $2\dfrac{1}{2}$ is the sum of 2 and $\dfrac{1}{2}$.

$$2\dfrac{1}{2} = 2 + \dfrac{1}{2}$$

Therefore, the sum of a whole number and a fraction is a mixed number.

$$2 + \dfrac{1}{2} = 2\dfrac{1}{2}$$

$$3 + \dfrac{4}{5} = 3\dfrac{4}{5}$$

$$8 + \dfrac{7}{9} = 8\dfrac{7}{9}$$

The sum of a whole number and a mixed number is a mixed number.

Add: $5 + 4\dfrac{2}{7}$

Add the whole numbers (5 and 4).

$$5 + 4\dfrac{2}{7} = 9\dfrac{2}{7}$$

Write the fraction.

To add two mixed numbers, first write the fractional parts as equivalent fractions with a common denominator. Then add the fractional parts and add the whole numbers.

Add: $3\frac{5}{8} + 4\frac{7}{12}$

Write the fractions as equivalent fractions with a common denominator. The common denominator is the LCM of 8 and 12 (24).

$$3\frac{5}{8} + 4\frac{7}{12} = 3\frac{15}{24} + 4\frac{14}{24}$$

Add the fractional parts and add the whole numbers.

$$= 7\frac{29}{24}$$

Write the sum in simplest form.

$$= 7 + \frac{29}{24}$$

$$= 7 + 1\frac{5}{24}$$

$$= 8\frac{5}{24}$$

Evaluate $x + y$ when $x = 2\frac{3}{4}$ and $y = 7\frac{5}{6}$.

$$x + y$$

Replace x with $2\frac{3}{4}$ and y with $7\frac{5}{6}$.

$$2\frac{3}{4} + 7\frac{5}{6}$$

Write the fractions as equivalent fractions with a common denominator.

$$= 2\frac{9}{12} + 7\frac{10}{12}$$

Add the fractional parts and add the whole numbers.

$$= 9\frac{19}{12}$$

Write the sum in simplest form.

$$= 10\frac{7}{12}$$

Example 1 Add: $\frac{9}{16} + \frac{5}{12}$

Solution $\frac{9}{16} + \frac{5}{12} = \frac{27}{48} + \frac{20}{48}$

$$= \frac{27 + 20}{48} = \frac{47}{48}$$

You Try It 1 Add: $\frac{7}{12} + \frac{3}{8}$

Your Solution

Solution on p. A11

Example 2

Add: $\frac{4}{5} + \frac{3}{4} + \frac{5}{8}$

Solution

$\frac{4}{5} + \frac{3}{4} + \frac{5}{8} = \frac{32}{40} + \frac{30}{40} + \frac{25}{40} = \frac{87}{40} = 2\frac{7}{40}$

You Try It 2

Add: $\frac{3}{5} + \frac{2}{3} + \frac{5}{6}$

Your Solution

Example 3

Find the sum of $12\frac{4}{7}$ and 19.

Solution

$12\frac{4}{7} + 19 = 31\frac{4}{7}$

You Try It 3

What is the sum of 16 and $8\frac{5}{9}$?

Your Solution

Example 4

Add: $-\frac{3}{8} + \frac{3}{4} + \left(-\frac{5}{6}\right)$

Solution

$-\frac{3}{8} + \frac{3}{4} + \left(-\frac{5}{6}\right) = \frac{-3}{8} + \frac{3}{4} + \frac{-5}{6}$

$= \frac{-9}{24} + \frac{18}{24} + \frac{-20}{24}$

$= \frac{-9 + 18 + (-20)}{24}$

$= \frac{-11}{24} = -\frac{11}{24}$

You Try It 4

Add: $-\frac{5}{12} + \frac{5}{8} + \left(-\frac{1}{6}\right)$

Your Solution

Example 5

Evaluate $x + y + z$ when $x = 2\frac{1}{6}$, $y = 4\frac{3}{8}$, and $z = 7\frac{5}{9}$.

Solution

$x + y + z$

$2\frac{1}{6} + 4\frac{3}{8} + 7\frac{5}{9} = 2\frac{12}{72} + 4\frac{27}{72} + 7\frac{40}{72}$

$= 13\frac{79}{72}$

$= 14\frac{7}{72}$

You Try It 5

Evaluate $x + y + z$ when $x = 3\frac{5}{6}$, $y = 2\frac{1}{9}$, and $z = 5\frac{5}{12}$.

Your Solution

Solutions on p. A11 | | | | |

OBJECTIVE B
Subtraction of fractions

In the last objective, it was stated that to add fractions, the fractions must have the same denominator. The same is true for subtracting fractions: The two fractions must have the same denominator.

Subtraction of Fractions

To subtract fractions with the same denominator, subtract the numerators and place the difference over the common denominator.

$$\frac{a}{b} - \frac{c}{b} = \frac{a - c}{b}, \quad \text{where} \quad b \neq 0$$

Subtract: $\frac{5}{8} - \frac{3}{8}$

The denominators are the same. Subtract the numerators and place the difference over the common denominator.

$$\frac{5}{8} - \frac{3}{8} = \frac{5 - 3}{8}$$

Write the answer in simplest form.

$$= \frac{2}{8} = \frac{1}{4}$$

To subtract fractions with different denominators, first rewrite the fractions as equivalent fractions with a common denominator. The common denominator is the least common multiple (LCM) of the denominators of the fractions.

Subtract: $\frac{5}{12} - \frac{3}{8}$

The common denominator is the LCM of 12 and 8.

The LCM of 12 and 8 is 24.

Write the fractions as equivalent fractions with the common denominator.

$$\frac{5}{12} - \frac{3}{8} = \frac{10}{24} - \frac{9}{24}$$

Subtract the fractions.

$$= \frac{10 - 9}{24} = \frac{1}{24}$$

To subtract fractions with negative signs, first rewrite the fractions with the negative signs in the numerators.

Simplify: $-\frac{2}{9} - \frac{5}{12}$

Rewrite with the negative sign in the numerator.

$$-\frac{2}{9} - \frac{5}{12} = \frac{-2}{9} - \frac{5}{12}$$

Write the fractions as equivalent fractions with a common denominator.

$$= \frac{-8}{36} - \frac{15}{36}$$

Subtract the numerators and place the difference over the common denominator.

$$= \frac{-8 - 15}{36} = \frac{-23}{36}$$

Write the negative sign in front of the fraction.

$$= -\frac{23}{36}$$

Subtract: $\dfrac{2}{3} - \left(-\dfrac{4}{5}\right)$

Rewrite subtraction as addition of the opposite.

$$\dfrac{2}{3} - \left(-\dfrac{4}{5}\right) = \dfrac{2}{3} + \dfrac{4}{5}$$

Write the fractions as equivalent fractions with a common denominator.

$$= \dfrac{10}{15} + \dfrac{12}{15}$$

$$= \dfrac{10 + 12}{15}$$

$$= \dfrac{22}{15} = 1\dfrac{7}{15}$$

To subtract mixed numbers when borrowing is not necessary, subtract the fractional parts and then subtract the whole numbers.

Find the difference between $5\dfrac{8}{9}$ and $2\dfrac{5}{6}$.   The LCM of 9 and 6 is 18.

Write the fractions as equivalent fractions with the LCM as the common denominator.

$$5\dfrac{8}{9} - 2\dfrac{5}{6} = 5\dfrac{16}{18} - 2\dfrac{15}{18}$$

Subtract the fractional parts and subtract the whole numbers.

$$= 3\dfrac{1}{18}$$

As in subtraction with whole numbers, subtraction of mixed numbers may involve borrowing.

Subtract: $7 - 4\dfrac{2}{3}$

Borrow 1 from 7. Write the 1 as a fraction with the same denominator as is in the fractional part of the mixed number (3).

$$7 - 4\dfrac{2}{3} = 6\dfrac{3}{3} - 4\dfrac{2}{3}$$

Note: $7 = 6 + 1 = 6 + \dfrac{3}{3} = 6\dfrac{3}{3}$

Subtract the fractional parts and subtract the whole numbers.

$$= 2\dfrac{1}{3}$$

Subtract: $9\dfrac{1}{8} - 2\dfrac{5}{6}$

Write the fractions as equivalent fractions with a common denominator.

$$9\dfrac{1}{8} - 2\dfrac{5}{6} = 9\dfrac{3}{24} - 2\dfrac{20}{24}$$

$3 < 20$. Borrow 1 from 9. Add the 1 to $\dfrac{3}{24}$.

Note: $9\dfrac{3}{24} = 9 + \dfrac{3}{24} = 8 + 1 + \dfrac{3}{24}$

$$= 8 + \dfrac{24}{24} + \dfrac{3}{24} = 8 + \dfrac{27}{24} = 8\dfrac{27}{24}$$

$$= 8\dfrac{27}{24} - 2\dfrac{20}{24}$$

Subtract.

$$= 6\dfrac{7}{24}$$

Evaluate $x - y$ when $x = 7\frac{2}{9}$ and $y = 3\frac{5}{12}$.

$$x - y$$

Replace x with $7\frac{2}{9}$ and y with $3\frac{5}{12}$.

$$7\frac{2}{9} - 3\frac{5}{12}$$

Write the fractions as equivalent fractions with a common denominator.

$$= 7\frac{8}{36} - 3\frac{15}{36}$$

$8 < 15$. Borrow 1 from 7. Add the 1 to $\frac{8}{36}$.

Note: $7\frac{8}{36} = 6 + \frac{36}{36} + \frac{8}{36} = 6\frac{44}{36}$

$$= 6\frac{44}{36} - 3\frac{15}{36}$$

Subtract.

$$= 3\frac{29}{36}$$

Example 6

Subtract: $-\frac{5}{6} - \left(-\frac{3}{8}\right)$

Solution

$-\frac{5}{6} - \left(-\frac{3}{8}\right) = -\frac{5}{6} + \frac{3}{8} = \frac{-20}{24} + \frac{9}{24}$

$\qquad = \frac{-20 + 9}{24}$

$\qquad = \frac{-11}{24} = -\frac{11}{24}$

You Try It 6

Subtract: $-\frac{5}{6} - \frac{7}{9}$

Your Solution

Example 7

Find the difference between $8\frac{5}{6}$ and $2\frac{3}{4}$.

Solution

$8\frac{5}{6} - 2\frac{3}{4} = 8\frac{10}{12} - 2\frac{9}{12} = 6\frac{1}{12}$

You Try It 7

Find the difference between $9\frac{7}{8}$ and $5\frac{2}{3}$.

Your Solution

Example 8

Subtract: $7 - 3\frac{5}{13}$

Solution

$7 - 3\frac{5}{13} = 6\frac{13}{13} - 3\frac{5}{13} = 3\frac{8}{13}$

You Try It 8

Subtract: $6 - 4\frac{2}{11}$

Your Solution

Solutions on p. A11

Example 9

Is $\frac{3}{8}$ a solution of the equation $\frac{2}{3} = w - \frac{5}{6}$?

Solution

$$\frac{2}{3} = w - \frac{5}{6}$$

$\frac{2}{3}$	$\frac{3}{8} - \frac{5}{6}$
$\frac{2}{3}$	$\frac{9}{24} - \frac{20}{24}$
$\frac{2}{3}$	$\frac{-11}{24}$

$$\frac{2}{3} \neq -\frac{11}{24}$$

No, $\frac{3}{8}$ is not a solution of the equation.

You Try It 9

Is $-\frac{1}{4}$ a solution of the equation $\frac{2}{3} - v = \frac{11}{12}$?

Your Solution

Solution on p. A12

OBJECTIVE C

Applications and formulas

Example 10

The length of a regulation NCAA football must be no less than $10\frac{7}{8}$ in. and no more than $11\frac{7}{16}$ in. What is the difference between the minimum and maximum lengths of an NCAA regulation football?

Strategy

To find the difference, subtract the minimum length $\left(10\frac{7}{8}\right)$ from the maximum length $\left(11\frac{7}{16}\right)$.

Solution

$$11\frac{7}{16} - 10\frac{7}{8} = 11\frac{7}{16} - 10\frac{14}{16} = 10\frac{23}{16} - 10\frac{14}{16} = \frac{9}{16}$$

The difference is $\frac{9}{16}$ in.

You Try It 10

In the 1996 presidential election, approximately $\frac{49}{100}$ of the voters voted for Bill Clinton, $\frac{41}{100}$ voted for Bob Dole, and $\frac{2}{25}$ voted for Ross Perot. What fraction of the voters did not vote for any of these three candidates?

Your Strategy

Your Solution

Solution on p. A12

Example 11

A chef has $1\frac{1}{2}$ c of granulated sugar and wants to make both of the recipes below. Does the chef have enough granulated sugar for both recipes?

Creme Brulee

$\frac{2}{3}$ c cream

$\frac{2}{3}$ c milk

$\frac{2}{3}$ c granulated sugar

1 vanilla bean

6 egg yokes

Chocolate Chip Cookies

$2\frac{1}{4}$ c flour

1 tsp salt

1 tsp baking soda

$\frac{3}{4}$ c granulated sugar

$\frac{3}{4}$ c brown sugar

8 oz chocolate chips

$\frac{1}{4}$ tsp vanilla

$\frac{1}{2}$ lb butter

2 eggs

You Try It 11

The dimensions of finished lumber are $\frac{1}{4}$ of an inch less in thickness and $\frac{1}{2}$ of an inch less in width than the given dimensions. For instance, a finished 2 by 4 is actually $1\frac{3}{4}$ inches thick and $3\frac{1}{2}$ inches wide. Nail sizes are measured in *pennys*. A 2-penny nail is 1 inch long. The length of the nail increases by $\frac{1}{4}$ of an inch for each 1-penny increase in size. For instance, a 3-penny nail is $1\frac{1}{4}$ in. long; a 4-penny nail is $1\frac{1}{2}$ in. long. What size penny nail is needed to nail four 1 by 6 pieces of lumber together so that the nail extends $\frac{1}{2}$ in. into the fourth board?

Strategy

To determine if the chef has enough sugar:

▶ Add the number of cups of sugar from the creme brulee recipe $\left(\frac{2}{3}\right)$ to the number of cups of sugar from the chocolate chip cookies recipe $\left(\frac{3}{4}\right)$.

▶ Compare the sum to the number of cups of sugar the chef has $\left(1\frac{1}{2}\right)$.

Your Strategy

Solution

$$\frac{2}{3} + \frac{3}{4} = \frac{8}{12} + \frac{9}{12} = \frac{17}{12} = 1\frac{5}{12}$$

$$1\frac{1}{2} = 1\frac{6}{12}$$

$$1\frac{6}{12} > 1\frac{5}{12}$$

The chef has enough sugar for both recipes.

Your Solution

Solution on p. A12

3.3 EXERCISES

OBJECTIVE A

Add.

1. $\dfrac{4}{11} + \dfrac{5}{11}$

2. $\dfrac{3}{7} + \dfrac{2}{7}$

3. $\dfrac{2}{3} + \dfrac{1}{3}$

4. $\dfrac{1}{2} + \dfrac{1}{2}$

5. $\dfrac{5}{6} + \dfrac{5}{6}$

6. $\dfrac{3}{8} + \dfrac{7}{8}$

7. $\dfrac{7}{18} + \dfrac{13}{18} + \dfrac{1}{18}$

8. $\dfrac{8}{15} + \dfrac{2}{15} + \dfrac{11}{15}$

9. $\dfrac{7}{b} + \dfrac{9}{b}$

10. $\dfrac{3}{y} + \dfrac{6}{y}$

11. $\dfrac{5}{c} + \dfrac{4}{c}$

12. $\dfrac{2}{a} + \dfrac{8}{a}$

13. $\dfrac{1}{x} + \dfrac{4}{x} + \dfrac{6}{x}$

14. $\dfrac{8}{n} + \dfrac{5}{n} + \dfrac{3}{n}$

15. $\dfrac{1}{4} + \dfrac{2}{3}$

16. $\dfrac{2}{3} + \dfrac{1}{2}$

17. $\dfrac{7}{15} + \dfrac{9}{20}$

18. $\dfrac{4}{9} + \dfrac{1}{6}$

19. $\dfrac{2}{3} + \dfrac{1}{12} + \dfrac{5}{6}$

20. $\dfrac{3}{8} + \dfrac{1}{2} + \dfrac{5}{12}$

21. $\dfrac{7}{12} + \dfrac{3}{4} + \dfrac{4}{5}$

22. $\dfrac{7}{11} + \dfrac{1}{2} + \dfrac{5}{6}$

23. $-\dfrac{3}{4} + \dfrac{2}{3}$

24. $-\dfrac{7}{12} + \dfrac{5}{8}$

25. $\dfrac{2}{5} + \left(-\dfrac{11}{15}\right)$

26. $\dfrac{1}{4} + \left(-\dfrac{1}{7}\right)$

27. $\dfrac{3}{8} + \left(-\dfrac{1}{2}\right) + \dfrac{7}{12}$

28. $-\dfrac{7}{12} + \dfrac{2}{3} + \left(-\dfrac{4}{5}\right)$

29. $\dfrac{2}{3} + \left(-\dfrac{5}{6}\right) + \dfrac{1}{4}$

30. $-\dfrac{5}{8} + \dfrac{3}{4} + \dfrac{1}{2}$

31. $8 + 7\dfrac{2}{3}$

32. $6 + 9\dfrac{3}{5}$

Add.

33. $2\frac{1}{6} + 3\frac{1}{2}$

34. $1\frac{3}{10} + 4\frac{3}{5}$

35. $8\frac{3}{5} + 6\frac{9}{20}$

36. $7\frac{5}{12} + 3\frac{7}{9}$

37. $5\frac{5}{12} + 4\frac{7}{9}$

38. $2\frac{11}{12} + 3\frac{7}{15}$

39. $2\frac{1}{4} + 3\frac{1}{2} + 1\frac{2}{3}$

40. $1\frac{2}{3} + 2\frac{5}{6} + 4\frac{7}{9}$

Solve.

41. What is $-\frac{5}{6}$ added to $\frac{4}{9}$?

42. What is $\frac{7}{12}$ added to $-\frac{11}{16}$?

43. Find the total of $\frac{2}{7}, \frac{3}{14}$, and $\frac{1}{4}$.

44. Find the total of $\frac{1}{3}, \frac{5}{18}$, and $\frac{2}{9}$.

45. What is $-\frac{2}{3}$ more than $-\frac{5}{6}$?

46. What is $-\frac{7}{12}$ more than $-\frac{5}{9}$?

47. Find $3\frac{7}{12}$ plus $2\frac{5}{8}$.

48. Find $5\frac{4}{9}$ plus $6\frac{5}{6}$.

49. Find $\frac{7}{8}$ increased by $1\frac{1}{3}$.

50. Find the sum of $7\frac{11}{15}, 2\frac{7}{10}$, and $5\frac{2}{5}$.

Evaluate the variable expression $x + y$ for the given values of x and y.

51. $x = \frac{3}{5}, y = \frac{4}{5}$

52. $x = \frac{5}{8}, y = \frac{3}{8}$

53. $x = \frac{2}{3}, y = -\frac{3}{4}$

54. $x = -\frac{3}{8}, y = \frac{2}{9}$

55. $x = \frac{5}{6}, y = \frac{8}{9}$

56. $x = \frac{3}{10}, y = -\frac{7}{15}$

57. $x = -\frac{5}{8}, y = -\frac{1}{6}$

58. $x = -\frac{3}{8}, y = -\frac{5}{6}$

Evaluate the variable expression $x + y + z$ for the given values of x, y, and z.

59. $x = \dfrac{3}{8}, y = \dfrac{1}{4}, z = \dfrac{7}{12}$

60. $x = \dfrac{5}{6}, y = \dfrac{2}{3}, z = \dfrac{7}{24}$

61. $x = 1\dfrac{1}{2}, y = 3\dfrac{3}{4}, z = 6\dfrac{5}{12}$

62. $x = 7\dfrac{2}{3}, y = 2\dfrac{5}{6}, z = 5\dfrac{4}{9}$

63. $x = 4\dfrac{3}{5}, y = 8\dfrac{7}{10}, z = 1\dfrac{9}{20}$

64. $x = 2\dfrac{3}{14}, y = 5\dfrac{5}{7}, z = 3\dfrac{1}{2}$

65. Is $-\dfrac{3}{5}$ a solution of the equation $z + \dfrac{1}{4} = -\dfrac{7}{20}$?

66. Is $\dfrac{3}{8}$ a solution of the equation $\dfrac{3}{4} = t + \dfrac{3}{8}$?

67. Is $-\dfrac{5}{6}$ a solution of the equation $\dfrac{1}{4} + x = -\dfrac{7}{12}$?

68. Is $-\dfrac{4}{5}$ a solution of the equation $0 = q + \dfrac{4}{5}$?

The figure at the right shows the approximate contribution in 1995 of various segments to J. C. Penney's total sales. Use this graph for Exercises 69 and 70.

69. Business   What fractional part of total sales in 1995 was due to women's and men's apparel?

70. Business   What fractional part of total sales in 1995 was due to women's, men's and children's apparel?

Home furnishings

Children's apparel

Men's apparel

Women's apparel

Total Sales for J. C. Penney, 1995
Source: J. C. Penney Annual Report

OBJECTIVE B

Subtract.

71. $\dfrac{7}{12} - \dfrac{5}{12}$

72. $\dfrac{17}{20} - \dfrac{9}{20}$

73. $\dfrac{11}{24} - \dfrac{7}{24}$

74. $\dfrac{39}{48} - \dfrac{23}{48}$

75. $\dfrac{8}{d} - \dfrac{3}{d}$

76. $\dfrac{12}{y} - \dfrac{7}{y}$

77. $\dfrac{5}{n} - \dfrac{10}{n}$

78. $\dfrac{6}{c} - \dfrac{13}{c}$

Subtract.

79. $\dfrac{3}{7} - \dfrac{5}{14}$

80. $\dfrac{7}{8} - \dfrac{5}{16}$

81. $\dfrac{2}{3} - \dfrac{1}{6}$

82. $\dfrac{5}{21} - \dfrac{1}{6}$

83. $\dfrac{11}{12} - \dfrac{2}{3}$

84. $\dfrac{9}{20} - \dfrac{1}{30}$

85. $-\dfrac{1}{2} - \dfrac{3}{8}$

86. $-\dfrac{5}{6} - \dfrac{1}{9}$

87. $-\dfrac{3}{10} - \dfrac{4}{5}$

88. $-\dfrac{7}{15} - \dfrac{3}{10}$

89. $-\dfrac{5}{12} - \left(-\dfrac{2}{3}\right)$

90. $-\dfrac{3}{10} - \left(-\dfrac{5}{6}\right)$

91. $-\dfrac{5}{9} - \left(-\dfrac{11}{12}\right)$

92. $-\dfrac{5}{8} - \left(-\dfrac{7}{12}\right)$

93. $4\dfrac{11}{18} - 2\dfrac{5}{18}$

94. $3\dfrac{7}{12} - 1\dfrac{1}{12}$

95. $8\dfrac{3}{4} - 2$

96. $6\dfrac{5}{9} - 4$

97. $8\dfrac{5}{6} - 7\dfrac{3}{4}$

98. $5\dfrac{7}{8} - 3\dfrac{2}{3}$

99. $7 - 3\dfrac{5}{8}$

100. $6 - 2\dfrac{4}{5}$

101. $10 - 4\dfrac{8}{9}$

102. $5 - 2\dfrac{7}{18}$

103. $7\dfrac{3}{8} - 4\dfrac{5}{8}$

104. $11\dfrac{1}{6} - 8\dfrac{5}{6}$

105. $12\dfrac{5}{12} - 10\dfrac{17}{24}$

106. $16\dfrac{1}{3} - 11\dfrac{5}{12}$

107. $6\dfrac{2}{3} - 1\dfrac{7}{8}$

108. $7\dfrac{7}{12} - 2\dfrac{5}{6}$

109. $10\dfrac{2}{5} - 8\dfrac{7}{10}$

110. $5\dfrac{5}{6} - 4\dfrac{7}{8}$

Solve.

111. What is $-\frac{7}{12}$ minus $\frac{7}{9}$?

112. What is $\frac{3}{5}$ decreased by $-\frac{7}{10}$?

113. What is $-\frac{2}{3}$ less than $-\frac{7}{8}$?

114. Find the difference between $-\frac{1}{6}$ and $-\frac{8}{9}$.

115. Find 8 less $1\frac{7}{12}$.

116. Find 9 minus $5\frac{3}{20}$.

Evaluate the variable expression $x - y$ for the given values of x and y.

117. $x = \frac{8}{9}, y = \frac{5}{9}$

118. $x = \frac{5}{6}, y = \frac{1}{6}$

119. $x = -\frac{11}{12}, y = \frac{5}{12}$

120. $x = -\frac{15}{16}, y = \frac{5}{16}$

121. $x = -\frac{2}{3}, y = -\frac{3}{4}$

122. $x = -\frac{5}{12}, y = -\frac{5}{9}$

123. $x = -\frac{3}{10}, y = -\frac{7}{15}$

124. $x = -\frac{5}{6}, y = -\frac{2}{15}$

125. $x = 5\frac{7}{9}, y = 4\frac{2}{3}$

126. $x = 9\frac{5}{8}, y = 2\frac{3}{16}$

127. $x = 7\frac{9}{10}, y = 3\frac{1}{2}$

128. $x = 6\frac{4}{9}, y = 1\frac{1}{6}$

129. $x = 5, y = 2\frac{7}{9}$

130. $x = 8, y = 4\frac{5}{6}$

131. $x = 10\frac{1}{2}, y = 5\frac{7}{12}$

132. $x = 9\frac{2}{15}, y = 6\frac{11}{15}$

133. Is $-\frac{3}{4}$ a solution of the equation $\frac{4}{5} = \frac{31}{20} - y$?

134. Is $\frac{5}{8}$ a solution of the equation $-\frac{1}{4} = x - \frac{7}{8}$?

135. Is $-\frac{3}{5}$ a solution of the equation $x - \frac{1}{4} = -\frac{17}{20}$?

136. Is $-\frac{2}{3}$ a solution of the equation $\frac{2}{3} - x = 0$?

The approximate market share of various manufacturers of cold breakfast cereal is represented in the figure on the right. Use this graph for Exercises 137 and 138.

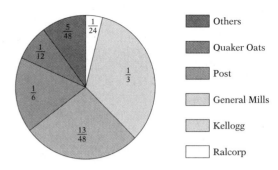

Distribution of Cold Cereal Sales, June 1996
Source: North Country Times, 8/15/96

137. *Business* What is the difference between the market share of Post and Quaker Oats?

138. *Business* How much greater is Kellogg's market share than that of General Mills?

OBJECTIVE C

Solve.

139. *Real Estate* You purchased $3\frac{1}{4}$ acres of land and then sold $1\frac{1}{2}$ acres of the property. How many acres of the property do you own now?

140. *Consumerism* You purchased two lobsters, one weighing $2\frac{1}{2}$ lb and one weighing $3\frac{1}{4}$ lb. Find the total weight of the two lobsters.

141. *Carpentry* A $2\frac{3}{4}$-foot piece is cut from a 6-foot board. Find the length of the remaining piece of board.

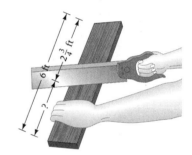

142. *Community Service* You are required to contribute 20 h of community service to the town in which your college is located. After you have contributed $12\frac{1}{4}$ h, how many more hours of community service are still required of you?

143. *Horse Racing* The 3-year-olds in the Kentucky Derby run $1\frac{1}{4}$ mi. The horses in the Belmont Stakes run $1\frac{1}{2}$ mi, and they run $1\frac{3}{16}$ mi in the Preakness Stakes. How much farther do the horses run in the Kentucky Derby than in the Preakness Stakes? How much farther do they run in the Belmont Stakes than in the Preakness Stakes?

144. *Sports* In the running high jump in the 1948 Summer Olympic Games, Alice Coachman's distance was $66\frac{1}{8}$ in. In the same event, Mildred McDaniel jumped $69\frac{1}{4}$ in. in the 1956 Summer Olympics, and Louise Ritter jumped 80 in. in the 1988 Olympic Games. Find the difference between Ritter's distance and Coachman's distance. Find the difference between Ritter's distance and McDaniel's distance.

145. *Sports* A boxer is put on a diet to gain 15 lb in four weeks. The boxer gains $4\frac{1}{2}$ lb the first week and $3\frac{3}{4}$ lb the second week. How much weight must the boxer gain during the third and fourth weeks in order to gain a total of 15 lb?

Solve.

146. *Health*   A patient is put on a diet to lose 25 lb in three months. The patient loses $10\frac{1}{2}$ lb during the first month and $9\frac{3}{8}$ lb during the second month. Find the amount of weight the patient must lose the third month in order to achieve the goal.

147. *Wages*   A student worked $4\frac{1}{3}$ h, 5 h, and $3\frac{2}{3}$ h this week at a part-time job. The student is paid $7 an hour. How much did the student earn this week?

148. *Wages*   An electrician worked $2\frac{3}{4}$ h of overtime on Tuesday, $1\frac{1}{2}$ h of overtime on Thursday, and $5\frac{3}{4}$ h of overtime on Saturday. If the electrician earns an overtime hourly wage of $24 an hour, how much overtime pay does the electrician earn for the week?

149. *Industry*   Two inlet pipes are being used to fill an oil tank. After one hour, the larger pipe has filled $\frac{2}{5}$ of the tank and the smaller pipe has filled $\frac{1}{4}$ of the tank. How much of the tank remains to be filled? Can the two pipes complete the job within another hour?

150. *Construction*   A roofer and an apprentice are roofing a newly constructed house. In one day, the roofer completes $\frac{1}{3}$ of the job and the apprentice completes $\frac{1}{4}$ of the job. How much of the job remains to be done? Working at the same rate, can the roofer and the apprentice complete the job in one more day?

The table at the right shows a 52-week high and low in 1996 of five automotive stocks. Use this table for Exercises 151 and 152.

Company	High	Low
Ford	$37\frac{1}{8}$	$27\frac{1}{2}$
GM	$58\frac{1}{8}$	$43\frac{3}{4}$
Chrysler	35	$23\frac{1}{2}$
Honda	$55\frac{1}{4}$	$31\frac{3}{8}$
Mercedes	56	$47\frac{3}{8}$

151. *Investments*   (a) What was the difference between the high and the low for Chrysler's stock? (b) How much greater was GM's high than Ford's high?

152. *Investments*   Using estimation, determine which company had (a) the smallest difference between the high and the low, and (b) the greatest difference between the high and the low.

153. *Investments*   Find the gain per share of a stock that an investor purchased for $14\frac{3}{8}$ per share and later sold for $18\frac{1}{4}$ per share. Use the formula $G = S - P$, where G is the profit or gain per share of stock, S is the selling price, and P is the purchase price.

Solve.

154. *Investments* Find the gain per share of a stock that an investor purchased for $23\frac{1}{8}$ per share and later sold for $29\frac{3}{4}$ per share. Use the formula $G = S - P$, where G is the profit or gain per share of stock, S is the selling price, and P is the purchase price.

155. *Investments* What is the loss per share of a stock that was purchased for $9\frac{1}{4}$ per share and was later sold for $6\frac{7}{8}$ per share? Use the formula $L = P - S$, where L is the loss per share of stock, P is the purchase price, and S is the selling price.

156. *Investments* What is the loss per share of a stock that was purchased for $37\frac{3}{4}$ per share and was later sold for $31\frac{5}{8}$ per share? Use the formula $L = P - S$, where L is the loss per share of stock, P is the purchase price, and S is the selling price.

CRITICAL THINKING

157. The figure at the right is divided into 5 parts. Is each part of the figure $\frac{1}{5}$ of the figure? Why or why not?

158. Draw a diagram that illustrates the addition of two fractions with the same denominator.

159. Use the diagram at the right to illustrate the sum of $\frac{1}{8}$ and $\frac{5}{6}$. Why does the figure contain 24 squares? Would it be possible to illustrate the sum of $\frac{1}{8}$ and $\frac{5}{6}$ if there were 48 squares in the figure? What if there were 16 squares? Make a list of the possible number of squares that could be used to illustrate the sum of $\frac{1}{8}$ and $\frac{5}{6}$.

160. A researcher completed a study of the ages of students at a college. The fraction of the total number of students enrolled in the college in various age groups was then recorded in the figure at the right. Are the results that are displayed in the circle graph possible? Explain your answer.

161. A local humane society reported that $\frac{3}{5}$ of the households in the city owned some type of pet. The report went on to say that $\frac{1}{6}$ of the households had a bird, $\frac{2}{5}$ had a dog, $\frac{3}{10}$ had a cat, and $\frac{1}{20}$ of the households had a different animal as a pet. The sum of $\frac{1}{6}$, $\frac{2}{5}$, $\frac{3}{10}$, and $\frac{1}{20}$ is $\frac{11}{12}$, which is more than $\frac{3}{5}$. Is this possible? Explain your answer.

162. Consult a history of mathematics text and prepare a report on unit fractions and how they were used by early Egyptians.

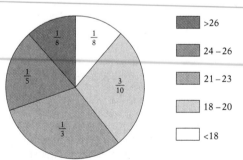

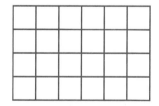

Age Distribution of Enrolled College Students

>26
24 – 26
21 – 23
18 – 20
<18

SECTION 3.4 **Multiplication and Division of Fractions**

OBJECTIVE A

Multiplication of fractions

To multiply two fractions, multiply the numerators and multiply the denominators.

Multiplication of Fractions

The product of two fractions is the product of the numerators over the product of the denominators.

$$\frac{a}{b} \cdot \frac{c}{d} = \frac{ac}{bd}, \qquad \text{where} \quad b \neq 0 \quad \text{and} \quad d \neq 0$$

Note that fractions do not need to have the same denominator in order to be multiplied.

After multiplying two fractions, write the product in simplest form.

Multiply: $\frac{2}{5} \cdot \frac{1}{3}$

Multiply the numerators.
Multiply the denominators.
$$\frac{2}{5} \cdot \frac{1}{3} = \frac{2 \cdot 1}{5 \cdot 3} = \frac{2}{15}$$

The product $\frac{2}{5} \cdot \frac{1}{3}$ can be read "$\frac{2}{5}$ times $\frac{1}{3}$" or "$\frac{2}{5}$ of $\frac{1}{3}$."

Reading the times sign as "of" is useful in diagraming the product of two fractions.

$\frac{1}{3}$ of the bar at the right is shaded.

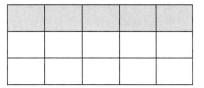

Shade $\frac{2}{5}$ of the $\frac{1}{3}$ already shaded.

$\frac{2}{15}$ of the bar is now shaded.

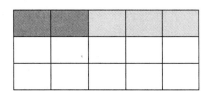

$$\frac{2}{5} \text{ of } \frac{1}{3} = \frac{2}{5} \cdot \frac{1}{3} = \frac{2}{15}$$

If a is a natural number, then $\frac{1}{a}$ is called the **reciprocal** or **multiplicative inverse** of a. Note that $a \cdot \frac{1}{a} = \frac{a}{1} \cdot \frac{1}{a} = \frac{a}{a} = 1$.

The product of a number and its multiplicative inverse is 1.
$$\frac{1}{8} \cdot 8 = 8 \cdot \frac{1}{8} = 1$$

Multiply: $\dfrac{3}{8} \cdot \dfrac{4}{9}$

Multiply the numerators. Multiply the denominators.	$\dfrac{3}{8} \cdot \dfrac{4}{9} = \dfrac{3 \cdot 4}{8 \cdot 9}$
Express the fraction in simplest form by first writing the prime factorization of each number.	$= \dfrac{3 \cdot 2 \cdot 2}{2 \cdot 2 \cdot 2 \cdot 3 \cdot 3}$
Divide by the common factors and write the product in simplest form.	$= \dfrac{1}{6}$

The sign rules for multiplying positive and negative fractions are the same rules used to multiply integers.

The product of two numbers with the same sign is positive.
The product of two numbers with different signs is negative.

Multiply: $-\dfrac{3}{4} \cdot \dfrac{8}{15}$

The signs are different. The product is negative.	$-\dfrac{3}{4} \cdot \dfrac{8}{15} = -\left(\dfrac{3}{4} \cdot \dfrac{8}{15}\right)$
Multiply the numerators. Multiply the denominators.	$= -\dfrac{3 \cdot 8}{4 \cdot 15}$
Write the product in simplest form.	$= -\dfrac{3 \cdot 2 \cdot 2 \cdot 2}{2 \cdot 2 \cdot 3 \cdot 5}$
	$= -\dfrac{2}{5}$

Multiply: $-\dfrac{3}{8}\left(-\dfrac{2}{5}\right)\left(-\dfrac{10}{21}\right)$

$$-\dfrac{3}{8}\left(-\dfrac{2}{5}\right)\left(-\dfrac{10}{21}\right)$$

Use the Order of Operations Agreement. Multiply the first two fractions. The product is positive.	$= \left(\dfrac{3}{8} \cdot \dfrac{2}{5}\right)\left(-\dfrac{10}{21}\right)$
The product of the first two fractions and the third fraction is negative.	$= -\left(\dfrac{3}{8} \cdot \dfrac{2}{5} \cdot \dfrac{10}{21}\right)$
Multiply the numerators. Multiply the denominators.	$= -\dfrac{3 \cdot 2 \cdot 10}{8 \cdot 5 \cdot 21}$
Write the product in simplest form.	$= -\dfrac{3 \cdot 2 \cdot 2 \cdot 5}{2 \cdot 2 \cdot 2 \cdot 5 \cdot 3 \cdot 7}$
	$= -\dfrac{1}{14}$

Thus, the product of three negative fractions is negative. We can modify the rule for multiplying positive and negative fractions to say that **the product of an odd number of negative fractions is negative and the product of an even number of negative fractions is positive.**

POINT OF INTEREST

Try this: What is the result if you take one-third of a half-dozen and add to it one-fourth of the product of the result and 8?

To multiply a whole number by a fraction or a mixed number, first write the whole number as a fraction with a denominator of 1.

Multiply: $3 \cdot \dfrac{5}{8}$

Write the whole number 3 as the fraction $\dfrac{3}{1}$.

$$3 \cdot \frac{5}{8} = \frac{3}{1} \cdot \frac{5}{8}$$

Multiply the fractions.
There are no common factors in the numerator and denominator.

$$= \frac{3 \cdot 5}{1 \cdot 8}$$

Write the improper fraction as a mixed number.

$$= \frac{15}{8} = 1\frac{7}{8}$$

Multiply: $\dfrac{x}{7} \cdot \dfrac{y}{5}$

Multiply the numerators.
Multiply the denominators.

$$\frac{x}{7} \cdot \frac{y}{5} = \frac{x \cdot y}{7 \cdot 5}$$

Write the product in simplest form.

$$= \frac{xy}{35}$$

When a factor is a mixed number, first write the mixed number as an improper fraction. Then multiply.

Find the product of $-4\dfrac{1}{6}$ and $2\dfrac{7}{10}$.

The signs are different.
The product is negative.

$$-4\frac{1}{6} \cdot 2\frac{7}{10} = -\left(4\frac{1}{6} \cdot 2\frac{7}{10}\right)$$

Write each mixed number as an improper fraction.

$$= -\left(\frac{25}{6} \cdot \frac{27}{10}\right)$$

Multiply the fractions.

$$= -\frac{25 \cdot 27}{6 \cdot 10}$$

$$= -\frac{5 \cdot 5 \cdot 3 \cdot 3 \cdot 3}{2 \cdot 3 \cdot 2 \cdot 5}$$

Write the product in simplest form.

$$= -\frac{45}{4} = -11\frac{1}{4}$$

Is $-\frac{2}{3}$ a solution of the equation $\frac{3}{4}x = -\frac{1}{2}$?

$$\frac{3}{4}x = -\frac{1}{2}$$

Replace x by $-\frac{2}{3}$ and then simplify.

$$
\begin{array}{c|c}
\frac{3}{4}\left(-\frac{2}{3}\right) & -\frac{1}{2} \\[2mm]
-\dfrac{3 \cdot 2}{4 \cdot 3} & -\dfrac{1}{2} \\[2mm]
-\dfrac{3 \cdot 2}{2 \cdot 2 \cdot 3} & -\dfrac{1}{2}
\end{array}
$$

The results are equal.

$$-\frac{1}{2} = -\frac{1}{2}$$

Yes, $-\frac{2}{3}$ is a solution of the equation.

Example 1   Multiply: $\frac{7}{9} \cdot \frac{3}{14} \cdot \frac{2}{5}$

Solution $\dfrac{7}{9} \cdot \dfrac{3}{14} \cdot \dfrac{2}{5} = \dfrac{7 \cdot 3 \cdot 2}{9 \cdot 14 \cdot 5}$

$$= \frac{7 \cdot 3 \cdot 2}{3 \cdot 3 \cdot 2 \cdot 7 \cdot 5} = \frac{1}{15}$$

You Try It 1   Multiply: $\frac{5}{12} \cdot \frac{9}{35} \cdot \frac{7}{8}$

Your Solution

Example 2   Multiply: $\frac{6}{x} \cdot \frac{8}{y}$

Solution $\dfrac{6}{x} \cdot \dfrac{8}{y} = \dfrac{6 \cdot 8}{x \cdot y}$

$$= \frac{48}{xy}$$

You Try It 2   Multiply: $\frac{y}{10} \cdot \frac{z}{7}$

Your Solution

Example 3   Multiply: $-\frac{3}{4}\left(\frac{1}{2}\right)\left(-\frac{8}{9}\right)$

Solution $-\dfrac{3}{4}\left(\dfrac{1}{2}\right)\left(-\dfrac{8}{9}\right)$ The product of two negative fractions is positive.

$$= \frac{3}{4} \cdot \frac{1}{2} \cdot \frac{8}{9}$$

$$= \frac{3 \cdot 1 \cdot 8}{4 \cdot 2 \cdot 9}$$

$$= \frac{3 \cdot 1 \cdot 2 \cdot 2 \cdot 2}{2 \cdot 2 \cdot 2 \cdot 3 \cdot 3} = \frac{1}{3}$$

You Try It 3   Multiply: $-\frac{1}{3}\left(-\frac{5}{12}\right)\left(\frac{8}{15}\right)$

Your Solution

Solutions on p. A12

Example 4 What is the product of $\frac{7}{12}$ and 4?

Solution $\frac{7}{12} \cdot 4 = \frac{7}{12} \cdot \frac{4}{1}$

$$= \frac{7 \cdot 4}{12 \cdot 1}$$

$$= \frac{7 \cdot 2 \cdot 2}{2 \cdot 2 \cdot 3 \cdot 1}$$

$$= \frac{7}{3}$$

$$= 2\frac{1}{3}$$

You Try It 4 Find the product of $\frac{8}{9}$ and 6.

Your Solution

Example 5 Multiply: $-7\frac{1}{2} \cdot 4\frac{2}{5}$

Solution $-7\frac{1}{2} \cdot 4\frac{2}{5} = -\left(\frac{15}{2} \cdot \frac{22}{5}\right)$

$$= -\frac{15 \cdot 22}{2 \cdot 5}$$

$$= -\frac{3 \cdot 5 \cdot 2 \cdot 11}{2 \cdot 5}$$

$$= -\frac{33}{1} = -33$$

You Try It 5 Multiply: $3\frac{6}{7} \cdot 2\frac{4}{9}$

Your Solution

Example 6 Evaluate the variable expression xy when $x = 1\frac{4}{5}$ and $y = -\frac{5}{6}$.

Solution xy

$$1\frac{4}{5}\left(-\frac{5}{6}\right) = -\left(\frac{9}{5} \cdot \frac{5}{6}\right)$$

$$= -\frac{9 \cdot 5}{5 \cdot 6}$$

$$= -\frac{3 \cdot 3 \cdot 5}{5 \cdot 2 \cdot 3}$$

$$= -\frac{3}{2} = -1\frac{1}{2}$$

You Try It 6 Evaluate the variable expression xy when $x = 5\frac{1}{8}$ and $y = \frac{2}{3}$.

Your Solution

Solutions on pp. A12–A13

OBJECTIVE B
Division of fractions

The **reciprocal** of a fraction is the fraction with the numerator and denominator interchanged.

$$\text{The reciprocal of } \frac{3}{4} \text{ is } \frac{4}{3}.$$

$$\text{The reciprocal of } \frac{a}{b} \text{ is } \frac{b}{a}.$$

The process of interchanging the numerator and denominator of a fraction is called **inverting** the fraction.

To find the reciprocal of a whole number, first rewrite the whole number as a fraction with a denominator of 1. Then invert the fraction.

$$6 = \frac{6}{1}$$

The reciprocal of 6 is $\frac{1}{6}$.

Reciprocals are used to rewrite division problems as related multiplication problems. Look at the following two problems:

$$6 \div 2 = 3 \qquad\qquad 6 \cdot \frac{1}{2} = 3$$

6 divided by 2 **equals** 3.   6 times the reciprocal of 2 **equals** 3.

Division is defined as multiplication by the reciprocal. Therefore, "divided by 2" is the same as "times $\frac{1}{2}$." Fractions are divided by making this substitution.

Division of Fractions

To divide two fractions, multiply by the reciprocal of the divisor.

$$\frac{a}{b} \div \frac{c}{d} = \frac{a}{b} \cdot \frac{d}{c}, \qquad \text{where} \quad b \neq 0, \quad c \neq 0, \quad \text{and} \quad d \neq 0$$

Divide: $\frac{2}{5} \div \frac{3}{4}$

Rewrite the division as multiplication by the reciprocal.

$$\frac{2}{5} \div \frac{3}{4} = \frac{2}{5} \cdot \frac{4}{3}$$

Multiply the fractions.

$$= \frac{2 \cdot 4}{5 \cdot 3}$$

$$= \frac{2 \cdot 2 \cdot 2}{5 \cdot 3} = \frac{8}{15}$$

The sign rules for dividing positive and negative fractions are the same rules used to divide integers.

The quotient of two numbers with the same sign is positive.
The quotient of two numbers with different signs is negative.

POINT OF INTEREST

Try this: What number, when multiplied by the square of its reciprocal, equals one-fourth of the original number?

Simplify: $-\dfrac{7}{10} \div \left(-\dfrac{14}{15}\right)$

The signs are the same.
The quotient is positive.

$$-\frac{7}{10} \div \left(-\frac{14}{15}\right) = \frac{7}{10} \div \frac{14}{15}$$

Rewrite the division as multiplication by the reciprocal.

$$= \frac{7}{10} \cdot \frac{15}{14}$$

Multiply the fractions.

$$= \frac{7 \cdot 15}{10 \cdot 14}$$

$$= \frac{7 \cdot 3 \cdot 5}{2 \cdot 5 \cdot 2 \cdot 7}$$

$$= \frac{3}{4}$$

To divide a fraction and a whole number, first write the whole number as a fraction with a denominator of 1.

Find the quotient of $\dfrac{2}{3}$ and 4.

Write the whole number 4 as the fraction $\dfrac{4}{1}$.

$$\frac{2}{3} \div 4 = \frac{2}{3} \div \frac{4}{1}$$

Rewrite the division as multiplication by the reciprocal.

$$= \frac{2}{3} \cdot \frac{1}{4}$$

Multiply the fractions.

$$= \frac{2 \cdot 1}{3 \cdot 4}$$

$$= \frac{2 \cdot 1}{3 \cdot 2 \cdot 2} = \frac{1}{6}$$

When a number in a quotient is a mixed number, first write the mixed number as an improper fraction. Then divide the fractions.

Divide: $\dfrac{2}{3} \div 1\dfrac{1}{4}$

Write the mixed number $1\dfrac{1}{4}$ as an improper fraction.

$$\frac{2}{3} \div 1\frac{1}{4} = \frac{2}{3} \div \frac{5}{4}$$

Rewrite the division as multiplication by the reciprocal.

$$= \frac{2}{3} \cdot \frac{4}{5}$$

Multiply the fractions.

$$= \frac{2 \cdot 4}{3 \cdot 5} = \frac{8}{15}$$

POINT OF INTEREST

Try this: What number when multiplied by its reciprocal is equal to 1?

Evaluate $-x \div y$ when $x = -\frac{3}{8}$ and $y = -\frac{5}{12}$.

$$-x \div y$$

Replace x with $-\frac{3}{8}$ and y with $-\frac{5}{12}$.

$$-\left(-\frac{3}{8}\right) \div \left(-\frac{5}{12}\right)$$

Simplify $-\left(-\frac{3}{8}\right)$.

$$= \frac{3}{8} \div \left(-\frac{5}{12}\right)$$

The signs are different.
The quotient is negative.

$$= -\left(\frac{3}{8} \div \frac{5}{12}\right)$$

Rewrite the division as multiplication by the reciprocal.

$$= -\left(\frac{3}{8} \cdot \frac{12}{5}\right)$$

Multiply the fractions.

$$= -\frac{3 \cdot 12}{8 \cdot 5}$$

$$= -\frac{3 \cdot 2 \cdot 2 \cdot 3}{2 \cdot 2 \cdot 2 \cdot 5}$$

$$= -\frac{9}{10}$$

Example 7   Divide: $\frac{4}{5} \div \frac{8}{15}$

Solution $\frac{4}{5} \div \frac{8}{15} = \frac{4}{5} \cdot \frac{15}{8}$

$$= \frac{4 \cdot 15}{5 \cdot 8}$$

$$= \frac{2 \cdot 2 \cdot 3 \cdot 5}{5 \cdot 2 \cdot 2 \cdot 2}$$

$$= \frac{3}{2} = 1\frac{1}{2}$$

You Try It 7   Divide: $\frac{5}{6} \div \frac{10}{27}$

Your Solution

Example 8   Divide: $\frac{x}{2} \div \frac{y}{4}$

Solution $\frac{x}{2} \div \frac{y}{4} = \frac{x}{2} \cdot \frac{4}{y}$

$$= \frac{x \cdot 4}{2 \cdot y}$$

$$= \frac{x \cdot 2 \cdot 2}{2 \cdot y} = \frac{2x}{y}$$

You Try It 8   Divide: $\frac{x}{8} \div \frac{y}{6}$

Your Solution

Solutions on p. A13

Example 9　What is the quotient of 6 and $-\dfrac{3}{5}$?

Solution　$6 \div \left(-\dfrac{3}{5}\right) = -\left(\dfrac{6}{1} \div \dfrac{3}{5}\right)$

$$= -\left(\dfrac{6}{1} \cdot \dfrac{5}{3}\right)$$

$$= -\dfrac{6 \cdot 5}{1 \cdot 3}$$

$$= -\dfrac{2 \cdot 3 \cdot 5}{1 \cdot 3}$$

$$= -\dfrac{10}{1} = -10$$

You Try It 9　Find the quotient of 4 and $-\dfrac{6}{7}$.

Your Solution

Example 10　Divide: $3\dfrac{4}{15} \div 2\dfrac{1}{10}$

Solution　$3\dfrac{4}{15} \div 2\dfrac{1}{10} = \dfrac{49}{15} \div \dfrac{21}{10}$

$$= \dfrac{49}{15} \cdot \dfrac{10}{21}$$

$$= \dfrac{49 \cdot 10}{15 \cdot 21}$$

$$= \dfrac{7 \cdot 7 \cdot 2 \cdot 5}{3 \cdot 5 \cdot 3 \cdot 7}$$

$$= \dfrac{14}{9} = 1\dfrac{5}{9}$$

You Try It 10　Divide: $4\dfrac{3}{8} \div 3\dfrac{1}{2}$

Your Solution

Example 11　Evaluate $x \div y$ when $x = 3\dfrac{1}{8}$ and $y = 5$.

Solution　$x \div y$

$$3\dfrac{1}{8} \div 5 = \dfrac{25}{8} \div \dfrac{5}{1}$$

$$= \dfrac{25}{8} \cdot \dfrac{1}{5}$$

$$= \dfrac{25 \cdot 1}{8 \cdot 5}$$

$$= \dfrac{5 \cdot 5 \cdot 1}{2 \cdot 2 \cdot 2 \cdot 5} = \dfrac{5}{8}$$

You Try It 11　Evaluate $x \div y$ when $x = 2\dfrac{1}{4}$ and $y = 9$.

Your Solution

Solutions on p. A13

OBJECTIVE C
Applications and formulas

Example 12

A 12-foot board is cut into pieces $2\frac{1}{2}$ ft long for use as bookshelves. What is the length of the remaining piece after as many shelves as possible are cut?

Strategy

To find the length of the remaining piece:

▶ Divide the total length (12) by the length of each shelf $\left(2\frac{1}{2}\right)$. The quotient is the number of shelves cut, with a certain fraction of a shelf left over.
▶ Multiply the fraction left over by the length of a shelf.

Solution

$$12 \div 2\frac{1}{2} = \frac{12}{1} \div \frac{5}{2} = \frac{12 \cdot 2}{1 \cdot 5} = \frac{24}{5} = 4\frac{4}{5}$$

4 shelves, each $2\frac{1}{2}$ ft long, can be cut from the board. The piece remaining is $\frac{4}{5}$ of $2\frac{1}{2}$ ft long.

$$\frac{4}{5} \cdot 2\frac{1}{2} = \frac{4}{5} \cdot \frac{5}{2} = \frac{4 \cdot 5}{5 \cdot 2} = 2$$

The length of the remaining piece is 2 ft.

You Try It 12

The Booster Club is making 22 sashes for the high school band members. Each sash requires $1\frac{3}{8}$ yd of material at a cost of $8 per yard. Find the total cost of the material.

Your Strategy

Your Solution

Example 13

The formula $C = \frac{5}{9}(F - 32)$, where C is Celcius and F is Fahrenheit, is used to convert Fahrenheit to Celsius. Use this formula to find the temperature in degrees Celsius when the Fahrenheit temperature is 86°.

Strategy

To find the Celsius temperature, replace F by 86 in the given formula and solve for C.

Solution

$$C = \frac{5}{9}(F - 32)$$

$$C = \frac{5}{9}(86 - 32) = \frac{5}{9}(54) = \frac{5}{9} \cdot \frac{54}{1} = 30$$

The Celsius temperature is 30°.

You Try It 13

The formula $C = \frac{5}{9}(F - 32)$, where C is Celsius and F is Fahrenheit, is used to convert Fahrenheit to Celsius. Use this formula to find the temperature in degrees Celsius when the Fahrenheit temperature is 68°.

Your Strategy

Your Solution

Solutions on p. A13

3.4 EXERCISES

OBJECTIVE A

Multiply.

1. $\dfrac{2}{3} \cdot \dfrac{9}{10}$

2. $\dfrac{3}{8} \cdot \dfrac{4}{5}$

3. $-\dfrac{6}{7} \cdot \dfrac{11}{12}$

4. $\dfrac{5}{6} \cdot \left(-\dfrac{2}{5}\right)$

5. $\dfrac{14}{15} \cdot \dfrac{6}{7}$

6. $\dfrac{15}{16} \cdot \dfrac{4}{9}$

7. $-\dfrac{6}{7} \cdot \dfrac{0}{10}$

8. $\dfrac{5}{12} \cdot \dfrac{3}{0}$

9. $\left(-\dfrac{4}{15}\right) \cdot \left(-\dfrac{3}{8}\right)$

10. $\left(-\dfrac{3}{4}\right) \cdot \left(-\dfrac{2}{9}\right)$

11. $-\dfrac{3}{4} \cdot \dfrac{1}{2}$

12. $-\dfrac{8}{15} \cdot \dfrac{5}{12}$

13. $\dfrac{9}{x} \cdot \dfrac{7}{y}$

14. $\dfrac{4}{c} \cdot \dfrac{8}{d}$

15. $-\dfrac{y}{5} \cdot \dfrac{z}{6}$

16. $-\dfrac{a}{10} \cdot \left(-\dfrac{b}{6}\right)$

17. $\dfrac{2}{3} \cdot \dfrac{3}{8} \cdot \dfrac{4}{9}$

18. $\dfrac{5}{7} \cdot \dfrac{1}{6} \cdot \dfrac{14}{15}$

19. $-\dfrac{7}{12} \cdot \dfrac{5}{8} \cdot \dfrac{16}{25}$

20. $\dfrac{5}{12} \cdot \left(-\dfrac{1}{3}\right) \cdot \left(-\dfrac{8}{15}\right)$

21. $\left(-\dfrac{3}{5}\right) \cdot \dfrac{1}{2} \cdot \left(-\dfrac{5}{8}\right)$

22. $\dfrac{5}{6} \cdot \left(-\dfrac{2}{3}\right) \cdot \dfrac{3}{25}$

23. $6 \cdot \dfrac{1}{6}$

24. $\dfrac{1}{10} \cdot 10$

25. $\dfrac{3}{4} \cdot 8$

26. $\dfrac{5}{7} \cdot 14$

27. $12 \cdot \left(-\dfrac{5}{8}\right)$

28. $24 \cdot \left(-\dfrac{3}{8}\right)$

29. $-16 \cdot \dfrac{7}{30}$

30. $-9 \cdot \dfrac{7}{15}$

31. $\dfrac{6}{7} \cdot 0$

32. $0 \cdot \dfrac{9}{11}$

Multiply.

33. $\dfrac{5}{22} \cdot 2\dfrac{1}{5}$

34. $\dfrac{4}{15} \cdot 1\dfrac{7}{8}$

35. $3\dfrac{1}{2} \cdot 5\dfrac{3}{7}$

36. $2\dfrac{1}{4} \cdot 1\dfrac{1}{3}$

37. $3\dfrac{1}{3} \cdot \left(-\dfrac{7}{10}\right)$

38. $2\dfrac{1}{4} \cdot \left(-\dfrac{7}{9}\right)$

39. $-1\dfrac{2}{3} \cdot \left(-\dfrac{3}{5}\right)$

40. $-2\dfrac{1}{8} \cdot \left(-\dfrac{4}{17}\right)$

41. $3\dfrac{1}{3} \cdot 2\dfrac{1}{3}$

42. $3\dfrac{1}{4} \cdot 2\dfrac{2}{3}$

43. $3\dfrac{1}{3} \cdot (-9)$

44. $-2\dfrac{1}{2} \cdot 4$

45. $8 \cdot 5\dfrac{1}{4}$

46. $3 \cdot 2\dfrac{1}{9}$

47. $3\dfrac{1}{2} \cdot 1\dfrac{5}{7} \cdot \dfrac{11}{12}$

48. $2\dfrac{2}{3} \cdot \dfrac{8}{9} \cdot 1\dfrac{5}{16}$

Solve.

49. Find the product of $\dfrac{3}{4}$ and $\dfrac{14}{15}$.

50. Find the product of $\dfrac{12}{25}$ and $\dfrac{5}{16}$.

51. Find $-\dfrac{9}{16}$ multiplied by $\dfrac{4}{27}$.

52. Find $\dfrac{3}{7}$ multiplied by $-\dfrac{14}{15}$.

53. What is the product of $-\dfrac{7}{24}, \dfrac{8}{21}$, and $\dfrac{3}{7}$?

54. What is the product of $-\dfrac{5}{13}, -\dfrac{26}{75}$, and $\dfrac{5}{8}$?

55. What is $4\dfrac{4}{5}$ times $\dfrac{3}{8}$?

56. What is $5\dfrac{1}{3}$ times $\dfrac{3}{16}$?

57. Find the product of $-2\dfrac{2}{3}$ and $-1\dfrac{11}{16}$.

58. Find the product of $1\dfrac{3}{11}$ and $5\dfrac{1}{2}$.

The distribution of the sale of 700 vehicles at a car dealership is shown in the figure at the right. Use this graph for Exercises 59 and 60.

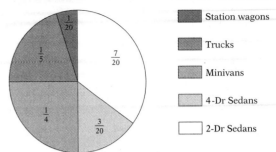

59.   ✍ *Business*   (a) How many more 4-door sedans were sold than station wagons? (b) How many more trucks were sold than station wagons?

60.   ✍ *Business*   (a) How many fewer minivans were sold than 2-door sedans? (b) How many fewer 4-door sedans were sold than trucks?

Distribution of the Sale of 700 Vehicles By a Car Dealership

Evaluate the variable expression xy for the given values of x and y.

61. $x = -\dfrac{5}{16}, y = \dfrac{7}{15}$

62. $x = -\dfrac{2}{5}, y = -\dfrac{5}{6}$

63. $x = \dfrac{4}{7}, y = 6\dfrac{1}{8}$

64. $x = 6\dfrac{3}{5}, y = 3\dfrac{1}{3}$

65. $x = -49, y = \dfrac{5}{14}$

66. $x = -\dfrac{3}{10}, y = -35$

67. $x = 1\dfrac{3}{13}, y = -6\dfrac{1}{2}$

68. $x = -3\dfrac{1}{2}, y = -2\dfrac{2}{7}$

Evaluate the variable expression xyz for the given values of x, y, and z.

69. $x = \dfrac{3}{8}, y = \dfrac{2}{3}, z = \dfrac{4}{5}$

70. $x = 4, y = \dfrac{0}{8}, z = 1\dfrac{5}{9}$

71. $x = 2\dfrac{3}{8}, y = -\dfrac{3}{19}, z = -\dfrac{4}{9}$

72. $x = \dfrac{4}{5}, y = -15, z = \dfrac{7}{8}$

73. $x = \dfrac{5}{6}, y = -3, z = 1\dfrac{7}{15}$

74. $x = 4\dfrac{1}{2}, y = 3\dfrac{5}{9}, z = 1\dfrac{7}{8}$

75.   Is $-\dfrac{1}{3}$ a solution of the equation $\dfrac{3}{4}y = -\dfrac{1}{4}$?

76.   Is $\dfrac{2}{5}$ a solution of the equation $-\dfrac{5}{6}z = \dfrac{1}{3}$?

77.   Is $\dfrac{3}{4}$ a solution of the equation $\dfrac{4}{5}x = \dfrac{5}{3}$?

78.   Is $\dfrac{1}{2}$ a solution of the equation $\dfrac{3}{4}p = \dfrac{3}{2}$?

79.   Is $-\dfrac{1}{6}$ a solution of the equation $6x = 1$?

80.   Is $-\dfrac{4}{5}$ a solution of the equation $\dfrac{5}{4}n = -1$?

OBJECTIVE B

Divide.

81. $\dfrac{5}{7} \div \dfrac{2}{5}$

82. $\dfrac{3}{8} \div \dfrac{2}{3}$

83. $\dfrac{4}{7} \div \left(-\dfrac{4}{7}\right)$

84. $-\dfrac{5}{7} \div \left(-\dfrac{5}{6}\right)$

85. $0 \div \dfrac{7}{9}$

86. $0 \div \dfrac{4}{5}$

87. $\left(-\dfrac{1}{3}\right) \div \dfrac{1}{2}$

88. $\left(-\dfrac{3}{8}\right) \div \dfrac{7}{8}$

89. $-\dfrac{5}{16} \div \left(-\dfrac{3}{8}\right)$

90. $\left(-\dfrac{3}{4}\right) \div \left(-\dfrac{5}{6}\right)$

91. $\dfrac{0}{1} \div \dfrac{1}{9}$

92. $\dfrac{1}{2} \div \left(-\dfrac{8}{0}\right)$

93. $6 \div \dfrac{3}{4}$

94. $8 \div \dfrac{2}{3}$

95. $\dfrac{3}{4} \div (-6)$

96. $-\dfrac{2}{3} \div 8$

97. $\dfrac{9}{10} \div 0$

98. $\dfrac{2}{11} \div 0$

99. $\dfrac{5}{12} \div \left(-\dfrac{15}{32}\right)$

100. $\dfrac{3}{8} \div \left(-\dfrac{5}{12}\right)$

101. $\left(-\dfrac{2}{3}\right) \div (-4)$

102. $\left(-\dfrac{4}{9}\right) \div (-6)$

103. $\dfrac{8}{x} \div \left(-\dfrac{y}{4}\right)$

104. $-\dfrac{9}{m} \div \dfrac{n}{7}$

105. $\dfrac{b}{6} \div \dfrac{5}{d}$

106. $\dfrac{y}{10} \div \dfrac{4}{z}$

107. $3\dfrac{1}{3} \div \dfrac{5}{8}$

108. $5\dfrac{1}{2} \div \dfrac{1}{4}$

109. $5\dfrac{3}{5} \div \left(-\dfrac{7}{10}\right)$

110. $6\dfrac{8}{9} \div \left(-\dfrac{31}{36}\right)$

111. $-1\dfrac{1}{2} \div 1\dfrac{3}{4}$

112. $-1\dfrac{3}{5} \div 3\dfrac{1}{10}$

Divide.

113. $5\frac{1}{2} \div 11$

114. $4\frac{2}{3} \div 7$

115. $5\frac{2}{7} \div 1$

116. $9\frac{5}{6} \div 1$

117. $-16 \div 1\frac{1}{3}$

118. $-9 \div \left(-3\frac{3}{5}\right)$

119. $2\frac{4}{13} \div 1\frac{5}{26}$

120. $3\frac{3}{8} \div 2\frac{7}{16}$

Solve.

121. Find the quotient of $\frac{9}{10}$ and $\frac{3}{4}$.

122. Find the quotient of $\frac{3}{5}$ and $\frac{12}{25}$.

123. What is $-\frac{15}{24}$ divided by $\frac{3}{5}$?

124. What is $\frac{5}{6}$ divided by $-\frac{10}{21}$?

125. Find $\frac{7}{8}$ divided by $3\frac{1}{4}$.

126. Find $-\frac{3}{8}$ divided by $2\frac{1}{4}$.

127. What is the quotient of $-3\frac{5}{11}$ and $3\frac{4}{5}$?

128. What is the quotient of $-10\frac{1}{5}$ and $-1\frac{7}{10}$?

Evaluate the variable expression $x \div y$ for the given values of x and y.

129. $x = \frac{2}{5}, y = \frac{4}{7}$

130. $x = \frac{3}{8}, y = \frac{5}{12}$

131. $x = -\frac{5}{8}, y = -\frac{15}{2}$

132. $x = -\frac{14}{3}, y = -\frac{7}{9}$

133. $x = -18, y = \frac{3}{8}$

134. $x = 20, y = -\frac{5}{6}$

135. $x = \frac{1}{7}, y = 0$

136. $x = \frac{4}{0}, y = 12$

Evaluate the variable expression $x \div y$ for the given values of x and y.

137. $x = 1\frac{2}{3}, y = \frac{7}{9}$ **138.** $x = -\frac{9}{10}, y = 3\frac{3}{5}$ **139.** $x = -\frac{1}{2}, y = -3\frac{5}{8}$ **140.** $x = 4\frac{3}{8}, y = 7$

141. $x = 6\frac{2}{5}, y = -4$ **142.** $x = -2\frac{5}{8}, y = 1\frac{3}{4}$ **143.** $x = -3\frac{2}{5}, y = -1\frac{7}{10}$ **144.** $x = -5\frac{2}{5}, y = -9$

OBJECTIVE C

Solve.

145. *Sports* A chukker is one period of play in a polo match. A chukker lasts $7\frac{1}{2}$ min. Find the length of time in four chukkers.

146. *History* The Assyrian calendar was based on the phases of the moon. One lunation was $29\frac{1}{2}$ days long. There were 12 lunations in one year. Find the number of days in one year in the Assyrian calendar.

147. *Aviation* The first polar flight, flown in 1926, lasted $15\frac{1}{2}$ h. How many minutes did the flight last?

148. *Measurement* One rod is equal to $5\frac{1}{2}$ yd. How many feet are in one rod? How many inches are in one rod?

149. *Cooking* A recipe for chocolate chip cookies calls for $1\frac{3}{4}$ c flour. If you are halving the recipe, how much flour do you need?

150. *Travel* A car used $12\frac{1}{2}$ gal of gasoline on a 275-mile trip. How many miles can this car travel on one gallon of gasoline?

151. *Housework* According to a national survey, the average couple spends $4\frac{1}{2}$ h cleaning house each weekend. How many hours does the average couple spend cleaning house each year?

Solve.

152. *Business*   A factory worker can assemble a product in $7\frac{1}{2}$ min. How many products can the worker assemble in one hour?

153. *Real Estate*   A developer purchases $25\frac{1}{2}$ acres of land and plans to set aside 3 acres for an entranceway to a housing development to be built on the property. Each house will be built on a $\frac{3}{4}$-acre plot of land. How many houses does the developer plan to build on the property?

154. *Consumerism*   You are planning a barbecue for 25 people. You want to serve $\frac{1}{4}$-pound hamburger patties to your guests and you estimate each person will eat two hamburgers. How much hamburger meat should you buy for the barbecue?

155. *Board Games*   A wooden travel game board has hinges which allow the board to be folded in half. If the dimensions of the open board are 14 in. by 14 in. by $\frac{7}{8}$ in., what are the dimensions of the board when it is closed?

156. *Carpentry*   A 16-foot board is cut into pieces $2\frac{1}{2}$ ft long for use as bookshelves. What is the length of the remaining piece after as many shelves as possible are cut?

157. *Investments*   Find the cost of purchasing 150 shares of stock selling for $\$22\frac{1}{8}$ per share.

158. *Investments*   Find the cost of purchasing 250 shares of stock selling for $\$18\frac{3}{4}$ per share.

159. *Wages*   Find the total wages of an employee who worked $26\frac{1}{2}$ h this week and who earns an hourly wage of $12.

160. *Wages*   Find the total wages of an employee who worked $18\frac{3}{4}$ h this week and who earns an hourly wage of $8.

161. *Sports*   The pressure on a submerged object is given by $P = 15 + \frac{1}{2}D$, where D is the depth in feet and P is the pressure measured in pounds per square inch. Find the pressure on a diver who is at a depth of $12\frac{1}{2}$ ft.

Solve.

162. *Sports* Find the rate of a hiker who walked $4\frac{2}{3}$ mi in $1\frac{1}{3}$ h. Use the equation $r = \frac{d}{t}$, where r is the rate in miles per hour, d is the distance, and t is the time.

163. *Sports* A hiker covers a distance of $7\frac{1}{2}$ mi in 3 h. Use the equation $r = \frac{d}{t}$, where r is the rate in miles per hour, d is the distance, and t is the time, to find the rate at which the hiker walked.

164. *Physics* Find the amount of force necessary to push a 75-pound crate across a floor where the coefficient of friction is $\frac{3}{8}$. Use the equation $F = \mu N$, where F is the force, μ is the coefficient of friction, and N is the weight of the crate. Force is measured in pounds.

CRITICAL THINKING

165. On a map, two cities are $3\frac{1}{8}$ in. apart. If $\frac{1}{8}$ in. on the map represents 50 mi, what is the number of miles in the distance between the two cities?

166. (a) A number increased by $\frac{1}{3}$ of the number is equal to 24. Find the number.

    (b) The product of $\frac{3}{10}$ and a number is $\frac{1}{2}$. Find the number.

167. Determine whether the statement is always true, sometimes true, or never true.
    a. Let n be an even number. Then $\frac{1}{2}n$ is a whole number.

    b. Let n be an odd number. Then $\frac{1}{2}n$ is an improper fraction.

168. Show by example that each of the following properties of multiplication of fractions is *not* satisfied by division of fractions: (a) Commutative Property, (b) Associative Property, (c) Inverse Property.

169. ✎ A box of stationery is to contain pieces of notepaper that measure $5\frac{3}{4}$ in. by $7\frac{3}{4}$ in. The notepaper is to be folded in half for mailing. What size envelopes would you design for this stationery? Explain how you arrived at your decision.

170. ✎ On page 196, Exercise 146 describes the Assyrian calendar. Our calendar is based on the solar year. One solar year is $365\frac{1}{4}$ days. Use this fact to explain leap years.

171. ✎ Draw a floor plan of your home or apartment.

SECTION 3.5   Solving Equations with Fractions

OBJECTIVE A
Solving equations

Earlier in the text, you solved equations using the addition, subtraction, and division properties of equations. These properties are reviewed below.

> **The same number can be subtracted from each side of an equation without changing the solution of the equation.**

> **The same number can be added to each side of an equation without changing the solution of the equation.**

> **Each side of an equation can be divided by the same nonzero number without changing the solution of the equation.**

A fourth property of equations involves *multiplying* each side of an equation by the same nonzero number.

As shown at the right, the solution of the equation $3x = 12$ is 4.

$$3x = 12$$
$$3 \cdot 4 = 12$$
$$12 = 12$$

If each side of the equation $3x = 12$ is multiplied by 2, the resulting equation is $6x = 24$. The solution of this equation is also 4.

$$3x = 12$$
$$2 \cdot 3x = 2 \cdot 12$$
$$6x = 24 \qquad 6 \cdot 4 = 24$$

This illustrates the multiplication property of equations.

> **Each side of an equation can be multiplied by the same nonzero number without changing the solution of the equation.**

Solve: $\dfrac{x}{5} = 2$

The variable is divided by 5. Multiply each side of the equation by 5.

$$\frac{x}{5} = 2$$

Note that $5 \cdot \dfrac{x}{5} = \dfrac{5}{1} \cdot \dfrac{x}{5} = \dfrac{5x}{5} = x$.

$$5 \cdot \frac{x}{5} = 5 \cdot 2$$

The variable x is alone on the left side of the equation. The number on the right side is the solution.

$$x = 10$$

Check the solution.

Check: $\dfrac{x}{5} = 2$

$$\frac{\dfrac{10}{5} \quad \bigg| \quad 2}{}$$

The solution checks.

$$2 = 2$$

The solution is 10.

Recall that the product of a number and its reciprocal is 1. For instance,

$$\frac{3}{4} \cdot \frac{4}{3} = 1 \quad \text{and} \quad \left(-\frac{5}{2}\right)\left(-\frac{2}{5}\right) = 1$$

Multiplying each side of an equation by the reciprocal of a number is useful when solving equations in which the variable is multiplied by a fraction.

Solve: $\frac{3}{4}a = 12$

$\frac{3}{4}$ multiplies the variable a. Note the effect of multiplying each side of the equation by $\frac{4}{3}$, the reciprocal of $\frac{3}{4}$.

Note: $\frac{4}{3} \cdot \frac{3}{4} \cdot a = 1 \cdot a = a$.

The result is an equation with the variable alone on the left side of the equation. The number on the right side is the solution.

$$\frac{3}{4}a = 12$$

$$\frac{4}{3} \cdot \frac{3}{4}a = \frac{4}{3} \cdot 12$$

$$a = 16$$

The solution is 16.

Solve: $6 = -\frac{3c}{5}$

$$-\frac{3c}{5} = -\left(\frac{3}{5} \cdot \frac{c}{1}\right) = -\frac{3}{5} \cdot c$$

$-\frac{3}{5}$ multiplies the variable c. Multiply each side of the equation by $-\frac{5}{3}$, the reciprocal of $-\frac{3}{5}$.

c is alone on the right side of the equation. The number on the left side is the solution.

Check your solution.

$$6 = -\frac{3c}{5}$$

$$6 = -\frac{3}{5} \cdot c$$

$$-\frac{5}{3} \cdot 6 = -\frac{5}{3}\left(-\frac{3}{5} \cdot c\right)$$

$$-10 = c$$

Check: $6 = -\frac{3c}{5}$

$\begin{array}{c|c} 6 & \dfrac{-3(-10)}{5} \\ \\ 6 & \dfrac{30}{5} \\ \\ & 6 = 6 \end{array}$

The solution is −10.

As shown in the following Example 1 and You Try It 1, the addition and subtraction properties of equations can be used to solve equations that contain fractions.

Example 1 Solve: $y + \frac{2}{3} = \frac{3}{4}$

You Try It 1 Solve: $-\frac{1}{5} = z - \frac{5}{6}$

Solution

$$y + \frac{2}{3} = \frac{3}{4}$$

$$y + \frac{2}{3} - \frac{2}{3} = \frac{3}{4} - \frac{2}{3}$$ $\frac{2}{3}$ is added to y.

Subtract $\frac{2}{3}$.

$$y = \frac{9}{12} - \frac{8}{12}$$

$$y = \frac{1}{12}$$

The solution is $\frac{1}{12}$.

Your Solution

Example 2 Solve: $-\frac{3}{5} = \frac{6}{7}c$

You Try It 2 Solve: $26 = 4x$

Solution

$$-\frac{3}{5} = \frac{6}{7}c$$

$$\frac{7}{6}\left(-\frac{3}{5}\right) = \frac{7}{6} \cdot \frac{6}{7}c$$

$$-\frac{7}{10} = c$$

The solution is $-\frac{7}{10}$.

Your Solution

Solutions on p. A13

OBJECTIVE **B**
Applications

Example 3

Three-eighths times a number is equal to negative one-fourth. Find the number.

Solution

The unknown number: y

Three-eighths times a number	is equal to	negative one-fourth

$$\frac{3}{8}y = -\frac{1}{4}$$

$$\frac{8}{3} \cdot \frac{3}{8}y = \frac{8}{3}\left(-\frac{1}{4}\right)$$

$$y = -\frac{2}{3}$$

The number is $-\frac{2}{3}$.

You Try It 3

Negative five-sixths is equal to ten-thirds of a number. Find the number.

Your Solution

Solution on p. A13

Example 4

One-third of all of the sugar produced by Sucor, Inc. is brown sugar. This year Sucor produced 250,000 lb of brown sugar. How many pounds of sugar were produced by Sucor?

Strategy

To find the number of pounds of sugar produced, write and solve an equation using x to represent the number of pounds of sugar produced.

Solution

| One-third of the sugar produced | is | brown sugar |

$$\frac{1}{3}x = 250,000$$

$$3 \cdot \frac{1}{3}x = 3 \cdot 250,000$$

$$x = 750,000$$

Sucor produced 750,000 lb of sugar.

Example 5

The average score on exams taken during a semester is given by $A = \frac{T}{N}$, where A is the average score, T is the total number of points scored on all tests, and N is the number of tests. Find the total number of points scored by a student whose average score for 6 tests was 84.

Strategy

To find the total number of points scored, replace A with 84 and N with 6 in the given formula and solve for T.

Solution

$$A = \frac{T}{N}$$

$$84 = \frac{T}{6}$$

$$6 \cdot 84 = 6 \cdot \frac{T}{6}$$

$$504 = T$$

The total number of points scored was 504.

You Try It 4

The number of computer software games sold by BAL Software in January was three-fifths of all the software products sold by the company. BAL Software sold 450 computer software games in January. Find the total number of software products sold in January.

Your Strategy

Your Solution

You Try It 5

The average score on exams taken during a semester is given by $A = \frac{T}{N}$, where A is the average score, T is the total number of points scored on all tests, and N is the number of tests. Find the total number of points scored by a student whose average score for 5 tests was 73.

Your Strategy

Your Solution

Solutions on p. A14

3.5 EXERCISES

OBJECTIVE A

Solve.

1. $\dfrac{x}{4} = 9$

2. $8 = \dfrac{y}{2}$

3. $-3 = \dfrac{m}{4}$

4. $\dfrac{n}{5} = -2$

5. $\dfrac{2}{5}x = 10$

6. $\dfrac{3}{4}z = 12$

7. $-\dfrac{5}{6}w = 10$

8. $-\dfrac{1}{2}x = 3$

9. $\dfrac{1}{4} + y = \dfrac{3}{4}$

10. $\dfrac{5}{9} = t - \dfrac{1}{9}$

11. $x + \dfrac{1}{4} = \dfrac{5}{6}$

12. $\dfrac{7}{8} = y - \dfrac{1}{6}$

13. $-\dfrac{2x}{3} = -\dfrac{1}{2}$

14. $-\dfrac{4a}{5} = \dfrac{2}{3}$

15. $\dfrac{5n}{6} = -\dfrac{2}{3}$

16. $\dfrac{7z}{8} = -\dfrac{5}{16}$

17. $-\dfrac{3}{8}t = -\dfrac{1}{4}$

18. $-\dfrac{3}{4}t = -\dfrac{7}{8}$

19. $4a = 6$

20. $6z = 10$

21. $-9c = 12$

22. $-10z = 28$

23. $-2x = \dfrac{8}{9}$

24. $-5y = -\dfrac{15}{16}$

OBJECTIVE B

Solve.

25. A number minus one-third equals one-half. Find the number.

26. The sum of a number and one-fourth is one-sixth. Find the number.

27. Three-fifths times a number is nine-tenths. Find the number.

28. The product of negative two-thirds and a number is five-sixths. Find the number.

29. The quotient of a number and negative four is three-fourths. Find the number.

30. A number divided by negative two equals two-fifths. Find the number.

31. Negative three-fourths of a number is equal to one-sixth. Find the number.

32. Negative three-eighths equals the product of two-thirds and some number. Find the number.

Solve.

33. *Construction* Three-fourths of the total number of square feet in a two-story house is on the first floor. There are 1,200 square feet on the first floor. Find the total number of square feet in this two-story house.

34. *Nutrition* The number of calories from fat in a serving of potato chips is one-half of the total number of calories in one serving. There are 112 calories from fat in one serving of these potato chips. What is the total number of calories in one serving?

35. *Catering* The number of quarts of orange juice in a fruit punch recipe is three-fifths of the total number of quarts in the punch. The number of quarts of orange juice in the punch is fifteen. Find the total number of quarts in the punch.

36. *The Electorate* The number of people who voted in an election for mayor of a city was two-thirds of the total number of eligible voters. There were 24,416 people who voted in the election. Find the number of eligible voters.

37. *Cost of Living* The amount of rent paid by a mechanic is $\frac{2}{5}$ of the mechanic's monthly income. Using the figure at the right, determine the mechanic's monthly income.

38. *Travel* The average number of miles per gallon for a car is calculated using the formula $a = \frac{m}{g}$, where a is the average number of miles per gallon and m is the number of miles traveled on g gallons of gas. Use this formula to find the number of miles a car can travel on 16 gal of gas if the car averages 26 mi per gallon.

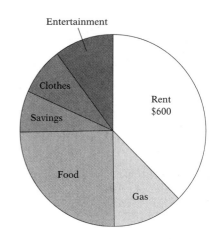

39. *Travel* The average number of miles per gallon for a truck is calculated using the formula $a = \frac{m}{g}$, where a is the average number of miles per gallon and m is the number of miles traveled on g gallons of gas. Use this formula to find the number of miles a truck can travel on 38 gal of diesel fuel if the truck averages 14 mi per gallon.

CRITICAL THINKING

40. If $\frac{3}{8}x = -\frac{1}{4}$, is $6x$ greater than -1 or less than -1?

41. Given $-\frac{x}{2} = \frac{2}{3}$, select the best answer from the choices below.
    a. $-9x > 10$ b. $-6x < 8$ c. $-9x > 10$ and $-6x < 8$

42. Explain why dividing each side of $3x = 6$ by 3 is the same as multiplying each side of the equation by $\frac{1}{3}$.

SECTION 3.6 Exponents, Complex Fractions, and the Order of Operations Agreement

OBJECTIVE A

Exponents

Recall that an exponent indicates the repeated multiplication of the same factor. For example,

$$3^5 = 3 \cdot 3 \cdot 3 \cdot 3 \cdot 3$$

The exponent, 5, indicates how many times the base, 3, occurs as a factor in the multiplication.

The base of an exponential expression can be a fraction, for example, $\left(\frac{2}{3}\right)^4$. To evaluate this expression, write the factor as many times as indicated by the exponent and then multiply.

$$\left(\frac{2}{3}\right)^4 = \frac{2}{3} \cdot \frac{2}{3} \cdot \frac{2}{3} \cdot \frac{2}{3} = \frac{2 \cdot 2 \cdot 2 \cdot 2}{3 \cdot 3 \cdot 3 \cdot 3} = \frac{16}{81}$$

Evaluate $\left(-\frac{3}{5}\right)^2 \cdot \left(\frac{5}{6}\right)^3$.

$$\left(-\frac{3}{5}\right)^2 \cdot \left(\frac{5}{6}\right)^3$$

Write each factor as many times as indicated by the exponent.

$$= \left(-\frac{3}{5}\right) \cdot \left(-\frac{3}{5}\right) \cdot \frac{5}{6} \cdot \frac{5}{6} \cdot \frac{5}{6}$$

Multiply. The product of two negative numbers is positive.

$$= \frac{3}{5} \cdot \frac{3}{5} \cdot \frac{5}{6} \cdot \frac{5}{6} \cdot \frac{5}{6}$$

$$= \frac{3 \cdot 3 \cdot 5 \cdot 5 \cdot 5}{5 \cdot 5 \cdot 6 \cdot 6 \cdot 6}$$

Write the product in simplest form.

$$= \frac{5}{24}$$

Evaluate x^3 when $x = 2\frac{1}{2}$.

$$x^3$$

Replace x with $x = 2\frac{1}{2}$.

$$\left(2\frac{1}{2}\right)^3$$

Write the mixed number as an improper fraction.

$$= \left(\frac{5}{2}\right)^3$$

Write the base as many times as indicated by the exponent.

$$= \frac{5}{2} \cdot \frac{5}{2} \cdot \frac{5}{2}$$

Multiply.

$$= \frac{125}{8}$$

Write the improper fraction as a mixed number.

$$= 15\frac{5}{8}$$

Example 1   Evaluate $\left(-\frac{3}{4}\right)^3 \cdot 8^2$.

Solution $\left(-\frac{3}{4}\right)^3 \cdot 8^2$

$$= \left(-\frac{3}{4}\right)\left(-\frac{3}{4}\right)\left(-\frac{3}{4}\right) \cdot 8 \cdot 8$$

$$= -\left(\frac{3}{4} \cdot \frac{3}{4} \cdot \frac{3}{4} \cdot \frac{8}{1} \cdot \frac{8}{1}\right)$$

$$= -\frac{3 \cdot 3 \cdot 3 \cdot 8 \cdot 8}{4 \cdot 4 \cdot 4 \cdot 1 \cdot 1} = -27$$

You Try It 1   Evaluate $\left(\frac{2}{9}\right)^2 \cdot (-3)^4$.

Your Solution

Example 2   Evaluate $x^2 y^2$
when $x = 1\frac{1}{2}$ and $y = \frac{2}{3}$.

Solution $x^2 y^2$

$$\left(1\frac{1}{2}\right)^2 \cdot \left(\frac{2}{3}\right)^2 = \left(\frac{3}{2}\right)^2 \cdot \left(\frac{2}{3}\right)^2$$

$$= \frac{3}{2} \cdot \frac{3}{2} \cdot \frac{2}{3} \cdot \frac{2}{3}$$

$$= \frac{3 \cdot 3 \cdot 2 \cdot 2}{2 \cdot 2 \cdot 3 \cdot 3} = 1$$

You Try It 2   Evaluate $x^4 y^3$ when
$x = 2\frac{1}{3}$ and $y = \frac{3}{7}$.

Your Solution

Solutions on p. A14

Objective B
Complex fractions

A **complex fraction** is a fraction whose numerator or denominator contains one or more fractions. Examples of complex fractions are shown below.

Main fraction bar ⟶ $\dfrac{\frac{3}{4}}{\frac{7}{8}}$ $\dfrac{4}{3 - \frac{1}{2}}$ $\dfrac{\frac{9}{10} + \frac{3}{5}}{\frac{5}{6}}$ $\dfrac{3\frac{1}{2} \cdot 2\frac{5}{8}}{\left(4\frac{2}{3}\right) \div \left(3\frac{1}{5}\right)}$

Look at the first example given above and recall that the fraction bar can be read "divided by."

Therefore, $\dfrac{\frac{3}{4}}{\frac{7}{8}}$ can be read "$\frac{3}{4}$ divided by $\frac{7}{8}$" and can be written $\frac{3}{4} \div \frac{7}{8}$. This is the division of two fractions and can be simplified by multiplying by the reciprocal.

$$\frac{\dfrac{3}{4}}{\dfrac{7}{8}} = \frac{3}{4} \div \frac{7}{8} = \frac{3}{4} \cdot \frac{8}{7} = \frac{3 \cdot 8}{4 \cdot 7} = \frac{6}{7}$$

To simplify a complex fraction, first simplify the expression above the main fraction bar and the expression below the main fraction bar; the result is one number in the numerator and one number in the denominator. Then rewrite the complex fraction as a division problem by reading the main fraction bar as "divided by."

Simplify: $\dfrac{4}{3 - \dfrac{1}{2}}$

The numerator (4) is already simplified. Simplify the expression in the denominator.

Note: $3 - \dfrac{1}{2} = \dfrac{6}{2} - \dfrac{1}{2} = \dfrac{5}{2}$

$$\frac{4}{3 - \dfrac{1}{2}} = \frac{4}{\dfrac{5}{2}}$$

Rewrite the complex fraction as division.

$$= 4 \div \frac{5}{2}$$

Divide.

$$= \frac{4}{1} \div \frac{5}{2}$$

$$= \frac{4}{1} \cdot \frac{2}{5}$$

Write the answer in simplest form.

$$= \frac{8}{5} = 1\frac{3}{5}$$

Simplify: $\dfrac{-\dfrac{9}{10} + \dfrac{3}{5}}{1\dfrac{1}{4}}$

Simplify the expression in the numerator.

Note: $-\dfrac{9}{10} + \dfrac{3}{5} = \dfrac{-9}{10} + \dfrac{6}{10} = \dfrac{-3}{10} = -\dfrac{3}{10}$

$$\frac{-\dfrac{9}{10} + \dfrac{3}{5}}{1\dfrac{1}{4}} = \frac{-\dfrac{3}{10}}{\dfrac{5}{4}}$$

Write the mixed number in the denominator as an improper fraction.

Rewrite the complex fraction as division. The quotient will be negative.

$$= -\left(\frac{3}{10} \div \frac{5}{4} \right)$$

Divide by multiplying by the reciprocal.

$$= -\left(\frac{3}{10} \cdot \frac{4}{5} \right)$$

$$= -\frac{6}{25}$$

Evaluate $\dfrac{wx}{yz}$ when $w = 1\frac{1}{3}$, $x = 2\frac{5}{8}$, $y = 4\frac{1}{2}$, and $z = 3\frac{1}{3}$.

$$\dfrac{wx}{yz}$$

Replace each variable with its given value.

$$\dfrac{1\frac{1}{3} \cdot 2\frac{5}{8}}{4\frac{1}{2} \cdot 3\frac{1}{3}}$$

Simplify the numerator.

Note: $1\frac{1}{3} \cdot 2\frac{5}{8} = \frac{4}{3} \cdot \frac{21}{8} = \frac{7}{2}$

Simplify the denominator.

Note: $4\frac{1}{2} \cdot 3\frac{1}{3} = \frac{9}{2} \cdot \frac{10}{3} = 15$

$$= \dfrac{\frac{7}{2}}{15}$$

Rewrite the complex fraction as division.

$$= \frac{7}{2} \div 15$$

Divide by multiplying by the reciprocal.

Note: $15 = \frac{15}{1}$; the reciprocal of $\frac{15}{1}$ is $\frac{1}{15}$.

$$= \frac{7}{2} \cdot \frac{1}{15} = \frac{7}{30}$$

Example 3   Is $\frac{2}{3}$ a solution of $\dfrac{x + \frac{1}{2}}{x} = \frac{7}{4}$?

You Try It 3   Is $-\frac{1}{2}$ a solution of $\dfrac{2y - 3}{y} = -2$?

Solution

$$\dfrac{x + \frac{1}{2}}{x} = \frac{7}{4}$$

$$\dfrac{\frac{2}{3} + \frac{1}{2}}{\frac{2}{3}} \,\bigg|\, \frac{7}{4}$$

$$\dfrac{\frac{7}{6}}{\frac{2}{3}} \,\bigg|\, \frac{7}{4}$$

$$\frac{7}{6} \div \frac{2}{3} \,\bigg|\, \frac{7}{4}$$

$$\frac{7}{6} \cdot \frac{3}{2} \,\bigg|\, \frac{7}{4}$$

$$\frac{7}{4} = \frac{7}{4}$$

Yes, $\frac{2}{3}$ is a solution of the equation.

Your Solution

Solution on p. A14

Example 4

Evaluate the variable expression $\dfrac{x - y}{z}$ when

$x = 4\frac{1}{8}$, $y = 2\frac{5}{8}$, and $z = \frac{3}{4}$.

Solution

$\dfrac{x - y}{z}$

$$\dfrac{4\frac{1}{8} - 2\frac{5}{8}}{\frac{3}{4}} = \dfrac{\frac{3}{2}}{\frac{3}{4}} = \frac{3}{2} \div \frac{3}{4} = \frac{3}{2} \cdot \frac{4}{3} = 2$$

You Try It 4

Evaluate the variable expression $\dfrac{x}{y - z}$ when

$x = 2\frac{4}{9}$, $y = 3$, and $z = 1\frac{1}{3}$.

Your Solution

Solution on p. A14

OBJECTIVE C

The Order of Operations Agreement

The Order of Operations Agreement applies in simplifying expressions containing fractions.

> ### The Order of Operations Agreement
>
> **Step 1** Do all operations inside parentheses.
> **Step 2** Simplify any numerical expressions containing exponents.
> **Step 3** Do multiplication and division as they occur from left to right.
> **Step 4** Do addition and subtraction as they occur from left to right.

Simplify: $\left(\dfrac{1}{2}\right)^2 + \left(\dfrac{2}{3} \div \dfrac{5}{9}\right) \cdot \dfrac{5}{6}$

$$\left(\dfrac{1}{2}\right)^2 + \left(\dfrac{2}{3} \div \dfrac{5}{9}\right) \cdot \dfrac{5}{6}$$

Do the operation inside the parentheses (Step 1).

$$= \left(\dfrac{1}{2}\right)^2 + \left(\dfrac{6}{5}\right) \cdot \dfrac{5}{6}$$

Simplify the exponential expression (Step 2).

$$= \dfrac{1}{4} + \left(\dfrac{6}{5}\right) \cdot \dfrac{5}{6}$$

Do the multiplication (Step 3).

$$= \dfrac{1}{4} + 1$$

Do the addition (Step 4).

$$= 1\dfrac{1}{4}$$

A fraction bar acts like parentheses. Therefore, simplify the numerator and denominator of a fraction as part of Step 1 in the Order of Operations Agreement.

Simplify: $6 - \dfrac{2+1}{15-8} \div \dfrac{3}{14}$

$$6 - \frac{2+1}{15-8} \div \frac{3}{14}$$

Perform operations above and below the fraction bar.

$$= 6 - \frac{3}{7} \div \frac{3}{14}$$

Do the division.

$$= 6 - \left(\frac{3}{7} \cdot \frac{14}{3}\right)$$

$$= 6 - 2$$

Do the subtraction.

$$= 4$$

Evaluate $\dfrac{w+x}{y} - z$ when $w = \dfrac{3}{4}$, $x = \dfrac{1}{4}$, $y = 2$, and $z = \dfrac{1}{3}$.

$$\frac{w+x}{y} - z$$

Replace each variable with its given value.

$$\frac{\dfrac{3}{4} + \dfrac{1}{4}}{2} - \frac{1}{3}$$

Simplify the numerator of the complex fraction.

$$= \frac{1}{2} - \frac{1}{3}$$

Do the subtraction.

$$= \frac{1}{6}$$

Example 5 Simplify: $\left(-\dfrac{2}{3}\right)^2 \div \dfrac{7-2}{13-4} - \dfrac{1}{3}$

Solution

$$\left(-\frac{2}{3}\right)^2 \div \frac{7-2}{13-4} - \frac{1}{3}$$

$$= \left(-\frac{2}{3}\right)^2 \div \frac{5}{9} - \frac{1}{3}$$

$$= \frac{4}{9} \div \frac{5}{9} - \frac{1}{3}$$

$$= \frac{4}{9} \cdot \frac{9}{5} - \frac{1}{3}$$

$$= \frac{4}{5} - \frac{1}{3} = \frac{7}{15}$$

You Try It 5 Simplify: $\left(-\dfrac{1}{2}\right)^3 \cdot \dfrac{7-3}{4-9} + \dfrac{4}{5}$

Your Solution

Solution on p. A14

3.6 EXERCISES

OBJECTIVE A

Evaluate.

1. $\left(\dfrac{3}{4}\right)^2$ **2.** $\left(\dfrac{5}{8}\right)^2$ **3.** $\left(-\dfrac{1}{6}\right)^3$ **4.** $\left(-\dfrac{2}{7}\right)^3$

5. $\left(2\dfrac{1}{4}\right)^2$ **6.** $\left(3\dfrac{1}{2}\right)^2$ **7.** $\left(\dfrac{5}{8}\right)^3 \cdot \left(\dfrac{2}{5}\right)^2$ **8.** $\left(\dfrac{3}{5}\right)^3 \cdot \left(\dfrac{1}{3}\right)^2$

9. $\left(\dfrac{18}{25}\right)^2 \cdot \left(\dfrac{5}{9}\right)^3$ **10.** $\left(\dfrac{2}{3}\right)^3 \cdot \left(\dfrac{5}{6}\right)^2$ **11.** $\left(\dfrac{4}{5}\right)^4 \cdot \left(-\dfrac{5}{8}\right)^3$ **12.** $\left(-\dfrac{9}{11}\right)^2 \cdot \left(\dfrac{1}{3}\right)^4$

13. $7^2 \cdot \left(\dfrac{2}{7}\right)^3$ **14.** $4^3 \cdot \left(\dfrac{5}{12}\right)^2$ **15.** $4 \cdot \left(\dfrac{4}{7}\right)^2 \cdot \left(-\dfrac{3}{4}\right)^3$ **16.** $3 \cdot \left(\dfrac{2}{5}\right)^2 \cdot \left(-\dfrac{1}{6}\right)^2$

Evaluate the variable expression for the given values of x and y.

17. x^4, when $x = \dfrac{2}{3}$ **18.** y^3, when $y = -\dfrac{3}{4}$

19. $x^4 y^2$, when $x = \dfrac{5}{6}$ and $y = -\dfrac{3}{5}$ **20.** $x^5 y^3$, when $x = -\dfrac{5}{8}$ and $y = \dfrac{4}{5}$

21. $x^3 y^2$ when $x = \dfrac{2}{3}$ and $y = 1\dfrac{1}{2}$ **22.** $x^2 y^4$ when $x = 2\dfrac{1}{3}$ and $y = \dfrac{3}{7}$

OBJECTIVE B

Simplify.

23. $\dfrac{\frac{9}{16}}{\frac{3}{4}}$ **24.** $\dfrac{\frac{7}{24}}{\frac{3}{8}}$ **25.** $\dfrac{-\frac{5}{6}}{\frac{15}{16}}$ **26.** $\dfrac{\frac{7}{12}}{-\frac{5}{18}}$

Simplify.

27. $\dfrac{\dfrac{2}{3} + \dfrac{1}{2}}{7}$

28. $\dfrac{-5}{\dfrac{3}{8} - \dfrac{1}{4}}$

29. $\dfrac{2 + \dfrac{1}{4}}{\dfrac{3}{8}}$

30. $\dfrac{1 - \dfrac{3}{4}}{\dfrac{5}{12}}$

31. $\dfrac{\dfrac{9}{25}}{\dfrac{4}{5} - \dfrac{1}{10}}$

32. $\dfrac{-\dfrac{5}{7}}{\dfrac{4}{7} - \dfrac{3}{14}}$

33. $\dfrac{\dfrac{1}{3} - \dfrac{3}{4}}{\dfrac{1}{6} + \dfrac{2}{3}}$

34. $\dfrac{\dfrac{9}{14} - \dfrac{1}{7}}{\dfrac{9}{14} + \dfrac{1}{7}}$

35. $\dfrac{3 + 2\dfrac{1}{3}}{5\dfrac{1}{6} - 1}$

36. $\dfrac{4 - 3\dfrac{5}{8}}{2\dfrac{1}{2} - \dfrac{3}{4}}$

37. $\dfrac{5\dfrac{2}{3} - 1\dfrac{1}{6}}{3\dfrac{5}{8} - 2\dfrac{1}{4}}$

38. $\dfrac{3\dfrac{1}{4} - 2\dfrac{1}{2}}{4\dfrac{3}{4} + 1\dfrac{1}{2}}$

Evaluate the expression for the given values of the variables.

39. $\dfrac{x + y}{z}$, when $x = \dfrac{2}{3}$, $y = \dfrac{3}{4}$, and $z = \dfrac{1}{12}$

40. $\dfrac{x}{y + z}$, when $x = \dfrac{8}{15}$, $y = \dfrac{3}{5}$, and $z = \dfrac{2}{3}$

41. $\dfrac{xy}{z}$, when $x = \dfrac{3}{4}$, $y = -\dfrac{2}{3}$, and $z = \dfrac{5}{8}$

42. $\dfrac{x}{yz}$, when $x = -\dfrac{5}{12}$, $y = \dfrac{8}{9}$, and $z = -\dfrac{3}{4}$

43. $\dfrac{x - y}{z}$, when $x = 2\dfrac{5}{8}$, $y = 1\dfrac{1}{4}$, and $z = 1\dfrac{3}{8}$

44. $\dfrac{x}{y - z}$, when $x = 2\dfrac{3}{10}$, $y = 3\dfrac{2}{5}$, and $z = 1\dfrac{4}{5}$

45. Is $-\dfrac{3}{4}$ a solution of the equation $\dfrac{4x}{x + 5} = -\dfrac{4}{3}$?

46. Is $-\dfrac{4}{5}$ a solution of the equation

$$\dfrac{15y}{\dfrac{3}{10} + y} = -24?$$

OBJECTIVE C

Simplify.

47. $\dfrac{3}{7} \cdot \dfrac{14}{15} + \dfrac{4}{5}$

48. $\dfrac{3}{5} \div \dfrac{6}{7} + \dfrac{4}{5}$

49. $\left(\dfrac{5}{6}\right)^2 - \dfrac{5}{9}$

50. $\left(\dfrac{3}{5}\right)^2 - \dfrac{3}{10}$

51. $\dfrac{3}{4} \cdot \left(\dfrac{11}{12} - \dfrac{7}{8}\right) + \dfrac{5}{16}$

52. $\dfrac{7}{18} + \dfrac{5}{6} \cdot \left(\dfrac{2}{3} - \dfrac{1}{6}\right)$

53. $\dfrac{11}{16} - \left(\dfrac{3}{4}\right)^2 + \dfrac{7}{8}$

54. $\left(-\dfrac{2}{3}\right)^2 - \dfrac{7}{18} + \dfrac{5}{6}$

55. $\left(1\dfrac{1}{3} - \dfrac{5}{6}\right) + \dfrac{7}{8} \div \left(-\dfrac{1}{2}\right)^2$

56. $\left(\dfrac{1}{4}\right)^2 \div \left(2\dfrac{1}{2} - \dfrac{3}{4}\right) + \dfrac{5}{7}$

57. $\left(\dfrac{2}{3}\right)^2 + \dfrac{8-7}{3-9} \div \dfrac{3}{8}$

58. $\left(\dfrac{1}{3}\right)^2 \cdot \dfrac{14-5}{6-10} + \dfrac{3}{4}$

59. $\dfrac{1}{2} + \dfrac{\dfrac{13}{25}}{4 - \dfrac{3}{4}} \div \dfrac{1}{5}$

60. $\dfrac{4}{5} + \dfrac{3 - \dfrac{7}{9}}{\dfrac{5}{6}} \cdot \dfrac{3}{8}$

61. $\left(\dfrac{2}{3}\right)^2 + \dfrac{\dfrac{5}{8} - \dfrac{1}{4}}{\dfrac{2}{3} - \dfrac{1}{6}} \cdot \dfrac{8}{9}$

Evaluate the expression for the given values of the variables.

62. $x^2 + \dfrac{y}{z}$, when $x = -\dfrac{2}{3}$, $y = \dfrac{5}{8}$, and $z = \dfrac{3}{4}$

63. $\dfrac{x}{y} - z^2$, when $x = \dfrac{5}{6}$, $y = \dfrac{1}{3}$, and $z = -\dfrac{3}{4}$

64. $x - y^3 z$, when $x = \dfrac{5}{6}$, $y = \dfrac{1}{2}$, and $z = \dfrac{8}{9}$

65. $xy^3 + z$, when $x = \dfrac{9}{10}$, $y = \dfrac{1}{3}$, and $z = \dfrac{7}{15}$

Evaluate the expression for the given values of the variables.

66. $\dfrac{wx}{y} + z$, when $w = \dfrac{4}{5}$, $x = \dfrac{5}{8}$, $y = \dfrac{3}{4}$, and $z = \dfrac{2}{3}$

67. $\dfrac{w}{xy} - z$, when $w = 2\dfrac{1}{2}$, $x = 4$, $y = \dfrac{3}{8}$, and $z = \dfrac{2}{3}$

68. Is $-\dfrac{1}{2}$ a solution of the equation $\dfrac{-8z}{z + \dfrac{5}{6}} - 4z = -14$?

69. Is $-\dfrac{1}{3}$ a solution of the equation $\dfrac{12w}{\dfrac{1}{6} - w} = -7$?

CRITICAL THINKING

70. Simplify: $\dfrac{\dfrac{3}{x} + \dfrac{2}{x}}{\dfrac{5}{6}}$

71. A computer can perform 600,000 operations in one second. To the nearest minute, how many minutes will it take for the computer to perform 10^8 operations?

72. Given x is a whole number, for what value of x will the expression $\left(\dfrac{3}{4}\right)^2 + x^5 \div \dfrac{7}{8}$ have a minimum value? What is the minimum value?

73. Which of the variables u, v, w, x and y can be doubled so that $\dfrac{u + \dfrac{v}{w}}{\dfrac{x}{y}}$ is

(a) halved or (b) doubled?

74. Prepare a report on the Munsell system of designating colors. Are the "fractions" used in this system the same as those presented in this chapter? Explain.

75. A farmer died and left 17 horses to be divided among 3 children. The first child was to receive one-half of the horses, the second child one-third of the horses, and the third child one-ninth of the horses. The executor for the family's estate realized that 17 horses could not be divided by halves, thirds, or ninths and so added a neighbor's horse to the farmer's. With 18 horses, the executor gave 9 horses to the first child, 6 horses to the second child, and 2 horses to the third child. This accounted for the 17 horses, so the executor returned the borrowed horse to the neighbor. Explain why this worked.

PROJECTS IN MATHEMATICS

Music

In musical notation, notes are printed on a **staff**, which is a set of five horizontal lines and the spaces between them. The notes of a musical composition are grouped into **measures**, or **bars**. Vertical lines separate measures on a staff. The shape of a note indicates how long it should be held. The whole note has the longest time value of any note. Each time value is divided by two in order to find the next smallest note value.

The **time signature** is a fraction that appears at the beginning of a piece of music. The numerator of the fraction indicates the number of beats in a measure. The denominator indicates what kind of note receives one beat. For example, music written in $\frac{2}{4}$ time has 2 beats to a measure, and a quarter note receives one beat. One measure in $\frac{2}{4}$ time may have 1 half note, 2 quarter notes, 4 eighth notes, or any other combination of notes totaling 2 beats. Other common time signatures include $\frac{4}{4}$, $\frac{3}{4}$, and $\frac{6}{8}$.

Explain the meaning of the 6 and the 8 in the time signature $\frac{6}{8}$. Give some possible combinations of notes in one measure of a piece written in $\frac{4}{4}$ time.

What does a dot at the right of a note indicate? What is the effect of a dot at the right of a half note? a quarter note? an eighth note?

Symbols called rests are used to indicate periods of silence in a piece of music. What symbols are used to indicate the different time values of rests?

Find some examples of musical compositions written in different time signatures. Use a few measures from each to show that the sum of the time values of the notes and rests in each measure equals the numerator of the time signature.

Construction

Suppose you are involved in building your own home. Design a stairway from the first floor of the house to the second floor. Some of the questions you will need to answer include:

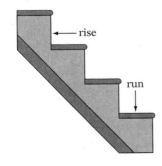

> What is the distance from the floor of the first story to the floor of the second story?
> Typically, what is the number of steps in a stairway?
> What is a reasonable length for the run of each step?
> What width wood is being used to build the staircase?

In designing the stairway, remember that each riser should be the same height and each run should be the same length. And the width of the wood used for the steps will have to be incorporated in the calculation.

The Trend of a Stock

A financial analyst tries to determine the *trend* of a stock (whether the stock is increasing or decreasing in price over a period of time). One way to do this is to make a line graph of the stock price. For each day of the week, the analyst records the closing price of the stock. Figure A, which is for Ford Motor Company, is an example of such a graph.

a. Select a stock from the New York Stock Exchange and graph on Figure B the daily closing price of the stock for three weeks. You must enter some scale numbers along the vertical axis that are appropriate for your stock.

b. After three weeks, on the basis of your graph, what is the trend of the stock? Explain your answer.

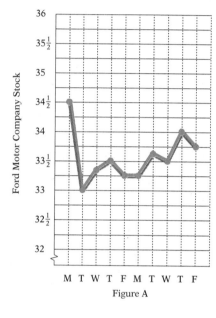

Figure A

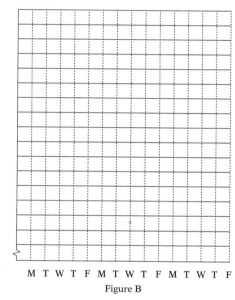

Figure B

*Errors in Operations
with Fractions*

For each of the following, explain the error that has been made.

a. $\dfrac{29}{97} = \dfrac{2}{7}; \dfrac{34}{48} = \dfrac{3}{8}; \dfrac{53}{57} = \dfrac{3}{7}$

b. $\dfrac{2}{3} + \dfrac{3}{4} = \dfrac{5}{7}; \dfrac{1}{5} + \dfrac{6}{7} = \dfrac{7}{12}; \dfrac{5}{9} + \dfrac{4}{7} = \dfrac{9}{16}$

c. $\dfrac{2}{3} \div \dfrac{5}{7} = \dfrac{15}{14}; \dfrac{3}{2} \div \dfrac{7}{9} = \dfrac{14}{27}; \dfrac{5}{9} \div \dfrac{2}{5} = \dfrac{18}{25}$

d. $\dfrac{2+3}{7+3} = \dfrac{2+\cancel{3}}{7+\cancel{3}} = \dfrac{3}{8}; \dfrac{\overset{1}{6}+4}{8+4} = \dfrac{6+\cancel{4}}{8+\cancel{4}} = \dfrac{7}{9}; \dfrac{\overset{1}{6}-5}{12-5} = \dfrac{6-\cancel{5}}{12-\cancel{5}} = \dfrac{5}{11}$

Puzzle from the Middle Ages

From the Middle Ages through the nineteenth century, it was quite common for the aristocracy to support court mathematicians and challenge other courts to contests between the mathematicians. In one such contest, Fibonacci (1170–1230), also known as Leonardo of Pisa, was given the following problem.

Find a fraction that is a square $\left(\text{for instance, } \frac{16}{25} \text{ because } \frac{16}{25} = \left(\frac{4}{5}\right)^2\right)$ and has the following property: If 5 is added to the fraction, the new fraction is still a square, and if 5 is subtracted from the original fraction, the new fraction is still a square. *Hint:* One of the fractions is $\frac{2{,}401}{144} = \left(\frac{49}{12}\right)^2$.

Tracking the Dow Jones Industrial Average

The Dow Jones Industrial Average is a gauge of the performance of the stock market. When the stock market is going up, it is referred to as a bullish market. When the stock market is going down, it is referred to as a bearish market. One way to measure the bullish or bearish tendency of the market is by determining how much a Dow Jones Industrial stock goes up or down.

$+1\frac{1}{2}$ or more			$-\frac{1}{8}$
$+1\frac{3}{8}$			$-\frac{1}{4}$
$+1\frac{1}{4}$			$-\frac{3}{8}$
$+1\frac{1}{8}$			$-\frac{1}{2}$
$+1$			$-\frac{5}{8}$
$+\frac{7}{8}$			$-\frac{3}{4}$
$+\frac{3}{4}$			$-\frac{7}{8}$
$+\frac{5}{8}$			-1
$+\frac{1}{2}$			$-1\frac{1}{8}$
$+\frac{3}{8}$			$-1\frac{1}{4}$
$+\frac{1}{4}$			$-1\frac{3}{8}$
$+\frac{1}{8}$			$-1\frac{1}{2}$ or less
Unchanged			

a. Find the names of the 30 stocks that compose the Dow Jones Industrial Average.

b. Using the business section of the newspaper, for each stock of the Dow Jones Industrial, place a tally mark (/) that indicates the change in the stock for that day. Update the table daily for a one-week period (Monday through Friday).

c. On the basis of the data you collected, does it appear that the Dow Jones Industrials are moving higher, moving lower, or staying about the same? Explain your answer.

CHAPTER SUMMARY

Key Words

A number that is a multiple of two or more numbers is a *common multiple* of those numbers. The *least common multiple (LCM)* is the smallest common multiple of two or more numbers.

A number that is a factor of two or more numbers is a *common factor* of those numbers. The *greatest common factor (GCF)* is the largest common factor of two or more numbers.

A *fraction* can represent the number of equal parts of a whole. In a fraction, the *fraction bar* separates the *numerator* and the *denominator*.

In a *proper fraction,* the numerator is smaller than the denominator; a proper fraction is a number less than 1. In an *improper fraction,* the numerator is greater than or equal to the denominator; an improper fraction is a number greater than or equal to 1. A *mixed number* is a number greater than 1 with a whole number part and a fractional part.

Equal fractions with different denominators are called *equivalent fractions.*

A fraction is in *simplest form* when the numerator and denominator have no common factors other than 1.

The *reciprocal* of a fraction is the fraction with the numerator and denominator interchanged.

A *complex fraction* is a fraction whose numerator or denominator contains one or more fractions.

Essential Rules

To find the LCM of two or more numbers, write the prime factorization of each number and circle the highest power of each prime factor. The LCM is the product of the circled factors.

To find the GCF of two or more numbers, write the prime factorization of each number and circle the lowest power of each prime factor that occurs in both factorizations. The GCF is the product of the circled factors.

To add fractions with the same denominators, add the numerators and place the sum over the common denominator.

To subtract fractions with the same denominators, subtract the numerators and place the difference over the common denominator.

To add or subtract fractions with different denominators, first rewrite the fractions as equivalent fractions with a common denominator. The common denominator is the least common multiple (LCM) of the denominators of the fractions. Then add or subtract the fractions.

To multiply two fractions, multiply the numerators and place the product over the product of the denominators.

To divide two fractions, multiply by the reciprocal of the divisor.

Multiplication Property of Equations
Each side of an equation can be multiplied by the same nonzero number without changing the solution of the equation.

To simplify a complex fraction, simplify the expression above the main fraction bar and simplify the expression below the main fraction bar. Then rewrite the complex fraction as a division problem by reading the main fraction bar as "divided by."

The Order of Operations Agreement
Step 1 Do all operations inside parentheses.
Step 2 Simplify any numerical expressions containing exponents.
Step 3 Do multiplication and division as they occur from left to right.
Step 4 Do addition and subtraction as they occur from left to right.

CHAPTER REVIEW EXERCISES

1. Write $\dfrac{19}{2}$ as a mixed number.

2. Subtract: $6\dfrac{2}{9} - 3\dfrac{7}{18}$

3. Evaluate $x \div y$ when $x = 2\dfrac{5}{8}$ and $y = 1\dfrac{3}{4}$.

4. Multiply: $\left(-2\dfrac{1}{3}\right) \cdot \dfrac{3}{7}$

5. Divide: $3\dfrac{3}{4} \div 1\dfrac{7}{8}$

6. Find the product of 3 and $\dfrac{8}{9}$.

7. Evaluate $\dfrac{x}{y+z}$ when $x = \dfrac{7}{8}$, $y = \dfrac{4}{5}$, and $z = -\dfrac{1}{2}$.

8. Place the correct symbol, $<$ or $>$, between the two numbers.

$\dfrac{3}{5} \quad \dfrac{7}{15}$

9. Find the LCM of 50 and 75.

10. Add: $6\dfrac{11}{15} + 4\dfrac{7}{10}$

11. Evaluate xy when $x = 8$ and $y = \dfrac{5}{12}$.

12. Express the shaded portion of the circles as an improper fraction and as a mixed number.

13. Place the correct symbol, $<$ or $>$, between the two numbers.

$\dfrac{7}{8} \quad \dfrac{17}{20}$

14. Simplify: $\dfrac{\dfrac{5}{8} - \dfrac{1}{4}}{\dfrac{1}{2} + \dfrac{1}{8}}$

15. Write a fraction that is equivalent to $\dfrac{4}{9}$ and has a denominator of 72.

16. Evaluate $x^2 y^3$ when $x = \dfrac{8}{9}$ and $y = -\dfrac{3}{4}$.

17. Evaluate $ab^2 - c$ when $a = 4$, $b = \dfrac{1}{2}$, and $c = \dfrac{5}{7}$.

18. Find the GCF of 42 and 63.

19. Write $2\dfrac{5}{14}$ as an improper fraction.

20. Evaluate $x + y + z$ when $x = \dfrac{5}{8}$, $y = -\dfrac{3}{4}$, and $z = \dfrac{1}{2}$.

21. Find the quotient of $\dfrac{5}{9}$ and $-\dfrac{2}{3}$.

22. Simplify: $\dfrac{2}{5} \div \dfrac{4}{7} + \dfrac{3}{8}$

23. Multiply: $5\frac{1}{4} \cdot \frac{8}{9} \cdot (-3)$

24. Find the difference between $\frac{2}{3}$ and $\frac{11}{18}$.

25. Subtract: $\frac{7}{8} - \left(-\frac{5}{6}\right)$

26. Evaluate $\left(-\frac{3}{8}\right)^2 \cdot 4^2$.

27. Find the sum of $3\frac{7}{12}$ and $5\frac{1}{2}$.

28. Write $\frac{30}{105}$ in the simplest form.

29. Evaluate $a - b$ when $a = 7$ and $b = 2\frac{3}{10}$.

30. Solve: $-\frac{5}{9} = \frac{1}{6} + p$

Solve.

31. *Time* What fractional part of an hour is 40 min?

32. *Business* Each day copies of the local newspaper are delivered to three different stores that have ordered 80, 120, and 160 copies, respectively. How many copies should be packaged together so that no package needs to be opened during delivery?

33. *Health* A wrestler is put on a diet to gain 12 lb in four weeks. The wrestler gains $3\frac{1}{2}$ lb the first week and $2\frac{1}{4}$ lb the second week. How much weight must the wrestler gain during the third and fourth weeks in order to gain a total of 12 lb?

34. *Business* An employee hired for piecework can assemble a unit in $2\frac{1}{2}$ min. How many units can this employee assemble during an 8-hour day?

35. *Wages* Find the overtime pay due an employee who worked $6\frac{1}{4}$ h of overtime this week. The employee's overtime rate is $24 an hour.

36. *Physics* What is the final velocity, in feet per second, of an object dropped from a plane with a starting velocity of 0 ft/s and a fall of $15\frac{1}{2}$ s? Use the formula $V = S + 32t$, where V is the final velocity of a falling object, S is its starting velocity, and t is the time of the fall.

CUMULATIVE REVIEW EXERCISES

1. Evaluate $3a + (a - b)^3$ when $a = 4$ and $b = 1$.

2. Find the product of 4 and $\frac{7}{8}$.

3. Add: $4\frac{7}{9} + 3\frac{5}{6}$

4. Subtract: $-42 - (-27)$

5. Find the GCF of 72 and 108.

6. Multiply: $3\frac{1}{13} \cdot 5\frac{1}{5}$

7. Find the quotient of $\frac{8}{9}$ and $-\frac{4}{5}$.

8. Subtract: $-\frac{2}{3} - \left(-\frac{2}{5}\right)$

9. Simplify: $\dfrac{\frac{1}{5} + \frac{1}{4}}{\frac{1}{4} - \frac{1}{5}}$

10. Place the correct symbol, $<$ or $>$, between the two numbers.

$\qquad \dfrac{7}{11} \qquad \dfrac{4}{5}$

11. Divide: $-2\frac{1}{3} \div 1\frac{2}{7}$

12. Multiply: $-\frac{3}{8} \cdot \frac{2}{5} \cdot \left(-\frac{4}{9}\right)$

13. Evaluate abc when $a = \frac{4}{7}$, $b = 1\frac{1}{6}$, and $c = 3$.

14. Subtract: $8\frac{3}{4} - 1\frac{5}{7}$

15. Subtract $-\frac{3}{8}$ from $\frac{7}{12}$.

16. Simplify: $\frac{2}{5} \div \frac{9 - 6}{3 + 7} + \left(-\frac{1}{2}\right)^2$

17. Evaluate $a - b$ when $a = \frac{3}{4}$ and $b = -\frac{7}{8}$.

18. Find the sum of $1\frac{9}{16}$ and $4\frac{5}{8}$.

19. Solve: $28 = -7y$

20. Write $\frac{41}{9}$ as a mixed number.

21. Find the difference between $\frac{5}{14}$ and $\frac{9}{42}$.

22. Evaluate $x^3 y^4$ when $x = \frac{7}{12}$ and $y = \frac{6}{7}$.

23. Evaluate $2a - (b - a)^2$ when $a = 2$ and $b = -3$.

24. Add: $6{,}847 + 3{,}501 + 924$

25. Evaluate $(x - y)^3 + 5x$ when $x = 8$ and $y = 6$.

26. Solve: $x + \dfrac{4}{5} = \dfrac{1}{4}$

27. Estimate the difference between 89,357 and 66,042.

28. Simplify: $-8 - (-12) - (-15) - 32$

29. Write $7\dfrac{3}{4}$ as an improper fraction.

30. Find the prime factorization of 140.

Solve.

31. _Health_   The chart at the right shows the calories burned per hour as a result of different aerobic activities. Suppose you weigh 150 lb. According to the chart, how many more calories would you burn by bicycling at 12 mph for 4 h than by walking at a rate of 3 mph for 5 h?

Activity	100 lb	150 lb
Bicycling, 6 mph	160	240
Bicycling, 12 mph	270	410
Jogging, 5 1/2 mph	440	660
Jogging, 7 mph	610	920
Jumping rope	500	750
Tennis, singles	265	400
Walking, 2 mph	160	240
Walking, 3 mph	210	320
Walking, 4 1/2 mph	295	440

32. _Demography_   It is projected that the population along the eastern seaboard, from Boston, Massachusetts, to Brunswick, Maine, will increase to 700,000 in 2010 from 370,000 in 1980. Find the projected increase in the population in that area during the 30-year period.

33. _Investments_   A stockholder sold 150 shares of a stock for $\$38\dfrac{1}{4}$ per share. The stockholder had purchased the stock at $\$27\dfrac{3}{8}$ per share. Find the stockholder's gain per share.

34. _Travel_   A bicyclist rode for $\dfrac{3}{4}$ h at a rate of $5\dfrac{1}{2}$ mph. Use the equation $d = rt$, where d is the distance traveled, r is the rate of travel, and t is the time, to find the distance traveled by the bicyclist.

35. _Travel_   Use the formula $d = rt$, where d is the distance traveled, r is the rate of speed, and t is the time, to find the speed of a car that travels 450 mi in 10 h.

36. _Sports_   The pressure on a submerged object is given by $P = 15 + \dfrac{1}{2}D$, where D is the depth in feet and P is the pressure measured in pounds per square inch. Find the pressure on a diver who is at a depth of $14\dfrac{3}{4}$ ft.

CHAPTER

Decimals and Real Numbers

FOCUS On Problem Solving

As you progress in your study of algebra, you will find that the problems become less concrete and more abstract. Problems that are concrete provide information pertaining to a specific instance. Abstract problems are theoretical; they are stated without reference to a specific instance. Let's look at an example of an abstract problem.

How many cents are in d dollars?

How can you solve this problem? Are you able to solve the same problem if the information given is concrete?

How many cents are in 5 dollars?

You know that there are 100 cents in 1 dollar. To find the number of cents in 5 dollars, multiply 5 by 100.

$100 \cdot 5 = 500$ There are 500 cents in 5 dollars.

Use the same procedure to find the number of cents in d dollars: multiply d by 100.

$100 \cdot d = 100d$ There are $100d$ cents in d dollars.

This problem might be taken a step further.

If one pen costs c cents, how many pens can be purchased with d dollars?

Consider the same problem using numbers in place of variables.

If one pen costs 25 cents, how many pens can be purchased with 2 dollars?

To solve this problem, you need to calculate the number of cents in 2 dollars (multiply 2 by 100), and divide the result by the cost per pen (25 cents).

$\dfrac{100 \cdot 2}{25} = \dfrac{200}{25} = 8$ If one pen costs 25 cents, 8 pens can be purchased with 2 dollars.

Use the same procedure to solve the related abstract problem. Calculate the number of cents in d dollars (multiply d by 100), and divide the result by the cost per pen (c cents).

$\dfrac{100 \cdot d}{c} = \dfrac{100d}{c}$ If one pen costs c cents, $\frac{100d}{c}$ pens can be purchased with d dollars.

At the heart of the study of algebra is the use of variables. It is the variables in the problems above that make them abstract. But it is variables that allow us to generalize situations and state rules about mathematics.

Try each of the following problems.

1. If you travel m miles on one gallon of gasoline, how far can you travel on g gallons of gasoline?

2. If you walk a mile in x minutes, how far can you walk in h hours?

3. If one photocopy costs n nickels, how many photocopies can you make for q quarters?

Answers:

1. If you travel m miles on one gallon of gasoline, you can travel gm miles on g gallons of gasoline.

2. If you walk a mile in x minutes, you can walk $\frac{60h}{x}$ miles in h hours.

3. If one photocopy costs n nickels, you can make $\frac{5q}{n}$ photocopies for q quarters.

SECTION 4.1   Introduction to Decimals

OBJECTIVE A
Place value

The price tag on a sweater reads $31.88. The number 31.88 is in **decimal notation.** A number written in decimal notation is often called simply a **decimal.**

A number written in decimal notation has three parts.

31	.	88
Whole number part	Decimal point	Decimal part

The decimal part of the number represents a number less than one. For example, $.88 is less than one dollar. The decimal point (.) separates the whole number part from the decimal part.

The position of a digit in a decimal determines the digit's place value. The place-value chart is extended to the right to show the place value of digits to the right of a decimal point.

In the decimal 458.302719, the position of the digit 7 determines that its place value is ten-thousandths.

Note the relationship between fractions and numbers written in decimal notation.

seven tenths	seven hundredths	seven thousandths
$\frac{7}{10} = 0.7$	$\frac{7}{100} = 0.07$	$\frac{7}{1,000} = 0.007$
1 zero in 10	2 zeros in 100	3 zeros in 1,000
1 decimal place in 0.7	2 decimal places in 0.07	3 decimal places in 0.007

To write a decimal in words, write the decimal part of the number as if it were a whole number, then name the place value of the last digit.

0.9684   nine thousand six hundred eighty-four ten-thousandths

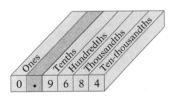

The decimal point in a decimal is read as "and."

372.516   three hundred seventy-two and five hundred sixteen thousandths

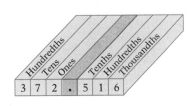

POINT OF INTEREST

The idea that all fractions should be represented in tenths, hundredths, and thousandths was presented in 1585 in Simon Stevin's publication *De Thiende* and its French translation, *La Disme*, which was well read and accepted by the French. This may help to explain why the French accepted the metric system so easily two hundred years later.

In *De Thiende*, Stevin argued in favor of his notation by including examples for astronomers, tapestry makers, surveyors, tailors, and the like. He stated that using decimals would enable calculations to be "performed. . . with as much ease as counterreckoning."

To write a decimal in standard form when it is written in words, write the whole number part, replace the word "and" with a decimal point, and write the decimal part so that the last digit is in the given place-value position.

four and twenty-three <u>hundredths</u>

3 is in the hundredths' place.   4.23

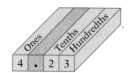

When writing a decimal in standard form, you may need to insert zeros after the decimal point so that the last digit is in the given place-value position.

ninety-one and eight <u>thousandths</u>

8 is in the thousandths' place.   91.008
Insert 2 zeros so that the 8 is in
the thousandths' place.

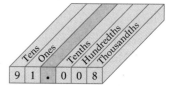

sixty-five <u>ten-thousandths</u>

5 is in the ten-thousandths' place.   0.0065
Insert 2 zeros so that the 5 is in
the ten-thousandths' place.

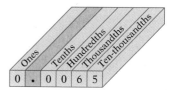

Example 1 Name the place value of the digit 8 in the number 45.687.

Solution The digit 8 is in the hundredths' place.

You Try It 1 Name the place value of the digit 4 in the number 907.1342.

Your Solution

Example 2 Write $\frac{43}{100}$ as a decimal.

Solution $\frac{43}{100} = 0.43$
[forty-three hundredths]

You Try It 2 Write $\frac{501}{1,000}$ as a decimal.

Your Solution

Example 3 Write 0.289 as a fraction.

Solution $0.289 = \frac{289}{1,000}$
[289 thousandths]

You Try It 3 Write 0.67 as a fraction.

Your Solution

Example 4 Write 293.50816 in words.

Solution two hundred ninety-three and fifty thousand eight hundred sixteen hundred-thousandths

You Try It 4 Write 55.6083 in words.

Your Solution

Solutions on pp. A14–A15

| | | | | |

Example 5   Write twenty-three and two
hundred forty-seven millionths
in standard form.

You Try It 5   Write eight hundred six and
four hundred ninety-one
hundred-thousandths in
standard form.

Solution   23.000247

Your Solution

Solution on p. A15 | | | | | |

OBJECTIVE B

Order relations between decimals

A whole number can be written as a decimal by writing a decimal point to the right of the last digit. For example:

62 = 62.   497 = 497.

You know that $62 and $62.00 both represent sixty-two dollars. Any number of zeros may be written to the right of the decimal point in a whole number without changing the value of the number.

62 = 62.00 = 62.0000   497 = 497.0 = 497.000

Also, any number of zeros may be written to the right of the last digit in a decimal without changing the value of the number.

0.8 = 0.80 = 0.800   1.35 = 1.350 = 1.3500 = 1.35000 = 1.350000

This fact is used to find the order relation between two decimals.

To compare two decimals, write the decimal part of each number so that each has the same number of decimal places. Then compare the two numbers.

Place the correct symbol, < or >, between the two numbers 0.693 and 0.71.

0.693 has 3 decimal places.
0.71 has 2 decimal places.
Write 0.71 with 3 decimal places.   0.71 = 0.710

Compare 0.693 and 0.710.
693 thousandths < 710 thousandths   0.693 < 0.710

Remove the zero written in 0.710.   0.693 < 0.71

Place the correct symbol, < or >, between the two numbers 5.8 and 5.493.

Write 5.8 with 3 decimal places.   5.8 = 5.800

Compare 5.800 and 5.493.
The whole number part (5) is the same.
800 thousandths > 493 thousandths   5.800 > 5.493

Remove the extra zeros written in 5.800.   5.8 > 5.493

POINT OF INTEREST

The decimal point did not make its appearance until the early 1600s. Stevin's notation used subscripts with circles around them after each digit: 0 for ones, 1 for tenths (which he called "primes"), 2 for hundredths (called "seconds"), 3 for thousandths ("thirds"), and so on. For example, 1.375 would be written:

1   3   7   5
⓪   ①   ②   ③

Example 6 Place the correct symbol, < or >, between the two numbers.

0.039   0.1001

Solution 0.039 = 0.0390

0.0390 < 0.1001
0.039 < 0.1001

You Try It 6 Place the correct symbol, < or >, between the two numbers.

0.065   0.0802

Your Solution

Example 7 Write the given numbers in order from smallest to largest.

1.01, 1.2, 1.002, 1.1, 1.12

Solution 1.010, 1.200, 1.002, 1.100, 1.120
1.002, 1.010, 1.100, 1.120, 1.200

1.002, 1.01, 1.1, 1.12, 1.2

You Try It 7 Write the given numbers in order from smallest to largest.

3.03, 0.33, 0.3, 3.3, 0.03

Your Solution

Solutions on p. A15 | | | | |

OBJECTIVE C
Rounding

In general, rounding decimals is similar to rounding whole numbers except that the digits to the right of the given place value are dropped instead of being replaced by zeros.

If the digit to the right of the given place value is less than 5, that digit and all digits to the right are dropped.

Round 6.9237 to the nearest hundredth.

Given place value (hundredths)

6.9237

3 < 5   Drop the digits 3 and 7.

6.9237 rounded to the nearest hundredth is 6.92.

If the digit to the right of the given place value is greater than or equal to 5, increase the digit in the given place value by 1, and drop all digits to its right.

Round 12.385 to the nearest tenth.

Given place value (tenths)

12.385

8 > 5   Increase 3 by 1 and drop all digits to the right of 3.

12.385 rounded to the nearest tenth is 12.4.

Round 0.46972 to the nearest thousandth.

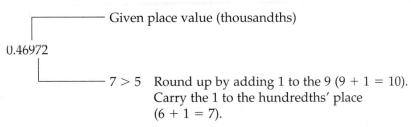

Given place value (thousandths)

0.46972

7 > 5   Round up by adding 1 to the 9 (9 + 1 = 10).
Carry the 1 to the hundredths' place
(6 + 1 = 7).

0.46972 rounded to the nearest thousandth is 0.470.

Note that in this example, the zero in the given place value is not dropped. This indicates that the number is rounded to the nearest thousandth. If we dropped the zero and wrote 0.47, it would indicate that the number was rounded to the nearest hundredth.

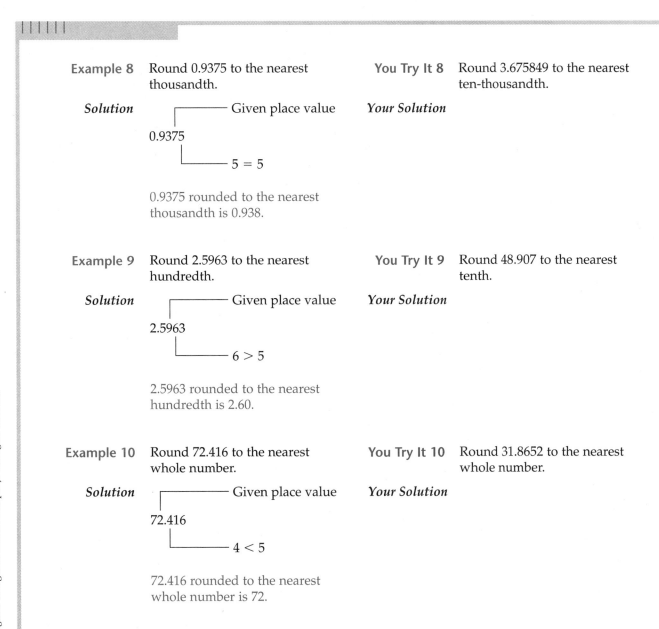

Example 8   Round 0.9375 to the nearest thousandth.

Solution

Given place value

0.9375

5 = 5

0.9375 rounded to the nearest thousandth is 0.938.

You Try It 8   Round 3.675849 to the nearest ten-thousandth.

Your Solution

Example 9   Round 2.5963 to the nearest hundredth.

Solution

Given place value

2.5963

6 > 5

2.5963 rounded to the nearest hundredth is 2.60.

You Try It 9   Round 48.907 to the nearest tenth.

Your Solution

Example 10   Round 72.416 to the nearest whole number.

Solution

Given place value

72.416

4 < 5

72.416 rounded to the nearest whole number is 72.

You Try It 10   Round 31.8652 to the nearest whole number.

Your Solution

Solutions on p. A15

OBJECTIVE D
Applications

6 CT

The table below shows the number of home runs hit, for every 100 times at bat, by four Major League baseball players. Use this table for Example 11 and You Try It 11.

Home Runs Hit for Every 100 At-Bats	
Harmon Killebrew	7.03
Ralph Kiner	7.09
Babe Ruth	8.05
Ted Williams	6.76

Source: Major League Baseball

Example 11

According to the table above, who had more home runs for every 100 times at bat, Ted Williams or Babe Ruth?

Strategy

To determine who had more home runs for every 100 times at bat, compare the numbers 6.76 and 8.05.

Solution

8.05 > 6.76

Babe Ruth had more home runs for every 100 at-bats.

You Try It 11

According to the table above, who had more home runs for every 100 times at bat, Harmon Killebrew or Ralph Kiner?

Your Strategy

Your Solution

Example 12

On average, an American goes to the movies 4.56 times per year. To the nearest whole number, how many times per year does an American go to the movies?

Strategy

To find the number, round 4.56 to the nearest whole number.

Solution

4.56 rounded to the nearest whole number is 5.

An American goes to the movies about 5 times per year.

You Try It 12

In the 1996 Summer Olympics, Michael Johnson ran the 200 m race in 19.32 s. To the nearest second, how fast did Johnson run the 200 m?

Your Strategy

Your Solution

Solutions on p. A15

4.1 EXERCISES

OBJECTIVE A

Name the place value of the digit 5.

1. 76.31587

2. 291.508

3. 432.09157

4. 0.0006512

5. 38.2591

6. 0.0000853

Write the fraction as a decimal.

7. $\dfrac{3}{10}$

8. $\dfrac{9}{10}$

9. $\dfrac{21}{100}$

10. $\dfrac{87}{100}$

11. $\dfrac{461}{1,000}$

12. $\dfrac{853}{1,000}$

13. $\dfrac{93}{1,000}$

14. $\dfrac{61}{1,000}$

Write the decimal as a fraction.

15. 0.1

16. 0.3

17. 0.47

18. 0.59

19. 0.289

20. 0.601

21. 0.09

22. 0.013

Write the number in words.

23. 0.37

24. 25.6

25. 9.4

26. 1.004

27. 0.0053

28. 41.108

29. 0.045

30. 3.157

31. 26.04

Write the number in standard form.

32. six hundred seventy-two thousandths

33. three and eight hundred six ten-thousandths

34. nine and four hundred seven ten-thousandths

35. four hundred seven and three hundredths

36. six hundred twelve and seven hundred four thousandths

37. two hundred forty-six and twenty-four thousandths

38. two thousand sixty-seven and nine thousand two ten-thousandths

39. seventy-three and two thousand six hundred eighty-four hundred-thousandths

OBJECTIVE B

Place the correct symbol, < or >, between the two numbers.

40. 0.16 0.6

41. 0.7 0.56

42. 5.54 5.45

43. 3.605 3.065

44. 0.047 0.407

45. 9.004 9.04

46. 1.0008 1.008

47. 9.31 9.031

48. 7.6005 7.605

49. 4.6 40.6

50. 0.31502 0.3152

51. 0.07046 0.07036

Write the given numbers in order from smallest to largest.

52. 0.39, 0.309, 0.399

53. 0.66, 0.699, 0.696, 0.609

54. 0.24, 0.024, 0.204, 0.0024

55. 1.327, 1.237, 1.732, 1.372

56. 0.06, 0.059, 0.061, 0.0061

57. 21.87, 21.875, 21.805, 21.78

OBJECTIVE C

Round the number to the given place value.

58. 6.249   Tenths

59. 5.398   Tenths

60. 21.007   Tenths

61. 30.0092   Tenths

62. 18.40937   Hundredths

63. 413.5972   Hundredths

64. 72.4983   Hundredths

65. 6.061745   Thousandths

66. 936.2905   Thousandths

67. 96.8027   Whole number

68. 47.3192   Whole number

69. 5,439.83   Whole number

70. 7,014.96   Whole number

71. 0.023591   Ten-thousandths

72. 2.975268   Hundred-thousandths

OBJECTIVE D

Solve.

73. *Measurement*   A nickel weighs about 0.1763668 oz. Find the weight of a nickel to the nearest hundredth of an ounce.

74. *Business*   The total cost of a parka, including sales tax, is $83.7188. Round the total cost to the nearest cent to find the amount a customer pays for the parka.

75. *Sports*   Runners in the Boston Marathon run a distance of 26.21875 mi. To the nearest tenth of a mile, find the distance an entrant in the Boston Marathon runs.

76. *Sports*   The table at the right lists National Football League leaders in passing efficiency. Who had the greater average gain, Sonny Jurgensen or Roger Staubach?

Football Player	Average Gain
Sonny Jurgensen	7.65
Bart Star	7.85
Roger Staubach	7.67
Fran Tarkenton	7.27

Source: 1996 Information Please Sports Almanac

77. *Health*   The average life expectancy in Great Britain is 75.3 years. The average life expectancy in Italy is 75.5 years. In which country is the average life expectancy higher, Great Britain or Italy?

78. *Economics*   On Thursday of a recent week, a British pound would have been exchanged for 1.5030 U.S. dollars. On Friday of the same week, a British pound would have been exchanged for 1.5067 U.S. dollars. On which day was the British pound worth more money in U.S. dollars, Thursday or Friday?

Solve.

79. *Economics* In the first quarter of 1996, the gross domestic product (GDP) was $6,823.6 billion. What was the GDP to the nearest billion dollars?

80. *Consumerism* Charge accounts generally require a minimum payment on the balance in the account each month. Use the minimum payment schedule shown below to determine the minimum payment due on the given account balances.

a. $187.93
b. $342.55
c. $261.48
d. $16.99
e. $310.00
f. $158.32
g. $200.10

If the New Balance Is:	The Minimum Required Payment Is:
Up to $20.00	The new balance
$20.01 to $200.00	$20.00
$200.01 to $250.00	$25.00
$250.01 to $300.00	$30.00
$300.01 to $350.00	$35.00
$350.01 to $400.00	$40.00

81. *Consumerism* Shipping and handling charges on catalog mail orders generally are based on the dollar amount of the order. Use the table shown below to determine the cost of shipping each order.

a. $12.42
b. $23.56
c. $47.80
d. $66.91
e. $35.75
f. $20.00
g. $18.25

If the Amount Ordered Is:	The Shipping and Handling Charge Is:
$10.00 and under	$1.60
$10.01 to $20.00	$2.40
$20.01 to $30.00	$3.60
$30.01 to $40.00	$4.70
$40.01 to $50.00	$6.00
$50.01 and up	$7.00

CRITICAL THINKING

82. Indicate which digits of the number, if any, need not be entered on a calculator.
   a. 1.500   b. 0.908   c. 60.07   d. 0.0032

83. Find a number between (a) 0.1 and 0.2, (b) 1 and 1.1, and (c) 0 and 0.005.

84. To what place value are timed events in the Olympics recorded? Provide some specific examples of events and the winning times in each.

85. Provide an example of a situation in which a decimal is always rounded up, even if the digit to the right is less than 5. Provide an example of a situation in which a decimal is always rounded down, even if the digit to the right is 5 or greater than 5. (*Hint:* Think about situations in which money changes hands.)

86. Prepare a report on the Richter scale. Include in your report the magnitudes that classify an earthquake as strong or moderate, the magnitudes that classify an earthquake as a microearthquake, and the largest known recorded shocks.

SECTION 4.2   Operations on Decimals

OBJECTIVE A

Addition and subtraction of decimals

To add decimals, write the numbers so that the decimal points are on a vertical line. Add as you would with whole numbers. Then write the decimal point in the sum directly below the decimal points in the addends.

Add: 0.326 + 4.8 + 57.23

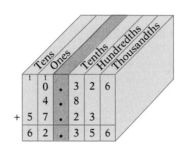

Note that by placing the decimal points on a vertical line, digits of the same place value are added.

> **POINT OF INTEREST**
>
> **Try this: Six different numbers are added together and their sum is 11. Four of the six numbers are 4, 3, 2, and 1. Find the other two numbers.**

Find the sum of 0.64, 8.731, 12, and 5.9.

Arrange the numbers vertically, placing the decimal points on a vertical line.

Add the numbers in each column.

Write the decimal point in the sum directly below the decimal points in the addends.

$$
\begin{array}{r}
1\,2 \\
0.64 \\
8.731 \\
12. \\
+\ 5.9 \\
\hline
27.271
\end{array}
$$

To subtract decimals, write the numbers so that the decimal points are on a vertical line. Subtract as you would with whole numbers. Then write the decimal point in the difference directly below the decimal point in the subtrahend.

Subtract and check: 31.642 − 8.759

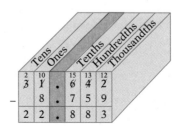

Note that by placing the decimal points on a vertical line, digits of the same place value are subtracted.

Check: Subtrahend   8.759
         + Difference   + 22.883
         = Minuend   31.642

Subtract and check: 5.4 − 1.6832

Insert zeros in the minuend so that it has the same number of decimal places as the subtrahend.

$$\begin{array}{r} 5.4000 \\ -\ 1.6832 \end{array}$$

Subtract and then check.

$$\begin{array}{r} {\scriptstyle 4\ 13\ 9\ 9\ 10} \\ \cancel{5.4000} \\ -\ 1.6832 \\ \hline 3.7168 \end{array} \qquad \text{Check:} \quad \begin{array}{r} 1.6832 \\ +\ 3.7168 \\ \hline 5.4000 \end{array}$$

Figure 4.1 shows the combined federal and state gasoline taxes, per gallon, for Connecticut, Rhode Island, Montana, Nebraska, and West Virginia. What is the difference between the taxes per gallon in Connecticut and West Virginia?

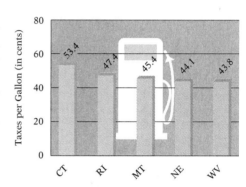

Figure 4.1 Combined Federal and State Gasoline Taxes *Source:* Tax Foundation

Subtract the taxes per gallon in West Virginia (43.8) from the taxes per gallon in Connecticut (53.4).

$53.4 - 43.8 = 9.6$

The difference is 9.6¢ per gallon.

The sign rules for adding and subtracting decimals are the same rules used to add and subtract integers.

Simplify: $-36.087 + 54.29$

The signs of the addends are different. Subtract the smaller absolute value from the larger absolute value.

$54.29 - 36.087 = 18.203$

Attach the sign of the number with the larger absolute value.

$|54.29| > |-36.087|$

The sum is positive.

$-36.087 + 54.29 = 18.203$

Recall that the opposite or additive inverse of n is $-n$ and the opposite of $-n$ is n. To find the opposite of a number, change the sign of the number.

Simplify: $-2.86 - 10.3$

Rewrite subtraction as addition of the opposite. The opposite of 10.3 is −10.3.

$-2.86 - 10.3$

$= -2.86 + (-10.3)$

The signs of the addends are the same. Add the absolute values of the numbers. Attach the sign of the addends.

$= -13.16$

Evaluate $c - d$ when $c = 6.731$ and $d = -2.48$.

Replace c with 6.731 and d with -2.48.

Rewrite subtraction as addition of the opposite.

Add.

$$c - d$$
$$6.731 - (-2.48)$$
$$= 6.731 + 2.48$$
$$= 9.211$$

Recall that to estimate the answer to a calculation, round each number to the highest place value of the number; the first digit of each number will be nonzero and all other digits will be zero. Perform the calculation using the rounded numbers.

Estimate the sum of 23.037 and 16.7892.

Round each number to the nearest ten.

$$23.037 \longrightarrow \quad 20$$
$$16.7892 \longrightarrow \underline{+ 20}$$

Add the rounded numbers.

$$40$$

40 is an estimate of the sum of 23.037 and 16.7892. Note that 40 is very close to the actual sum of 39.8262.

$$23.037$$
$$\underline{+ 16.7892}$$
$$39.8262$$

When a number in an estimation is a decimal less than one, round the decimal so that there is one nonzero digit.

Estimate the difference between 4.895 and 0.6193.

Round 4.895 to the nearest one.
Round 0.6193 to the nearest tenth.
Subtract the rounded numbers.

$$4.895 \longrightarrow \quad 5.0$$
$$0.6193 \longrightarrow \underline{- 0.6}$$
$$4.4$$

4.4 is an estimate of the difference between 4.895 and 0.6193.
It is close to the actual difference of 4.2757.

$$4.8950$$
$$\underline{- 0.6193}$$
$$4.2757$$

Example 1   Add: $35.8 + 182.406 + 71.0934$

Solution
$$\begin{array}{r} \overset{1\ \ 1}{} \\ 35.8 \\ 182.406 \\ \underline{+\ 71.0934} \\ 289.2994 \end{array}$$

You Try It 1   Add: $8.64 + 52.7 + 0.39105$

Your Solution

Example 2   What is -251.49 more than -638.7?

Solution $-638.7 + (-251.49) = -890.19$

You Try It 2   What is 4.002 minus 9.378?

Your Solution

Solutions on p. A15

Example 3 Subtract and check: $73 - 8.16$

Solution

_{6 12 9 10}
$$\begin{array}{r} 73.00 \\ -\ 8.16 \\ \hline 64.84 \end{array}$$

Check:
$$\begin{array}{r} 8.16 \\ +64.84 \\ \hline 73.00 \end{array}$$

You Try It 3 Subtract and check: $25 - 4.91$

Your Solution

Example 4 Estimate the sum of 0.3927, 0.4856, and 0.2104.

Solution

$$\begin{array}{r} 0.3927 \longrightarrow\ \ 0.4 \\ 0.4856 \longrightarrow\ \ 0.5 \\ 0.2104 \longrightarrow +0.2 \\ \hline 1.1 \end{array}$$

You Try It 4 Estimate the sum of 6.514, 8.903, and 2.275.

Your Solution

Example 5 Evaluate $x + y + z$ when $x = -1.6$, $y = 7.9$, and $z = -4.8$.

Solution $x + y + z$
$$-1.6 + 7.9 + (-4.8) = 6.3 + (-4.8)$$
$$= 1.5$$

You Try It 5 Evaluate $x + y + z$ when $x = -7.84$, $y = -3.05$, and $z = 2.19$.

Your Solution

Example 6 Is -4.3 a solution of the equation $9.7 - b = 5.4$?

Solution

$$\begin{array}{c|c} \multicolumn{2}{c}{9.7 - b = 5.4} \\ \hline 9.7 - (-4.3) & 5.4 \\ 9.7 + 4.3 & 5.4 \\ 14.0 \neq 5.4 \end{array}$$

No, -4.3 is not a solution of the equation.

You Try It 6 Is -23.8 a solution of the equation $-m + 16.9 = 40.7$?

Your Solution

Solutions on p. A15

OBJECTIVE B

Multiplication of decimals

Decimals are multiplied as if they were whole numbers; then the decimal point is placed in the product. Writing the decimals as fractions shows where to write the decimal point in the product.

$$0.4 \cdot 2 = \frac{4}{10} \cdot \frac{2}{1} = \frac{8}{10} = 0.8$$

1 decimal place in 0.4   1 decimal place in 0.8

$$0.4 \cdot 0.2 = \frac{4}{10} \cdot \frac{2}{10} = \frac{8}{100} = 0.08$$

1 decimal place in 0.4   2 decimal places in 0.08
1 decimal place in 0.2

$$0.4 \cdot 0.02 = \frac{4}{10} \cdot \frac{2}{100} = \frac{8}{1,000} = 0.008$$

1 decimal place in 0.4   3 decimal places in 0.008
2 decimal places in 0.02

To multiply decimals, multiply the numbers as you would whole numbers. Then write the decimal point in the product so that the number of decimal places in the product is the sum of the numbers of decimal places in the factors.

Multiply: (32.41)(7.6)

$$
\begin{array}{rl}
32.41 & \text{2 decimal places} \\
\times \quad 7.6 & \text{1 decimal place} \\
\hline
19446 & \\
22687 \quad\;\; & \\
\hline
246.316 & \text{3 decimal places}
\end{array}
$$

Estimating the product of 32.41 and 7.6 shows that the decimal point has been correctly placed.

Round 32.41 to the nearest ten.   32.41 $\longrightarrow$ 30
Round 7.6 to the nearest one.   7.6 $\longrightarrow$ $\times$ 8
Multiply the two numbers.   240

240 is an estimate of (32.41)(7.6). It is close to the actual product 246.316.

Multiply: 0.061(0.08)

$$
\begin{array}{rl}
0.061 & \text{3 decimal places} \\
\times \;\; 0.08 & \text{2 decimal places} \\
\hline
0.00488 & \text{5 decimal places}
\end{array}
$$

Insert two zeros between the 4 and the decimal point so that there are 5 decimal places in the product.

Figure 4.2 shows the classified advertising rates for two magazines, *Family Circle* and *McCall's*. What would it cost you to place a one-time ad consisting of 25 words in *McCall's*?

Multiply the rate per word (29.95) by the number of words (25).

29.95(25) = 748.75

The cost would be $748.75.

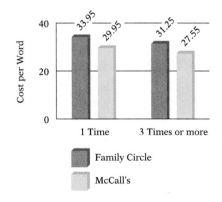

Figure 4.2 Classified Advertising Rates

To multiply a decimal by a power of 10 (10, 100, 1,000, . . .), move the decimal point to the right the same number of places as there are zeros in the power of 10.

$2.7935 \cdot \underline{10}$ = 27.935

1 zero   1 decimal place

$2.7935 \cdot \underline{100}$ = 279.35

2 zeros   2 decimal places

$2.7935 \cdot \underline{1,000}$ = 2,793.5

3 zeros   3 decimal places

$2.7935 \cdot \underline{10,000}$ = 27,935.

4 zeros   4 decimal places

$2.7935 \cdot \underline{100,000}$ = 279,350.   A zero must be inserted before the decimal point.

5 zeros   5 decimal places

Note that if the power of 10 is written in exponential notation, the exponent indicates how many places to move the decimal point.

$2.7935 \cdot 10^1 = 27.935$

1 decimal place

$2.7935 \cdot 10^2 = 279.35$

2 decimal places

$2.7935 \cdot 10^3 = 2,793.5$

3 decimal places

$2.7935 \cdot 10^4 = 27,935.$

4 decimal places

$2.7935 \cdot 10^5 = 279,350.$

5 decimal places

Find the product of 64.18 and 10^3.

The exponent on 10 is 3. Move the decimal point in 64.18 three places to the right. $64.18 \cdot 10^3 = 64,180$

Evaluate $100x$ when $x = 5.714$.

$100x$

Replace x with 5.714. $100(5.714)$

Multiply. There are two zeros in 100. Move the decimal point two places to the right. $= 571.4$

The sign rules for multiplying decimals are the same rules used to multiply integers.

The product of two numbers with the same sign is positive.
The product of two numbers with different signs is negative.

Multiply: $(-3.2)(-0.008)$

The signs are the same.
The product is positive.
Multiply the absolute values of
the numbers. $(-3.2)(-0.008) = 0.0256$

Is -0.6 a solution of the equation $4.3a = -2.58$?

$$4.3a = -2.58$$

Replace a by -0.6 and then simplify. $4.3(-0.6) \mid -2.58$
The results are equal. $-2.58 = -2.58$

Yes, -0.6 is a solution of the equation.

Example 7   Multiply: $0.00073(0.052)$

Solution $\begin{array}{r} 0.00073 \\ \times\ \ 0.052 \\ \hline 146 \\ 365 \\ \hline 0.00003796 \end{array}$

You Try It 7   Multiply: $0.000081(0.025)$

Your Solution

Example 8   Estimate the product of 0.7639 and 0.2188.

Solution $\begin{array}{r} 0.7639 \longrightarrow 0.8 \\ 0.2188 \longrightarrow \times\ 0.2 \\ \hline 0.16 \end{array}$

You Try It 8   Estimate the product of 6.407 and 0.959.

Your Solution

Example 9   What is 835.294 multiplied by 1,000?

Solution $835.294 \cdot 1,000 = 835,294$

You Try It 9   Find the product of 1.756 and 10^4.

Your Solution

Example 10   Multiply: $-3.42(6.1)$

Solution $-3.42(6.1) = -20.862$

You Try It 10   Multiply: $(-0.7)(-5.8)$

Your Solution

Solutions on p. A15

Example 11 Evaluate $50ab$ when $a = -0.9$ and $b = -0.2$.

Solution $50ab$
$50(-0.9)(-0.2) = -45(-0.2)$
$= 9$

You Try It 11 Evaluate $25xy$ when $x = -0.8$ and $y = 0.6$.

Your Solution

Solution on p. A15

OBJECTIVE C
Division of decimals

To divide decimals, move the decimal point in the divisor to the right so that the divisor is a whole number. Move the decimal point in the dividend the same number of places to the right. Place the decimal point in the quotient directly above the decimal point in the dividend. Then divide as you would with whole numbers.

Divide: $29.585 \div 4.85$

$$4.85.\overline{)29.58.5}$$

Move the decimal point 2 places to the right in the divisor. Move the decimal point 2 places to the right in the dividend. Place the decimal point in the quotient. Then divide as shown at the right.

$$
\begin{array}{r}
6.1 \\
485.\overline{)2958.5} \\
-2910 \\
\hline
485 \\
-485 \\
\hline
0
\end{array}
$$

Moving the decimal point the same number of places in the divisor and the dividend does not change the quotient because the process is the same as multiplying the numerator and denominator of a fraction by the same number. For the last example,

$$4.85\overline{)29.585} = \frac{29.585}{4.85} = \frac{29.585 \cdot 100}{4.85 \cdot 100} = \frac{2958.5}{485} = 485\overline{)2958.5}$$

In division of decimals, rather than writing the quotient with a remainder, the quotient is usually rounded off to a specified place value. The symbol $\approx$, which is read "is approximately equal to," is used to indicate that the quotient is an approximate value after being rounded off.

Divide and round to the nearest tenth: $0.86 \div 0.7$.

$$
\begin{array}{r}
1.22 \approx 1.2 \\
0.7.\overline{)0.8.60} \\
-7 \\
\hline
1\,6 \\
-1\,4 \\
\hline
20 \\
-14 \\
\hline
6
\end{array}
$$

⟵ To round the quotient to the nearest tenth, the division must be carried to the hundredths' place. Therefore, zeros must be inserted in the dividend so that the quotient has a digit in the hundredths' place.

Figure 4.3 shows average hourly earnings in the United States. How many times greater were the average hourly earnings in 1990 than in 1970? Round to the nearest whole number.

Divide the 1990 average hourly earnings (10.02) by those in 1970 (3.23).

$10.02 \div 3.23 \approx 3$

The average hourly earnings in 1990 were about 3 times greater than in 1970.

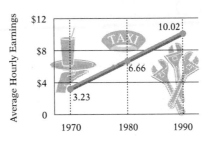

Figure 4.3   Average Hourly Earnings in Current Dollars by Private Industry
Source: 1994 Statistical Abstract

To divide a decimal by a power of 10 (10, 100, 1,000, 10,000, . . .), move the decimal point to the left the same number of places as there are zeros in the power of 10.

$462.81 \div 1\underline{0}$ $= 46.281$

       1 zero   1 decimal place

$462.81 \div 1\underline{00}$ $= 4.6281$

       2 zeros   2 decimal places

$462.81 \div 1\underline{,000}$ $= 0.46281$

       3 zeros   3 decimal places

$462.81 \div 1\underline{0,000}$ $= 0.046281$

       4 zeros   4 decimal places

A zero must be inserted between the decimal point and the 4.

$462.81 \div 1\underline{00,000}$ $= 0.0046281$

       5 zeros   5 decimal places

Two zeros must be inserted between the decimal point and the 4.

If the power of 10 is written in exponential notation, the exponent indicates how many places to move the decimal point.

$462.81 \div 10^1 = 46.281$

       1 decimal place

$462.81 \div 10^2 = 4.6281$

       2 decimal places

$462.81 \div 10^3 = 0.46281$

       3 decimal places

$462.81 \div 10^4 = 0.046281$

       4 decimal places

$462.81 \div 10^5 = 0.0046281$

       5 decimal places

Find the quotient of 3.59 and 100.

There are two zeros in 100. Move the decimal point in 3.59 two places to the left. $3.59 \div 100 = 0.0359$

What is the quotient of 64.79 and 10^4?

The exponent on 10 is 4. Move the decimal point in 64.79 four places to the left. $64.79 \div 10^4 = 0.006479$

The sign rules for dividing integers are the same rules used to divide decimals.

The quotient of two numbers with the same sign is positive.
The quotient of two numbers with different signs is negative.

Divide: $-1.16 \div 2.9$

The signs are different.
The quotient is negative.
Divide the absolute values of the numbers. $-1.16 \div 2.9 = -0.4$

Evaluate $c \div d$ when $c = -8.64$ and $d = -0.4$.

Replace c with -8.64 and d with -0.4. $c \div d$
$(-8.64) \div (-0.4)$

The signs are the same. The quotient is positive.
Divide the absolute values of the numbers. $= 21.6$

Example 12 Divide: $431.97 \div 7.26$

Solution

$$
\begin{array}{r}
5\,9.5 \\
7.26\,\overline{)4\,3\,1.9\,7\,0} \\
-3\,6\,3\,0 \\
\hline
6\,8\,9\,7 \\
-6\,5\,3\,4 \\
\hline
3\,6\,3\,0 \\
-3\,6\,3\,0 \\
\hline
0
\end{array}
$$

You Try It 12 Divide: $314.746 \div 6.53$

Your Solution

Example 13 Estimate the quotient of 8.37 and 0.219.

Solution   8.37 $\longrightarrow$ 8
0.219 $\longrightarrow$ 0.2

$8 \div 0.2 = 40$

You Try It 13 Estimate the quotient of 62.7 and 3.45.

Your Solution

Example 14   Divide and round to the nearest hundredth: $448.2 \div 53$

You Try It 14   Divide and round to the nearest thousandth: $519.37 \div 86$

Solution

$$
\begin{array}{r}
8.4\,5\,6 \approx 8.46 \\
53\overline{)4\,4\,8.2\,0\,0} \\
-4\,2\,4 \\
\hline
2\,4\,2 \\
-2\,1\,2 \\
\hline
3\,0\,0 \\
-2\,6\,5 \\
\hline
3\,5\,0 \\
-3\,1\,8 \\
\hline
3\,2
\end{array}
$$

Your Solution

Example 15   Find the quotient of 592.4 and 10^4.

You Try It 15   What is 63.7 divided by 100?

Solution $592.4 \div 10^4 = 0.05924$

Your Solution

Example 16   Divide and round to the nearest tenth: $-6.94 \div -1.5$

You Try It 16   Divide and round to the nearest tenth: $-25.7 \div 0.31$

Solution   The quotient is positive.

$$-6.94 \div (-1.5) \approx 4.6$$

Your Solution

Example 17   Evaluate $\frac{x}{y}$ when $x = -76.8$ and $y = 0.8$.

You Try It 17   Evaluate $\frac{x}{y}$ when $x = -40.6$ and $y = -0.7$.

Solution $\dfrac{x}{y}$

$$\dfrac{-76.8}{0.8} = -96$$

Your Solution

Example 18   Is -0.4 a solution of the equation $\frac{8}{x} = -20$?

You Try It 18   Is -1.2 a solution of the equation $-2 = \frac{d}{-0.6}$?

Solution

$$\dfrac{8}{x} = -20$$

$$
\begin{array}{c|c}
\dfrac{8}{-0.4} & -20 \\
\end{array}
$$

$$-20 = -20$$

Yes, -0.4 is a solution of the equation.

Your Solution

Solutions on p. A16

OBJECTIVE D
Fractions and decimals

Since the fraction bar can be read "divided by," any fraction can be written as a decimal. To write a fraction as a decimal, divide the numerator of the fraction by the denominator.

Convert $\frac{3}{4}$ to a decimal.

$$
\begin{array}{r}
0.75 \\
4\overline{)3.00} \\
-2\,8 \\
\hline
20 \\
-20 \\
\hline
0
\end{array}
$$

← This is a **terminating decimal.**

← The remainder is zero.

$$\frac{3}{4} = 0.75$$

Convert $\frac{5}{11}$ to a decimal.

$$
\begin{array}{r}
0.4545 \\
11\overline{)5.0000} \\
-4\,4 \\
\hline
60 \\
-55 \\
\hline
50 \\
-44 \\
\hline
60 \\
-55 \\
\hline
5
\end{array}
$$

← This is a **repeating decimal.**

← The remainder is never zero.

$\frac{5}{11} = 0.\overline{45}$ The bar over the digits 45 is used to show that these digits repeat.

Convert $2\frac{4}{9}$ to a decimal.

Write the fractional part of the mixed number as a decimal. Divide the numerator by the denominator.

$$
\begin{array}{r}
0.444 \\
9\overline{)4.000}
\end{array} = 0.\overline{4}
$$

The whole number part of the mixed number is the whole number part of the decimal.

$$2\frac{4}{9} = 2.\overline{4}$$

To convert a decimal to a fraction, remove the decimal point and place the decimal part over a denominator equal to the place value of the last digit in the decimal.

⤷ hundredths

$$0.57 = \frac{57}{100}$$

⤷ hundredths

$$7.65 = 7\frac{65}{100} = 7\frac{13}{20}$$

⤷ tenths

$$8.6 = 8\frac{6}{10} = 8\frac{3}{5}$$

Convert 4.375 to a fraction.

The 5 in 4.375 is in the thousandths' place.
Write 0.375 as a fraction with a denominator
of 1,000.

$$4.375 = 4\frac{375}{1,000}$$

Simplify the fraction.

$$= 4\frac{3}{8}$$

To find the order relation between a fraction and a decimal, first rewrite the fraction as a decimal. Then compare the two decimals.

Find the order relation between $\frac{6}{7}$ and 0.855.

Write the fraction as a decimal. Round to one more place value than the given decimal. (0.855 has 3 decimal places; round to 4 decimal places.)

$$\frac{6}{7} \approx 0.8571$$

Compare the two decimals.

$$0.8571 > 0.8550$$

Replace the decimal approximation of $\frac{6}{7}$ with $\frac{6}{7}$.

$$\frac{6}{7} > 0.855$$

CALCULATOR NOTE

Some calculators *truncate* a decimal number that exceeds the calculator display. This means that the digits beyond the calculator's display are not shown. For this type of calculator, $\frac{2}{3}$ would be shown as 0.66666666. Other calculators *round* a decimal number when the calculator display is exceeded. For this type of calculator, $\frac{2}{3}$ would be shown as 0.66666667.

Example 19   Convert $\frac{5}{8}$ to a decimal.

Solution $\begin{array}{r} 0.625 \\ 8)\overline{5.000} \end{array}$ $\frac{5}{8} = 0.625$

You Try It 19   Convert $\frac{4}{5}$ to a decimal.

Your Solution

Example 20   Convert $3\frac{1}{3}$ to a decimal.

Solution   Write $\frac{1}{3}$ as a decimal.

$\begin{array}{r} 0.333 \\ 3)\overline{1.000} \end{array} = 0.\overline{3}$

$3\frac{1}{3} = 3.\overline{3}$

You Try It 20   Convert $1\frac{5}{6}$ to a decimal.

Your Solution

Example 21   Convert 7.25 to a fraction.

Solution $7.25 = 7\frac{25}{100} = 7\frac{1}{4}$

You Try It 21   Convert 6.2 to a fraction.

Your Solution

Example 22 Place the correct symbol, < or >, between the two numbers.

$$0.845 \qquad \frac{5}{6}$$

Solution $\dfrac{5}{6} \approx 0.8333$

$$0.8450 > 0.8333$$

$$0.845 > \frac{5}{6}$$

You Try It 22 Place the correct symbol, < or >, between the two numbers.

$$0.588 \qquad \frac{7}{12}$$

Your Solution

Solution on p. A16

Objective E
Applications and formulas

Example 23

A one-year subscription to a monthly magazine costs $21. The price of each issue at the newsstand is $2.25. How much would you save per issue by buying a year's subscription rather than buying each issue at the newsstand?

Strategy

To find the amount saved:

▶ Find the subscription price per issue by dividing the cost of the subscription (21) by the number of issues (12).

▶ Subtract the subscription price per issue from the newsstand price (2.25).

Solution

$$
\begin{array}{r}
1.7\,5 \\
12\overline{)2\,1.0\,0} \\
-1\,2 \\
\hline
9\,0 \\
-8\,4 \\
\hline
6\,0 \\
-6\,0 \\
\hline
0
\end{array}
\qquad
\begin{array}{r}
2.2\,5 \\
-1.7\,5 \\
\hline
0.5\,0
\end{array}
$$

The savings would be $.50 per issue.

You Try It 23

You hand a postal clerk a five-dollar bill to pay for the purchase of twelve 32¢ stamps. How much change do you receive?

Your Strategy

Your Solution

Solution on p. A16

Figure 4.4 shows the point-of-sale volume for Visa, Master Card, and American Express for the years 1990 and 1995. Use Figure 4.4 for Example 24 and You Try It 24.

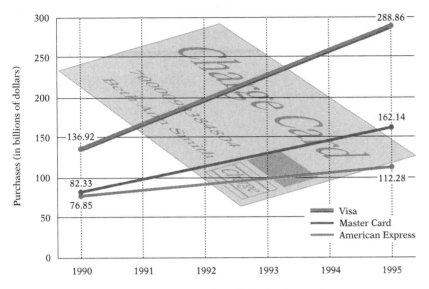

Figure 4.4   Point-of-Sale Volume for Three U.S. Credit Cards
Source: The Nilson Report

Example 24

In 1995, was the total amount charged on both Master Card and American Express greater than or less than the amount charged on Visa? Use Figure 4.4.

Strategy

To make the comparison:

▶ Add the purchases made on Master Card in 1995 (162.14 billion) and the purchases made on American Express in 1995 (112.28 billion).

▶ Compare the sum with the purchases made on Visa in 1995 (288.86).

Solution

162.14 billion + 112.28 billion = 274.42 billion

274.42 billion < 288.86 billion

The total amount changed on both Master Card and American Express was less than the amount charged on Visa.

You Try It 24

For the credit card with the greatest volume, find the increase from 1990 to 1995 in the amount of the annual purchases made with the credit card. Is this amount less than or greater than the annual purchases made in 1990? Use Figure 4.4.

Your Strategy

Your Solution

Solution on p. A16

Example 25

An overseas flight charges $6.40 for each kilogram or part of a kilogram over 50 kg of luggage weight. How much extra must be paid for three pieces of luggage weighing 21.4 kg, 19.3 kg, and 16.8 kg?

Strategy

To find the extra charge:

▶ Add the three weights (21.4, 19.3, and 16.8) to find the total weight of the luggage.
▶ Subtract 50 kg from the total weight of the luggage to find the excess weight.
▶ Round the difference up to the nearest whole number.
▶ Multiply the charge per kilogram of excess weight (6.40) by the excess weight.

Solution

$21.4 + 19.3 + 16.8 = 57.5$

$57.5 - 50 = 7.5$

7.5 rounded up to the nearest whole number is 8.

$6.40(8) = 51.20$

The extra charge for the luggage is $51.20.

You Try It 25

A health food store buys nuts in 100-pound containers and repackages the nuts for resale. The store packages the nuts in 2-pound bags, costing $.04 each, and sells them for $8.50 per bag. Find the profit on a 100-pound container of nuts costing $325.

Your Strategy

Your Solution

Example 26

Use the formula $P = BF$, where P is the insurance premium, B is the base rate, and F is the rating factor, to find the insurance premium due on an insurance policy with a base rate of $342.50 and a rating factor of 2.2.

Strategy

To find the insurance premium due, replace B by 342.50 and F by 2.2 in the given formula and solve for P.

Solution

$P = BF$
$P = 342.50(2.2)$
$P = 753.50$

The insurance premium due is $753.50.

You Try It 26

Use the formula $P = BF$, where P is the insurance premium, B is the base rate, and F is the rating factor, to find the insurance premium due on an insurance policy with a base rate of $276.25 and a rating factor of 1.8.

Your Strategy

Your Solution

Solutions on pp. A16–A17

4.2 EXERCISES

OBJECTIVE A

Add or subtract.

1. $1.864 + 39 + 25.0781$

2. $2.04 + 35.6 + 4.918$

3. $35.9 + 8.217 + 146.74$

4. $12 + 73.59 + 6.482$

5. $36.47 - 15.21$

6. $85.69 - 2.13$

7. $28 - 6.74$

8. $5 - 1.386$

9. $6.02 - 3.252$

10. $0.92 - 0.0037$

11. $-42.1 - 8.6$

12. $-6.57 - 8.933$

13. $5.73 - 9.042$

14. $-31.894 + 7.5$

15. $-9.37 + 3.465$

16. $1.09 - (-8.3)$

17. $-19 - (-2.65)$

18. $3.18 - 5.72 - 6.4$

19. $-12.3 - 4.07 + 6.82$

20. $-8.9 + 7.36 - 14.2$

21. $-5.6 - (-3.82) - 17.409$

Solve.

22. Find the sum of 2.536, 14.97, 8.014, and 21.67.

23. Find the total of 6.24, 8.573, 19.06, and 22.488.

24. What is 6.9217 decreased by 3.4501?

25. What is 8.9 less than 62.57?

26. How much larger is 5 than 1.63?

27. What is the sum of -65.47 and -32.91?

28. Find 382.9 more than -430.6.

29. Find -138.72 minus 510.64.

30. What is 4.793 less than -6.82?

31. How much larger is -31 than -62.09?

Estimate by rounding. Then find the exact answer.

32. 45.06 + 80.71

33. 6.408 + 5.917

34. 0.24 + 0.38 + 0.96

35. 56.87 − 23.24

36. 6.272 − 1.848

37. 0.931 − 0.628

38. 5.37 + 26.49

39. 87.65 − 49.032

40. 387.6 − 54.92

Solve.

41. *Business* The figure at the right shows the breakdown of Microsoft's revenue in 1996. The segment entitled Operating Systems includes revenue from Windows 95. The segment entitled Business Systems includes revenue from Windows NT. (a) What is the total revenue from the Business Systems and Operating Systems segments? (b) Find the total revenue from all three segments.

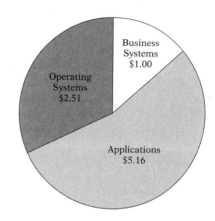

Microsoft's 1996 Revenue (in billions)
Sources: Microsoft, Montgomery Securities

Evaluate the variable expression $x + y$ for the given values of x and y.

42. $x = 62.97; y = -43.85$

43. $x = 5.904; y = -7.063$

44. $x = -125.41; y = 361.55$

45. $x = -6.175; y = -19.49$

Evaluate the variable expression $x + y + z$ for the given values of x, y, and z.

46. $x = 41.33; y = -26.095; z = 70.08$

47. $x = -6.059; y = 3.884; z = 15.71$

48. $x = 81.72; y = 36.067; z = -48.93$

49. $x = -16.219; y = 47; z = -2.3885$

Evaluate the variable expression $x - y$ for the given values of x and y.

50. $x = 43.29; y = 18.76$

51. $x = 6.029; y = -4.708$

52. $x = -16.329; y = 4.54$

53. $x = -21.073; y = 6.48$

54. $x = -3.69; y = -1.527$

55. $x = -8.21; y = -6.798$

56. Is -1.2 a solution of the equation $6.4 = 5.2 + a$?

57. Is -2.8 a solution of the equation $0.8 - p = 3.6$?

58. Is -0.5 a solution of the equation $x - 0.5 = 1$?

59. Is 36.8 a solution of the equation $27.4 = y - 9.4$?

OBJECTIVE B

Multiply.

60. $0.9(0.3)$

61. $(3.4)(0.5)$

62. $(0.72)(3.7)$

63. $8.29(0.004)$

64. $-5.2(0.8)$

65. $(-6.3)(-2.4)$

66. $(1.9)(-3.7)$

67. $-1.3(4.2)$

68. $-8.1(-7.5)$

69. $1.31(-0.006)$

70. $-10(0.59)$

71. $(-100)(4.73)$

Solve.

72. What is the product of 5.92 and 100?

73. What is 1,000 times 4.25?

74. Find 0.82 times 10^2.

75. Find the product of 6.71 and 10^4.

76. Find the product of 2.7, -16, and 3.04.

77. What is the product of 0.06, -0.4, and -1.5?

Estimate by rounding. Then find the exact answer.

78. 86.4(4.2)

79. (9.81)(0.77)

80. 0.238(8.2)

81. (6.88)(9.97)

82. (8.432)(0.043)

83. 28.45(1.13)

The table at the right shows currency exchange rates for several foreign countries. To determine how many French francs would be exchanged for 1,000 U.S. dollars, multiply the number of francs exchanged for 1 U.S. dollar (5.0315) by $1,000: $1,000(5.0315) = $5,031.50. Use this table for Exercises 84 and 85.

Country and Monetary Unit	Number of Units Exchanged for 1 U.S. Dollar
Britain (Pound)	0.64342
France (Franc)	5.0315
Germany (Mark)	1.4823
Italy (Lira)	1,530.0
Japan (Yen)	108.38
Mexico (Peso)	7.6050

84. *Exchange Rates* How many Italian lira would be exchanged for 5,000 U.S. dollars?

85. *Exchange Rates* How many British pounds would be exchanged for 15,000 U.S. dollars?

Evaluate the expression for the given values of the variables.

86. xy, when $x = 5.68$ and $y = 0.2$

87. ab, when $a = 6.27$ and $b = 8$

88. $40c$, when $c = 2.5$

89. $10t$, when $t = -4.8$

90. xy, when $x = -3.71$ and $y = 2.9$

91. ab, when $a = 0.379$ and $b = -0.22$

92. ab, when $a = 452$ and $b = -0.86$

93. cd, when $c = -2.537$ and $d = -9.1$

94. cd, when $c = -4.259$ and $d = -6.3$

95. Is -8 a solution of the equation $1.6 = -0.2z$?

96. Is -1 a solution of the equation $-7.9c = -7.9$?

97. Is -10 a solution of the equation $-83.25r = 8.325$?

98. Is -3.6 a solution of the equation $32.4 = -9w$?

OBJECTIVE C

Divide.

99. $16.15 \div 0.5$

100. $7.02 \div 3.6$

101. $27.08 \div (-0.4)$

102. $-8.919 \div 0.9$

103. $(-3.312) \div (-0.8)$

104. $84.66 \div (-1.7)$

105. $-2.501 \div 0.41$

106. $1.003 \div (-0.59)$

Divide. Round to the nearest tenth.

107. $55.63 \div 8.8$

108. $1.873 \div 1.4$

109. $(-52.8) \div (-9.1)$

110. $-6.824 \div 0.053$

Divide. Round to the nearest hundredth.

111. $6.457 \div 8$

112. $19.07 \div 0.54$

113. $0.0416 \div (-0.53)$

114. $(-31.792) \div (-0.86)$

Solve.

115. Find the quotient of 52.78 and 10.

116. What is 37,942 divided by 1,000?

117. What is the quotient of 48.05 and 10^2?

118. Find 9.407 divided by 10^3.

Solve. Round to the nearest tenth.

119. Find the quotient of -19.04 and 0.75.

120. What is the quotient of -21.892 and -0.96?

121. Find 27.735 divided by -60.3.

122. What is -13.97 divided by 28.4?

Estimate by rounding. Then divide and round to the nearest hundredth.

123. $42.43 \div 3.8$

124. $678 \div 0.71$

125. $6.398 \div 5.5$

126. $0.994 \div 0.456$

127. $1.237 \div 0.021$

128. $421.093 \div 4.087$

129. $33.14 \div 4.6$

130. $129.38 \div 4.47$

Solve.

131. *Business* The figure at the right shows revenue passenger miles for four airlines. A revenue passenger mile is one paying passenger flown one mile. How many times greater were the number of American Trans Air's revenue passenger miles than Reno Airlines'?

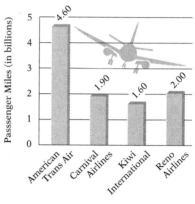

1995 Revenue Passenger Miles for Selected Low-Cost Carriers
Source: Transportation Department

Evaluate the variable expression $\frac{x}{y}$ for the given values of x and y.

132. $x = 52.8; y = 0.4$

133. $x = 3.542; y = 0.7$

134. $x = -2.436; y = 0.6$

135. $x = 0.648; y = -2.7$

136. $x = 26.22; y = -6.9$

137. $x = -8.034; y = -3.9$

138. $x = -64.05; y = -6.1$

139. $x = -2.501; y = 0.41$

140. $x = 1.003; y = -0.59$

141. Is 24.8 a solution of the equation $\frac{q}{-8} = -3.1$?

142. Is 0.48 a solution of the equation $\frac{-6}{z} = -12.5$?

143. Is -8.4 a solution of the equation $21 = \frac{t}{0.4}$?

144. Is -0.9 a solution of the equation $\frac{-2.7}{a} = \frac{a}{-0.3}$?

Objective D

Convert the fraction to a decimal. Place a bar over repeating digits of a repeating decimal.

145. $\dfrac{3}{8}$ **146.** $\dfrac{7}{15}$ **147.** $\dfrac{8}{11}$ **148.** $\dfrac{9}{16}$ **149.** $\dfrac{7}{12}$

150. $\dfrac{5}{3}$ **151.** $\dfrac{7}{4}$ **152.** $2\dfrac{3}{4}$ **153.** $1\dfrac{1}{2}$ **154.** $3\dfrac{2}{9}$

155. $4\dfrac{1}{6}$ **156.** $\dfrac{3}{25}$ **157.** $2\dfrac{1}{4}$ **158.** $6\dfrac{3}{5}$ **159.** $3\dfrac{8}{9}$

Convert the decimal to a fraction.

160. 0.6   **161.** 0.2   **162.** 0.25   **163.** 0.75   **164.** 0.48

165. 0.125   **166.** 0.325   **167.** 2.5   **168.** 3.4   **169.** 4.55

170. 9.95   **171.** 1.72   **172.** 5.68   **173.** 0.045   **174.** 0.085

Place the correct symbol, $<$ or $>$, between the two numbers.

175. $\dfrac{9}{10}$ 0.89   **176.** $\dfrac{7}{20}$ 0.34   **177.** $\dfrac{4}{5}$ 0.803   **178.** $\dfrac{3}{4}$ 0.706

179. 0.444 $\dfrac{4}{9}$ **180.** 0.72 $\dfrac{5}{7}$ **181.** 0.13 $\dfrac{3}{25}$ **182.** 0.25 $\dfrac{13}{50}$

183. $\dfrac{5}{16}$ 0.312   **184.** $\dfrac{7}{18}$ 0.39   **185.** $\dfrac{10}{11}$ 0.909   **186.** $\dfrac{8}{15}$ 0.543

OBJECTIVE E

Solve.

187. *Finances* If you earn an annual salary of $41,619, what is your monthly salary?

188. *Finances* You pay $947.60 a year in car insurance. The insurance is paid in four equal payments. Find the amount of each payment.

189. *Temperature* On January 22, 1943, in Spearfish, South Dakota, the temperature fell from 12.22°C at 9:00 A.M. to −20°C at 9:27 A.M. How many degrees did the temperature fall during the 27-minute period?

190. *Temperature* On January 10, 1911, in Rapid City, South Dakota, the temperature fell from 12.78°C at 7:00 A.M. to −13.33°C at 7:15 A.M. How many degrees did the temperature fall during the 15-minute period?

191. *Consumerism* A case of diet cola costs $6.79. If there are 24 cans in a case, find the cost per can. Round to the nearest cent.

192. *Travel* You travel 295 mi on 12.5 gal of gasoline. How many miles can you travel on one gallon of gasoline?

193. *Consumerism* It costs $.038 an hour to operate an electric motor. How much does it cost to operate the motor for 90 h?

194. *Investments* During one day, the Dow Jones Industrial Average rose 6.14 points to close at 5,575.22 points. What did the Dow Jones Industrial Average close at the day before?

195. *Consumerism* Using the catalog information shown below, estimate the cost of ordering 4 boxes of 3.5" DS/DD diskettes.

Catalog Number	Description	Cost
DC-4392	5.25" DS/DD Diskettes, Box of 10	$3.90
DC-4397	5.25" DS/HD Diskettes, Box of 50	$19.50
DC-5108	3.5" DS/DD Diskettes, Box of 10	$5.90
DC-5260	3.5" SS/DD Diskettes, Box of 25	$8.75
DC-5914	3.5" DS/HD Diskettes, Box of 25	$14.75

196. *Consumerism* Using the catalog information shown above, estimate the cost of ordering 6 boxes of 5.25" DS/HD diskettes.

Solve.

197. *Consumerism*   You make a down payment of $225 on a camcorder and agree to make payments of $34.17 a month for the next 18 months. Find the total cost of the camcorder.

198. *Finances*   You have a monthly budget of $810. This month you have already spent $22.78 for the telephone bill, $64.93 for food, $15.50 for gasoline, $160 for rent, and $91.62 for a loan repayment. How much money do you have left in the budget for the remainder of the month?

199. *Finances*   You had a balance of $347.08 in your checking account. You then made a deposit of $189.53 and wrote a check for $62.89. Find the new balance in your checking account.

200. *Finances*   You earn a salary of $327.25 per week. You have deductions of $24.54 for social security, $16.80 for medical insurance, and $6.78 for union dues. Find your take-home pay.

201. *Finances*   A bookkeeper earns a salary of $340 for a 40-hour week. This week the bookkeeper worked 6 h of overtime at a rate of $12.75 for each hour of overtime worked. Find the bookkeeper's total income for the week.

202. *Consumerism*   Using the menu shown below, estimate the bill for the following order: 1 soup, 1 cheese sticks, 1 blackened swordfish, 1 chicken divan, and 1 carrot cake.

Appetizers
Soup of the Day.................$2.75
Cheese Sticks....................$3.25
Potato Skins.....................$3.50

Entrees
Roast Prime Rib.................$18.95
Blackened Swordfish............$16.95
Chicken Divan...................$14.95

Desserts
Carrot Cake......................$4.25
Ice Cream Pie...................$5.50
Cheese Cake.....................$6.75

203. *Consumerism*   Using the menu shown above, estimate the bill for the following order: 1 potato skins, 1 cheese sticks, 1 roast prime rib, 1 chicken divan, 1 ice cream pie, and 1 cheese cake.

204. *Consumerism*   A customer purchases a product that sells for $39.88. The sales tax is $2.39. How much change should the customer receive from a $50 bill?

The following figure shows the life expectancy at birth for males and females. Use this graph for Exercise 205.

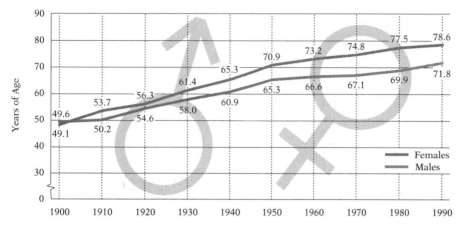

Life Expectancy of Males and Females in the United States

205. *Demography* (a) Has life expectancy increased for both males and females between each 10-year period shown in the graph? (b) Did males or females have a longer life expectancy in 1990? How much longer? (c) During which year was the difference between male and female life expectancy greatest?

206. *Business* For $65, a druggist purchases 5 L of cough syrup and repackages it in 250-milliliter bottles. Each bottle costs the druggist $.25. Each bottle of cough syrup is sold for $5.89. Find the profit on the 5 L of cough syrup.

207. *Environmental Science* In the United States today, on average each person throws away 3.6 lb of garbage per day. On average, how many pounds per year does a family of four discard?

208. *Investment* The figure at the right shows the gains and losses in the technology stock funds during the four quarters of 1995. Find the difference between the performance of these funds in the third and fourth quarters of 1995.

209. *Business* Use the formula $M = S - C$, where M is the markup on a consumer product, S is the selling price, and C is the cost of the product to the business, to find the markup on a product that cost a business $1,653.19 and has a selling price of $2,231.81.

210. *Business* Use the formula $M = S - C$, where M is the markup on a consumer product, S is the selling price, and C is the cost of the product to the business, to find the markup on a product that cost a business $30.73 and has a selling price of $87.80.

211. *Accounting* The amount of an employee's earnings that is subject to federal withholding is called federal earnings. Find the federal earnings for an employee who earns $694.89 and has a withholding allowance of $132.69. Use the formula $F = E - W$, where F is the federal earnings, E is the employee's earnings, and W is the withholding allowance.

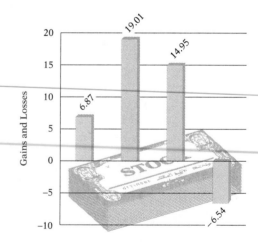

Technology Stock Funds Performance
Source: Lipper Analytical Services, Inc.

Solve.

212. *Accounting* The amount of an employee's earnings that is subject to federal withholding is called federal earnings. Find the federal earnings for an employee who earns $572.45 and has a withholding allowance of $88.46. Use the formula $F = E - W$, where F is the federal earnings, E is the employee's earnings, and W is the withholding allowance.

213. *Consumerism* Use the formula $M = \dfrac{C}{N}$, where M is the cost per mile for a rental car, C is the total cost, and N is the number of miles driven, to find the cost per mile when the total cost of renting a car is $260.16 and you drive the car 542 mi.

214. *Consumerism* Use the formula $M = \dfrac{C}{N}$, where M is the cost per mile for a rental car, C is the total cost, and N is the number of miles driven, to find the cost per mile when the total cost of renting a car is $311.88 and you drive the car 678 mi.

215. *Physics* Find the force exerted on a falling object that has a mass of 4.25 kg. Use the formula $F = ma$, where F is the force exerted by gravity on a falling object, m is the mass of the object and a is the acceleration of gravity. The acceleration of gravity is -9.80 m/s^2 (meters per second squared). The force is measured in newtons.

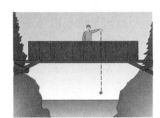

216. *Physics* Find the force exerted on a falling object that has a mass of 6.75 kg. Use the formula $F = ma$, where F is the force exerted by gravity on a falling object, m is the mass of the object, and a is the acceleration of gravity. The acceleration of gravity is -9.80 m/s^2 (meters per second squared). The force is measured in newtons.

217. *Finances* Find the equity on a home that is valued at $125,000 when the homeowner has $67,853.25 in loans on the property. Use the formula $E = V - L$, where E is the equity, V is the value of the home, and L is the loan amount on the property.

218. *Finances* Find the equity on a home that is valued at $240,000 when the homeowner has $142,976.80 in loans on the property. Use the formula $E = V - L$, where E is the equity, V is the value of the home, and L is the loan amount on the property.

219. *Utilities* Find the cost of operating a 1800-watt TV set for 5 h at a cost of $.06 per kilowatt-hour. Use the formula $c = \dfrac{1}{1,000} wtk$, where c is the cost of operating an appliance, w is the number of watts, t is the time in hours, and k is the cost per kilowatt-hour.

220. *Utilities* Find the cost of operating a 200-watt stereo for 3 h at a cost of $.10 per kilowatt-hour. Use the formula $c = \dfrac{1}{1,000} wtk$, where c is the cost of operating an appliance, w is the number of watts, t is the time in hours, and k is the cost per kilowatt-hour.

CRITICAL THINKING

221. Using the method, presented in this section, of estimating the sum of two decimals, what is the largest amount by which the estimate of the sum of two decimals with tenths', hundredths', and thousandths' places could differ from the exact sum? Assume the number in the thousandths' place is not zero.

222. Use a calculator to order the numbers $-\frac{67}{131}$, $-\frac{103}{197}$, $-\frac{199}{379}$, and $-\frac{211}{409}$ from smallest to largest.

223. Place the correct symbol between the two numbers.
   a. $(1.1)^3$ 1.31   **b.** $(0.9)^3$ 1^5 **c.** $(1.2)^3$ $(0.8)^3$

224. Find the product of 1.0035 and 1.00079 without using a calculator. Then find the product using a calculator and compare the two numbers. If your calculator has an eight-digit display, what number did the calculator display? Some calculators truncate the product, which means that the digits that cannot be displayed are discarded. Other calculators round the answer to the rightmost place value in the calculator's display. Determine which method your calculator uses to handle approximate answers. If the decimal places in a negative number are truncated, is the resulting number greater than, less than, or equal to the original number?

225. A ballpoint pen priced at 50¢ was not selling. When the price was reduced to a different whole number of cents, the entire stock sold for $31.93. How many cents were charged per pen when the price was reduced?

226. Determine whether the statement is always true, sometimes true, or never true.
   a. The product of an even number of negative factors is a negative number.
   b. The sum of an odd number of negative addends is a negative number.
   c. If $a \geq 0$, then $|a| = a$.
   d. If $a \leq 0$, then $|a| = -a$.

227. Convert $\frac{1}{9}$, $\frac{2}{9}$, $\frac{3}{9}$, and $\frac{4}{9}$ to decimals. Describe the pattern. Use the pattern to convert $\frac{5}{9}$, $\frac{7}{9}$, and $\frac{8}{9}$ to decimals.

228. Explain how baseball batting averages are determined.

229. What does the term *population density* mean? How is population density determined? What is the population density of the state you live in? How does this compare with the population density of the country as a whole?

230. What is the national debt of the United States of America? Divide the national debt by the number of U.S. citizens. What do these numbers mean?

231. Prepare a report on the Kelvin scale. The report should include a definition of absolute zero and an explanation of how to convert from Kelvin to Celsius and from Celsius to Kelvin.

SECTION 4.3 Solving Equations with Decimals

OBJECTIVE A
Solving equations

The properties of equations discussed earlier are restated here.

The same number can be added to each side of an equation without changing the solution of the equation.

The same number can be subtracted from each side of an equation without changing the solution of the equation.

Each side of an equation can be multiplied by the same nonzero number without changing the solution of the equation.

Each side of an equation can be divided by the same nonzero number without changing the solution of the equation.

Solve: $3.4 = a - 3.56$

3.56 is subtracted from the variable a. Add 3.56 to each side of the equation.

a is alone on the right side of the equation. The number on the left side is the solution.

$$3.4 = a - 3.56$$
$$3.4 + 3.56 = a - 3.56 + 3.56$$
$$6.96 = a$$

The solution is 6.96.

> **TAKE NOTE**
> Remember to check the solutions for all equations.
>
> Check:
>
> $3.4 = a - 3.56$
>
> $3.4 \mid 6.96 - 3.56$
>
> $3.4 = 3.4$

Solve: $-1.25y = 3.875$

The variable is multiplied by -1.25. Divide each side of the equation by -1.25.

y is alone on the left side of the equation. The number on the right side is the solution.

$$-1.25y = 3.875$$
$$\frac{-1.25y}{-1.25} = \frac{3.875}{-1.25}$$
$$y = -3.1$$

The solution is -3.1.

Example 1 Solve: $4.56 = 9.87 + z$

Solution
$$4.56 = 9.87 + z$$
$$4.56 - 9.87 = 9.87 - 9.87 + z$$
$$-5.31 = z$$

The solution is -5.31.

You Try It 1 Solve: $a - 1.23 = -6$

Your Solution

Example 2 Solve: $\dfrac{x}{2.45} = -0.3$

Solution
$$\frac{x}{2.45} = -0.3$$
$$2.45 \cdot \frac{x}{2.45} = 2.45(-0.3)$$
$$x = -0.735$$

The solution is -0.735.

You Try It 2 Solve: $-2.13 = -0.71c$

Your Solution

Solutions on p. A17

OBJECTIVE B
Applications

Example 3

The cost of operating an electrical appliance is given by the formula $c = \frac{wtk}{1,000}$, where c is the cost of operating the appliance, w is the number of watts, t is the number of hours, and k is the cost per kilowatt-hour. Find the cost per kilowatt-hour if it costs $.60 to operate a 2,000-watt television for 5 h.

Strategy

To find the cost per kilowatt-hour, replace c by 0.60, w by 2,000, and t by 5 in the given formula and solve for k.

Solution

$$c = \frac{wtk}{1,000} \qquad 0.60 = \frac{2,000(5)k}{1,000}$$

$$0.60 = 10k$$

$$\frac{0.60}{10} = \frac{10k}{10}$$

$$0.06 = k$$

It costs $.06 per kilowatt-hour.

You Try It 3

The net worth of a business is given by the equation $N = A - L$, where N is the net worth, A is the assets of the business (the amount owned) and L is the liabilities of the business (the amount owed). Use the net worth equation to find the assets of a business that has a net worth of $24.3 billion and liabilities of $17.9 billion.

Your Strategy

Your Solution

Example 4

The total of the monthly payments for an installment loan is the product of the number of months of the loan and the monthly payment. The total of the monthly payments for a 48-month new-car loan is $10,433.28. What is the monthly payment?

Strategy

To find the monthly payment, write and solve an equation using m to represent the amount of the monthly payment.

Solution

The total of the monthly payments	is	the product of the number of months of the loan and the monthly payment

$$10,433.28 = 48m$$

$$217.36 = m \qquad \text{Divide each side by 48.}$$

The monthly payment is $217.36.

You Try It 4

The selling price of a product is the sum of the amount paid by the store for the product and the amount of the markup. The selling price of a titanium golf club is $295.50, and the amount paid by the store for the golf club is $223.75. Find the markup.

Your Strategy

Your Solution

Solutions on p. A17

4.3 EXERCISES

OBJECTIVE A

Solve. Write the answer as a decimal.

1. $y + 3.96 = 8.45$

2. $x - 2.8 = 1.34$

3. $-9.3 = c - 15$

4. $-28 = x - 3.27$

5. $7.3 = -\dfrac{n}{1.1}$

6. $-5.1 = \dfrac{y}{3.2}$

7. $-7x = 8.4$

8. $1.44 = -0.12t$

9. $y - 0.234 = -0.09$

10. $9 = z + 0.98$

11. $6.21r = -1.863$

12. $-78.1a = 85.91$

13. $-0.001 = x + 0.009$

14. $5 = 43.5 + c$

15. $\dfrac{x}{2} = -0.93$

16. $-1.03 = -\dfrac{z}{3}$

17. $-9.85y = 2.0685$

18. $7w = -0.014$

19. $-6v = 15$

20. $-55 = -40x$

21. $0.908 = 2.913 + x$

22. $-76.51 = y - 43.9$

23. $\dfrac{t}{-2.1} = -7.8$

24. $\dfrac{w}{0.02} = -9.64$

OBJECTIVE B

Solve.

25. *Business*   Xlint Office Supplies wants to make a profit of $3.50 on the sale of each desk calendar that costs the store $5.23. Use the equation $P = S - C$, where P is the profit on an item, S is the selling price, and C is the cost, to find the selling price of the calendar.

26. *Cost of Living*   The average cost per mile to operate a car is given by the equation $M = \dfrac{C}{N}$, where M is the average cost per mile, C is the total cost of operating the car, and N is the number of miles the car is driven. Use this formula to find the total cost of operating a car for 25,000 mi when the average cost per mile is $.23.

Solve.

27. *Physics* The average acceleration of an object is given by $a = \frac{v}{t}$, where a is the average acceleration, v is the velocity, and t is the time. Find the velocity after 6.3 s of an object whose acceleration is 16 feet/second² (feet per second squared).

28. *Accounting* The fundamental accounting equation is $A = L + S$, where A is the assets of a company, L is the liabilities of the company, and S is the stockholders' equity. Find the stockholders' equity in a company whose assets are $34.8 million and whose liabilities are $29.9 million.

29. *Cost of Living* The cost of operating an electrical appliance is given by the formula $c = \frac{wtk}{1,000}$, where c is the cost of operating the appliance, w is the number of watts, t is the number of hours, and k is the cost per kilowatt-hour. Find the cost per kilowatt-hour if it costs $.01 to operate a 1,000-watt microwave for 8 h.

30. *Business* The markup on an item in a store equals the difference between the selling price of the item and the cost of the item. Find the selling price of a package of golf balls for which the cost is $3.27 and the markup is $1.73.

31. *Consumerism* The total of the monthly payments for a car lease is the product of the number of months of the lease and the monthly lease payment. The total of the monthly payments for a 60-month car lease is $15,387. Find the monthly lease payment.

32. *Business* A company's revenue for a language tutorial is the price of each tutorial times the number of tutorials sold. If the company has revenues of $4.2 million for this tutorial and sells 25,000 tutorials, what is the price per tutorial?

The revenue	is	the price per tutorial times number sold

33. *Physics* The distance a spring will stretch is equal to the spring constant (a measure of the stiffness of the spring) times the weight of the object placed on the spring. An object that is placed on a spring with a spring constant of 16 pounds per inch stretches the spring a distance of 3.2 in. Find the weight of the object.

The distance	is equal to	the spring constant times the object's weight

---

CRITICAL THINKING

34. Solve: $0.\overline{33}x = 7$

35. **a.** Make up an equation of the form $x - b = c$ for which $x = -0.96$.
    b. Make up an equation of the form $ax = b$ for which $x = 2.1$.

36. For the equation $0.375x = 0.6$, a student offered the solution shown at the right. Is this a correct method of solving the equation? Explain your answer.

37. Consider the equation $12 = \frac{x}{a}$, where a is any positive number. Explain how increasing values of a affect the solution, x, of the equation.

$0.375x = 0.6$

$\frac{375}{1,000}x = \frac{6}{10}$

$\frac{3}{8}x = \frac{3}{5}$

$\frac{8}{3} \cdot \frac{3}{8}x = \frac{8}{3} \cdot \frac{3}{5}$

$x = \frac{8}{5} = 1.6$

SECTION 4.4   Radical Expressions

OBJECTIVE A
Square roots of perfect squares

Recall that the square of a number is equal to the number multiplied times itself.

$$3^2 = 3 \cdot 3 = 9$$

The square of an integer is called a **perfect square.**

9 is a perfect square because 9 is the square of 3: $3^2 = 9$.

The numbers 1, 4, 9, 16, 25, 36, 49, 64, 81, and 100 are perfect squares.

$$1^2 = 1$$
$$2^2 = 4$$
$$3^2 = 9$$
$$4^2 = 16$$
$$5^2 = 25$$
$$6^2 = 36$$
$$7^2 = 49$$
$$8^2 = 64$$

Larger perfect squares can be found by squaring 11, squaring 12, squaring 13, and so on.

$$9^2 = 81$$
$$10^2 = 100$$

Note that squaring the negative integers results in the same list of numbers.

$$(-1)^2 = 1$$
$$(-2)^2 = 4$$
$$(-3)^2 = 9$$
$$(-4)^2 = 16, \text{ and so on.}$$

Perfect squares are used in simplifying square roots. The symbol for square root is $\sqrt{}$.

Square Root

A square root of a positive number x is a number whose square is x.

If $a^2 = x$, then $\sqrt{x} = a$.

The expression $\sqrt{9}$, read "the square root of 9," is equal to the number that when squared is equal to 9.

Since $3^2 = 9$, $\sqrt{9} = 3$.

Every positive number has two square roots, one a positive number and one a negative number. The symbol $\sqrt{}$ is used to indicate the positive square root of a number. When the negative square root of a number is to be found, a negative sign is placed in front of the square root symbol. For example:

$$\sqrt{9} = 3 \quad \text{and} \quad -\sqrt{9} = -3$$

POINT OF INTEREST

The radical symbol was first used in 1525, when it was written as √. Some historians suggest that the radical symbol also developed into the symbols for "less than" and "greater than." Because typesetters of that time did not want to make additional symbols, the radical was rotated to the position ⟩ and used as a "greater than" symbol and rotated to ⟨ and used for the "less than" symbol. Other evidence, however, suggests that the "less than" and "greater than" symbols were developed independently of the radical symbol.

The square root symbol, $\sqrt{}$, is also called a **radical.** The number under the radical is called the **radicand.** In the radical expression $\sqrt{9}$, 9 is the radicand.

Simplify: $\sqrt{49}$

$\sqrt{49}$ is equal to the number that when squared equals 49. $7^2 = 49$.

$$\sqrt{49} = 7$$

Simplify: $-\sqrt{49}$

The negative sign in front of the square root symbol indicates the negative square root of 49. $(-7)^2 = 49$.

$$-\sqrt{49} = -7$$

Simplify: $\sqrt{25} + \sqrt{81}$

Simplify each radical expression.

Since $5^2 = 25$, $\sqrt{25} = 5$.

Since $9^2 = 81$, $\sqrt{81} = 9$.

Add.

$$\sqrt{25} + \sqrt{81} = 5 + 9$$
$$= 14$$

Simplify: $5\sqrt{64}$

The expression $5\sqrt{64}$ means 5 times $\sqrt{64}$.

Simplify $\sqrt{64}$.

Multiply.

$$5\sqrt{64} = 5 \cdot 8$$
$$= 40$$

Simplify: $6 + 4\sqrt{9}$

Simplify $\sqrt{9}$.

Use the Order of Operations Agreement.

$$6 + 4\sqrt{9} = 6 + 4 \cdot 3$$
$$= 6 + 12$$
$$= 18$$

Simplify: $\sqrt{\dfrac{1}{9}}$

$\sqrt{\dfrac{1}{9}}$ is equal to the number that when squared equals $\dfrac{1}{9}$. $\left(\dfrac{1}{3}\right)^2 = \dfrac{1}{9}$.

$$\sqrt{\frac{1}{9}} = \frac{1}{3}$$

Note that the square root of $\dfrac{1}{9}$ is equal to the square root of the numerator ($\sqrt{1} = 1$) over the square root of the denominator ($\sqrt{9} = 3$).

Evaluate $\sqrt{xy}$ when $x = 5$ and $y = 20$.

$$\sqrt{xy}$$

Replace x with 5 and y with 20.

$$\sqrt{5 \cdot 20}$$

Simplify under the radical.

$$= \sqrt{100}$$

Take the square root of 100. $10^2 = 100$.

$$= 10$$

Example 1   Simplify: $\sqrt{121}$

Solution   Since $11^2 = 121$, $\sqrt{121} = 11$.

You Try It 1   Simplify: $-\sqrt{144}$

Your Solution

Example 2   Simplify: $\sqrt{\dfrac{4}{25}}$

Solution   Since $\left(\dfrac{2}{5}\right)^2 = \dfrac{4}{25}$, $\sqrt{\dfrac{4}{25}} = \dfrac{2}{5}$.

You Try It 2   Simplify: $\sqrt{\dfrac{81}{100}}$

Your Solution

Example 3   Simplify: $\sqrt{36} - 9\sqrt{4}$

Solution $\sqrt{36} - 9\sqrt{4} = 6 - 9 \cdot 2$
$\qquad\qquad\qquad = 6 - 18$
$\qquad\qquad\qquad = 6 + (-18)$
$\qquad\qquad\qquad = -12$

You Try It 3   Simplify: $4\sqrt{16} - \sqrt{9}$

Your Solution

Example 4   Evaluate $6\sqrt{ab}$ when $a = 2$ and $b = 8$.

Solution $6\sqrt{ab}$
$6\sqrt{2 \cdot 8} = 6\sqrt{16}$
$\qquad\qquad = 6(4)$
$\qquad\qquad = 24$

You Try It 4   Evaluate $5\sqrt{a + b}$ when $a = 17$ and $b = 19$.

Your Solution

Solutions on p. A17

═══════ ⁄⁄⁄ ═══════

OBJECTIVE B
Square roots of whole numbers

In the last objective, the radicand in each radical expression was a perfect square. Since the square root of a perfect square is an integer, the exact value of each radical expression could be found.

If the radicand is not a perfect square, the square root can only be approximated. For example, the radicand in the radical expression $\sqrt{2}$ is 2, and 2 is not a perfect square. The square root of 2 can be approximated to any desired place value.

To the nearest tenth:	$\sqrt{2} \approx 1.4$	$(1.4)^2 = 1.96$
To the nearest hundredth:	$\sqrt{2} \approx 1.41$	$(1.41)^2 = 1.9881$
To the nearest thousandth:	$\sqrt{2} \approx 1.414$	$(1.414)^2 = 1.999396$
To the nearest ten-thousandth:	$\sqrt{2} \approx 1.4142$	$(1.4142)^2 = 1.99996164$

The square of each approximation gets closer and closer to 2 as the number of place values in the decimal approximation increases. But no matter how many place values are used to approximate $\sqrt{2}$, the digits never terminate or repeat. In general, the square root of any number that is not a perfect square can only be approximated.

Approximate $\sqrt{11}$ to the nearest ten-thousandth.

11 is not a perfect square.

Use a calculator to approximate $\sqrt{11}$. $\qquad\qquad \sqrt{11} \approx 3.3166$

Approximate $3\sqrt{5}$ to the nearest ten-thousandth.

$3\sqrt{5}$ means 3 times $\sqrt{5}$. $\qquad\qquad 3\sqrt{5} \approx 6.7082$

Between what two whole numbers is the value of $\sqrt{41}$?

Since the number 41 is between the perfect squares 36 and 49, the value of $\sqrt{41}$ is between $\sqrt{36}$ and $\sqrt{49}$.

$\sqrt{36} = 6$ and $\sqrt{49} = 7$,

so the value of $\sqrt{41}$ is between the whole numbers 6 and 7.

This can be written using inequality symbols as $6 < \sqrt{41} < 7$, which is read

"the square root of 41 is greater than 6 and less than 7."

Use a calculator to verify that $\sqrt{41} \approx 6.4$, which is between 6 and 7.

Sometimes we are not interested in an approximation of the square root of a number, but rather the exact value in simplest form.

A radical expression is in simplest form when the radicand contains no factor, other than 1, that is a perfect square. The Product Property of Square Roots is used to simplify radical expressions.

> ### Product Property of Square Roots
> If a and b are positive numbers, then $\sqrt{a \cdot b} = \sqrt{a} \cdot \sqrt{b}$.

The Product Property of Square Roots states that the square root of a product is equal to the product of the square roots. For example:

$$\sqrt{4 \cdot 9} = \sqrt{4} \cdot \sqrt{9}$$

Note that $\sqrt{4 \cdot 9} = \sqrt{36} = 6$ and $\sqrt{4} \cdot \sqrt{9} = 2 \cdot 3 = 6$.

Simplify: $\sqrt{50}$

Think: What perfect square is a factor of 50?

Begin with a perfect square that is larger than 50.

Then test each successively smaller perfect square.

$8^2 = 64$; 64 is too big.
$7^2 = 49$; 49 is not a factor of 50.
$6^2 = 36$; 36 is not a factor of 50.
$5^2 = 25$; 25 is a factor of 50. $(50 = 25 \cdot 2)$

Write $\sqrt{50}$ as $\sqrt{25 \cdot 2}$.

$$\sqrt{50} = \sqrt{25 \cdot 2}$$

Use the Product Property of Square Roots.

$$= \sqrt{25} \cdot \sqrt{2}$$

Simplify $\sqrt{25}$.

$$= 5 \cdot \sqrt{2}$$

The radicand 2 contains no factor other than 1 that is a perfect square. The radical expression $5\sqrt{2}$ is in simplest form.

$$= 5\sqrt{2}$$

Remember that $5\sqrt{2}$ means 5 times $\sqrt{2}$. Using a calculator, $5\sqrt{2} \approx 5(1.4142) = 7.071$ and $\sqrt{50} \approx 7.071$.

> **CALCULATOR NOTE**
> The keystrokes used to evaluate $5\sqrt{2}$ on a calculator, are either:
> 1. 5 × 2 $\boxed{\sqrt{}}$ $\boxed{=}$
> or
> 2. 5 × $\boxed{\sqrt{}}$ 2 $\boxed{\text{ENTER}}$

Example 5 Approximate $4\sqrt{17}$ to the nearest ten-thousandth.

Solution $4\sqrt{17} \approx 16.4924$

You Try It 5 Approximate $5\sqrt{23}$ to the nearest ten-thousandth.

Your Solution

Solution on p. A17

Example 6 Between what two whole numbers is the value of $\sqrt{79}$?

Solution 79 is between the perfect squares 64 and 81.

$\sqrt{64} = 8$ and $\sqrt{81} = 9$.

$8 < \sqrt{79} < 9$

You Try It 6 Between what two whole numbers is the value of $\sqrt{57}$?

Your Solution

Example 7 Simplify: $\sqrt{32}$

Solution $6^2 = 36$; 36 is too big.
$5^2 = 25$; 25 is not a factor of 32.
$4^2 = 16$; 16 is a factor of 32.

$\sqrt{32} = \sqrt{16 \cdot 2}$

$= \sqrt{16} \cdot \sqrt{2}$

$= 4 \cdot \sqrt{2}$

$= 4\sqrt{2}$

You Try It 7 Simplify: $\sqrt{80}$

Your Solution

Solutions on p. A17

Objective C

Applications and formulas

Example 8

Find the range of a submarine periscope that is 8 ft above the surface of the water. Use the formula $R = 1.4\sqrt{h}$, where R is the range in miles and h is the height in feet of the periscope above the surface of the water. Round to the nearest hundredth.

Strategy

To find the range, replace h by 8 in the given formula and solve for R.

Solution

$R = 1.4\sqrt{h}$
$R = 1.4\sqrt{8}$
$R \approx 3.96$

The range of the periscope is 3.96 mi.

You Try It 8

Find the range of a submarine periscope that is 6 ft above the surface of the water. Use the formula $R = 1.4\sqrt{h}$, where R is the range in miles and h is the height in feet of the periscope above the surface of the water. Round to the nearest hundredth.

Your Strategy

Your Solution

Solution on p. A17

4.4 EXERCISES

OBJECTIVE A

Simplify.

1. $\sqrt{36}$

2. $\sqrt{1}$

3. $-\sqrt{9}$

4. $-\sqrt{1}$

5. $\sqrt{169}$

6. $\sqrt{196}$

7. $\sqrt{225}$

8. $\sqrt{81}$

9. $-\sqrt{25}$

10. $-\sqrt{64}$

11. $-\sqrt{100}$

12. $-\sqrt{4}$

13. $\sqrt{8 + 17}$

14. $\sqrt{40 + 24}$

15. $\sqrt{49} + \sqrt{9}$

16. $\sqrt{100} + \sqrt{16}$

17. $\sqrt{121} - \sqrt{4}$

18. $\sqrt{144} - \sqrt{25}$

19. $3\sqrt{81}$

20. $8\sqrt{36}$

21. $-2\sqrt{49}$

22. $-6\sqrt{121}$

23. $5\sqrt{16} - 4$

24. $7\sqrt{64} + 9$

25. $3 + 10\sqrt{1}$

26. $14 - 3\sqrt{144}$

27. $\sqrt{4} - 2\sqrt{16}$

28. $\sqrt{144} + 3\sqrt{9}$

29. $5\sqrt{25} + \sqrt{49}$

30. $20\sqrt{1} - \sqrt{36}$

31. $\sqrt{\dfrac{1}{100}}$

32. $\sqrt{\dfrac{1}{81}}$

33. $\sqrt{\dfrac{9}{16}}$

34. $\sqrt{\dfrac{25}{49}}$

35. $\sqrt{\dfrac{1}{4}} + \sqrt{\dfrac{1}{64}}$

36. $\sqrt{\dfrac{1}{36}} - \sqrt{\dfrac{1}{144}}$

Evaluate the expression for the given values of the variables.

37. $-4\sqrt{xy}$, where $x = 3$ and $y = 12$

38. $-3\sqrt{xy}$, where $x = 20$ and $y = 5$

39. $8\sqrt{x + y}$, where $x = 19$ and $y = 6$

40. $7\sqrt{x + y}$, where $x = 34$ and $y = 15$

41. $5 + 2\sqrt{ab}$, where $a = 27$ and $b = 3$

42. $6\sqrt{ab} - 9$, where $a = 2$ and $b = 32$

43. $\sqrt{a^2 + b^2}$, where $a = 3$ and $b = 4$

44. $\sqrt{c^2 - a^2}$, where $a = 6$ and $c = 10$

45. $\sqrt{c^2 - b^2}$, where $b = 12$ and $c = 13$

46. $\sqrt{b^2 - 4ac}$, where $a = 1$, $b = -4$, and $c = -5$

Solve.

47. What is the sum of five and the square root of nine?

48. Find eight more than the square root of four.

49. Find the difference between six and the square root of twenty-five.

50. What is seven decreased by the square root of sixteen?

51. What is negative four times the square root of eighty-one?

52. Find the product of negative three and the square root of forty-nine.

OBJECTIVE B

Approximate to the nearest ten-thousandth.

53. $\sqrt{3}$

54. $\sqrt{7}$

55. $\sqrt{10}$

56. $\sqrt{19}$

57. $2\sqrt{6}$

58. $10\sqrt{21}$

59. $3\sqrt{14}$

60. $6\sqrt{15}$

61. $-4\sqrt{2}$

62. $-5\sqrt{13}$

63. $-8\sqrt{30}$

64. $-12\sqrt{53}$

Between what two whole numbers is the value of the radical expression?

65. $\sqrt{23}$
      66. $\sqrt{47}$
      67. $\sqrt{29}$
      68. $\sqrt{71}$

69. $\sqrt{62}$
      70. $\sqrt{103}$
      71. $\sqrt{130}$
      72. $\sqrt{95}$

Simplify.

73. $\sqrt{8}$
    74. $\sqrt{12}$
    75. $\sqrt{45}$
    76. $\sqrt{18}$
    77. $\sqrt{20}$

78. $\sqrt{44}$
    79. $\sqrt{27}$
    80. $\sqrt{56}$
    81. $\sqrt{48}$
    82. $\sqrt{28}$

83. $\sqrt{75}$
    84. $\sqrt{96}$
    85. $\sqrt{63}$
    86. $\sqrt{72}$
    87. $\sqrt{98}$

88. $\sqrt{108}$
    89. $\sqrt{112}$
    90. $\sqrt{200}$
    91. $\sqrt{175}$
    92. $\sqrt{180}$

OBJECTIVE C

Solve.

93. *Earth Science* A tsunami is a great sea wave produced by underwater earthquakes or volcanic eruption. Find the velocity of a tsunami when the depth of the water is 100 ft. Use the formula $v = 3\sqrt{d}$, where v is the velocity in feet per second of a tsunami as it approaches land and d is the depth in feet of the water.

94. *Earth Science* A tsunami is a great sea wave produced by underwater earthquakes or volcanic eruption. Find the velocity of a tsunami when the depth of the water is 144 ft. Use the formula $v = 3\sqrt{d}$, where v is the velocity in feet per second of a tsunami as it approaches land and d is the depth in feet of the water.

95. *Physics* If an object is dropped from a plane, how long will it take for the object to fall 144 ft? Use the formula $t = \sqrt{\dfrac{d}{16}}$, where t is the time in seconds that the object falls and d is the distance in feet that the object falls.

Solve.

96. *Physics* If an object is dropped from a bridge, how long will it take for the object to fall 64 ft? Use the formula $t = \sqrt{\dfrac{d}{16}}$, where t is the time in seconds that the object falls and d is the distance in feet that the object falls.

97. *Astronautics* The weight of an object is related to the distance the object is above the surface of the earth. A formula for this relationship is $d = 4{,}000\sqrt{\dfrac{E}{S}} - 4{,}000$, where E is the object's weight on the surface of the earth and S is the object's weight at a distance of d miles above the earth's surface. A space explorer, who weighs 144 lb on the surface of the earth, weighs 36 lb in space. How far above the earth's surface is the space explorer?

98. *Astronautics* The weight of an object is related to the distance the object is above the surface of the earth. A formula for this relationship is $d = 4{,}000\sqrt{\dfrac{E}{S}} - 4{,}000$, where E is the object's weight on the surface of the earth and S is the object's weight at a distance of d miles above the earth's surface. A space explorer, who weighs 189 lb on the surface of the earth, weighs 21 lb in space. How far above the earth's surface is the space explorer?

CRITICAL THINKING

99. List the whole numbers between $\sqrt{4}$ and $\sqrt{100}$.

100. Simplify. **a.** $\sqrt{0.81}$ **b.** $-\sqrt{0.64}$ **c.** $\sqrt{2\dfrac{7}{9}}$ **d.** $-\sqrt{3\dfrac{1}{16}}$

101. List the expressions $\sqrt{\dfrac{1}{4} + \dfrac{1}{8}}$, $\sqrt{\dfrac{1}{3} + \dfrac{1}{9}}$, and $\sqrt{\dfrac{1}{5} + \dfrac{1}{6}}$ in order from smallest to largest.

102. a. Use the expressions $\sqrt{16 + 9}$ and $\sqrt{16} + \sqrt{9}$ to show that $\sqrt{a + b} \neq \sqrt{a} + \sqrt{b}$.
   b. Use the expressions $\sqrt{16 - 9}$ and $\sqrt{16} - \sqrt{9}$ to show that $\sqrt{a - b} \neq \sqrt{a} - \sqrt{b}$.

103. a. Find the two-digit perfect square that has exactly nine factors.
   b. Find two whole numbers such that their difference is 10, the smaller number is a perfect square, and the larger number is two less than a perfect square.

104. Describe in your own words (a) how to find the square root of a perfect square and (b) how to simplify the square root of a number that is not a perfect square.

105. Explain why $2\sqrt{2}$ is in simplest form and $\sqrt{8}$ is not in simplest form.

106. Locate definitions of a perfect cube and of a cube root. Describe how to find the cube root of a perfect cube.

SECTION 4.5   Real Numbers

OBJECTIVE A
Real numbers and the real number line

A **rational number** is the quotient of two integers.

> ### Rational Numbers
>
> A rational number is a number that can be written in the form $\frac{a}{b}$, where a and b are integers and $b \neq 0$.

Each of the three numbers shown at the right is a rational number.

$$\frac{3}{4} \qquad \frac{-2}{9} \qquad \frac{13}{-5}$$

An integer can be written as the quotient of the integer and 1. Therefore, every integer is a rational number.

$$6 = \frac{6}{1} \qquad -8 = \frac{-8}{1}$$

A mixed number can be written as the quotient of two integers. Therefore, every mixed number is a rational number.

$$1\frac{4}{7} = \frac{11}{7} \qquad 3\frac{2}{5} = \frac{17}{5}$$

Recall from Section 4.2 that a fraction can be written as a decimal by dividing the numerator of the fraction by the denominator. The result is either a terminating decimal or a repeating decimal.

To convert $\frac{3}{8}$ to a decimal, read the fraction bar as "divided by."

$$\frac{3}{8} = 3 \div 8 = 0.375.$$ This is an example of a terminating decimal.

To convert $\frac{6}{11}$ to a decimal, divide 6 by 11.

$$\frac{6}{11} = 6 \div 11 = 0.\overline{54}.$$ This is an example of a repeating decimal.

Every rational number can be written either as a terminating decimal or as a repeating decimal. All terminating and repeating decimals are rational numbers.

Some numbers have decimal representations that never terminate or repeat, for example,

$$0.12122122212222 \ldots$$

The pattern in this number is one more 2 following each successive 1 in the number. There is no repeating block of digits. This number is an **irrational number.** Other examples of irrational numbers include π (which is presented in Chapter 9) and square roots of integers that are not perfect squares.

> **TAKE NOTE**
>
> Rational numbers are fractions such as $-\frac{4}{5}$ or $\frac{10}{7}$ where the numerator and denominator are integers. Rational numbers are also represented by repeating decimals such as 0.2626262 . . . or terminating decimals such as 1.83. An irrational number is neither a repeating decimal nor a terminating decimal. For instance, 1.45445444544445 . . . is an irrational number.

Irrational Numbers

An **irrational number** is a number whose decimal representation never terminates or repeats.

The rational numbers and the irrational numbers taken together are called the **real numbers**.

Real Numbers

The **real numbers** are all the rational numbers together with all the irrational numbers.

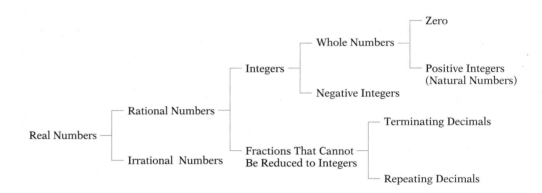

The number line is also called the **real number line.** Every real number corresponds to a point on the real number line, and every point on the real number line corresponds to a real number.

Graph $3\frac{1}{2}$ on the real number line.

$3\frac{1}{2}$ is a positive number and is therefore to the right of zero on the number line. Draw a solid dot three and one-half units to the right of zero on the number line.

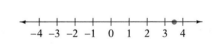

Graph -2.5 on the real number line.

-2.5 is a negative number and is therefore to the left of zero on the number line. Draw a solid dot two and one-half units to the left of zero on the number line.

Graph the real numbers greater than 2.

To graph the real numbers greater than 2 would mean that a solid dot should be placed above every number to the right of 2 on the number line. It is not possible to list all the real numbers greater than 2. It is not even possible to list all the real numbers between 2 and 3, or even to give the smallest real number greater than 2. The number 2.0000000001 is greater than 2 and is certainly very close to 2, but even smaller numbers greater than 2 can be written by inserting more and more zeros after the decimal point. Therefore, the graph of the real numbers greater than 2 is shown by drawing a heavy line to the right of 2.

The arrow indicates that the heavy line continues without end. The real numbers greater than 2 do not include the number 2. The circle on the graph indicates that 2 is not included in the graph.

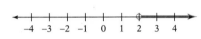

Graph the real numbers between 1 and 3.

The real numbers between 1 and 3 do not include the number 1 or the number 3; thus circles are drawn at 1 and 3. Draw a heavy line between 1 and 3 to indicate all the real numbers between these two numbers.

||||||

Example 1 Graph 0.5 on the real number line.	**You Try It 1** Graph $-1\frac{1}{2}$ on the real number line.
Solution Draw a solid dot one-half unit to the right of zero on the number line. 	**Your Solution** 
Example 2 Graph the real numbers less than -1.	**You Try It 2** Graph the real numbers greater than -2.
Solution The real numbers less than -1 are to the left of -1 on the number line. Draw a circle at -1. Draw a heavy line to the left of -1. Draw an arrow at the left of the line. 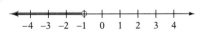	**Your Solution** 

Example 3 Graph the real numbers between −3 and 0.

Solution Draw a circle at −3 and a circle at 0. Draw a heavy line between −3 and 0.

You Try It 3 Graph the real numbers between −1 and 4.

Your Solution

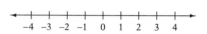

Solution on p. A17

OBJECTIVE B

Inequalities in one variable

Recall that the symbol for "is greater than" is >, and the symbol for "is less than" is <. The symbol ≥ means "is greater than or equal to." The symbol ≤ means "is less than or equal to."

The statement 5 < 5 is a false statement because 5 is not less than 5.

$5 < 5$ False

The statement 5 ≤ 5 is a true statement because 5 is "less than <u>or</u> equal to" 5; 5 is equal to 5.

$5 ≤ 5$ True

An **inequality** contains the symbol >, <, ≥, or ≤, and expresses the relative order of two mathematical expressions.

$$\left. \begin{array}{l} 4 > -3 \\ -9.7 < 0 \\ 6 + 2 \geq 1 \\ x \leq 5 \end{array} \right\} \text{Inequalities}$$

The inequality $x \leq 5$ is read "x is less than or equal to 5."

For the inequality $x > -3$, which values of the variable listed below make the inequality true?

a. −6   **b.** −3.9   **c.** 0   **d.** $\sqrt{7}$

Replace x in $x > -3$ with each number, and determine whether each inequality is true.

a.	**b.**	**c.**	**d.**
$x > -3$	$x > -3$	$x > -3$	$x > -3$
$-6 > -3$	$-3.9 > -3$	$0 > -3$	$\sqrt{7} > -3$
False	False	True	True

The numbers 0 and $\sqrt{7}$ make the inequality true.

There are many values of the variable x that will make the inequality $x > -3$ true; any number greater than −3 makes the inequality true. Replacing x with any number less than −3 will result in a false statement.

What values of the variable x make the inequality $x \leq 4$ true?

All real numbers less than or equal to 4 make the inequality true.

The numbers that make an inequality true can be graphed on the real number line.

Graph $x > 1$.

The numbers that, when substituted for x, make this inequality true are all the real numbers greater than 1. The numbers greater than 1 are all the numbers to the right of 1 on the number line. The circle on the graph indicates that 1 is not included in the numbers greater than 1.

Graph $x \geq 1$.

The numbers that make this inequality true are all the real numbers greater than or equal to 1. The solid dot at 1 indicates that 1 is included in the numbers greater than or equal to 1.

Note: for $<$ or $>$, draw a circle on the graph. For $\leq$ or $\geq$, draw a solid dot.

Example 4   For the inequality $x \leq -6$, which values of the variable listed below make the inequality true?

a. -12 **b.** -6 **c.** 0 **d.** $\sqrt{5}$

Solution   **a.** $x \leq -6$
$-12 \leq -6$ True

b. $x \leq -6$
$-6 \leq -6$ True

c. $x \leq -6$
$0 \leq -6$ False

d. $x \leq -6$
$\sqrt{5} \leq -6$ False

The numbers -12 and -6 make the inequality true.

You Try It 4   For the inequality $x \geq 4$, which values of the variable listed below make the inequality true?

a. -1 **b.** 0 **c.** 4 **d.** $\sqrt{26}$

Your Solution

Example 5   What values of the variable x make the inequality $x < 8$ true?

Solution   All real numbers less than 8 make the inequality true.

You Try It 5   What values of the variable x make the inequality $x > -7$ true?

Your Solution

Solutions on p. A17

Example 6 Graph $x \leq 3$.

Solution Draw a solid dot at 3.
Draw an arrow to the left of 3.

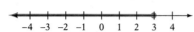

You Try It 6 Graph $x \geq -4$.

Your Solution

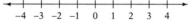

Solution on p. A17

OBJECTIVE C
Applications

Solving application problems requires recognition of the verbal phrases that translate into mathematical symbols. Below is a partial list of the phrases used to indicate each of the four inequality symbols.

$<$ is less than

$>$ is greater than
is more than
exceeds

$\leq$ is less than or equal to
maximum
at most
or less

$\geq$ is greater than or equal to
minimum
at least
or more

Example 7

The minimum wage at the company you work for is $5.25 an hour. Write an inequality for the wages at the company. Is it possible for an employee to earn $5.15 an hour?

Strategy

▶ To write the inequality, let w represent the wages. Since $5.25 is a minimum wage, all wages are greater than or equal to $5.25.

▶ To determine whether a wage of $5.15 is possible, replace w in the inequality by 5.15. If the inequality is true, it is possible. If the inequality is false, it is not possible.

Solution

$w \geq 5.25$

$5.15 \geq 5.25$ False

It is not possible for an employee to earn $5.15 an hour.

You Try It 7

On the highway near your home, motorists who exceed a speed of 55 mph are ticketed. Write an inequality for the speeds at which a motorist is ticketed. Will a motorist traveling at 58 mph be ticketed?

Your Strategy

Your Solution

Solution on p. A18

4.5 EXERCISES

OBJECTIVE A

Graph the number on the real number line.

1. $2\dfrac{1}{2}$

<!-- number line -6 to 6 -->

2. $-2\dfrac{1}{2}$

<!-- number line -6 to 6 -->

3. -3.5

<!-- number line -6 to 6 -->

4. -0.5

<!-- number line -6 to 6 -->

5. $-4\dfrac{1}{2}$

<!-- number line -6 to 6 -->

6. $\dfrac{1}{2}$

<!-- number line -6 to 6 -->

7. 1.5

<!-- number line -6 to 6 -->

8. 5.5

<!-- number line -6 to 6 -->

Graph.

9. the real numbers greater than 6

<!-- number line -6 to 6 -->

10. the real numbers greater than 1

<!-- number line -6 to 6 -->

11. the real numbers less than 0

<!-- number line -6 to 6 -->

12. the real numbers less than 2

<!-- number line -6 to 6 -->

13. the real numbers greater than -1

<!-- number line -6 to 6 -->

14. the real numbers greater than -4

<!-- number line -6 to 6 -->

Graph.

15. the real numbers less than -5

16. the real numbers less than -3

17. the real numbers between 2 and 5

18. the real numbers between 4 and 6

19. the real numbers between -4 and 0

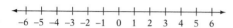

20. the real numbers between 0 and 3

21. the real numbers between -2 and 6

22. the real numbers between -1 and 5

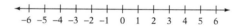

23. the real numbers between -6 and 1

24. the real numbers between -5 and 0

OBJECTIVE B

Solve.

25. For the inequality $x > 9$, which numbers listed below make the inequality true?
   a. -3.8 **b.** 0   **c.** 9   **d.** $\sqrt{101}$

26. For the inequality $x \leq 5$, which numbers listed below make the inequality true?
   a. $-\sqrt{11}$ **b.** 0   **c.** 5   **d.** 5.01

27. For the inequality $x \geq -2$, which numbers listed below make the inequality true?
   a. -6 **b.** -2 **c.** 0.4   **d.** $\sqrt{17}$

28. For the inequality $x \leq -7$, which numbers listed below make the inequality true?
   a. -14 **b.** -7 **c.** -1.3 **d.** $-\sqrt{2}$

What values of the variable x make the inequality true?

29. $x < 3$ **30.** $x > -6$ **31.** $x \geq -1$ **32.** $x \leq 5$

Graph the inequality on the real number line.

33. $x < -2$

34. $x > 4$

35. $x \geq 0$

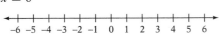

36. $x \leq -3$

37. $x > -5$

38. $x < -1$

39. $x \leq 2$

40. $x \geq 6$

OBJECTIVE C

Solve.

41. *Business*   Each sales representative for a company must sell at least 50,000 units per year. Write an inequality for the number of units a sales representative must sell. Has a representative who sold 49,000 units this past year met the sales goal?

42. *Health*   A health official recommends a cholesterol level of less than 220 units. Write an inequality for the acceptable cholesterol levels. Is a cholesterol level of 238 within the recommended levels?

43. *Education*   A part-time student can take a maximum of 9 credit hours per semester. Write an inequality for the number of credit hours a part-time student can take. Does a student taking 8.5 credit hours fulfill the requirement for being a part-time student?

44. *Community Service*   A service organization will receive a bonus of $200 for collecting more than 1,750 lb of aluminum cans during a collection drive. Write an inequality for the number of cans that must be collected in order to earn the bonus. If 1,705.5 lb of aluminum cans are collected, will the organization receive the bonus?

45. *Finances*   Your monthly budget allows you to spend at most $1,200 per month. Write an inequality for the amount of money you can spend per month. Have you kept within your budget during a month in which you spent $1,190.50?

Solve.

46. *Education* In order to get a B in a history course, you must earn more than 80 points on the final exam. Write an inequality for the number of points you need to score on the final exam. Will a score of $80\frac{1}{2}$ earn you a B in the course?

47. *Computers* Computer disks should be stored at temperatures greater than 50°F. Write an inequality for the temperatures at which computer disks should be stored. Is it safe to store a computer disk at a temperature of 47.5°F?

48. *Sports* According to NCAA rules, the diameter of the ring on a basketball hoop is to be $\frac{5}{8}$ in. or less. Write an inequality for the diameter of the ring on a basketball hoop. Does a ring with a diameter of $\frac{9}{16}$ in. meet the NCAA regulations?

CRITICAL THINKING

49. Classify each number as a whole number, an integer, a positive integer, a negative integer, a rational number, an irrational number, and/or a real number.

    a. -2 b. 18 c. $-\frac{9}{37}$ d. -6.606 e. $4.\overline{56}$ f. $3.050050005\ldots$

50. Using the variable x, write an inequality to represent the graph.
    a.
    b.

51. For the given inequality, which of the numbers in parentheses make the inequality true?
    a. $|x| < 9$ $(-2.5, 0, 9, 15.8)$ b. $|x| > -3$ $(-6.3, -3, 0, 6.7)$
    c. $|x| \geq 4$ $(-1.5, 0, 4, 13.6)$ d. $|x| \leq 5$ $(-4.9, 0, 2.1, 5)$

52. Given that a, b, c, and d are real numbers, which will ensure that $a + c < b + d$?
    a. $a < b$ and $c < d$ b. $a > b$ and $c > d$
    c. $a < b$ and $c > d$ d. $a > b$ and $c < d$

53. Determine whether the statement is always true, sometimes true, or never true.
    a. Given that $a > 0$ and $b > 0$, then $ab > 0$.
    b. Given that $a < 0$, then $a^2 > 0$.
    c. Given that $a > 0$ and $b > 0$, then $a^2 > b$.

54. Enter -4 on your calculator and then press the square root key. What is in the calculator's display? Explain why.

55. In your own words, define (a) a rational number, (b) an irrational number, and (c) a real number.

PROJECTS IN MATHEMATICS

Sequences

Suppose you are offered a 30-day job that pays $.01 the first day, $.02 the second day, $.04 the third day, and so on. Each day you work, your earnings are twice your earnings for the previous day. Would you accept this job over a 30-day job that pays $50,000 per day?

Day 1

Day 2

For the job in which earnings double each day, make a guess as to your earnings on the 30th day of work, and your total earnings over the 30-day period. Then calculate these figures. You may be surprised at the results!

The list of numbers that indicates your earnings each day is an ordered list of numbers, called a **sequence**.

Day 3

$$0.01, 0.02, 0.04, 0.08, 0.16, 0.32, 0.64, 1.28, 2.56, 5.12, \ldots$$

This list is ordered because the position of a number in the list indicates the day on which that amount was earned. For example, the 8th term of the sequence is 1.28, and $1.28 is earned on the 8th day.

Each of the numbers of a sequence is called a **term** of the sequence. A formula can be used to find a specific term of the sequence given above.

term t = (first term)(2^{t-1})

For example:

$$\text{term } 5 = (0.01)(2^{5-1}) = 0.01(2^4) = 0.01(16) = 0.16$$

The amount earned on day 5 is $.16.

Use the formula to find the amount earned on day 15 and on day 20. What amount is earned on the 30th day?

A formula can be used to find the total amount earned after any given day.

$$\textbf{day } d = \frac{\textbf{(first term)}(1 - 2^d)}{1 - 2}$$

For example:

$$\text{day } 5 = \frac{(0.01)(1 - 2^5)}{1 - 2} = \frac{(0.01)(1 - 32)}{1 - 2}$$

$$= \frac{(0.01)(-31)}{-1} = 0.31$$

[Note that the Order of Operations Agreement is used to simplify this expression.]

The total amount earned during the first 5 days is $.31.

Use the formula to find the total amount earned by the end of day 15 and by the end of day 20. What is the total amount earned after 30 days? How does this compare with being paid $50,000 per day?

Small Business and Gross Income

Do some research on your local newspaper. What is the newspaper's circulation? How often is an edition of the newspaper published? What is the cost per issue?

Is there a special rate for subscribers to the newspaper? What is the cost of a subscription? What is the length of time for which a subscription is paid? How many of the newspaper's readers are subscribers? Use this figure and the newspaper's total circulation to determine the number of copies sold at newsstands.

Use the figures you have gathered in answering the questions above to determine the total annual income, or **gross income,** derived from sales of the newspaper.

Demographics

Demography is the statistical study of human populations. The chart below includes statistics on a portion of the population in the United States. It shows the number of 30- to 34-year-olds who are or will be in the job market in the United States in 1992, 1999, and 2005.

The Twenty-First Century: An Easier Ladder to Climb		
Number of 30- to 34-year-olds competing for jobs		
1992	1999	2005
Men 10.3 million	9.2 million	8.5 million
Women 8.3 million	7.9 million	7.8 million
Total 18.6 million	17.1 million	16.3 million

Discuss the significance of the data. Consider including the following in your discussion: an interpretation of the title of the chart, The Baby Boomers, the Baby Bust. What other factors influence competition for jobs?

Averages

We often discuss temperature in terms of average high or average low temperature. Temperatures collected over a period of time are analyzed to determine, for example, the average high temperature for a given month in your city or state. The following activity is planned to help you better understand the concept of "average."

1. Choose two cities in the United States. We will refer to them as City X and City Y. Over an eight-day period, record the daily high temperature each day in each city.

2. Determine the average high temperature for City *X* for the eight-day period. (Add the eight numbers, and then divide the sum by 8.) Do not round your answer.

3. Subtract the average high temperature for City *X* from each of the eight daily high temperatures for City *X*. You should have a list of eight numbers; the list should include positive numbers, negative numbers, and possibly zero.

4. Find the sum of the list of eight differences recorded in Step 3.

5. Repeat Steps 2 through 4 for City *Y*.

6. Compare the two sums found in Step 5 for City *X* and City *Y*.

7. If you were to conduct this activity again, what would you expect the outcome to be? Use the results to explain what an average high temperature means. In your own words, explain what "average" means.

CHAPTER SUMMARY

Key Words

A number written in *decimal notation* has three parts: a whole number part, a decimal point, and a decimal part. The *decimal part* of a number represents a number less than one. A number written in decimal notation is often simply called a *decimal*.

The square of an integer is called a *perfect square*.

A *square root* of a positive number *x* is a number whose square is *x*. The symbol for square root is $\sqrt{}$, which is called a *radical sign*. The number under the radical is called the *radicand*. A radical expression is in *simplest form* when the radicand contains no factor, other than 1, that is a perfect square.

A *rational number* is a number that can be written in the form $\frac{a}{b}$, where *a* and *b* are integers and $b \neq 0$. Every rational number can be written either as a terminating or a repeating decimal.

An *irrational number* is a number whose decimal representation never terminates or repeats.

The *real numbers* are all the rational numbers together with all the irrational numbers.

An *inequality* contains the symbol $>$, $<$, $\geq$, or $\leq$, and expresses the relative order of two mathematical expressions.

Essential Rules

To write a decimal in words, write the decimal part as if it were a whole number. Then name the place value of the last digit.

To write a decimal in standard form when it is written in words, write the whole number part, replace the word *and* with a decimal point, and write the decimal part so that the last digit is in the given place-value position.

To compare two decimals, write the decimal part of each number so that each has the same number of decimal places. Then compare the two numbers.

To round a decimal, use the same rules used with whole numbers, except drop the digits to the right of the given place value instead of replacing them with zeros.

To add or subtract decimals, write the decimals so that the decimal points are on a vertical line. Add or subtract as you would with whole numbers. Then write the decimal point in the answer directly below the decimal points in the given numbers.

To estimate the answer to a calculation, round each number to the highest place value of the number; the first digit of each number will be nonzero and all other digits will be zero. If a number is a decimal less than one, round the decimal so that there is one nonzero digit. Perform the calculation using the rounded numbers.

To multiply decimals, multiply the numbers as you would whole numbers. Then write the decimal point in the product so that the number of decimal places in the product is the sum of the decimal places in the factors.

To multiply a decimal by a power of 10, move the decimal point to the right the same number of places as there are zeros in the power of 10. If the power of 10 is written in exponential notation, the exponent indicates how many places to move the decimal point.

To divide decimals, move the decimal point in the divisor to the right so that it is a whole number. Move the decimal point in the dividend the same number of places to the right. Place the decimal point in the quotient directly above the decimal point in the dividend. Then divide as you would with whole numbers.

To divide a decimal by a power of 10, move the decimal point to the left the same number of places as there are zeros in the power of 10. If the power of 10 is written in exponential notation, the exponent indicates how many places to move the decimal point.

To write a fraction as a decimal, divide the numerator of the fraction by the denominator.

To convert a decimal to a fraction, remove the decimal point and place the decimal part over a denominator equal to the place value of the last digit in the decimal.

To find the order relation between a decimal and a fraction, first rewrite the fraction as a decimal. Then compare the two decimals.

Square Root   For $x > 0$, if $a^2 = x$, then $\sqrt{x} = a$.

Product Property of Square Roots   If a and b are positive numbers, then $\sqrt{a \cdot b} = \sqrt{a} \cdot \sqrt{b}$.

CHAPTER REVIEW EXERCISES

1. Approximate $3\sqrt{47}$ to the nearest ten-thousandth.

2. Find the product of 0.918 and 10^5.

3. Simplify: $-\sqrt{121}$

4. Subtract: $-3.981 - 4.32$

5. Evaluate $a + b + c$ when $a = 80.59$, $b = -3.647$, and $c = 12.3$.

6. Write five and thirty-four thousandths in standard form.

7. Simplify: $\sqrt{100} - 2\sqrt{49}$

8. Find the quotient of 14.2 and 10^3.

9. Solve: $4.2z = -1.428$

10. Place the correct symbol, $<$ or $>$, between the two numbers.

8.039   8.31

11. Evaluate $\dfrac{x}{y}$ when $x = 0.396$ and $y = 3.6$.

12. Multiply: $(9.47)(0.26)$

13. For the inequality $x \geq -1$, what numbers listed below make the inequality true?
a. -6 **b.** -1 **c.** -0.5 **d.** $\sqrt{10}$

14. Place the correct symbol, $<$ or $>$, between the two numbers.

$\dfrac{3}{7}$ 0.429

15. Convert 0.28 to a fraction.

16. Divide and round to the nearest tenth: $-6.8 \div 47.92$

17. *Labor*   In July 1996, how many more hours per week, on average, did employees in the manufacturing sector work than those in private nonfarm jobs? Use the figure at the right.

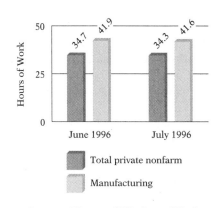

Average Hours of Work per Week
Source: Labor Department

18. Graph all the real numbers between −6 and −2.

19. Graph $x \geq -3$.

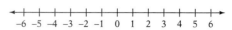

20. Find the sum of −247.8 and −193.4.

21. Find the quotient of 614.3 and 100.

22. Evaluate $a - b$ when $a = 80.32$ and $b = 29.577$.

23. Simplify: $\sqrt{90}$

24. Evaluate $60st$ when $s = 5$ and $t = -3.7$.

25. Estimate the difference between 506.81 and 64.1.

Solve.

26. *Education* A student must have a grade point average of at least 3.5 to qualify for a certain scholarship. Write an inequality for the grade point average a student must have in order to qualify for the scholarship. Does a student who has a grade point average of 3.48 qualify for the scholarship?

27. *Chemistry* The boiling point of mercury is 356.58°C. The melting point of mercury is −38.87°C. Find the difference between the boiling point and the melting point of mercury.

28. *History* The figure at the right shows the monetary cost of four wars. (a) What is the difference between the monetary costs of the two World Wars? (b) How many times greater was the monetary cost of the Vietnam War than World War I?

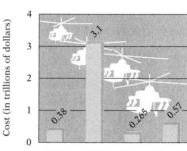

Cost (in trillions of dollars)

WWI WWII Korea Vietnam

Monetary Cost of War
Source: Congressional Research Service Using Numbers from the *Statistical Abstract of the United States*

29. *Consumerism* A 7-ounce jar of instant coffee costs $5.89. Find the cost per ounce. Round to the nearest cent.

30. *Consumerism* The total of the monthly payments for a car lease is the product of the number of months of the lease and the monthly lease payment. The total of the monthly payments for a 24-month car lease is $4,988.88. Find the monthly lease payment.

31. *Business* Use the formula $P = C + M$, where P is the price of a product to a customer, C is the cost paid by a store for the product, and M is the markup, to find the price of a treadmill that costs a business $369.99 and has a markup of $129.50.

32. *Physics* The velocity of a falling object is given by the formula $v = \sqrt{64d}$, where v is the velocity in feet per second and d is the distance the object has fallen. Find the velocity of an object that has fallen a distance of 25 ft.

CUMULATIVE REVIEW EXERCISES

1. Find the quotient of 387.9 and 10^4.

2. Evaluate $(x + y)^2 - 2z$ when $x = -3$, $y = 2$, and $z = -5$.

3. Solve: $-9.8 = -0.49c$

4. Write eight million seventy-two thousand ninety-two in standard form.

5. Graph all the real numbers between -4 and 1.

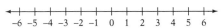

6. Graph $x \le -2$.

7. Find the difference between -23 and -19.

8. Estimate the sum of 372, 541, 608, and 429.

9. Simplify: $\sqrt{192}$

10. Evaluate $x \div y$ when $x = 3\frac{2}{3}$ and $y = 2\frac{4}{9}$.

11. What is -36.92 increased by 18.5?

12. Simplify: $\left(\frac{5}{9}\right)\left(-\frac{3}{10}\right)\left(-\frac{6}{7}\right)$

13. Evaluate x^4y^2 when $x = 2$ and $y = 10$.

14. Find the prime factorization of 260.

15. Convert $\frac{19}{25}$ to a decimal.

16. Approximate $10\sqrt{91}$ to the nearest ten-thousandth.

17. ✍ *Labor* The figure at the right shows the number of vacation days that are legally mandated in several countries. (a) Which country mandates more vacation days, Ireland or Finland? (b) How many times more vacation days does Germany mandate than Mexico?

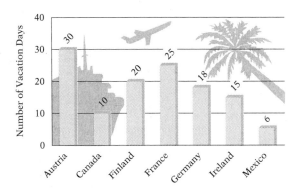

Number of Legally Mandated Vacation Days for Employees with One Year of Service
Source: Hewitt Associates

18. Divide: $\dfrac{-8}{0}$

19. Simplify: $-\dfrac{5}{7} + \dfrac{4}{21}$

20. Simplify: $4\sqrt{25} - \sqrt{81}$

21. Estimate the product of 62.8 and 0.47.

22. Simplify: $5(3 - 7) \div (-4) + 6(2)$

23. Evaluate $\dfrac{a}{b + c}$ when $a = \dfrac{3}{8}$, $b = \dfrac{1}{2}$, and $c = \dfrac{3}{4}$.

24. Evaluate $x - y + z$ when $x = \dfrac{5}{12}$, $y = -\dfrac{3}{8}$, and $z = -\dfrac{3}{4}$.

25. Divide and round to the nearest tenth: $2.617 \div 0.93$

Solve.

26. *Consumerism* Your cellular phone company charges $29.99 per month, which includes 10 min of free air time, and $.75 for each additional minute after the first 10. What is your cellular phone service bill for a month in which you had 37 min of calls?

27. *Temperature* On December 24, 1924, in Fairfield, Montana, the temperature fell from 17.22°C at noon to −29.4°C at midnight. How many degrees did the temperature fall in the 12-hour period?

28. *Consumerism* Use the formula $C = \dfrac{M}{N}$, where C is the cost per visit at a health club, M is the membership fee, and N is the number of visits to the club, to find the cost per visit when your annual membership fee at a health club is $195 and you visit the club 125 times during the year.

29. *Business* The figure at the right shows how the average salesperson spends the workweek. (a) On average, how many hours per week does a salesperson work? (b) Does the average salesperson spend more time face-to-face selling or doing both administrative work and placing service calls?

30. *Physics* The relationship between the velocity of a car and its braking distance is given by the formula $v = \sqrt{20d}$, where v is the velocity in miles per hour and d is its braking distance in feet. How fast is a car going when its braking distance is 45 ft?

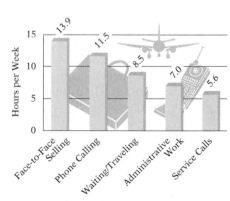

Average Salesperson's Work Week
Source: Dartnell s 28th Survey of Sales Force Compensation

Variable Expressions

FOCUS On Problem Solving

A very useful problem-solving strategy is to look for a pattern.

Problem A legend says that a peasant invented the game of chess and gave it to a very rich king as a present. The king so enjoyed the game that he gave the peasant the choice of anything in the kingdom. The peasant's request was simple. "Place one grain of wheat on the first square, 2 grains on the second square, 4 grains on the third square, 8 on the fourth square, and continue doubling the number of grains until the last square of the chessboard is reached." How many grains of wheat must the king give the peasant?

A chessboard consists of 64 squares. To find the total number of grains of wheat on the 64 squares, we begin by looking at the amount of wheat on the first few squares (see figure below.)

Square 1	Square 2	Square 3	Square 4
1	2	4	8
1	3	7	15

Square 5	Square 6	Square 7	Square 8
16	32	64	128
31	63	127	255

The bottom row of numbers represents the sum of the number of grains of wheat up to and including that square. For instance, the number of grains of wheat on the first 7 squares is

$$1 + 2 + 4 + 8 + 16 + 32 + 64 = 127.$$

One pattern to observe is that the number of grains of wheat on a square can be expressed by a power of 2.

The number of grains on square $n = 2^{n-1}$.

For example, the number of grains on square $7 = 2^{7-1} = 2^6 = 64$.

A second pattern of interest is that the number *below* a square (the total number of grains up to and including that square) is one less than the number of grains of wheat *on* the next square. For example, the number *below* square 7 is one less than the number *on* square 8 ($128 - 1 = 127$). From this observation, the number of grains of wheat on the first eight squares is the number on square 8 (128) plus one less than the number on square 8 (127); the total number of grains of wheat on the first eight squares is $128 + 127 = 255$.

From this observation,

Number of grains of wheat on the chessboard

$$= \begin{array}{c}\text{number of grains}\\\text{on square 64}\end{array} + \begin{array}{c}\text{one less than the number}\\\text{of grains on square 64}\end{array}$$

$$= 2^{64-1} + 2^{64-1} - 1$$

$$= 2^{63} + 2^{63} - 1 \approx 18{,}000{,}000{,}000{,}000{,}000{,}000$$

To give you an idea of the magnitude of this number, this is more wheat than has been produced in the world since chess was invented.

The same king decided to have a banquet in the long banquet room of the palace to celebrate the invention of chess. The king had 50 square tables, and each table could seat only one person on each side. The king pushed the tables together to form one long banquet table. How many people can sit at this table? *Hint:* Try constructing a pattern by using 2 tables, 3 tables, and 4 tables. [Answer: 102 people]

SECTION 5.1   **Properties of Real Numbers**

OBJECTIVE **A**

Application of the Properties of Real Numbers

The Properties of Real Numbers describe the way operations on numbers can be performed. These properties have been stated in previous chapters but are restated here for review. The properties are used to rewrite variable expressions.

PROPERTIES OF REAL NUMBERS

The Commutative Property of Addition If a and b are real numbers, then $a + b = b + a$.	$7 + 12 = 12 + 7$ $19 = 19$

The Commutative Property of Addition states that when adding two numbers, the numbers can be added in either order; the sum is the same.

The Commutative Property of Multiplication If a and b are real numbers, then $a \cdot b = b \cdot a$.	$7 \cdot (-2) = (-2) \cdot 7$ $-14 = -14$

The Commutative Property of Multiplication states that when multiplying two numbers, the numbers can be multiplied in either order; the product is the same.

The Associative Property of Addition If a, b, and c are real numbers, then $(a + b) + c = a + (b + c)$.	$(7 + 3) + 8 = 7 + (3 + 8)$ $10 + 8 = 7 + 11$ $18 = 18$

The Associative Property of Addition states that when adding three or more numbers, the numbers can be grouped in any order; the sum is the same.

The Associative Property of Multiplication If a, b, and c are real numbers, then $(a \cdot b) \cdot c = a \cdot (b \cdot c)$.	$(4 \cdot 5) \cdot 3 = 4 \cdot (5 \cdot 3)$ $20 \cdot 3 = 4 \cdot 15$ $60 = 60$

The Associative Property of Multiplication states that when multiplying three or more factors, the factors can be grouped in any order; the product is the same.

> ### The Addition Property of Zero
>
> If a is a real number, then
> $a + 0 = 0 + a = a.$

$$(-7) + 0 = 0 + (-7) = -7$$

The Addition Property of Zero states that the sum of a number and zero is the number.

> ### The Multiplication Property of Zero
>
> If a is a real number, then
> $a \cdot 0 = 0 \cdot a = 0.$

$$5 \cdot 0 = 0 \cdot 5 = 0$$

The Multiplication Property of Zero states that the product of a number and zero is zero.

> ### The Multiplication Property of One
>
> If a is a real number, then
> $a \cdot 1 = 1 \cdot a = a.$

$$9 \cdot 1 = 1 \cdot 9 = 9$$

The Multiplication Property of One states that the product of a number and one is the number.

> ### The Inverse Property of Addition
>
> If a is a real number, then
> $a + (-a) = (-a) + a = 0.$

$$2 + (-2) = (-2) + 2 = 0$$

The sum of a number and its opposite is zero.
$-a$ is the opposite of a. $-a$ is also called the **additive inverse of** a.
a is the opposite of $-a$, or a is the *additive inverse* of $-a$.
The sum of a number and its additive inverse is zero.

> ### The Inverse Property of Multiplication
>
> If a is a real number and $a \neq 0$, then
> $a \cdot \dfrac{1}{a} = \dfrac{1}{a} \cdot a = 1.$

$$4 \cdot \frac{1}{4} = \frac{1}{4} \cdot 4 = 1$$

The product of a nonzero number and its reciprocal is one.
$\dfrac{1}{a}$ is the reciprocal of a. $\dfrac{1}{a}$ is also called the **multiplicative inverse** of a.

a is the reciprocal of $\dfrac{1}{a}$, or a is the *multiplicative inverse* of $\dfrac{1}{a}$.
The product of a nonzero number and its multiplicative inverse is one.

The Properties of Real Numbers can be used to rewrite a variable expression in a simpler form. This process is referred to as *simplifying* the variable expression.

Simplify: $5 \cdot (4x)$

| Use the Associative Property of Multiplication. | $5 \cdot (4x) = (5 \cdot 4)x$ |
| Multiply 5 times 4. | $= 20x$ |

Simplify: $(6x) \cdot 2$

Use the Commutative Property of Multiplication.	$(6x) \cdot 2 = 2 \cdot (6x)$
Use the Associative Property of Multiplication.	$= (2 \cdot 6)x$
Multiply 2 times 6.	$= 12x$

Simplify: $(5y)(3y)$

Use the Commutative and Associative Properties of Multiplication.	$(5y)(3y) = 5 \cdot y \cdot 3 \cdot y$
	$= 5 \cdot 3 \cdot y \cdot y$
Write $y \cdot y$ in exponential form.	$= (5 \cdot 3)(y \cdot y)$
Multiply 5 times 3.	$= 15y^2$

By the Multiplication Property of One, the product of 1 and x is x.

$$1 \cdot x = x$$
$$1x = x$$

Just as the product of 1 and x is written x, the product of -1 and x is written $-x$.

$$-1 \cdot x = -x$$
$$-1x = -x$$

Simplify: $(-2)(-x)$

Write $-x$ as $-1x$.	$(-2)(-x) = (-2)(-1x)$
Use the Associative Property of Multiplication.	$= [(-2)(-1)]x$ ●
Multiply -2 times -1.	$= 2x$

> ● *TAKE NOTE*
>
> **Brackets, [ ], are used as a grouping symbol to group the factors -2 and -1 because parentheses have already been used in the expression to show that -2 and -1 are being multiplied. The expression $[(-2)(-1)]$ is considered easier to read than $((-2)(-1))$.**

Simplify: $-4t + 9 + 4t$

Use the Commutative Property of Addition.	$-4t + 9 + 4t = -4t + 4t + 9$
Use the Associative Property of Addition.	$= (-4t + 4t) + 9$
Use the Inverse Property of Addition.	$= 0 + 9$
Use the Addition Property of Zero.	$= 9$

Example 1 Simplify: $-5(7b)$

Solution $-5(7b) = (-5 \cdot 7)b$
$= -35b$

You Try It 1 Simplify: $-6(-3p)$

Your Solution

Example 2 Simplify: $(-4r)(-9t)$

Solution $(-4r)(-9t) = [(-4)(-9)](r \cdot t)$
$= 36rt$

You Try It 2 Simplify: $(-2m)(-8n)$

Your Solution

Example 3 Simplify: $(-8)(-z)$

Solution $(-8)(-z) = (-8)(-1z)$
$= [(-8)(-1)]z$
$= 8z$

You Try It 3 Simplify: $(-12)(-d)$

Your Solution

Example 4 Simplify: $-5y + 5y + 7$

Solution $-5y + 5y + 7 = 0 + 7$
$= 7$

You Try It 4 Simplify: $6n + 9 + (-6n)$

Your Solution

Solutions on p. A18 | | | | | |

OBJECTIVE **B**
The Distributive Property

Consider the numerical expression $6 \cdot (7 + 9)$.

This expression can be evaluated by applying the Order of Operations Agreement.

Simplify the expression inside the parentheses. $6 \cdot (7 + 9) = 6 \cdot 16$
Multiply. $= 96$

There is an alternate method of evaluating this expression.

Multiply each number inside the $6 \cdot (7 + 9) = 6 \cdot 7 + 6 \cdot 9$
parentheses by 6 and add the $= 42 + 54$
products. $= 96$

Each method produced the same result. The second method uses the **Distributive Property,** which is another of the Properties of Real Numbers.

The Distributive Property

If a, b, and c are real numbers, then $a(b + c) = ab + ac$.

The Distributive Property is used to remove parentheses from a variable expression.

Simplify $3(5a + 4)$ by using the Distributive Property.

Use the Distributive Property.	$3(5a + 4) = 3(5a) + 3(4)$
Simplify.	$= 15a + 12$

Simplify $-4(2a + 3)$ by using the Distributive Property.

Use the Distributive Property.	$-4(2a + 3) = -4(2a) + (-4)(3)$
Simplify.	$= -8a + (-12)$
Rewrite addition of the opposite as subtraction.	$= -8a - 12$

The Distributive Property can also be stated in terms of subtraction.

$a(b - c) = ab - ac$

Simplify $5(2x - 4y)$ by using the Distributive Property.

Use the Distributive Property.	$5(2x - 4y) = 5(2x) - 5(4y)$
Simplify.	$= 10x - 20y$

Simplify $-3(2x - 8)$ by using the Distributive Property.

Use the Distributive Property.	$-3(2x - 8) = -3(2x) - (-3)(8)$
Simplify.	$= -6x - (-24)$
Rewrite the subtraction as addition of the opposite.	$= -6x + 24$

The Distributive Property can be extended to more than two addends inside the parentheses. For example:

$$4(2a + 3b - 5c) = 4(2a) + 4(3b) - 4(5c)$$
$$= 8a + 12b - 20c$$

The Distributive Property is used to remove the parentheses from an expression that has a negative sign in front of the parentheses. Just as $-x = -1 \cdot x$, the expression $-(x + y) = -1(x + y)$. Therefore:

$$-(x + y) = -1(x + y) = -1x - 1y = -x - y$$

When a negative sign precedes parentheses, remove the parentheses and change the sign of *each* term inside the parentheses.

Rewrite the expression $-(4a - 3b + 7)$ without parentheses.

Remove the parentheses and change the sign of each term inside the parentheses.	$-(4a - 3b + 7) = -4a + 3b - 7$

Example 5

Simplify by using the Distributive Property:
$6(5c - 12)$

Solution

$6(5c - 12) = 6(5c) - 6(12)$
$\qquad\qquad\quad = 30c - 72$

You Try It 5

Simplify by using the Distributive Property:
$-7(2k - 5)$

Your Solution

Example 6

Simplify by using the Distributive Property:
$-4(-2a - b)$

Solution

$-4(-2a - b) = -4(-2a) - (-4)(b)$
$\qquad\qquad\qquad = 8a + 4b$

You Try It 6

Simplify by using the Distributive Property:
$-4(x - 2y)$

Your Solution

Example 7

Simplify by using the Distributive Property:
$-2(3m - 8n + 5)$

Solution

$-2(3m - 8n + 5)$
$= -2(3m) - (-2)(8n) + (-2)(5)$
$= -6m + 16n - 10$

You Try It 7

Simplify by using the Distributive Property:
$3(-2v + 3w - 7)$

Your Solution

Example 8

Simplify by using the Distributive Property:
$3(2a + 6b - 5c)$

Solution

$3(2a + 6b - 5c) = 3(2a) + 3(6b) - 3(5c)$
$\qquad\qquad\qquad\quad = 6a + 18b - 15c$

You Try It 8

Simplify by using the Distributive Property:
$-4(2x - 7y - z)$

Your Solution

Example 9

Rewrite $-(5x + 3y - 2z)$ without parentheses.

Solution

$-(5x + 3y - 2z) = -5x - 3y + 2z$

You Try It 9

Rewrite $-(c - 9d + 1)$ without parentheses.

Your Solution

Solutions on p. A18

5.1 EXERCISES

OBJECTIVE A

Identify the Property of Real Numbers that justifies the statement.

1. $3 \cdot (4 \cdot 7) = (3 \cdot 4) \cdot 7$

2. $a + 0 = a$

3. $x + 7 = 7 + x$

4. $12 \cdot a = a \cdot 12$

5. $4r + (-4r) = 0$

6. $5 + (a + 7) = (a + 7) + 5$

7. $\dfrac{2}{3} \cdot \dfrac{3}{2} = 1$

8. $-\dfrac{2}{3} + \dfrac{2}{3} = 0$

9. $a(bc) = (bc)a$

10. $1 \cdot x = x$

11. $\dfrac{1}{2}(2x) = \left(\dfrac{1}{2} \cdot 2\right)x$ **a.** _____

         $= 1 \cdot x$ **b.** _____

         $= x$ **c.** _____

12. $(5x + 6) + (-6) = 5x + [6 + (-6)]$ **a.** _____

                       $= 5x + 0$ **b.** _____

                       $= 5x$ **c.** _____

Use the given Property of Real Numbers to complete the statement.

13. The Associative Property of Addition
$x + (4 + y) = ?$

14. The Commutative Property of Multiplication
$v \cdot w = ?$

15. The Inverse Property of Multiplication
$5 \cdot ? = 1$

16. The Inverse Property of Multiplication
$? \cdot \dfrac{3}{4} = 1$

17. The Multiplication Property of Zero
$a \cdot ? = 0$

18. The Inverse Property of Multiplication
For $a \neq 0$, $a \cdot \dfrac{1}{a} = ?$

19. The Inverse Property of Addition
$-7y + ? = 0$

20. The Inverse Property of Addition
$\dfrac{2}{3}x + ? = 0$

21. The multiplicative inverse of $-\dfrac{2}{3}$ is _?_ .

22. For $a \neq 0$, the multiplicative inverse of $-\dfrac{2}{a}$ is _?_ .

Simplify the variable expression.

23. $6(2x)$

24. $3(4y)$

25. $-5(3x)$

26. $-3(6z)$

27. $(3t) \cdot 7$

28. $(9r) \cdot 5$

29. $(-3p) \cdot 7$

30. $(-4w) \cdot 6$

31. $(-2)(-6q)$

32. $(-3)(-5m)$

33. $\frac{1}{2}(4x)$

34. $\frac{2}{3}(6n)$

35. $-\frac{5}{3}(9w)$

36. $-\frac{2}{5}(10v)$

37. $-\frac{1}{2}(-2x)$

38. $-\frac{1}{3}(-3x)$

39. $(2x)(3x)$

40. $(4k)(6k)$

41. $(-3x)(9x)$

42. $(4b)(-12b)$

43. $\left(\frac{1}{2}x\right)(2x)$

44. $\left(\frac{1}{3}h\right)(3h)$

45. $\left(-\frac{2}{3}\right)(x)\left(-\frac{3}{2}\right)$

46. $\left(-\frac{4}{3}\right)(z)\left(-\frac{3}{4}\right)$

47. $6\left(\frac{1}{6}c\right)$

48. $9\left(\frac{1}{9}v\right)$

49. $-5\left(-\frac{1}{5}a\right)$

50. $-9\left(-\frac{1}{9}s\right)$

51. $\frac{4}{5}w \cdot 15$

52. $\frac{7}{5}y \cdot 30$

53. $2v \cdot 8w$

54. $3m \cdot 7n$

55. $(-4b)(7c)$

56. $(-3k)(-6m)$

57. $3x + (-3x)$

58. $7xy + (-7xy)$

Simplify the variable expression.

59. $-12h + 12h$

60. $5 + 8y + (-8y)$

61. $9 + 2m + (-2m)$

62. $12 - 3m + 3m$

63. $8x + 7 + (-8x)$

64. $13v + 12 + (-13v)$

65. $6t - 15 + (-6t)$

66. $10z - 4 + (-10z)$

67. $8 + (-8) - 5y$

68. $12 + (-12) - 7b$

69. $(-4) + 4 + 13b$

70. $-7 + 7 - 15t$

OBJECTIVE B

Simplify by using the Distributive Property.

71. $2(5z + 2)$

72. $3(4n + 5)$

73. $6(2y + 5z)$

74. $4(7a + 2b)$

75. $3(7x - 9)$

76. $9(3w - 7)$

77. $-(2x - 7)$

78. $-(3x + 4)$

79. $-(-4x - 9)$

80. $-(-5y - 12)$

81. $-5(y + 3)$

82. $-4(x + 5)$

83. $-6(2x - 3)$

84. $-3(7y - 4)$

85. $-5(4n - 8)$

86. $-4(3c - 2)$

87. $-8(-6z + 3)$

88. $-2(-3k + 9)$

89. $-6(-4p - 7)$

90. $-5(-8c - 5)$

91. $5(2a + 3b + 1)$

Simplify by using the Distributive Property.

92. $5(3x + 9y + 8)$ **93.** $4(3x - y - 1)$ **94.** $3(2x - 3y + 7)$

95. $9(4m - n + 2)$ **96.** $-4(3x + 2y - 5)$ **97.** $-6(-2v + 3w + 7)$

98. $-7(-2b - 4)$ **99.** $-4(-5x - 1)$ **100.** $-9(3x - 6y)$

101. $5(4a - 5b + c)$ **102.** $-4(-2m - n + 3)$ **103.** $-6(3p - 2r - 9)$

Rewrite without parentheses.

104. $-(4x + 6y - 8z)$ **105.** $-(5a - 9b + 7)$ **106.** $-(-6m + 3n + 1)$ **107.** $-(11p - 2q - r)$

CRITICAL THINKING

108. Determine whether the statement is true or false. If the statement is false, give an example that illustrates that it is false.
   a. Division is a commutative operation.
   b. Division is an associative operation.
   c. Subtraction is an associative operation.
   d. Subtraction is a commutative operation.

109. Is the statement "any number divided by itself is one" a true statement? If not, for what number or numbers is the statement not true?

110. Does every real number have an additive inverse? If not, which real numbers do not have an additive inverse?

111. Does every real number have a multiplicative inverse? If not, which real numbers do not have a multiplicative inverse?

112. ✎ In your own words, explain the Distributive Property.

113. ✎ Explain why division by zero is not allowed.

114. ✎ Give examples of two operations that occur in everyday experience that are not commutative (for example, putting on socks and then shoes).

SECTION 5.2   Variable Expressions in Simplest Form

OBJECTIVE A

Addition of like terms

A variable expression is shown at the right. The expression can be rewritten by writing subtraction as addition of the opposite. A **term** of a variable expression is one of the addends of the expression.

$$4y^3 - 3xy + x - 9$$

$$4y^3 + (-3xy) + x + (-9)$$

The variable expression has 4 terms: $4y^3$, $-3xy$, x, and -9.

The term -9 is a **constant term,** or simply a **constant.** The terms $4y^3$, $-3xy$, and x are **variable terms.**

Each variable term consists of a **numerical coefficient** and a **variable part.** The table at the right gives the numerical coefficient and the variable part of each variable term.

Term	Numerical Coefficient	Variable Part
$4y^3$	4	y^3
$-3xy$	-3	xy
x	1	x

For an expression such as x, the numerical coefficient is 1 ($x = 1x$). The numerical coefficient for $-x$ is -1 ($-x = -1x$). The numerical coefficient of $-xy$ is -1 ($-xy = -1xy$). Usually the 1 is not written.

For the variable expression at the right, state:

$$9x^2 - x - 7yz^2 + 8$$

a. the number of terms,
b. the coefficient of the second term,
c. the variable part of the third term, and
d. the constant term.

 a. There are 4 terms: $9x^2$, $-x$, $-7yz^2$, and 8.
 b. The coefficient of the second term is -1.
 c. The variable part of the third term is yz^2.
 d. The constant term is 8.

Like terms of a variable expression have the same variable part. Constant terms are also like terms.

> For the expression $13ab + 4 - 2ab - 10$, the terms $13ab$ and $-2ab$ are like variable terms, and 4 and -10 are like constant terms.

For the expression at the right, note that $5y^2$ and $-3y$ are not like terms because $y^2 = y \cdot y$, and $y \cdot y \neq y$. However, $6xy$ and $9yx$ are like variable terms because $xy = yx$ by the Commutative Property of Multiplication.

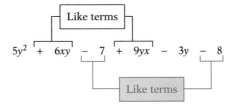

> For the variable expression $7 - 9x^2 - 8x - 9 + 4x$, state which terms are like terms.
>
> The terms $-8x$ and $4x$ are like variable terms.
>
> The terms 7 and -9 are like constant terms.

Variable expressions containing like terms are simplified by using an alternate form of the Distributive Property.

Alternate Form of the Distributive Property

If a, b, and c are real numbers, then $ac + bc = (a + b)c$.

Simplify: $6c + 7c$

$6c$ and $7c$ are like terms.
Use the Alternate Form of
the Distributive Property.
Then simplify.

$$6c + 7c = (6 + 7)c$$
$$= 13c$$

This example shows that to simplify a variable expression with like terms, add the coefficients of the like terms. Adding or subtracting the like terms of a variable expression is called **combining like terms.**

Simplify: $6a + 7 - 9a + 3$

Use the Commutative Property
of Addition to rearrange terms so
that like terms are together.

$$6a + 7 - 9a + 3$$
$$= 6a + 7 + (-9a) + 3$$
$$= 6a + (-9a) + 7 + 3$$

Use the Alternate Form of the
Distributive Property to add like
variable terms. Add the like constant terms.

$$= [6 + (-9)]a + (7 + 3)$$
$$= -3a + 10$$

Simplify: $4x^2 - 7x + x^2 - 12x$

Use the Commutative Property
of Addition to rearrange terms so
that like terms are together.

$$4x^2 - 7x + x^2 - 12x$$
$$= 4x^2 + (-7x) + x^2 + (-12x)$$
$$= 4x^2 + x^2 + (-7x) + (-12x)$$

Use the Alternate Form of the
Distributive Property to add like
terms.

$$= (4 + 1)x^2 + [-7 + (-12)]x$$
$$= 5x^2 + (-19)x$$
$$= 5x^2 - 19x$$

Example 1 Simplify:
$9y - 3z - 12y + 3z + 2$

Solution $9y - 3z - 12y + 3z + 2$
$$= 9y - 12y - 3z + 3z + 2$$
$$= -3y + 0z + 2$$
$$= -3y + 2$$

You Try It 1 Simplify:
$12a^2 - 8a + 3 - 16a^2 + 8a$

Your Solution

Solution on p. A18

Example 2 Simplify:
$$6b^2 - 9ab + 3b^2 - ab$$

Solution $6b^2 - 9ab + 3b^2 - ab$
$$= 6b^2 + 3b^2 - 9ab - ab$$
$$= 9b^2 - 10ab$$

You Try It 2 Simplify:
$$-7x^2 + 4xy + 8x^2 - 12xy$$

Your Solution

Example 3 Simplify:
$$6u + 7v - 8 + 9u - 12v + 14$$

Solution $6u + 7v - 8 + 9u - 12v + 14$
$$= 6u + 9u + 7v - 12v - 8 + 14$$
$$= 15u - 5v + 6$$

You Try It 3 Simplify:
$$-2r + 7s - 12 - 8r + s + 8$$

Your Solution

Example 4 Simplify:
$$5r^2t - 6rt^2 + 8rt^2 - 9r^2t$$

Solution $5r^2t - 6rt^2 + 8rt^2 - 9r^2t$
$$= 5r^2t - 9r^2t - 6rt^2 + 8rt^2$$
$$= -4r^2t + 2rt^2$$

You Try It 4 Simplify:
$$8x^2y - 15xy^2 + 12xy^2 - 7x^2y$$

Your Solution

Solutions on p. A18

Objective B
General variable expressions

General variable expressions are simplified by repeated use of the Properties of the Real Numbers.

Simplify: $7(2a - 4b) - 3(4a - 2b)$

Use the Distributive Property to remove parentheses.

$7(2a - 4b) - 3(4a - 2b)$
$= 14a - 28b - 12a + 6b$

Use the Commutative Property of Addition to rearrange terms.

$= 14a - 12a - 28b + 6b$

Use the Alternate Form of the Distributive Property to combine like terms.

$= 2a - 22b$

To simplify variable expressions that contain grouping symbols within other grouping symbols, simplify the inner grouping symbols first.

Simplify: $2x - 4[3 - 2(6x + 5)]$

Use the Distributive Property to remove the parentheses.	$2x - 4[3 - 2(6x + 5)]$ $= 2x - 4[3 - 12x - 10]$
Combine like terms inside the brackets.	$= 2x - 4[-12x - 7]$
Use the Distributive Property to remove the brackets.	$= 2x + 48x + 28$
Combine like terms.	$= 50x + 28$

Simplify: $2a^2 + 3[4(2a^2 - 5) - 4(3a - 1)]$

Use the Distributive Property to remove both sets of parentheses.	$2a^2 + 3[4(2a^2 - 5) - 4(3a - 1)]$ $= 2a^2 + 3[8a^2 - 20 - 12a + 4]$
Combine like terms inside the brackets.	$= 2a^2 + 3[8a^2 - 12a - 16]$
Use the Distributive Property to remove the brackets.	$= 2a^2 + 24a^2 - 36a - 48$
Combine like terms.	$= 26a^2 - 36a - 48$

Example 5   Simplify:
$4 - 3(2a - b) + 4(3a + 2b)$

Solution $4 - 3(2a - b) + 4(3a + 2b)$
$= 4 - 6a + 3b + 12a + 8b$
$= 6a + 11b + 4$

You Try It 5   Simplify:
$6 - 4(2x - y) + 3(x - 4y)$

Your Solution

Example 6   Simplify:
$7y - 4(2y - 3z) - (6y - 4z)$

Solution $7y - 4(2y - 3z) - (6y - 4z)$
$= 7y - 8y + 12z - 6y + 4z$
$= -7y + 16z$

You Try It 6   Simplify:
$8c - 4(3c - 8) - 5(c + 4)$

Your Solution

Example 7   Simplify:
$9v - 4[2(1 - 3v) - 5(2v + 4)]$

Solution $9v - 4[2(1 - 3v) - 5(2v + 4)]$
$= 9v - 4[2 - 6v - 10v - 20]$
$= 9v - 4[-16v - 18]$
$= 9v + 64v + 72$
$= 73v + 72$

You Try It 7   Simplify:
$6p + 5[3(2 - 3p) - 2(5 - 4p)]$

Your Solution

Solutions on p. A18

5.2 EXERCISES

OBJECTIVE A

List the terms of the variable expression. Then underline the constant term.

1. $3x^2 + 4x - 9$

2. $-7y^2 - 2y + 6$

3. $b + 5$

4. $8n^2 - 1$

List the variable terms of the expression. Then underline the variable part of each term.

5. $9a^2 - 12a + 4b^2$

6. $6x^2y + 7xy^2 + 11$

7. $3x^2 + 16$

8. $-2n^2 + 5n - 8$

State the coefficients of the variable terms.

9. $x^2 - 6x - 7$

10. $-x + 15$

11. $12a^2 + 4ab - 1$

12. $x^2y - x + y$

Simplify by combining like terms.

13. $7a + 9a$

14. $8c + 15c$

15. $12x + 15x$

16. $9b + 24b$

17. $9z - 6z$

18. $12h - 4h$

19. $9x - x$

20. $12y - y$

21. $8z - 15z$

22. $2p - 13p$

23. $w - 7w$

24. $y - 9y$

25. $12v - 12v$

26. $11c - 11c$

27. $9s - 8s$

28. $6n - 5n$

29. $4x - 3y + 2x$

30. $3m - 6n + 4m$

31. $4r + 8p - 2r + 5p$

32. $-12t - 6s + 9t + 4s$

33. $9w - 5v - 12w + 7v$

34. $3c - 8 + 7c - 9$

Simplify by combining like terms.

35. $-4p + 9 - 5p + 2$

36. $-6y - 17 + 4y + 9$

37. $8p + 7 - 6p - 7$

38. $9m - 12 + 2m + 12$

39. $7h + 15 - 7h - 9$

40. $7v^2 - 9v + v^2 - 8v$

41. $9y^2 - 8 + 4y^2 + 9$

42. $r^2 + 4r - 8r - 5r^2$

43. $3w^2 - 7 - 9 + 9w^2$

44. $4c - 7c^2 + 8c - 8c^2$

45. $9w^2 - 15w + w - 9w^2$

46. $12v^2 + 15v - 14v - 12v^2$

47. $7a^2b + 5ab^2 - 2a^2b + 3ab^2$

48. $3xy^2 + 2x^2y - 7xy^2 - 4x^2y$

49. $8a - 9b + 2 - 8a + 9b + 3$

50. $10v + 12w - 9 - v - 12w + 9$

51. $6x^2 - 7x + 1 + 5x^2 + 5x - 1$

52. $4y^2 + 7y + 1 + y^2 - 10y + 9$

53. $-3b^2 + 6b + 1 + 11b^2 - 8b - 1$

54. $-4z^2 - 6z + 1 - z^2 + 7z + 8$

OBJECTIVE B

Simplify.

55. $5x + 2(x + 1)$

56. $6y + 2(2y + 3)$

57. $9n - 3(2n - 1)$

Simplify.

58. $12x - 2(4x - 6)$

59. $7a - (3a - 4)$

60. $9m - 4(2m - 3)$

61. $7 + 2(2a - 3)$

62. $5 + 3(2y - 8)$

63. $6 + 4(2x + 9)$

64. $4 + 3(7d + 7)$

65. $8 - 4(3x - 5)$

66. $13 - 7(4y + 3)$

67. $2 - 9(2m + 6)$

68. $4 - 7(6w - 9)$

69. $3(6c + 5) + 2(c + 4)$

70. $7(2k - 5) + 3(4k - 3)$

71. $2(a - 2b) + 3(2a + 3b)$

72. $4(3x - 6y) + 5(2x - 3y)$

73. $6(7z - 5) - 3(9z - 6)$

74. $8(2t + 4) - 4(3t - 1)$

75. $-2(6y + 2) + 3(4y - 5)$

76. $-3(2a - 5) - 2(4a + 3)$

77. $-5(x - 2y) - 4(2x + 3y)$

78. $-6(-x - 3y) - 2(-3x + 9y)$

79. $2 - 3(2v - 1) + 2(2v + 4)$

80. $5 - 2(3x + 5) - 3(4x - 1)$

81. $2c - 3(c + 4) - 2(2c - 3)$

82. $5m - 2(3m + 2) - 4(m - 1)$

83. $8a + 3(2a - 1) + 6(4 - 2a)$

84. $9z - 2(2z - 7) + 4(3 - 5z)$

85. $3n - 2[5 - 2(2n - 4)]$

86. $6w + 4[3 - 5(6w - 2)]$

87. $9x - 3[8 - 2(5 - 3x)]$

Simplify.

88. $11y - 7[2(2y - 5) + 3(7 - 5y)]$

89. $-3v - 6[2(3 - 2v) - 5(3v - 7)]$

90. $8b - 3[2(3 - 5b) - 4(3b - 4)]$

91. $21r - 4[3(4 - 5r) - 3(2 - 7r)]$

92. $7y^2 - 2[3(2y - 4) + 3(2y^2)]$

93. $9z^2 - 3[4(2z + 3) - 3(2z^2 - 6)]$

CRITICAL THINKING

94. The square and the rectangle at the right can be used to illustrate algebraic expressions. Note at the right the expression for $2x + 1$. The expression below is $3(x + 2)$.

| x | 1 | 1 | x | 1 | 1 | x | 1 | 1 |

Rearrange these rectangles so that the x's are together and the 1's are together. Write a mathematical expression for the rearranged figure. Using similar squares and rectangles, draw figures that represent the expressions $2 + 3x$, $5x$, $2(2x + 3)$, $4x + 3$, and $4x + 6$. Does the figure for $2(2x + 3)$ equal the figure for $4x + 6$? How does this relate to the Distributive Property? Does the figure for $2 + 3x$ equal the figure for $5x$? How does this relate to combining like terms?

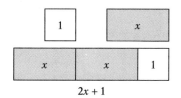

$2x + 1$

95. The procedure for multiplying whole numbers with more than one digit is based on the Distributive Property. Examine the following.

$$7 \cdot 435 = 7(400 + 30 + 5) = 7 \cdot 400 + 7 \cdot 30 + 7 \cdot 5$$
$$= 2,800 + 210 + 35 = 3,045$$

Use the Distributive Property to multiply 527 by 6.

$$\begin{array}{r} 2\ 3 \\ 4\ 3\ 5 \\ \times \quad\quad 7 \\ \hline 3,0\ 4\ 5 \end{array}$$

96. Explain why the simplification of the expression $2 + 3(2x + 4)$ shown at the right is incorrect. What is the correct simplification?

Why is this incorrect?

$$2 + 3(2x + 4) = 5(2x + 4)$$
$$= 10x + 20$$

97. Simplifying variable expressions requires combining like terms. Give some examples of how this applies to everyday experience.

98. It was stated in this section that the variable terms x^2 and x are not like terms. Use measurements of distance and area to show that these terms would not be combined as measurements.

SECTION 5.3 Addition and Subtraction of Polynomials

OBJECTIVE A
Addition of polynomials

A **monomial** is a number, a variable, or a product of numbers and variables. The expressions below are all monomials.

$$7 \qquad b \qquad \frac{2}{3}a \qquad 12xy^2$$

A number   A variable   A product of a   A product of a
                          number and a variable   number and variables

The expression $3\sqrt{x}$ is not a monomial, because $\sqrt{x}$ cannot be written as a product of variables.

The expression $\frac{2x}{y^2}$ is not a monomial, because it is a quotient of variables.

A **polynomial** is a variable expression in which the terms are monomials.

A polynomial of *one* term is a **monomial.** $-7x^2$ is a monomial.

A polynomial of *two* terms is a **binomial.** $4x + 2$ is a binomial.

A polynomial of *three* terms is a **trinomial.** $7x^2 + 5x - 7$ is a trinomial.

> **TAKE NOTE**
> The expression $x + y + z$ has 3 terms; it is a trinomial. The expression xyz has 1 term; it is a monomial.

To add polynomials, add the coefficients of the like terms. Either a horizontal format or a vertical format can be used.

Use a horizontal format to simplify $(6x^3 + 4x^2 - 7) + (-12x^2 + 4x - 8)$.

Use the Commutative and Associative Properties of Addition to rearrange the terms so that like terms are grouped together.

$$(6x^3 + 4x^2 - 7) + (-12x^2 + 4x - 8)$$
$$= 6x^3 + (4x^2 - 12x^2) + 4x + (-7 - 8)$$

Combine like terms.

$$= 6x^3 - 8x^2 + 4x - 15$$

Use a vertical format to simplify $(-4x^2 + 6x - 9) + (2x^3 - 8x + 12)$.

Rewrite the expression with like terms in the same columns.

Add the like terms in each column.

$$\begin{array}{r} -4x^2 + 6x - 9 \\ 2x^3 \qquad - 8x + 12 \\ \hline 2x^3 - 4x^2 - 2x + 3 \end{array}$$

Example 1

Use a horizontal format to simplify
$(8x^2 - 4x - 9) + (2x^2 + 9x - 9)$.

Solution

$(8x^2 - 4x - 9) + (2x^2 + 9x - 9)$
$= (8x^2 + 2x^2) + (-4x + 9x) + (-9 - 9)$
$= 10x^2 + 5x - 18$

You Try It 1

Use a horizontal format to simplify
$(-4x^3 + 2x^2 - 8) + (4x^3 + 6x^2 - 7x + 5)$.

Your Solution

Solution on p. A18

Example 2

Use a vertical format to simplify
$(-5x^3 + 4x^2 - 7x + 9) + (2x^3 + 5x - 11)$.

Solution

$$
\begin{array}{r}
-5x^3 + 4x^2 - 7x + 9 \\
2x^3 \qquad\; + 5x - 11 \\
\hline
-3x^3 + 4x^2 - 2x - 2
\end{array}
$$

You Try It 2

Use a vertical format to simplify
$(6x^3 + 2x + 8) + (-9x^3 + 2x^2 - 12x - 8)$.

Your Solution

Solution on p. A18

OBJECTIVE B

Subtraction of polynomials

The **opposite** of the polynomial $(3x^2 - 7x + 8)$ is $-(3x^2 - 7x + 8)$.

To simplify the opposite of a polynomial, change the sign of each term inside the parentheses. $-(3x^2 - 7x + 8) = -3x^2 + 7x - 8$

To subtract two polynomials, add the opposite of the second polynomial to the first. Polynomials are subtracted using either a horizontal or a vertical format.

Use a horizontal format to simplify $(5a^2 - a + 2) - (-2a^3 + 3a - 3)$.

Rewrite subtraction as addition $(5a^2 - a + 2) - (-2a^3 + 3a - 3)$
of the opposite polynomial. $= (5a^2 - a + 2) + (2a^3 - 3a + 3)$

Combine like terms. $= 2a^3 + 5a^2 - 4a + 5$

Use a vertical format to simplify $(3y^3 + 4y + 9) - (2y^2 + 4y - 21)$.

Find the opposite of $(2y^2 + 4y - 21)$. $3y^3 \qquad\;\; + 4y + 9$
The opposite polynomial is $(-2y^2 - 4y + 21)$. $\underline{\quad\; - 2y^2 - 4y + 21}$

Add this polynomial to the first polynomial. $3y^3 - 2y^2 \qquad + 30$

Example 3

Use a horizontal format to simplify
$(7c^2 - 9c - 12) - (9c^2 + 5c - 8)$.

Solution

$(7c^2 - 9c - 12) - (9c^2 + 5c - 8)$
$= (7c^2 - 9c - 12) + (-9c^2 - 5c + 8)$
$= -2c^2 - 14c - 4$

You Try It 3

Use a vertical format to simplify
$(13y^3 - 6y - 7) - (4y^2 - 6y - 9)$.

Your Solution

Solution on p. A18

5.3 EXERCISES

OBJECTIVE A

Add. Use a horizontal format.

1. $(5y^2 + 3y - 7) + (6y^2 - 7y + 9)$

2. $(-8x^2 - 11x - 15) + (4x^2 - 12x + 13)$

3. $(3w^3 + 8w^2 - 2w) + (5w^2 - 6w - 5)$

4. $(11p^3 - 9p^2 - 6p) + (10p^2 - 8p + 4)$

5. $(-9a^3 + 3a^2 + 2a - 7) + (7a^3 - 12a^2 - 10a + 8)$

6. $(7x^3 - 8x^2 + 9x - 12) + (-3x^3 - 7x^2 + 5x - 9)$

Add. Use a vertical format.

7. $(5k^2 - 7k - 8) + (6k^2 + 9k - 10)$

8. $(8v^2 - 9v + 12) + (12v^2 - 11v - 2)$

9. $(8x^3 - 9x^2 + 2) + (9x^3 + 9x - 7)$

10. $(13z^3 - 7z^2 + 4z) + (10z^2 + 5z - 9)$

11. $(12b^3 + 9b^2 + 5b - 10) + (4b^3 + 5b^2 - 5b + 11)$

12. $(5a^3 - a^2 + 4a - 19) + (-a^3 + a^2 - 7a + 19)$

Solve.

13. Find the sum of $6t^2 - 8t - 15$ and $7t^2 + 8t - 20$.

14. What is $8y^2 - 3y - 1$ plus $-6y^2 + 3y - 1$?

OBJECTIVE B

Subtract. Use a horizontal format.

15. $(3x^2 - 2x - 5) - (x^2 + 7x - 3)$

16. $(7y^2 - 8y - 10) - (3y^2 + 2y - 9)$

17. $(11b^3 - 2b^2 + 1) - (6b^2 - 12b - 13)$

18. $(13w^3 + 3w^2 - 9) - (7w^3 - 9w + 10)$

Subtract. Use a horizontal format.

19. $(8z^3 - 9z^2 + 4z + 12) - (10z^3 - z^2 + 4z - 9)$

20. $(15t^3 - 9t^2 + 8t + 11) - (17t^3 - 9t^2 - 8t + 6)$

21. $(-6r^3 + 9r^2 + 19) - (6r^3 - 16r + 19)$

22. $(-4v^3 + 8v - 2) - (6v^3 - 13v^2 + 7v + 1)$

Subtract. Use a vertical format.

23. $(4a^2 + 9a - 11) - (2a^2 - 3a - 9)$

24. $(8b^2 - 7b - 6) - (5b^2 + 8b + 12)$

25. $(6z^3 + 4z^2 + 1) - (3z^3 - 8z - 9)$

26. $(10y^3 - 8y - 13) - (6y^2 + 2y + 7)$

27. $(5n^3 + 8n^2 - 4n - 9) - (2n^3 + 8n^2 + 4n - 9)$

28. $(4q^3 + 7q^2 + 8q - 9) - (14q^3 + 7q^2 - 8q - 9)$

Solve.

29. Find the difference between $10b^2 - 7b + 4$ and $8b^2 + 5b - 14$.

30. What is $7m^2 - 3m - 6$ minus $2m^2 - m + 5$?

CRITICAL THINKING

31. The **degree of a monomial** is the sum of the exponents of the variables. For example, the degree of the monomial $4x^2y^4z^3$ is $2 + 4 + 3 = 9$. Find the degree of (a) $2a^2b^6$, (b) $-5xy^5z^3$, and (c) $9mn^2p$.

32. The **degree of a polynomial** is the greatest of the degrees of any of its terms. For example, the degree of the polynomial $7x^4 + 5x^2 - 6$ is 4. Find the degree of each of the following polynomials.

    a. $4x^3 - 6x^2 + 7$ **b.** $3 - 2x - 7x^2 - 9x^5$ **c.** $4x + 9x^3 - 6x^2 + 8$

33. In your own words, explain the meaning of *monomial, binomial, trinomial,* and *polynomial.* Give an example of each. Give an example of an expression that is not a polynomial.

34. Computer programming languages, such as BASIC, allow the programmer to enter polynomials. How would the polynomial $x^3 - 4x^2 + 5x - 3$ be entered in BASIC? What does BASIC stand for?

SECTION 5.4   Multiplication of Monomials

OBJECTIVE A
Multiplication of monomials

Recall that in the exponential expression 3^4, 3 is the base and 4 is the exponent. The exponential expression 3^4 means to multiply 3, the base, 4 times. Therefore, $3^4 = 3 \cdot 3 \cdot 3 \cdot 3 = 81$.

For the variable exponential expression x^6, x is the base and 6 is the exponent. The exponent indicates the number of times the base occurs as a factor. Therefore,

$$\text{Multiply } x \text{ 6 times.}$$
$$x^6 = \overbrace{x \cdot x \cdot x \cdot x \cdot x \cdot x}$$

The product of exponential expressions with the *same* base can be simplified by writing each expression in factored form and writing the result with an exponent.

$$x^3 \cdot x^2 = \overbrace{(x \cdot x \cdot x)}^{3 \text{ factors}} \cdot \overbrace{(x \cdot x)}^{2 \text{ factors}}$$
$$\underbrace{}_{5 \text{ factors}}$$
$$= x \cdot x \cdot x \cdot x \cdot x$$
$$= x^5$$

Note that adding the exponents results in the same product.

$$x^3 \cdot x^2 = x^{3+2} = x^5$$

This suggests the following rule for multiplying exponential expressions.

Rule for Multiplying Exponential Expressions

If m and n are positive integers, then $x^m \cdot x^n = x^{m+n}$.

Simplify: $a^4 \cdot a^5$

The bases are the same.
Add the exponents.

$$a^4 \cdot a^5 = a^{4+5}$$
$$= a^9$$

Simplify: $c^3 \cdot c^4 \cdot c$

The bases are the same.
Add the exponents. Note that $c = c^1$.

$$c^3 \cdot c^4 \cdot c = c^{3+4+1}$$
$$= c^8$$

Simplify: $x^5 y^3$

The bases are *not* the same. The exponential expression is in simplest form.

$x^5 y^3$ is in simplest form.

Simplify: $(4x^3)(2x^2)$

Use the Commutative and Associative Properties of Multiplication to rearrange and group like factors.

$$(4x^3)(2x^2) = (4 \cdot 2)(x^3 \cdot x^2)$$

Multiply variables with the same base by adding the exponents.

$$= 8x^{3+2}$$
$$= 8x^5$$

Simplify: $(a^3b^2)(a^4)$

Multiply variables with the same base by adding the exponents.

$$(a^3b^2)(a^4) = a^{3+4}b^2$$
$$= a^7b^2$$

Simplify: $(-2v^3z^5)(5v^2z^6)$

Multiply the coefficients of the monomials. Multiply variables with the same base by adding the exponents.

$$(-2v^3z^5)(5v^2z^6) = [(-2)5](v^{3+2})(z^{5+6})$$
$$= -10v^5z^{11}$$

Example 1 Simplify: $(-6c^5)(7c^8)$

Solution $(-6c^5)(7c^8) = [(-6)7](c^{5+8})$

$$= -42c^{13}$$

You Try It 1 Simplify: $(-7a^4)(4a^2)$

Your Solution

Example 2 Simplify: $(-5ab^3)(4a^5)$

Solution $(-5ab^3)(4a^5) = (-5 \cdot 4)(a \cdot a^5)b^3$

$$= -20a^{1+5}b^3$$
$$= -20a^6b^3$$

You Try It 2 Simplify: $(8m^3n)(-3n^5)$

Your Solution

Example 3 Simplify: $(6x^3y^2)(4x^4y^5)$

Solution $(6x^3y^2)(4x^4y^5)$

$$= (6 \cdot 4)(x^3 \cdot x^4)(y^2 \cdot y^5)$$
$$= 24x^{3+4}y^{2+5}$$
$$= 24x^7y^7$$

You Try It 3 Simplify: $(12p^4q^3)(-3p^5q^2)$

Your Solution

Solutions on p. A18

Objective B
Powers of monomials

The expression $(x^4)^3$ is an example of a *power of a monomial;* the monomial x^4 is raised to the third (3) power.

The power of a monomial can be simplified by writing the power in factored form and then using the Rule for Multiplying Exponential Expressions.

$$(x^4)^3 = x^4 \cdot x^4 \cdot x^4$$
$$= x^{4+4+4} = x^{12}$$

Note that multiplying the exponent inside the parentheses by the exponent outside the parentheses results in the same product.

$$(x^4)^3 = x^{4 \cdot 3} = x^{12}$$

This suggests the following rule for simplifying powers of monomials.

Rule for Simplifying the Power of an Exponential Expression

If m and n are positive integers, then $(x^m)^n = x^{m \cdot n}$.

Simplify: $(z^2)^5$

Use the Rule for Simplifying the Power of an exponential expression.

$$(z^2)^5 = z^{2 \cdot 5} = z^{10}$$

The expression $(a^2b^3)^2$ is the *power of the product* of the two exponential expressions a^2 and b^3. The power of a product of exponential expressions can be simplified by writing the product in factored form and then using the Rule for Multiplying Exponential Expressions.

Write the power of the product of the monomial in factored form.
Use the Rule for Multiplying Exponential Expressions.

$$(a^2b^3)^2 = (a^2b^3)(a^2b^3)$$
$$= a^{2+2}b^{3+3}$$
$$= a^4b^6$$

Note that multiplying each exponent inside the parentheses by the exponent outside the parentheses results in the same product.

$$(a^2b^3)^2 = a^{2 \cdot 2}b^{3 \cdot 2}$$
$$= a^4b^6$$

Rule for Simplifying Powers of Products

If m, n, and p are positive integers, then $(x^my^n)^p = x^{m \cdot p}y^{n \cdot p}$.

Simplify: $(x^4y)^6$

Multiply each exponent inside the parentheses by the exponent outside the parentheses. Remember that $y = y^1$.

$$(x^4y)^6 = x^{4\cdot6}y^{1\cdot6}$$
$$= x^{24}y^6$$

Simplify: $(5x^2)^3$

Multiply each exponent inside the parentheses by the exponent outside the parentheses. Note that $5 = 5^1$.

Evaluate 5^3.

$$(5x^2)^3 = 5^{1\cdot3}x^{2\cdot3}$$
$$= 5^3x^6$$
$$= 125x^6$$

Simplify: $(-a^5)^4$

Multiply each exponent inside the parentheses by the exponent outside the parentheses. Note that $-a^5 = (-1)a^5 = (-1)^1a^5$.

$$(-a^5)^4 = (-1)^{1\cdot4}a^{5\cdot4}$$
$$= (-1)^4a^{20}$$
$$= 1a^{20} = a^{20}$$

Simplify: $(3m^5p^2)^4$

Multiply each exponent inside the parentheses by the exponent outside the parentheses.

Evaluate 3^4.

$$(3m^5p^2)^4 = 3^{1\cdot4}m^{5\cdot4}p^{2\cdot4}$$
$$= 3^4m^{20}p^8$$
$$= 81m^{20}p^8$$

Example 4 Simplify: $(-2x^4)^3$

Solution $(-2x^4)^3 = (-2)^{1\cdot3}x^{4\cdot3}$
$$= (-2)^3x^{12}$$
$$= -8x^{12}$$

You Try It 4 Simplify: $(-y^4)^5$

Your Solution

Example 5 Simplify: $(-2p^3r)^4$

Solution $(-2p^3r)^4 = (-2)^{1\cdot4}p^{3\cdot4}r^{1\cdot4}$
$$= (-2)^4p^{12}r^4$$
$$= 16p^{12}r^4$$

You Try It 5 Simplify: $(-3a^4bc^2)^3$

Your Solution

Solutions on pp. A18–A19

5.4 Exercises

OBJECTIVE A

Multiply.

1. $a^4 \cdot a^5$

2. $y^5 \cdot y^8$

3. $z^3 \cdot z \cdot z^4$

4. $b \cdot b^2 \cdot b^6$

5. $(a^3b^2)(a^5b)$

6. $(xy^5)(x^3y^7)$

7. $(-m^3n)(m^6n^2)$

8. $(-r^4t^3)(r^2t^9)$

9. $(2x^3)(5x^4)$

10. $(6x^3)(9x)$

11. $(8x^2y)(xy^5)$

12. $(4a^3b^4)(3ab^5)$

13. $(-4m^3)(3m^4)$

14. $(6r^2)(-4r)$

15. $(7v^3)(-2w)$

16. $(-9a^3)(4b^2)$

17. $(ab^2c^3)(-2b^3c^2)$

18. $(4x^2y^3)(-5x^5)$

19. $(4b^4c^2)(6a^3b)$

20. $(3xy^5)(5y^2z)$

21. $(-8r^2t^3)(-5rt^4v)$

22. $(-4ab^3c^2)(b^3c)$

23. $(9mn^4p)(-3mp^2)$

24. $(-3v^2wz)(-4vz^4)$

25. $(2x)(3x^2)(4x^4)$

26. $(5a^2)(4a)(3a^5)$

27. $(3ab)(2a^2b^3)(a^3b)$

28. $(4x^2y)(3xy^5)(2x^2y^2)$

29. $(-xy^5)(3x^2)(5y^3)$

30. $(-6m^3n)(-mn^2)(m)$

31. $(8rt^3)(-2r^3v^2)(-3t^5v^2)$

32. $(-y^5z)(-2x^3z)(-3xy^4)$

33. $(-5ac^3)(-4b^3c)(-3a^2b^2)$

Solve.

34. Find the product of $7x^2y^3z^5$ and $3xy^4$.

35. What is $2ab^6$ times $-4a^5b^4$?

OBJECTIVE B

Simplify.

36. $(x^3)^5$

37. $(b^2)^4$

38. $(z^6)^3$

39. $(p^4)^7$

40. $(3x)^2$

41. $(2y)^3$

42. $(x^2y^3)^6$

43. $(m^4n^2)^3$

44. $(r^3t)^4$

45. $(a^2b)^5$

46. $(-y^2)^2$

47. $(-z^3)^2$

48. $(2x^4)^3$

49. $(3n^3)^3$

50. $(-2a^2)^3$

51. $(-3b^3)^2$

52. $(3x^2y)^2$

53. $(4a^4b^5)^3$

54. $(2a^3bc^2)^3$

55. $(4xy^3z^2)^2$

56. $(-mn^5p^3)^4$

CRITICAL THINKING

57. Evaluate $(2^3)^2$ and $2^{(3^2)}$. Are the results the same? If not, which expression has the larger value?

58. What is the Order of Operations Agreement for the expression x^{m^n}?

59. If n is a positive integer and $x^n = y^n$, does $x = y$? Explain your answer.

60. The distance a rock will fall in t seconds is $16t^2$ (neglecting air resistance). Find other examples of quantities that can be expressed in terms of an exponential expression, and explain the application of the expression.

Objective A
Multiplication of a polynomial by a monomial

Recall that the Distributive Property states that if a, b, and c are real numbers, then $a(b + c) = ab + ac$. The Distributive Property is used to multiply a polynomial by a monomial. Each term of the polynomial is multiplied by the monomial.

Multiply: $y^2(4y^2 + 3y - 7)$

Use the Distributive Property. Multiply each term of the polynomial by y^2.

$y^2(4y^2 + 3y - 7)$

$= y^2(4y^2) + y^2(3y) - y^2(7)$

Use the Rule for Multiplying Exponential Expressions.

$= 4y^4 + 3y^3 - 7y^2$

Multiply: $3x^3(4x^4 - 2x + 5)$

Use the Distributive Property. Multiply each term of the polynomial by $3x^3$.

$3x^3(4x^4 - 2x + 5)$

$= 3x^3(4x^4) - 3x^3(2x) + 3x^3(5)$

Use the Rule for Multiplying Exponential Expressions.

$= 12x^7 - 6x^4 + 15x^3$

Multiply: $-3a(6a^4 - 3a^2)$

Use the Distributive Property. Multiply each term of the polynomial by $-3a$.

$-3a(6a^4 - 3a^2)$

$= -3a(6a^4) - (-3a)(3a^2)$

Use the Rule for Multiplying Exponential Expressions.

$= -18a^5 - (-9a^3)$

Rewrite $-(-9a^3)$ as $+ 9a^3$.

$= -18a^5 + 9a^3$

Example 1 Multiply: $-2x(7x - 4y)$

Solution $-2x(7x - 4y)$
$= -2x(7x) - (-2x)(4y)$
$= -14x^2 + 8xy$

You Try It 1 Multiply: $-3a(-6a + 5b)$

Your Solution

Example 2 Multiply:
$2xy(3x^2 - xy + 2y^2)$

Solution $2xy(3x^2 - xy + 2y^2)$
$= 2xy(3x^2) - {}^2xy(xy) + 2xy(2y^2)$
$= 6x^3y - 2x^2y^2 + 4xy^3$

You Try It 2 Multiply:
$3mn^2(2m^2 - 3mn - 1)$

Your Solution

Solutions on p. A19

Objective B
Multiplication of two binomials

In the previous objective, a monomial and a polynomial were multiplied. Using the Distributive Property, each term of the polynomial was multiplied by the monomial. Two binomials are also multiplied by using the Distributive Property. Each term of one binomial is multiplied by the other binomial.

Multiply: $(x + 2)(x + 6)$

Use the Distributive Property. Multiply each term of $(x + 6)$ by $(x + 2)$.

Use the Distributive Property again to multiply $(x + 2)x$ and $(x + 2)6$.

Simplify by combining like terms.

$(x + 2)(x + 6)$
$= (x + 2)x + (x + 2)6$
$= x(x) + 2(x) + x(6) + 2(6)$
$= x^2 + 2x + 6x + 12$
$= x^2 + 8x + 12$

Because it is frequently necessary to multiply two binomials, the terms of the binomials are labeled as shown in the diagram below and the product is computed by using a method called FOIL. The letters of FOIL stand for **First, Outer, Inner,** and **Last.** The FOIL method is based on the Distributive Property and involves adding the product of the first terms, the outer terms, the inner terms, and the last terms.

The product $(2x + 3)(3x + 4)$ is shown below using FOIL.

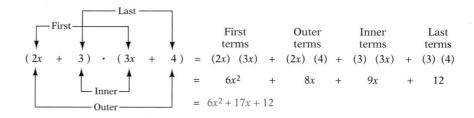

Multiply $(4x - 3)(2x + 3)$ using the FOIL method.

$(4x - 3)(2x + 3) = (4x)(2x) + (4x)(3) + (-3)(2x) + (-3)(3)$
$= 8x^2 + 12x - 6x - 9$
$= 8x^2 + 6x - 9$

Example 3

Multiply: $(2x - 3)(x + 2)$

Solution

$(2x - 3)(x + 2)$
$= (2x)(x) + (2x)(2) + (-3)(x) + (-3)(2)$
$= 2x^2 + 4x - 3x - 6$
$= 2x^2 + x - 6$

You Try It 3

Multiply: $(3c + 7)(3c - 7)$

Your Solution

Solution on p. A19

5.5 EXERCISES

OBJECTIVE A

Multiply.

1. $x(x^2 - 3x - 4)$

2. $y(3y^2 + 4y - 8)$

3. $4a(2a^2 + 3a - 6)$

4. $3b(6b^2 - 5b - 7)$

5. $-2a(3a^2 + 9a - 7)$

6. $-4x(x^2 - 3x - 7)$

7. $m^3(4m - 9)$

8. $r^2(2r^2 + 7)$

9. $2x^3(5x^2 - 6xy + 2y^2)$

10. $4b^4(3a^2 + 4ab - b^2)$

11. $-6r^5(r^2 - 2r - 6)$

12. $-5y^4(3y^2 - 6y^3 + 7)$

13. $4a^2(3a^2 + 6a - 7)$

14. $5b^3(2b^2 - 4b - 9)$

15. $-2n^2(3 - 4n^3 - 5n^5)$

16. $-4x^3(6 - 4x^2 - 5x^4)$

17. $ab^2(3a^2 - 4ab + b^2)$

18. $x^2y^3(5y^3 - 6xy - x^3)$

19. $-x^2y^3(4x^5y^2 - 5x^3y - 7x)$

20. $-a^2b^4(3a^6b^4 + 6a^3b^2 - 5a)$

21. $6r^2t^3(1 - rt - r^3t^3)$

Solve.

22. What is $3p$ times $4p^2 + 5p - 8$?

23. What is the product of $-4q$ and $-9q + 7$?

OBJECTIVE B

Multiply. Use the FOIL method.

24. $(x + 4)(x + 6)$

25. $(y + 9)(y + 3)$

26. $(a - 6)(a - 7)$

Multiply. Use the FOIL method.

27. $(x + 6)(x + 5)$

28. $(y + 4)(y + 3)$

29. $(a - 3)(a - 8)$

30. $(3c + 4)(2c + 3)$

31. $(5z + 2)(2z + 1)$

32. $(3v - 7)(4v + 3)$

33. $(8c - 7)(5c + 3)$

34. $(8x - 3)(5x - 4)$

35. $(5v - 3)(2v - 1)$

36. $(4n - 9)(4n - 5)$

37. $(7t - 2)(5t + 4)$

38. $(3y - 4)(4y + 7)$

39. $(8x + 5)(3x - 2)$

40. $(4a - 5)(4a + 5)$

41. $(5r + 2)(5r - 2)$

Solve.

42. What is $2b + 3$ multiplied by $3b + 8$?

43. Find the product of $7y + 5$ and $3y - 8$.

CRITICAL THINKING

44. State whether the statement is true or false. If the statement is false, change the statement to make it true.
  a. $(5 + x)^2 = 25 + x^2$ **b.** $(5x)^2 = 25x^2$ **c.** $(a - 4)^2 = a^2 - 16$

45. Simplify: $(x - a)(x - b)(x - c) \cdots (x - y)(x - z)$

46. Simplify: $(a - b)^2 - (a + b)^2$

47. Use the Distributive Property to multiply $(x + 4)(x^2 + 3x + 5)$.

48. Explain the similarities and differences between multiplying two binomials and multiplying a monomial and a binomial.

49. Explain the similarities and differences between simplifying the power of a monomial and simplifying the power of a binomial.

SECTION 5.6 Division of Monomials

Objective A

Division of monomials

The quotient of two exponential expressions with the *same* base can be simplified by writing each expression in factored form, dividing by the common factors, and then writing the result with an exponent.

$$\frac{x^6}{x^2} = \frac{\overset{1}{\cancel{x}} \cdot \overset{1}{\cancel{x}} \cdot x \cdot x \cdot x \cdot x}{\underset{1}{\cancel{x}} \cdot \underset{1}{\cancel{x}}} = x^4$$

Note that subtracting the exponents results in the same quotient.

$$\frac{x^6}{x^2} = x^{6-2} = x^4$$

This example suggests that to divide monomials with like bases, subtract the exponents.

Rule for Dividing Exponential Expressions

If m and n are positive integers and $x \neq 0$, then $\frac{x^m}{x^n} = x^{m-n}$.

Simplify: $\frac{c^8}{c^5}$

Use the Rule for Dividing Exponential Expressions.

$$\frac{c^8}{c^5} = c^{8-5} = c^3$$

Simplify: $\frac{x^5 y^7}{x^4 y^2}$

Use the Rule for Dividing Exponential Expressions by subtracting the exponents of the like bases. Note that $x^{5-4} = x^1$ but that the exponent 1 is not written.

$$\frac{x^5 y^7}{x^4 y^2} = x^{5-4} y^{7-2} = xy^5$$

The expression at the right has been simplified in two ways: dividing by common factors, and using the Rule for Dividing Exponential Expressions.

$$\frac{x^3}{x^3} = \frac{\overset{1}{\cancel{x}} \cdot \overset{1}{\cancel{x}} \cdot \overset{1}{\cancel{x}}}{\underset{1}{\cancel{x}} \cdot \underset{1}{\cancel{x}} \cdot \underset{1}{\cancel{x}}} = 1$$

$$\frac{x^3}{x^3} = x^{3-3} = x^0$$

Because $\frac{x^3}{x^3} = 1$ and $\frac{x^3}{x^3} = x^0$, 1 must equal x^0. Therefore, the following definition of zero as an exponent is used.

Zero as an Exponent

If $x \neq 0$, then $x^0 = 1$. The expression 0^0 is not defined.

Simplify: 15^0

Any nonzero expression to the zero power is 1.

$15^0 = 1$

Simplify: $(4t^3)^0$

Any nonzero expression to the zero power is 1.

$(4t^3)^0 = 1$

Simplify: $-(2r)^0$

Any nonzero expression to the zero power is 1. Because the negative sign is in front of the parentheses, the answer is -1.

$-(2r)^0 = -1$

The expression at the right has been simplified in two ways: dividing by common factors, and using the Rule for Dividing Exponential Expressions.

$$\frac{x^3}{x^5} = \frac{\overset{1}{\cancel{x}} \cdot \overset{1}{\cancel{x}} \cdot \overset{1}{\cancel{x}}}{\underset{1}{\cancel{x}} \cdot \underset{1}{\cancel{x}} \cdot \underset{1}{\cancel{x}} \cdot x \cdot x} = \frac{1}{x^2}$$

$$\frac{x^3}{x^5} = x^{3-5} = x^{-2}$$

Because $\frac{x^3}{x^5} = \frac{1}{x^2}$ and $\frac{x^3}{x^5} = x^{-2}$, $\frac{1}{x^2}$ must equal x^{-2}. Therefore, the following definition of a negative exponent is used.

Definition of Negative Exponents

If n is a positive integer and $x \neq 0$, then $x^{-n} = \frac{1}{x^n}$ and $\frac{1}{x^{-n}} = x^n$.

An exponential expression is in simplest form when there are no negative exponents in the expression.

Simplify: y^{-7}

Use the Definition of Negative Exponents to rewrite the expression with a positive exponent.

$y^{-7} = \frac{1}{y^7}$

Simplify: $\frac{1}{c^{-4}}$

Use the Definition of Negative Exponents to rewrite the expression with a positive exponent.

$\frac{1}{c^{-4}} = c^4$

A numerical expression with a negative exponent can be evaluated by first rewriting the expression with a positive exponent.

Evaluate: 2^{-3}

Use the Definition of Negative Exponents to write the expression with a positive exponent. Then simplify.

$$2^{-3} = \frac{1}{2^3} = \frac{1}{8}$$

TAKE NOTE

Note from the example at the left that 2^{-3} is a *positive* number. A negative exponent does not indicate a negative number.

Sometimes applying the Rule for Dividing Exponential Expressions results in a quotient that contains a negative exponent. If this happens, use the Definition of Negative Exponents to rewrite the expression with a positive exponent.

Simplify: $\dfrac{p^4}{p^7}$

Use the Rule for Dividing Exponential Expressions.

$$\frac{p^4}{p^7} = p^{4-7}$$

Use the Definition of Negative Exponents to rewrite the expression with a positive exponent.

$$= p^{-3}$$
$$= \frac{1}{p^3}$$

Example 1   Simplify: $\dfrac{1}{a^{-8}}$

Solution $\dfrac{1}{a^{-8}} = a^8$

You Try It 1   Simplify: $\dfrac{1}{d^{-6}}$

Your Solution

Example 2   Simplify: 3^{-4}

Solution $3^{-4} = \dfrac{1}{3^4} = \dfrac{1}{81}$

You Try It 2   Simplify: 4^{-2}

Your Solution

Example 3   Simplify: $\dfrac{b^2}{b^9}$

Solution $\dfrac{b^2}{b^9} = b^{2-9} = b^{-7} = \dfrac{1}{b^7}$

You Try It 3   Simplify: $\dfrac{n^6}{n^{11}}$

Your Solution

Solutions on p. A19

Objective B
Scientific notation

Very large and very small numbers are encountered in the natural sciences. For example, the mass of an electron is 0.00000000000000000000000000000911 kg. Numbers such as this are difficult to read, so a more convenient system called **scientific notation** is used. In scientific notation, a number is expressed as the product of two factors, one a number between 1 and 10, and the other a power of ten.

To express a number in scientific notation, write it in the form $a \times 10^n$, where a is a number between 1 and 10 and n is an integer.

<table>
<tr><td>For numbers greater than 10, move the decimal point to the right of the first digit. The exponent n is positive and equal to the number of places the decimal point has been moved.</td><td>$240{,}000 = 2.4 \times 10^5$

$93{,}000{,}000 = 9.3 \times 10^7$</td></tr>
</table>

For numbers greater than 10, move the decimal point to the right of the first digit. The exponent n is positive and equal to the number of places the decimal point has been moved.

$240{,}000 = 2.4 \times 10^5$

$93{,}000{,}000 = 9.3 \times 10^7$

For numbers less than 1, move the decimal point to the right of the first nonzero digit. The exponent n is negative. The absolute value of the exponent is equal to the number of places the decimal point has been moved.

$0.0003 = 3.0 \times 10^{-4}$

$0.0000832 = 8.32 \times 10^{-5}$

Changing a number written in scientific notation to decimal notation also requires moving the decimal point.

When the exponent is positive, move the decimal point to the right the same number of places as the exponent.

$3.45 \times 10^6 = 3{,}450{,}000$

$2.3 \times 10^8 = 230{,}000{,}000$

When the exponent is negative, move the decimal point to the left the same number of places as the absolute value of the exponent.

$8.1 \times 10^{-3} = 0.0081$

$6.34 \times 10^{-7} = 0.000000634$

TAKE NOTE

There are two steps in writing a number in scientific notation: (1) determine the number between 1 and 10, and (2) determine the exponent on 10.

Example 4 Write 824,300,000,000 in scientific notation.

Solution The number is greater than 10. Move the decimal point 11 places to the left. The exponent on 10 is 11.

$824{,}300{,}000{,}000 = 8.243 \times 10^{11}$

You Try It 4 Write 0.000000961 in scientific notation.

Your Solution

Example 5 Write 6.8×10^{-10} in decimal notation.

Solution The exponent on 10 is negative. Move the decimal point 10 places to the left.

$6.8 \times 10^{-10} = 0.00000000068$

You Try It 5 Write 7.329×10^6 in decimal notation.

Your Solution

Solutions on p. A19

5.6 EXERCISES

OBJECTIVE A

Simplify.

1. 27^0

2. $(3x)^0$

3. $-(17)^0$

4. $(2a)^0$

5. 3^{-2}

6. 4^{-3}

7. 2^{-3}

8. 5^{-2}

9. x^{-5}

10. v^{-3}

11. w^{-8}

12. m^{-9}

13. $\dfrac{1}{a^{-5}}$

14. $\dfrac{1}{c^{-6}}$

15. $\dfrac{1}{b^{-3}}$

16. $\dfrac{1}{y^{-7}}$

17. $\dfrac{a^8}{a^2}$

18. $\dfrac{c^{12}}{c^5}$

19. $\dfrac{q^5}{q}$

20. $\dfrac{r^{10}}{r}$

21. $\dfrac{m^4 n^7}{m^3 n^5}$

22. $\dfrac{a^5 b^6}{a^3 b^2}$

23. $\dfrac{t^4 u^8}{t^2 u^5}$

24. $\dfrac{b^{11} c^5}{b^4 c}$

25. $\dfrac{x^4}{x^9}$

26. $\dfrac{r^2}{r^5}$

27. $\dfrac{b}{b^5}$

28. $\dfrac{m^5}{m^8}$

OBJECTIVE B

Write the number in scientific notation.

29. 2,370,000

30. 75,000

31. 0.00045

32. 0.000076

33. 309,000

34. 819,000,000

35. 0.000000601

36. 0.00000000096

37. 57,000,000,000

38. 934,800,000,000

39. 0.000000017

40. 0.0000009217

Write the number in decimal notation.

41. 7.1×10^5 **42.** 2.3×10^7 **43.** 4.3×10^{-5} **44.** 9.21×10^{-7}

45. 6.71×10^8 **46.** 5.75×10^9 **47.** 7.13×10^{-6} **48.** 3.54×10^{-8}

49. 5×10^{12} **50.** 1.0987×10^{11} **51.** 8.01×10^{-3} **52.** 4.0162×10^{-9}

Solve.

53. *Physics* Light travels approximately 16,000,000,000 mi in one day. Write this number in scientific notation.

54. *Earth Science* Write the mass of Earth, which is approximately 5,980,000,000,000,000,000,000,000 kg, in scientific notation.

55. *Physics* The electric charge on an electron is 0.00000000000000000016 coulomb. Write this number in scientific notation.

56. *Physics* The length of an infrared light wave is approximately 0.0000037 m. Write this number in scientific notation.

57. *Computers* One unit used to measure the speed of a computer is the picosecond. One picosecond is 0.000000001 of a second. Write one picosecond in scientific notation.

58. *Economics* What was the United States' auto trade deficit with Japan in 1995? Use the figure on the right. Write the answer in scientific notation.

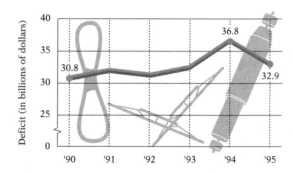

The U.S. Auto and Auto Parts Trade Deficit with Japan
Source: U.S. Census Bureau, Japan Automobile Importers Association, AAMA

CRITICAL THINKING

59. Place the correct symbol, $<$ or $>$, between the two numbers.
  a. 3.45×10^{-14} ? 6.45×10^{-15} **b.** 5.23×10^{18} ? 5.23×10^{17}
  c. 3.12×10^{12} ? 4.23×10^{11} **d.** -6.81×10^{-24} ? -9.37×10^{-25}

60. **a.** Evaluate 3^{-x} when $x = -2, -1, 0, 1,$ and 2.
  b. Evaluate 2^{-x} when $x = -2, -1, 0, 1,$ and 2.

61. In your own words, explain how to divide exponential expressions.

SECTION 5.7   Verbal Expressions and Variable Expressions

OBJECTIVE A

Translation of verbal expressions into variable expressions

One of the major skills required in applied mathematics is translating a verbal expression into a mathematical expression. Doing so requires recognizing the verbal phrases that translate into mathematical operations. Following is a partial list of the verbal phrases used to indicate the different mathematical operations.

Addition	more than	8 more than w	$w + 8$
	the sum of	the sum of z and 9	$z + 9$
	the total of	the total of r and s	$r + s$
	increased by	x increased by 7	$x + 7$
Subtraction	less than	12 less than b	$b - 12$
	the difference between	the difference between x and 1	$x - 1$
	decreased by	17 decreased by a	$17 - a$
Multiplication	times	negative 2 times c	$-2c$
	the product of	the product of x and y	xy
	of	three-fourths of m	$\dfrac{3}{4}m$
	twice	twice d	$2d$
Division	divided by	v divided by 15	$\dfrac{v}{15}$
	the quotient of	the quotient of y and 3	$\dfrac{y}{3}$
Power	the square of or the second power of	the square of x	x^2
	the cube of or the third power of	the cube of r	r^3
	the fifth power of	the fifth power of a	a^5

◆◆
POINT OF INTEREST

The way in which expressions are symbolized has changed over time. Here are some expressions as they may have appeared in the early sixteenth century.

R p. 9 for $x + 9$. The symbol R was used for a variable to the first power. The symbol p. was used for plus.

R m. 3 for $x - 3$. The symbol R is again the variable. The symbol m. is used for minus.

The square of a variable was designated by Q, and the cube was designated by C. The expression $x^3 + x^2$ was written C p. Q.

Translating a phrase that contains the word *sum, difference, product,* or *quotient* can sometimes cause a problem. In the examples at the right, note where the operation symbol is placed.

the *sum* of x and y $\qquad$ $x + y$

the *difference* between x and y $\qquad$ $x - y$

the *product* of x and y $\qquad$ $x \cdot y$

the *quotient* of x and y $\qquad$ $\dfrac{x}{y}$

Translate "3 times the sum of c and 5" into a variable expression.

Identify words that indicate the mathematical operations.

3 <u>times</u> the <u>sum of</u> c and 5

Use the identified words to write the variable expression. Note that the phrase <u>times the sum of</u> requires parentheses.

$3(c + 5)$

The sum of two numbers is 37. If x represents the smaller number, translate "twice the larger number" into a variable expression.

Write an expression for the larger number by subtracting the smaller number, x, from the sum.

larger number: $37 - x$

Identify the words that indicate the mathematical operations on the larger number.

<u>twice</u> the larger number

Use the identified words to write a variable expression.

$2(37 - x)$

Example 1

Translate "the quotient of r and the sum of r and four" into a variable expression.

Solution

the <u>quotient of</u> r and the <u>sum of</u> r and four

$$\frac{r}{r + 4}$$

You Try It 1

Translate "twice x divided by the difference between x and 7" into a variable expression.

Your Solution

Example 2

Translate "the sum of the square of y and six" into a variable expression.

Solution

the <u>sum of</u> the <u>square</u> of y and six

$y^2 + 6$

You Try It 2

Translate "the product of negative three and the square of d" into a variable expression.

Your Solution

Solutions on p. A19

OBJECTIVE B

Translation and simplification of verbal expressions

After translating a verbal expression into a variable expression, it may be possible to simplify the variable expression.

Translate "a number plus five less than the product of eight and the number" into a variable expression. Then simplify.

The letter x is chosen for the unknown number. Any letter could be used.	the unknown number: x
Identify words that indicate the mathematical operations.	x plus 5 less than the product of 8 and x
Use the identified words to write the variable expression.	$x + (8x - 5)$
Simplify the expression by adding like terms.	$x + 8x - 5$ $9x - 5$

Translate "five less than twice the difference between a number and seven" into a variable expression. Then simplify.

	the unknown number: x
Identify words that indicate the mathematical operations.	5 less than twice the difference between x and 7
Use the identified words to write the variable expression.	$2(x - 7) - 5$
Simplify the expression.	$2x - 14 - 5$ $2x - 19$

Example 3

The sum of two numbers is 28. Using x to represent the smaller number, translate "the sum of the smaller number and three times the larger number" into a variable expression. Then simplify.

Solution

The smaller number is x.
The larger number is $28 - x$.

the sum of the smaller number and three times the larger number

$x + 3(28 - x)$ ▶ This is the variable expression.

$x + 84 - 3x$ ▶ Simplify.

$-2x + 84$

You Try It 3

The sum of two numbers is 16. Using x to represent the smaller number, translate "the difference between the larger number and twice the smaller number" into a variable expression. Then simplify.

Your Solution

Solution on p. A19

Example 4

Translate "eight more than the product of four and the total of a number and twelve" into a variable expression. Then simplify.

Solution

Let the unknown number be x.

8 <u>more than</u> the <u>product</u> of 4 and the <u>total of</u> x and 12

$4(x + 12) + 8$ ▶ This is the variable expression.
$4x + 48 + 8$ ▶ Now simplify.
$4x + 56$

You Try It 4

Translate "the difference between fourteen and the sum of a number and seven" into a variable expression. Then simplify.

Your Solution

Solution on p. A19

OBJECTIVE C
Applications

Many applications of mathematics require that you identify the unknown quantity, assign a variable to that quantity, and then attempt to express other unknowns in terms of that quantity.

Thirty gallons of paint were poured into two containers of different sizes. Express the amount of paint poured into the smaller container in terms of the amount poured into the larger container.

Assign a variable to the amount of paint poured into the larger container.

gallons of paint poured into the larger container: g

Express the amount of paint in the smaller container in terms of g. (g gallons of paint were poured into the larger container.)

The number of gallons of paint in the smaller container is $30 - g$.

> **TAKE NOTE**
> Any variable can be used. For example, if the gallons of paint poured into the large container is p, then the number of gallons in the smaller container is $30 - p$.

Example 5

A cyclist is riding at twice the speed of a runner. Express the speed of the cyclist in terms of the speed of the runner.

Solution

the speed of the runner: r
the speed of the cyclist is twice r: $2r$

You Try It 5

A mixture of candy contains 3 lb more of milk chocolate than of caramel. Express the amount of milk chocolate in the mixture in terms of the amount of caramel in the mixture.

Your Solution

Solution on p. A19

5.7 EXERCISES

OBJECTIVE A

Translate into a variable expression.

1. three more than t

2. the total of twice q and five

3. five less than the product of six and m

4. seven subtracted from the product of eight and d

5. the difference between three times b and seven

6. the difference between six times c and twelve

7. the product of n and seven

8. the quotient of nine times k and seven

9. twice the sum of three and w

10. six times the difference between y and eight

11. four times the difference between twice r and five

12. seven times the total of p and ten

13. the quotient of v and the difference between v and four

14. x divided by the sum of x and one

15. four times the square of t

16. six times the cube of q

17. the sum of the square of m and the cube of the m

18. the difference between the square of d and d

19. The sum of two numbers is 31. Using s to represent the smaller number, translate "five more than the larger number" into a variable expression.

20. The sum of two numbers is 74. Using L to represent the larger number, translate "the quotient of the larger number and the smaller number" into a variable expression.

OBJECTIVE B

Translate into a variable expression. Then simplify.

21. a number decreased by the total of the number and twelve

22. a number decreased by the difference between six and the number

Translate into a variable expression. Then simplify.

23. the difference between two thirds of a number and three eighths of the number

24. two more than the total of a number and five

25. twice the sum of seven times a number and six

26. five times the product of seven and a number

27. the sum of eleven times a number and the product of three and the number

28. a number plus the product of the number and ten

29. nine times the sum of a number and seven

30. a number added to the product of four and the number

31. seven more than the sum of a number and five

32. a number minus the sum of the number and six

33. the product of seven and the difference between a number and four

34. six times the difference between a number and three

35. the difference between ten times a number and the product of three and the number

36. fifteen more than the difference between a number and seven

37. the sum of a number and twice the difference between the number and four

38. the difference between a number and the total of three times the number and five

39. seven times the difference between a number and fourteen

40. the product of three and the sum of a number and twelve

41. the product of eight and the sum of a number and ten

42. the difference between the square of a number and the total of twelve and the square of the number

43. a number increased by the difference between seven times the number and eight

44. the product of ten and the total of a number and one

45. five increased by twice the sum of a number and fifteen

46. eleven less than the difference between a number and eight

Translate into a variable expression. Then simplify.

47. fourteen decreased by the sum of a number and thirteen

48. eleven minus the sum of a number and six

49. the product of eight times a number and two

50. eleven more than a number added to the difference between the number and seventeen

51. a number plus nine added to the difference between four times the number and three

52. the sum of a number and ten added to the difference between the number and eleven

53. The sum of two numbers is 9. Using y to represent the smaller number, translate "five times the larger number" into a variable expression. Then simplify.

54. The sum of two numbers is 14. Using p to represent the smaller number, translate "eight less than the larger number" into a variable expression. Then simplify.

55. The sum of two numbers is 17. Using m to represent the larger number, translate "nine less than three times the smaller number" into a variable expression. Then simplify.

56. The sum of two numbers is 19. Using k to represent the larger number, translate "the difference between twice the smaller number and ten" into a variable expression. Then simplify.

OBJECTIVE C

Solve.

57. *Astronomy* The distance from Earth to the sun is approximately 390 times the distance from Earth to the moon. Express the distance from Earth to the sun in terms of the distance from Earth to the moon.

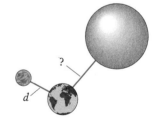

58. *Physics* The length of an infrared ray is twice the length of an ultraviolet ray. Express the length of the infrared ray in terms of the ultraviolet ray.

59. *Geography* Mt. Everest is 4,430 m higher than Mt. Whitney. Express the height of Mt. Everest in terms of the height of Mt. Whitney.

60. *Carpentry* The height of a door is five feet more than the width of the door. Express the height of the door in terms of the width of the door.

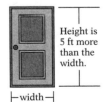

Height is 5 ft more than the width.

├─ width ─┤

61. *Food Mixtures* A mixture contains three times as many peanuts as cashews. Express the amount of peanuts in the mixture in terms of the amount of cashews in the mixture.

Solve.

62. *Computers* A "double-clocked" computer operates at twice its normally rated speed. Express the "double-clocked" speed in terms of the normally rated speed.

63. *Consumerism* The sale price of a suit is three fourths of the original price. Express the sale price in terms of the original price.

64. *Travel* One cyclist drives six miles per hour faster than another cyclist. Express the speed of the faster cyclist in terms of the speed of the slower cyclist.

65. *Sports* A fishing line three feet long is cut into two pieces, one shorter than the other. Express the length of the shorter piece in terms of the length of the longer piece.

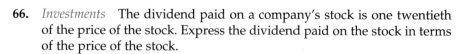

66. *Investments* The dividend paid on a company's stock is one twentieth of the price of the stock. Express the dividend paid on the stock in terms of the price of the stock.

67. *Carpentry* A twelve-foot board is cut into two pieces of different lengths. Express the length of the longer piece in terms of the length of the shorter piece.

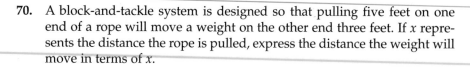

CRITICAL THINKING

68. A wire whose length is given as x inches is bent into a square. Express the length of a side of the square in terms of x.

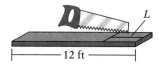

69. The chemical formula for water is H_2O. This formula means that there are two hydrogen atoms and one oxygen atom in each molecule of water. If x represents the number of atoms of oxygen in a glass of pure water, express the number of hydrogen atoms in the glass of water.

70. A block-and-tackle system is designed so that pulling five feet on one end of a rope will move a weight on the other end three feet. If x represents the distance the rope is pulled, express the distance the weight will move in terms of x.

71. A mechanical gear is designed so that a larger wheel makes four turns as a smaller wheel makes seven turns. Express the number of turns made by the larger wheel in terms of the number made by the smaller wheel.

72. ✎ Translate the expressions $3x + 4$ and $3(x + 4)$ into phrases.

73. ✎ In your own words, explain how variables are used.

74. ✎ Explain the similarities and differences between the expressions "the difference between x and 5" and "5 less than x."

PROJECTS IN MATHEMATICS

Multiplication of Polynomials

Section 5.5 introduced multiplying a polynomial by a monomial and multiplying two binomials. Here we are demonstrating multiplying a binomial times a polynomial of three or more terms. This skill is used below in the project entitled Pascal's Triangle.

Multiplying two polynomials requires the repeated application of the Distributive Property. Each term of one polynomial is multiplied by the other polynomial. For the product $(2y - 3)(y^2 + 2y + 5)$ shown below, note that the Distributive Property is used twice. The final result is simplified by combining like terms.

$$(2y - 3)(y^2 + 2y + 5) = (2y - 3)y^2 + (2y - 3)2y + (2y - 3)5$$
$$= 2y(y^2) - 3(y^2) + 2y(2y) - 3(2y) + 2y(5) - 3(5)$$
$$= 2y^3 - 3y^2 + 4y^2 - 6y + 10y - 15$$
$$= 2y^3 + y^2 + 4y - 15$$

A vertical format similar to that used for multiplication of whole numbers can also be used to multiply polynomials. The multiplication problem shown above is performed below in a vertical format.

$$
\begin{array}{r}
y^2 + 2y + 5 \\
2y - 3 \\
\hline
-3y^2 - 6y - 15 \\
2y^3 + 4y^2 + 10y \\
\hline
2y^3 + y^2 + 4y - 15
\end{array}
$$

This is $-3(y^2 + 2y + 5)$.
This is $2y(y^2 + 2y + 5)$. Like terms are placed in the same columns.

Add the terms in each column.

Note that each term of one polynomial is multiplied by each term of the other polynomial. Here is another example.

Multiply: $(a - 2)(a^3 + 4a^2 - 3a + 5)$

$$
\begin{array}{r}
a^3 + 4a^2 - 3a + 5 \\
a - 2 \\
\hline
-2a^3 - 8a^2 + 6a - 10 \\
a^4 + 4a^3 - 3a^2 + 5a \\
\hline
a^4 + 2a^3 - 11a^2 + 11a - 10
\end{array}
$$

Multiply -2 times $a^3 + 4a^2 - 3a + 5$.
Multiply a times $a^3 + 4a^2 - 3a + 5$. Put like terms in the same column so they can be added.

Multiply.

1. $(y + 6)(y^2 - 3y + 4)$

2. $(x - 3)(x^3 + 2x^2 - 4x - 5)$

3. $(2a^2 + 4a - 5)(2a + 1)$

4. $(c^3 + 3c^2 - 4c + 5)(2c - 3)$

Multiply.

5. $(2b^3 + 4b^2 - 5)(2b + 3)$

6. $(3z^3 - 5z + 7)(z - 3)$

Pascal's Triangle

Simplifying the power of a binomial is called **expanding the binomial.** The expansion of the first two powers of a binomial is shown below.

$$(a + b)^1 = a + b$$

$$(a + b)^2 = (a + b)(a + b) \quad \text{Squaring an expression means to multiply it times itself.}$$

$$= a^2 + 2ab + b^2 \quad \text{Use the FOIL method to multiply the two binomials.}$$

Find $(a + b)^3$. *Hint:* $(a + b)^3 = (a + b)^2(a + b) = (a^2 + 2ab + b^2)(a + b)$

Find $(a + b)^4$. *Hint:* $(a + b)^4 = (a + b)^3(a + b)$

If we continue in this manner, the results for $(a + b)^5$ and $(a + b)^6$ would be as follows:

$$(a + b)^5 = a^5 + 5a^4b + 10a^3b^2 + 10a^2b^3 + 5ab^4 + b^5$$

$$(a + b)^6 = a^6 + 6a^5b + 15a^4b^2 + 20a^3b^3 + 15a^2b^4 + 6ab^5 + b^6$$

Now expand $(a + b)^8$. Before you begin, you might think about the problem-solving strategy given at the beginning of this chapter. Is there a pattern that will assist you in finding $(a + b)^8$ without having to multiply it out? Here are some hints.

1. Write out the variable terms of each binomial expansion without the coefficients. Observe how the exponents on the variables change.

2. Write out the coefficients of each term without the variable part. It will be helpful to make a triangular table as shown at the left.

Note that each row begins and ends with a 1.

Also note from the two shaded regions that any number in a row is the sum of the two closest numbers above it. For instance, $1 + 5 = 6$ and $6 + 4 = 10$.

```
        1   1
      1   2   1
    1   3   3   1
  1   4   6   4   1
1   5   10   10   5   1
1   6   15   20   15   6   1
```

POINT OF INTEREST

Pascal did not invent the triangle of numbers known as Pascal's Triangle. It was known to mathematicians in China probably as early as A.D. 1050. But Pascal's *Traite du triangle arithmetique* (Treatise Concerning the Arithmetical Triangle) brought together all the different aspects of the numbers for the first time.

The triangle of numbers given above is called **Pascal s Triangle.** To find the expansion of $(a + b)^8$, you will need to find the eighth row of Pascal's Triangle. First find row seven. Now find row eight.

Use the patterns you have observed to write the expansion. The result is

$$(a + b)^8$$
$$= a^8 + 8a^7b + 28a^6b^2 + 56a^5b^3 + 70a^4b^4 + 56a^3b^5 + 28a^2b^6 + 8ab^7 + b^8$$

Pascal's Triangle has been the subject of extensive analysis, and many patterns have been found within it. See if you can find some of them.

CHAPTER SUMMARY

Key Words

The *additive inverse* of a number a is $-a$. The additive inverse of a number is also called the *opposite* number.

The *multiplicative inverse* of a nonzero number a is $\frac{1}{a}$. The multiplicative inverse of a number is also called the *reciprocal* of the number.

A *term* of a variable expression is one of the addends of the expression. A *variable term* consists of a *numerical coefficient* and a *variable part*. A *constant term* has no variable part.

A *monomial* is a number, a variable, or a product of numbers and variables.

A *polynomial* is a variable expression in which the terms are monomials.

A polynomial of one term is a *monomial.*

A polynomial of two terms is a *binomial.*

A polynomial of three terms is a *trinomial.*

Like terms of a variable expression have the same variable part.

Adding or subtracting the like terms of a variable expression is called *combining like terms.*

Essential Rules

Commutative Property of Addition	$a + b = b + a$
Commutative Property of Multiplication	$a \cdot b = b \cdot a$
Associative Property of Addition	$(a + b) + c = a + (b + c)$
Associative Property of Multiplication	$(a \cdot b) \cdot c = a \cdot (b \cdot c)$

Addition Property of Zero	$a + 0 = 0 + a = a$
Multiplication Property of Zero	$a \cdot 0 = 0 \cdot a = 0$
Multiplication Property of One	$a \cdot 1 = 1 \cdot a = a$
Inverse Property of Addition	$a + (-a) = (-a) + a = 0$
Inverse Property of Multiplication	For $a \neq 0$, $a \cdot \dfrac{1}{a} = \dfrac{1}{a} \cdot a = 1$.
Distributive Property	$a(b + c) = ab + ac$
Alternate Form of the Distributive Property	$ac + bc = (a + b)c$
Rule for Multiplying Exponential Expressions	$x^m \cdot x^n = x^{m+n}$
Rule for Simplifying the Power of an Exponential Expression	$(x^m)^n = x^{m \cdot n}$
Rule for Simplifying Powers of Products	$(x^m y^n)^p = x^{m \cdot p} y^{n \cdot p}$
Rule for Dividing Exponential Expressions	For $x \neq 0$, $\dfrac{x^m}{x^n} = x^{m-n}$.
Zero as an Exponent	For $x \neq 0$, $x^0 = 1$.
Definition of Negative Exponents	For $x \neq 0$, $x^{-n} = \dfrac{1}{x^n}$ and $\dfrac{1}{x^{-n}} = x^n$.
Addition of Polynomials	To add polynomials, add the coefficients of the like terms.
Subtraction of Polynomials	To subtract two polynomials, add the opposite of the second polynomial to the first.
The FOIL Method	Add the products of the First terms, the Outer terms, the Inner terms, and the Last terms.

Scientific Notation

To express a number in scientific notation, write it in the form $a \times 10^n$, where a is a number between 1 and 10 and n is an integer. If the number is greater than 10, the exponent on 10 will be positive. If the number is less than 1, the exponent on 10 will be negative.

$$367,000,000 = 3.67 \times 10^8$$

$$0.0000059 = 5.9 \times 10^{-6}$$

To change a number written in scientific notation to decimal notation, move the decimal point to the right if the exponent on 10 is positive and to the left if the exponent on 10 is negative. Move the decimal point the same number of places as the absolute value of the exponent on 10.

$$2.418 \times 10^7 = 24,180,000$$

$$9.06 \times 10^{-5} = 0.0000906$$

CHAPTER REVIEW EXERCISES

1. Simplify: $-3x - 7 + 5x - 9$

2. Multiply: $-2(9z + 1)$

3. Add: $(3z^2 + 4z - 7) + (7z^2 - 5z - 8)$

4. Multiply: $(2m^3n)(-4m^2n)$

5. Evaluate: 3^{-5}

6. Write the additive inverse of $\frac{3}{7}$.

7. Multiply: $\frac{2}{3}\left(\frac{3}{2}x\right)$

8. Simplify: $-5(2s - 5t) + 6(3t + s)$

9. Multiply: $(-5xy^4)(-3x^2y^3)$

10. Multiply: $(7a + 6)(3a - 4)$

11. Subtract:
$(6b^3 - 7b^2 + 5b - 9) - (9b^3 - 7b^2 + b + 9)$

12. Simplify: $(2z^4)^5$

13. Multiply: $-\frac{3}{4}(-8w)$

14. Multiply: $5xyz^2(-3x^2z + 6yz^2 - x^3y^4)$

15. Write the multiplicative inverse of $-\frac{9}{4}$.

16. Multiply: $-4(3c - 8)$

17. Simplify: $2m - 6n + 7 - 4m + 6n + 9$

18. Multiply: $(4a^3b^8)(-3a^2b^7)$

19. Identify the property that justifies the statement.
$a(b + c) = ab + ac$

20. Simplify: $(p^2q^3)^3$

21. Simplify: $\dfrac{a^4}{a^{11}}$

22. Write 0.0000397 in scientific notation.

23. Identify the property that justifies the statement.
$a + b = b + a$

24. Add: $(9y^3 + 8y^2 - 10) + (-6y^3 + 8y - 9)$

25. Simplify: $8(2c - 3d) - 4(c - 5d)$

26. Multiply: $7(2m - 6)$

27. Simplify: $\dfrac{x^3y^5}{xy}$

28. Simplify: $7a^2 + 9 - 12a^2 + 3a$

29. Multiply: $(3p - 9)(4p + 7)$

30. Multiply: $-2a^2b(4a^3 - 5ab^2 + 3b^4)$

31. Simplify: $-12x + 7y + 15x - 11y$

32. Simplify: $-7(3a - 4b) - 5(3b - 4a)$

33. Simplify: c^{-5}

34. Subtract:
$(12x^3 + 9x^2 - 5x - 1) - (6x^3 + 9x^2 + 5x - 1)$

35. Write 2.4×10^5 in decimal notation.

36. Simplify: $4z^2 + 3z - 9z + 2z^2$

Solve.

37. Translate "nine less than the quotient of four times a number and seven" into a variable expression.

38. Translate "the sum of three times a number and twice the difference between the number and seven" into a variable expression. Then simplify.

39. *Chemistry* Avogadro's number is used in chemistry, and its value is approximately 602,300,000,000,000,000,000,000. Express this number in scientific notation.

40. *Food Mixtures* Thirty pounds of a blend of coffee beans uses only mocha java and expresso beans. Express the number of pounds of expresso beans in the blend in terms of the number of pounds of mocha java beans in the blend.

CUMULATIVE REVIEW EXERCISES

1. Find the quotient of 4.712 and -0.38.

2. Simplify: $9v - 10 + 5v + 8$

3. Multiply: $(3x - 5)(2x + 4)$

4. Evaluate $-a - b$ when $a = \frac{11}{24}$ and $b = -\frac{5}{6}$.

5. Simplify: $\sqrt{81} + 3\sqrt{25}$

6. Graph the real numbers greater than -3.

7. Simplify: $\dfrac{1}{x^{-7}}$

8. Solve: $-4t = 36$

9. Write 0.00000084 in scientific notation.

10. Add: $(5x^2 - 3x + 2) + (4x^2 + x - 6)$

11. Evaluate $-5\sqrt{x + y}$ when $x = 18$ and $y = 31$.

12. Simplify: $\dfrac{\frac{5}{8} + \frac{3}{4}}{3 - \frac{1}{2}}$

13. Multiply: $(-3a^2b)(4a^5b^8)$

14. Simplify: $\dfrac{x^3}{x^5}$

15. Evaluate x^3y^2 when $x = \frac{2}{5}$ and $y = 2\frac{1}{2}$.

16. Simplify: $-8p(6)$

17. Estimate the difference between 829.43 and 567.109.

18. Multiply: $-3ab^2(4a^2b + 5ab - 2ab^2)$

19. Simplify: $6(5x - 4y) - 12(x - 2y)$

20. Evaluate $\dfrac{a}{-b}$ when $a = -56$ and $b = -8$.

21. Convert 0.5625 to a fraction.

22. Simplify: $6 \cdot (-2)^3 \div 12 - (-8)$

23. Simplify: $\sqrt{300}$

24. Subtract: $(8y^2 - 7y + 4) - (3y^2 - 5y + 9)$

25. Evaluate $-6cd$ when $c = -\frac{2}{9}$ and $d = \frac{3}{4}$.

26. Simplify: $-(3a^2)^0$

27. Simplify: $(2a^4b^3)^5$

28. Evaluate $(a - b)^2 + 5c$ when $a = -4$, $b = 6$, and $c = -2$.

29. Find the product of $2\frac{4}{5}$ and $\frac{6}{7}$.

30. Write 6.23×10^{-5} in decimal notation.

Solve.

31. Translate "the quotient of ten and the difference between a number and nine" into a variable expression.

32. Translate "two less than twice the sum of a number and four" into a variable expression. Then simplify.

33. *Meteorology* The average annual precipitation in Seattle, Washington, is 38.6 in. The average annual precipitation in El Paso, Texas, is 7.82 in. Find the difference between the average annual precipitation in Seattle and the average annual precipitation in El Paso.

34. *Education* The figure at the right shows the projected market in educational institutions for PC and Macintosh computers. Find the difference between the projected numbers for the years 2000 and 1996.

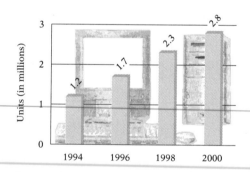

Projected Academic Market for PC and Macintosh
Source: Understanding Information Systems in Higher Education, 1994, by CCA Consulting Inc., Wellesley, MA

35. *Astronomy* The distance from Neptune to the sun is approximately 30 times the distance from Earth to the sun. Express the distance from Neptune to the sun in terms of the distance from Earth to the sun.

36. *Investments* The cost, C, of the shares of stock in a stock purchase is equal to the cost per share, S, times the number of shares purchased, N. Use the equation $C = SN$ to find the cost of purchasing 200 shares of stock selling for $\$15\frac{3}{8}$ per share.

First-Degree Equations

FOCUS On Problem Solving

In the Critical Thinking exercises in this text, you are sometimes asked to determine whether a statement is true or false. For instance, the statement "every real number has a reciprocal" is false because 0 is a real number and 0 does not have a reciprocal.

Finding an example, such as 0 has no reciprocal, to show that a statement is not always true is called *finding a counterexample.* A counterexample is an example that shows that a statement is not always true.

Consider the statement "the product of two numbers is greater than either factor." A counterexample to this statement is the numbers $\frac{2}{3}$ and $\frac{3}{4}$. The product of these numbers is $\frac{1}{2}$, and $\frac{1}{2}$ is *smaller* than $\frac{2}{3}$ or $\frac{3}{4}$. There are many other counterexamples to the given statement.

Here are some counterexamples to the statement that the square root of a number is smaller than the number.

$$\sqrt{\frac{1}{4}} = \frac{1}{2} \quad \text{but} \quad \frac{1}{2} > \frac{1}{4} \qquad \sqrt{1} = 1 \quad \text{but} \quad 1 = 1$$

For each of the next five statements, find at least one counterexample to show that the conjecture is false.

1. The product of two integers is always a positive number.

2. The sum of two prime numbers is never a prime number.

3. For all real numbers, $|x + y| = |x| + |y|$.

4. If x and y are nonzero real numbers and $x > y$, then $x^2 > y^2$.

5. The quotient of any two nonzero real numbers is less than either one of the numbers.

When a problem is posed, it may not be known whether the problem statement is true or false. For instance, Christian Goldbach (1690–1764) stated that every even integer greater than 2 can be written as the sum of two prime numbers. No one has been able to find a counterexample to this statement, but no one has been able to prove that it is always true.

In the next set of problems, answer true if the statement is always true or give a counterexample if there is an instance when the statement is false.

1. The reciprocal of a positive number is always smaller than the number.

2. If $x < 0$, then $|x| = -x$.

3. For any two real numbers x and y, $x + y > x - y$.

4. For any positive integer n, $n^2 + n + 17$ is a prime number.

5. The list of numbers, 1, 11, 111, 1111, 11111, . . . contains infinitely many composite numbers.
    Hint: A number is divisible by 3 if the sum of the digits of the number is divisible by 3.

SECTION 6.1   Equations of the Form $x + a = b$ and $ax = b$

OBJECTIVE A

Equations of the form $x + a = b$

Recall that an **equation** expresses the equality of two mathematical expressions. The display at the right shows some examples of equations.

$$3x - 7 = 4x + 9$$
$$y = 3x - 6$$
$$2z^2 - 5z + 10 = 0$$
$$\frac{3}{x} + 7 = 9$$

The first equation in the display above is a *first-degree equation in one variable*. The equation has one variable, x, and each instance of the variable is the first power (the exponent on x is 1). First-degree equations in one variable are the topic of Sections 1 through 4 of this chapter. The second equation is a *first-degree equation in two variables*. These are discussed in Section 5. The remaining equations are not first-degree equations and will not be discussed in this text.

Which of the equations shown at the right are first-degree equations in one variable?

1. $5x + 4 = 9 - 3(2x + 1)$
2. $\sqrt{x} + 9 = 10$
3. $p = -14$
4. $2x - 5 = x^2 - 9$

Equation 1 is a first-degree equation in one variable.

Equation 2 is not a first-degree equation in one variable. First-degree equations do not contain square roots of variable expressions.

Equation 3 is a first-degree equation in one variable.

Equation 4 is not a first-degree equation in one variable. First-degree equations in one variable do not have exponents greater than 1 on the variable.

Recall that a **solution** of an equation is a number that, when substituted for the variable, results in a true equation.

15 is a solution of the equation $x - 5 = 10$ because $15 - 5 = 10$ is a true equation.

20 is not a solution of $x - 5 = 10$ because $20 - 5 = 10$ is a false equation.

To solve an equation means to determine the solutions of the equation. The simplest equation to solve is an equation of the form **variable = constant**. The constant is the solution.

Consider the equation $x = 7$, which is in the form variable = constant. The solution is 7 because $7 = 7$ is a true equation.

Find the solution of the equation $y = 3 + 7$.

Simplify the right side of the equation.

$$y = 3 + 7$$
$$y = 10$$

The solution is 10.

Note that replacing x in $x + 8 = 12$ by 4 results in a true equation. The solution of the equation $x + 8 = 12$ is 4.

$$x + 8 = 12$$
$$4 + 8 = 12$$
$$12 = 12$$

If 5 is added to each side of $x + 8 = 12$, the solution is still 4.

$$x + 8 = 12$$
$$x + 8 + 5 = 12 + 5$$
$$x + 13 = 17$$

Check:
$$x + 13 = 17$$
$$\frac{4 + 13 \mid 17}{17 = 17}$$

If -3 is added to each side of $x + 8 = 12$, the solution is still 4.

$$x + 8 = 12$$
$$x + 8 + (-3) = 12 + (-3)$$
$$x + 5 = 9$$

Check:
$$x + 5 = 9$$
$$\frac{4 + 5 \mid 9}{9 = 9}$$

These examples suggest that adding the same number to each side of an equation does not change the solution of the equation. This is called the Addition Property of Equations.

Addition Property of Equations

The same number or variable expression can be added to each side of an equation without changing the solution of the equation.

This property is used in solving equations. Note the effect of adding, to each side of the equation $x + 8 = 12$, the opposite of the constant term 8. After simplifying, the equation is in the form *variable* = *constant*. The solution is the constant, 4.

$$x + 8 = 12$$
$$x + 8 + (-8) = 12 + (-8)$$
$$x + 0 = 4$$
$$x = 4$$

Check the solution.

Check:
$$x + 8 = 12$$
$$\frac{4 + 8 \mid 12}{12 = 12}$$

The solution checks.

The solution is 4.

The goal in solving an equation is to rewrite it in the form **variable = constant**. The Addition Property of Equations is used to remove a term from one side of an equation by adding the opposite of that term to each side of the equation. The resulting equation has the same solution as the original equation.

Solve: $m - 9 = 2$

Remove the constant term -9 from the left side of the equation by adding 9, the opposite of -9, to each side of the equation. Then simplify.

$$m - 9 = 2$$
$$m - 9 + 9 = 2 + 9$$
$$m + 0 = 11$$
$$m = 11$$

You should check the solution.

The solution is 11.

In each of the equations above, the variable appeared on the left side of the equation, and the equation was rewritten in the form *variable = constant*. For some equations, it may be more practical to work toward the goal of *constant = variable*, as shown in the example below.

Solve: $12 = n - 8$

The variable is on the right side of the equation. The goal is to rewrite the equation in the form *constant = variable*.

Remove the constant term from the right side of the equation by adding 8 to each side of the equation. Then simplify.

$$12 = n - 8$$
$$12 + 8 = n - 8 + 8$$
$$20 = n + 0$$
$$20 = n$$

You should check the solution.

The solution is 20.

Because subtraction is defined in terms of addition, the Addition Property of Equations allows the same number to be subtracted from each side of an equation without changing the solution of the equation.

Solve: $z + 9 = 6$

The goal is to rewrite the equation in the form *variable = constant*.

Add the opposite of 9 to each side of the equation. This is equivalent to subtracting 9 from each side of the equation. Then simplify.

$$z + 9 = 6$$
$$z + 9 - 9 = 6 - 9$$
$$z + 0 = -3$$
$$z = -3$$

The solution checks.

The solution is -3.

TAKE NOTE

Remember to check the solution.

Check: $z + 9 = 6$
          $\overline{-3 + 9 \mid 6}$
          $ 6 = 6$

Solve: $5 + x - 9 = -10$

Simplify the left side of the equation by combining the constant terms.

$$5 + x - 9 = -10$$
$$x - 4 = -10$$

Add 4 to each side of the equation.

$$x - 4 + 4 = -10 + 4$$
$$x + 0 = -6$$

Simplify.

$$x = -6$$

-6 checks as a solution.

The solution is -6.

Example 1

Solve: $6 + x = 4$

Solution

$$6 + x = 4$$
$$6 - 6 + x = 4 - 6$$
$$x = -2$$

The solution is -2.

You Try It 1

Solve: $7 + y = 12$

Your Solution

Example 2

Solve: $8 = y - 12$

Solution

$$8 = y - 12$$
$$8 + 12 = y - 12 + 12$$
$$20 = y$$

The solution is 20.

You Try It 2

Solve: $19 = b - 23$

Your Solution

Example 3

Solve: $7x - 4 - 6x = 3$

Solution

$$7x - 4 - 6x = 3$$
$$x - 4 = 3 \qquad \text{Combine like terms.}$$
$$x - 4 + 4 = 3 + 4$$
$$x = 7$$

The solution is 7.

You Try It 3

Solve: $-5r + 3 + 6r = 1$

Your Solution

Solutions on p. A19

OBJECTIVE **B**

Equations of the form $ax = b$

16   CT

Note that replacing x by 3 in $4x = 12$ results in a true equation. The solution of the equation is 3.

$$4x = 12$$
$$4(3) = 12$$
$$12 = 12$$

If each side of the equation $4x = 12$ is multiplied by 2, the solution is still 3.

$$4x = 12$$
$$2(4x) = 2(12)$$
$$8x = 24$$

Check:
$$8x = 24$$
$$\overline{8(3) \;|\; 24}$$
$$24 = 24$$

If each side of the equation $4x = 12$ is multiplied by -3, the solution is still 3.

$$4x = 12$$
$$-3(4x) = -3(12)$$
$$-12x = -36$$

Check:
$$-12x = -36$$
$$\overline{-12(3) \;|\; -36}$$
$$-36 = -36$$

These examples suggest that multiplying each side of an equation by the same nonzero number does not change the solution of the equation. This is called the Multiplication Property of Equations.

> ## *Multiplication Property of Equations*
>
> Each side of an equation can be multiplied by the same nonzero number without changing the solution of the equation.

This property is used in solving equations. Note the effect of multiplying each side of the equation $4x = 12$ by $\frac{1}{4}$, the reciprocal of the coefficient 4. After simplifying, the equation is in the form *variable = constant*.

$$4x = 12$$
$$\frac{1}{4} \cdot 4x = \frac{1}{4} \cdot 12$$
$$1 \cdot x = 3$$
$$x = 3$$

The solution is 3.

The Multiplication Property of Equations is used to remove a coefficient from a variable term of an equation by multiplying each side of the equation by the reciprocal of the coefficient. The resulting equation will have the same solution as the original equation.

Solve: $\frac{3}{4}x = -9$

The goal is to rewrite the equation in the form *variable = constant*.

Multiply each side of the equation by $\frac{4}{3}$, the reciprocal of $\frac{3}{4}$. After simplifying, the equation is in the form *variable = constant*.

$$\frac{3}{4}x = -9$$
$$\frac{4}{3} \cdot \frac{3}{4}x = \frac{4}{3} \cdot (-9)$$
$$1 \cdot x = -12$$
$$x = -12$$

You should check this solution.

The solution is -12.

Because division is defined in terms of multiplication, the Multiplication Property of Equations allows each side of an equation to be divided by the same nonzero number without changing the solution of the equation.

Solve: $-2x = 8$

Multiply each side of the equation by the reciprocal of -2. This is equivalent to dividing each side of the equation by -2.

$$-2x = 8$$
$$\frac{-2x}{-2} = \frac{8}{-2}$$
$$1 \cdot x = -4$$
$$x = -4$$

Check the solution.

Check:
$$\begin{array}{r|l} -2x = 8 \\ \hline -2(-4) & 8 \\ 8 = 8 \end{array}$$

The solution checks.

The solution is -4.

When using the Multiplication Property of Equations, multiply each side of the equation by the reciprocal of the coefficient when the coefficient is a fraction. Divide each side of the equation by the coefficient when the coefficient is an integer or a decimal.

Example 4

Solve: $48 = -12y$

Solution

$$48 = -12y$$

$$\frac{48}{-12} = \frac{-12y}{-12}$$

$$-4 = y$$

The solution is -4.

You Try It 4

Solve: $-60 = 5d$

Your Solution

Example 5

Solve: $\frac{2x}{3} = 12$

Solution

$$\frac{2x}{3} = 12$$

$$\frac{3}{2}\left(\frac{2}{3}x\right) = \frac{3}{2}(12) \quad \blacktriangleright \quad \frac{2x}{3} = \frac{2}{3}x$$

$$x = 18$$

The solution is 18.

You Try It 5

Solve: $10 = \frac{-2x}{5}$

Your Solution

Example 6

Solve and check: $3y - 7y = 8$

Solution

$$3y - 7y = 8$$

$$-4y = 8 \qquad \text{Combine like terms.}$$

$$\frac{-4y}{-4} = \frac{8}{-4}$$

$$y = -2$$

Check:

$$
\begin{array}{c|c}
3y - 7y = 8 & \\
\hline
3(-2) - 7(-2) & 8 \\
-6 - (-14) & 8 \\
-6 + 14 & 8 \\
8 = 8 &
\end{array}
$$

-2 checks as the solution.

The solution is -2.

You Try It 6

Solve and check: $\frac{1}{3}x - \frac{5}{6}x = 4$

Your Solution

Solutions on p. A19

6.1 EXERCISES

OBJECTIVE A

Solve.

1. $x + 3 = 9$

2. $y + 6 = 8$

3. $4 + x = 13$

4. $9 + y = 14$

5. $m - 12 = 5$

6. $n - 9 = 3$

7. $x - 3 = -2$

8. $y - 6 = -1$

9. $a + 5 = -2$

10. $b + 3 = -3$

11. $3 + m = -6$

12. $5 + n = -2$

13. $8 = x + 3$

14. $7 = y + 5$

15. $3 = w - 6$

16. $4 = y - 3$

17. $-7 = -7 + m$

18. $-9 = -9 + n$

19. $-3 = v + 5$

20. $-1 = w + 2$

21. $-5 = 1 + x$

22. $-3 = 4 + y$

23. $3 = -9 + m$

24. $4 = -5 + n$

25. $4 + x - 7 = 3$

26. $12 + y - 4 = 8$

27. $8t + 6 - 7t = -6$

28. $-5z + 5 + 6z = 12$

29. $y + \dfrac{4}{7} = \dfrac{6}{7}$

30. $z + \dfrac{3}{5} = \dfrac{4}{5}$

31. $x - \dfrac{3}{8} = \dfrac{1}{8}$

32. $a - \dfrac{1}{6} = \dfrac{5}{6}$

33. $c + \dfrac{2}{3} = \dfrac{3}{4}$

34. $n + \dfrac{1}{3} = \dfrac{2}{5}$

35. $w - \dfrac{1}{4} = \dfrac{3}{8}$

36. $t - \dfrac{1}{3} = \dfrac{1}{2}$

OBJECTIVE B

Solve.

37. $3x = 9$

38. $8a = 16$

39. $4c = -12$

40. $5z = -25$

41. $-2r = 16$

42. $-6p = 72$

43. $-4m = -28$

44. $-12x = -36$

Solve.

45. $-3y = 0$

46. $-7a = 0$

47. $12 = 2c$

48. $28 = 7x$

49. $-72 = 18v$

50. $35 = -5p$

51. $-68 = -17t$

52. $-60 = -15y$

53. $12x = 30$

54. $9v = 15$

55. $-6a = 21$

56. $-8c = 20$

57. $28 = -12y$

58. $36 = -16z$

59. $-52 = -18a$

60. $-40 = -30w$

61. $\dfrac{2}{3}x = 4$

62. $\dfrac{3}{4}y = 9$

63. $\dfrac{1}{3}a = -12$

64. $\dfrac{3y}{5} = -15$

65. $-\dfrac{4c}{7} = 16$

66. $-\dfrac{5n}{8} = 20$

67. $-\dfrac{z}{4} = -3$

68. $-\dfrac{3x}{8} = -15$

69. $8 = \dfrac{4}{5}y$

70. $10 = -\dfrac{5}{6}c$

71. $\dfrac{5y}{6} = \dfrac{7}{12}$

72. $\dfrac{-3v}{4} = -\dfrac{7}{8}$

73. $7y - 9y = 10$

74. $8w - 5w = 9$

75. $m - 4m = 21$

76. $2a - 6a = 10$

CRITICAL THINKING

77. (a) Solve the equation $x + a = b$ for x. Is the solution you have written valid for all real numbers a and b? (b) Solve the equation $ax = b$ for x. Is the solution you have written valid for all real numbers a and b?

78. Solve: **a.** $\dfrac{2}{\dfrac{1}{x}} = 8$ **b.** $\dfrac{3}{\dfrac{2}{x}} = 6$

79. ✎ Write out the steps for solving the equation $\dfrac{2}{3}x = 6$. Identify each Property of Real Numbers or Property of Equations as you use it.

80. ✎ Using your own words, state the Addition Property of Equations and the Multiplication Property of Equations.

SECTION 6.2   Equations of the Form $ax + b = c$

OBJECTIVE A

Equations of the form $ax + b = c$

To solve an equation such as $3w - 5 = 16$, both the Addition and Multiplication Properties of Equations are used.

$$3w - 5 = 16$$

First add the opposite of the constant term -5 to each side of the equation.

$$3w - 5 + 5 = 16 + 5$$
$$3w = 21$$

Divide each side of the equation by the coefficient of w.

$$\frac{3w}{3} = \frac{21}{3}$$

The equation is in the form *variable = constant*.

$$w = 7$$

Check the solution.

Check: $$\begin{array}{c|c} 3w - 5 = 16 \\ \hline 3(7) - 5 & 16 \\ 21 - 5 & 16 \\ & 16 = 16 \end{array}$$

> **TAKE NOTE**
>
> Note that the Order of Operations Agreement applies to evaluating the expression $3(7) - 5$.

7 checks as the solution.

The solution is 7.

Solve: $8 = 4 - \frac{2}{3}x$

The variable is on the right side of the equation. Work toward the goal of *constant = variable*.

$$8 = 4 - \frac{2}{3}x$$

Subtract 4 from each side of the equation.

$$8 - 4 = 4 - 4 - \frac{2}{3}x$$
$$4 = -\frac{2}{3}x$$

Multiply each side of the equation by $-\frac{3}{2}$.

$$-\frac{3}{2} \cdot 4 = \left(-\frac{3}{2}\right)\left(-\frac{2}{3}x\right)$$

The equation is in the form *constant = variable*.

$$-6 = x$$

You should check the solution.

The solution is -6.

> **TAKE NOTE**
>
> Always check the solution.
>
> Check: $$\begin{array}{c|c} 8 = 4 - \frac{2}{3}x \\ \hline 8 & 4 - \frac{2}{3}(-6) \\ 8 & 4 + 4 \\ 8 = 8 \end{array}$$

Example 1 Solve: $6 - 5x = 16$

Solution
$$6 - 5x = 16$$
$$6 - 6 - 5x = 16 - 6$$
$$-5x = 10$$
$$\frac{-5x}{-5} = \frac{10}{-5}$$
$$x = -2$$

The solution is -2.

You Try It 1 Solve: $-5 - 4t = 7$

Your Solution

Solution on p. A20

Example 2

Solve: $4m - 7 + m = 8$

Solution

$$4m - 7 + m = 8$$
$$5m - 7 = 8 \qquad \text{Combine like terms.}$$
$$5m - 7 + 7 = 8 + 7$$
$$5m = 15$$
$$m = 3 \qquad \text{Divide each side by 5.}$$

The solution is 3.

You Try It 2

Solve: $5v + 3 - 9v = 9$

Your Solution

Solution on p. A20

OBJECTIVE B

Applications

Some application problems can be solved by using a formula. For instance, the formula for the monthly car payment for a 60-month car loan at a 9 percent interest rate is **$P = 0.02076L$**, where P is the monthly car payment and L is the amount of the loan. A car buyer who can afford a maximum monthly car payment of $250 can use this formula to find the maximum loan amount.

CALCULATOR NOTE

To solve for *L*, use your calculator: 250 ÷ 0.02076. Then round the answer to the nearest cent.

$$P = 0.02076L$$
$$250 = 0.02076L \qquad \text{Replace } P \text{ by 250.}$$
$$\frac{250}{0.02076} = \frac{0.02076L}{0.02076} \qquad \text{Solve for } L.$$
$$12{,}042.39 \approx L \qquad \text{The maximum loan amount is \$12,042.39.}$$

Example 3

The formula $V = C - 4{,}500t$ is used to determine the value V, after t years, of a computer that originally cost C dollars. Determine in how many years the value of a computer that originally cost $39,000 will be worth $25,500.

Strategy

To find the number of years, replace each of the variables by their value and solve for t. $V = 25{,}500$, $C = 39{,}000$.

Solution

$$V = C - 4{,}500t$$
$$25{,}500 = 39{,}000 - 4{,}500t$$
$$25{,}500 - 39{,}000 = 39{,}000 - 39{,}000 - 4{,}500t$$
$$-13{,}500 = -4{,}500t$$
$$\frac{-13{,}500}{-4{,}500} = \frac{-4{,}500t}{-4{,}500}$$
$$3 = t$$

In 3 years, the value will be $25,500.

You Try It 3

The pressure P, in pounds per square inch, at a certain depth in the ocean is approximated by the formula $P = 15 + \frac{1}{2}D$, where D is the depth in feet. Find the depth when the pressure is 45 pounds per square inch.

Your Strategy

Your Solution

Solution on p. A20

6.2 EXERCISES

OBJECTIVE A

Solve.

1. $5y + 1 = 11$

2. $3x + 5 = 26$

3. $2z - 9 = 11$

4. $7p - 2 = 26$

5. $12 = 2 + 5a$

6. $29 = 1 + 7v$

7. $-5y + 8 = 13$

8. $-7p + 6 = -8$

9. $-12a - 1 = 23$

10. $-15y - 7 = 38$

11. $10 - c = 14$

12. $3 - x = 1$

13. $4 - 3x = -5$

14. $8 - 5x = -12$

15. $-33 = 3 - 4z$

16. $-41 = 7 - 8v$

17. $-4t + 16 = 0$

18. $-6p - 72 = 0$

19. $5a + 9 = 12$

20. $7c + 5 = 20$

21. $2t - 5 = 2$

22. $3v - 1 = 4$

23. $8x + 1 = 7$

24. $6y + 5 = 8$

25. $4z - 5 = 1$

26. $8 = 5 + 6p$

27. $25 = 11 + 8v$

Solve.

28. $-4 = 11 + 6z$

29. $-3 = 7 + 4y$

30. $9w - 4 = 17$

31. $8a - 5 = 31$

32. $5 - 8x = 5$

33. $7 - 12y = 7$

34. $-3 - 8z = 11$

35. $-9 - 12y = 5$

36. $5n - \dfrac{2}{9} = \dfrac{43}{9}$

37. $6z - \dfrac{1}{3} = \dfrac{5}{3}$

38. $7y - \dfrac{2}{5} = \dfrac{12}{5}$

39. $3p - \dfrac{5}{8} = \dfrac{19}{8}$

40. $\dfrac{3}{4}x - 1 = 2$

41. $\dfrac{4}{5}y + 3 = 11$

42. $\dfrac{5t}{6} + 4 = -1$

43. $\dfrac{3v}{7} - 2 = 10$

44. $\dfrac{2a}{5} - 5 = 7$

45. $\dfrac{4z}{9} + 23 = 3$

46. $\dfrac{x}{3} + 6 = 1$

47. $\dfrac{y}{4} + 5 = 2$

48. $17 = 20 + \dfrac{3}{4}x$

49. $\dfrac{2}{5}y - 3 = 1$

50. $\dfrac{7}{3}v + 2 = 8$

51. $5 - \dfrac{7}{8}y = 2$

52. $3 - \dfrac{5}{2}z = 6$

53. $\dfrac{3}{5}y + \dfrac{1}{4} = \dfrac{3}{4}$

54. $\dfrac{5}{6}x - \dfrac{2}{3} = \dfrac{5}{3}$

Solve.

55. $\dfrac{3}{5} = \dfrac{2}{7}t + \dfrac{1}{5}$

56. $\dfrac{10}{3} = \dfrac{9}{5}w - \dfrac{2}{3}$

57. $\dfrac{z}{3} - \dfrac{1}{2} - \dfrac{1}{4}$

58. $\dfrac{a}{6} + \dfrac{1}{4} = \dfrac{3}{8}$

59. $5.6t - 5.1 = 1.06$

60. $7.2 + 5.2z = 8.76$

61. $6.2 - 3.3t = -12.94$

62. $2.4 - 4.8v = 13.92$

63. $6c - 2 - 3c = 10$

64. $12t + 6 + 3t = 16$

65. $4y + 5 - 12y = -3$

66. $7m - 15 - 10m = 6$

67. $17 = 12p - 5 - 6p$

68. $29 = 4x + 5 - 9x$

69. $3 = 6n + 23 - 10n$

OBJECTIVE B

To determine the depreciated value of an X-ray machine, an accountant uses the formula $V = C - 5,500t$, where V is the depreciated value of the machine in t years and C is the original cost.

70. *Accounting*   An X-ray machine originally cost $70,000. In how many years will the depreciated value be $48,000?

71. *Accounting*   An X-ray machine originally cost $63,000. In how many years will the depreciated value be $47,500? Round to the nearest tenth.

The formula for the monthly car payment for a 60-month car loan at a 9 percent interest rate is $P = 0.02076L$, where P is the monthly car payment and L is the amount of the loan. Use this formula for Exercises 72 and 73.

72. *Consumerism*   If you can afford a maximum monthly car payment of $300, what is the maximum loan amount you can afford? Round to the nearest cent.

73. *Consumerism*   If you can afford a maximum of $325 for a monthly car payment, what is the largest loan amount you can afford? Round to the nearest cent.

The world record time for a 1-mile race can be approximated by the formula $t = 17.08 - 0.0067y$, where y is the year of the race and t is the time, in minutes, of the race.

74. *Sports*   Approximate the year in which the first "4-minute" mile was run. The actual year was 1954.

75. *Sports*   In 1985, the world record for a 1-mile race was 3.77 min. For what year does the equation predict this record time?

Black ice is an ice covering on roads that is especially difficult to see and therefore extremely dangerous for motorists. The distance a car traveling 30 mph will slide after its brakes are applied is related to the outside air temperature by the formula $C = \frac{1}{4}D - 45$, where C is the Celsius temperature and D is the distance in feet the car will slide.

76. *Physics*   Determine the distance a car will slide on black ice when the outside air temperature is $-3°C$.

77. *Physics*   Determine the distance a car will slide on black ice when the outside air temperature is $-11°C$.

CRITICAL THINKING

78. Solve: $x \div 28 = 1{,}481$ remainder 25

79. Make up an equation of the form $ax + b = c$ that has -3 as its solution.

80. If a and b are real numbers, is it always possible to solve the equation $ax + b = c$? If not, what values of a, b, or c must be excluded?

81. Does the sentence "Solve $2x - 3(4x + 1)$" make sense? Why or why not?

82. Explain in your own words the steps you would take to solve the equation $\frac{2}{3}x + 4 = 10$. State the Property of Real Numbers or the Property of Equations that is used at each step.

83. Explain the difference between the word *equation* and the word *expression*.

SECTION 6.3 General First-Degree Equations

Objective A
Equations of the form $ax + b = cx + d$

An equation that contains variable terms on both the left and the right side is solved by repeated application of the Addition Property of Equations. The Multiplication Property of Equations is then used to remove the coefficient of the variable and write the equation in the form *variable* = *constant*.

Solve: $5z - 4 = 8z + 5$

The goal is to rewrite the equation in the form *variable* = *constant*.

Use the Addition Property of Equations to remove $8z$ from the right side by subtracting $8z$ from each side of the equation. After simplifying, there is only one variable term in the equation.

$$5z - 4 = 8z + 5$$
$$5z - 8z - 4 = 8z - 8z + 5$$
$$-3z - 4 = 5$$

Solve this equation by following the procedure developed in the last section. Using the Addition Property of Equations, add 4 to each side of the equation.

$$-3z - 4 + 4 = 5 + 4$$
$$-3z = 9$$

Divide each side of the equation by -3. After simplifying, the equation is in the form *variable* = *constant*.

$$\frac{-3z}{-3} = \frac{9}{-3}$$
$$z = -3$$

Check the solution.

$$\begin{array}{c|c} \multicolumn{2}{c}{5z - 4 = 8z + 5} \\ \hline 5(-3) - 4 & 8(-3) + 5 \\ -15 - 4 & -24 + 5 \\ \multicolumn{2}{c}{-19 = -19} \end{array}$$

-3 checks as a solution.

The solution is -3.

Example 1

Solve: $2c + 5 = 8c + 2$

Solution

$$2c + 5 = 8c + 2$$
$$2c - 8c + 5 = 8c - 8c + 2$$
$$-6c + 5 = 2$$
$$-6c + 5 - 5 = 2 - 5$$
$$-6c = -3$$
$$\frac{-6c}{-6} = \frac{-3}{-6}$$
$$c = \frac{1}{2}$$

The solution is $\frac{1}{2}$.

You Try It 1

Solve: $r - 7 = 5 - 3r$

Your Solution

Solution on p. A20

Example 2

Solve: $6a + 3 - 9a = 3a + 7$

Solution

$$
\begin{aligned}
6a + 3 - 9a &= 3a + 7 \\
-3a + 3 &= 3a + 7 \\
-3a - 3a + 3 &= 3a - 3a + 7 \\
-6a + 3 &= 7 \\
-6a + 3 - 3 &= 7 - 3 \\
-6a &= 4 \\
\frac{-6a}{-6} &= \frac{4}{-6} \\
a &= -\frac{2}{3}
\end{aligned}
$$

The solution is $-\frac{2}{3}$.

Combine like terms.

▶ $\dfrac{4}{-6} = -\dfrac{2}{3}$

Remember to check the solution.

You Try It 2

Solve: $4a - 2 + 5a = 2a - 2 + 3a$

Your Solution

Solution on p. A20

OBJECTIVE B

Equations with parentheses

When an equation contains parentheses, one of the steps in solving the equation requires the use of the Distributive Property. The Distributive Property is used to remove parentheses from a variable expression.

Solve: $6 - 2(3x - 1) = 3(3 - x) + 5$

The goal is to rewrite the equation in the form *variable = constant*.

Use the Distributive Property to remove parentheses. Then combine like terms on each side of the equation.

$$
\begin{aligned}
6 - 2(3x - 1) &= 3(3 - x) + 5 \\
6 - 6x + 2 &= 9 - 3x + 5 \\
8 - 6x &= 14 - 3x
\end{aligned}
$$

Using the Addition Property of Equations, add $3x$ to each side of the equation. After simplifying, there is only one variable term in the equation.

$$
\begin{aligned}
8 - 6x + 3x &= 14 - 3x + 3x \\
8 - 3x &= 14
\end{aligned}
$$

Subtract 8 from each side of the equation. After simplifying, there is only one constant term in the equation.

$$
\begin{aligned}
8 - 8 - 3x &= 14 - 8 \\
-3x &= 6
\end{aligned}
$$

Divide each side of the equation by -3, the coefficient of x. The equation is in the form *variable = constant*.

$$
\begin{aligned}
\frac{-3x}{-3} &= \frac{6}{-3} \\
x &= -2
\end{aligned}
$$

-2 checks as a solution.

The solution is -2.

Example 3

Solve: $4 - 3(2t + 1) = 15$

Solution

$$4 - 3(2t + 1) = 15$$
$$4 - 6t - 3 = 15$$
$$-6t + 1 = 15$$
$$-6t + 1 - 1 = 15 - 1$$
$$-6t = 14$$
$$\frac{-6t}{-6} = \frac{14}{-6} \qquad \blacktriangleright \frac{14}{-6} = -\frac{7}{3}$$
$$t = -\frac{7}{3}$$

$-\frac{7}{3}$ checks as the solution.

The solution is $-\frac{7}{3}$.

You Try It 3

Solve: $6 - 5(3y + 2) = 26$

Your Solution

Example 4

Solve: $5x - 3(2x - 3) = 4(x - 2)$

Solution

$$5x - 3(2x - 3) = 4(x - 2)$$
$$5x - 6x + 9 = 4x - 8$$
$$-x + 9 = 4x - 8$$
$$-x - 4x + 9 = 4x - 4x - 8$$
$$-5x + 9 = -8$$
$$-5x + 9 - 9 = -8 - 9$$
$$-5x = -17$$
$$\frac{-5x}{-5} = \frac{-17}{-5}$$
$$x = \frac{17}{5}$$

$\frac{17}{5}$ checks as the solution.

The solution is $\frac{17}{5}$.

You Try It 4

Solve: $2w - 7(3w + 1) = 5(5 - 3w)$

Your Solution

Solutions on p. A20

OBJECTIVE C
Applications

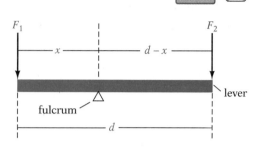

A lever system is shown at the right. It consists of a lever, or bar; a fulcrum; and two forces, F_1 and F_2. The distance d represents the length of the lever, x represents the distance from F_1 to the fulcrum, and $d - x$ represents the distance from F_2 to the fulcrum.

When a lever system balances, $F_1x = F_2(d - x)$. This is known as Archimedes' Principle of Levers.

Example 5

A lever 10 ft long is used to move a 100-pound rock. The fulcrum is placed 2 ft from the rock. What minimum force must be applied to the other end of the lever to move the rock?

You Try It 5

A screwdriver 9 in. long is used as a lever to open a can of paint. The tip of the screwdriver is placed under the lip of the can with the fulcrum 0.15 in. from the lip. A force of 30 lb is applied to the other end of the screwdriver. Find the force on the lip of the can.

Strategy

To find the minimum force needed, replace the variables F_1, d, and x by the given values, and solve for F_2.
$F_1 = 100, d = 10, x = 2$

F_2 (10 − 2) ft   2 ft   100 lb

Your Strategy

Solution

$$F_1x = F_2(d - x)$$
$$100 \cdot 2 = F_2 \cdot (10 - 2)$$
$$200 = 8F_2$$
$$\frac{200}{8} = \frac{8F_2}{8}$$
$$25 = F_2$$

Your Solution

Check:

$$F_1x = F_2(d - x)$$

$100 \cdot 2$	$25 \cdot (10 - 2)$
200	$25(8)$
$200 =$	200

25 checks as the solution.

The minimum force required is 25 lb.

Solution on p. A20

6.3 EXERCISES

OBJECTIVE A

Solve.

1. $4x + 3 = 2x + 9$

2. $6z + 5 = 3z + 20$

3. $7y - 6 = 3y + 6$

4. $8w - 5 = 5w + 10$

5. $12m + 11 = 5m + 4$

6. $8a + 9 = 2a - 9$

7. $7c - 5 = 2c - 25$

8. $7r - 1 = 5r - 13$

9. $2n - 3 = 5n - 18$

10. $4t - 7 = 10t - 25$

11. $3z + 5 = 19 - 4z$

12. $2m + 3 = 23 - 8m$

13. $5v - 3 = 4 - 2v$

14. $3r - 8 = 2 - 2r$

15. $7 - 4a = 2a$

16. $5 - 3x = 5x$

17. $12 - 5y = 3y - 12$

18. $8 - 3m = 8m - 14$

19. $7r = 8 + 2r$

20. $-2w = 4 - 5w$

21. $5a + 3 = 3a + 10$

22. $7y + 3 = 5y + 12$

Solve.

23. $9w - 2 = 5w + 4$

24. $7n - 3 = 3n + 6$

25. $x - 7 = 5x - 21$

26. $3y - 4 = 9y - 24$

27. $5n - 1 + 2n = 4n + 8$

28. $3y + 1 + y = 2y + 11$

29. $3z - 2 - 7z = 4z + 6$

30. $2a + 3 - 9a = 3a + 33$

31. $4t - 8 + 12t = 3 - 4t - 11$

32. $6x - 5 + 9x = 7 - 4x - 12$

OBJECTIVE B

Solve.

33. $3(4y + 5) = 25$

34. $5(3z - 2) = 8$

35. $-2(4x + 1) = 22$

36. $-3(2x - 5) = 30$

37. $5(2k + 1) - 7 = 28$

38. $7(3t - 4) + 8 = -6$

39. $3(3v - 4) + 2v = 10$

40. $4(3x + 1) - 5x = 25$

41. $3y + 2(y + 1) = 12$

42. $7x + 3(x + 2) = 33$

Solve.

43. $7v - 3(v - 4) = 20$

44. $15m - 4(2m - 5) = 34$

45. $6 + 3(3x - 3) = 24$

46. $9 + 2(4p - 3) = 24$

47. $9 - 3(4a - 2) = 9$

48. $17 - 8(x - 3) = 1$

49. $3(2z - 5) = 4z + 1$

50. $4(3z - 1) = 5z + 17$

51. $2 - 3(5x + 2) = 2(3 - 5x)$

52. $5 - 2(3y + 1) = 3(2 - 3y)$

53. $4r + 11 = 5 - 2(3r + 3)$

54. $3v + 6 = 9 - 4(2v - 2)$

55. $7n - 2 = 5 - (9 - n)$

56. $8x - 5 = 7 - 2(5 - x)$

OBJECTIVE C

Solve. Use the lever system equation $F_1 x = F_2(d - x)$.

57. *Physics*   Two people are sitting 15 ft apart on a seesaw. One person weighs 180 lb; the second person weighs 120 lb. How far from the 180-pound person should the fulcrum be placed so that the seesaw balances?

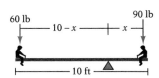

58. *Physics*   Two children are sitting on a seesaw that is 10 ft long. One child weighs 60 lb; the second child weighs 90 lb. How far from the 90-pound child should the fulcrum be placed so that the seesaw balances?

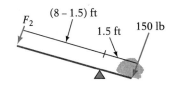

59. *Physics*   A metal bar 8 ft long is used to move a 150-pound rock. The fulcrum is placed 1.5 ft from the rock. What minimum force must be applied to the other end of the bar to move the rock? Round to the nearest tenth.

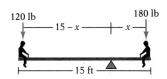

Solve. Use the lever system equation $F_1x = F_2(d - x)$.

60. *Physics* A screwdriver 8 in. long is used as a lever to open a can of paint. The tip of the screwdriver is placed under the lip of the can with the fulcrum 0.2 in. from the lip. A force of 50 lb is applied to the other end of the screwdriver. Find the force on the lip of the top of the can.

To determine the breakeven point, or the number of units that must be sold so that no profit or loss occurs, an economist uses the formula $Px = Cx + F$, where P is the selling price per unit, x is the number of units that must be sold to break even, C is the cost to make each unit, and F is the fixed cost.

61. *Business* A business analyst has determined that the selling price per unit for a laser printer is $1600. The cost to make the laser printer is $950, and the fixed cost is $211,250. Find the breakeven point.

62. *Business* An economist has determined that the selling price per unit for a gas barbecue is $325. The cost to make one gas barbecue is $175, and the fixed cost is $39,000. Find the breakeven point.

63. *Business* A manufacturer of thermostats determines that the cost per unit for a programmable thermostat is $38 and that the fixed cost is $24,400. The selling price for the thermostat is $99. Find the breakeven point.

64. *Business* A manufacturing engineer determines that the cost per unit for a desk lamp is $12 and that the fixed cost is $19,240. The selling price for the desk lamp is $49. Find the breakeven point.

CRITICAL THINKING

65. If $5a - 4 = 3a + 2$, what is the value of $4a^3$?

66. If $3 + 2(4a - 3) = 5$ and $4 - 3(2 - 3b) = 11$, which is larger, a or b?

67. Explain the problem with the demonstration shown at the right that suggests $2 = 3$.

$$2x + 5 = 3x + 5$$
$$2x + 5 - 5 = 3x + 5 - 5 \quad \text{Subtract 5 from each side of the equation.}$$
$$2x = 3x$$
$$\frac{2x}{x} = \frac{3x}{x} \quad \text{Divide each side of the equation by } x.$$
$$2 = 3$$

68. The equation $x = x + 1$ has no solution, whereas the solution of the equation $2x + 3 = 3$ is zero. Is there a difference between no solution and a solution of zero? Explain your answer.

69. Archimedes supposedly said, "Give me a long enough lever and I can move the world." Explain what Archimedes meant by that statement.

SECTION 6.4 Translating Sentences into Equations

OBJECTIVE A

Translate a sentence into an equation and solve

An equation states that two mathematical expressions are equal. Therefore, to translate a sentence into an equation requires recognition of the words or phrases that mean "equals." Some of these words and phrases are listed below.

equals   *is*   *represents*
amounts to   *totals*   *is the same as*

Translate "five less than four times a number is four more than the number" into an equation and solve.

Assign a variable to the unknown number.	the unknown number: n

Find two verbal expressions for the same value.

Five less than four times a number	is	four more than the number

Write an equation. $4n - 5 = n + 4$

Solve the equation.
Subtract n from each side. $3n - 5 = 4$
Add 5 to each side. $3n = 9$
Divide each side by 3. $n = 3$
The solution checks.   The number is 3.

TAKE NOTE

You can check the solution to a translation problem.

Check:

5 less than 4 times 3	4 more than 3
$4 \cdot 3 - 5$	$3 + 4$
$12 - 5$	7
$7 =$	7

Example 1

Translate "eight less than three times a number equals five times the number" into an equation and solve.

Solution

the unknown number: x

Eight less than three times a number	equals	five times the number

$$3x - 8 = 5x$$
$$3x - 3x - 8 = 5x - 3x$$
$$-8 = 2x$$
$$\frac{-8}{2} = \frac{2x}{2}$$
$$-4 = x$$

-4 checks as the solution.

The number is -4.

You Try It 1

Translate "six more than one-half a number is the total of the number and nine" into an equation and solve.

Your Solution

Solution on p. A20

Example 2

Translate "four more than five times a number is six less than three times the number" into an equation and solve.

Solution

the unknown number: m

Four more than five times a number	is	six less than three times the number

$$5m + 4 = 3m - 6$$
$$5m - 3m + 4 = 3m - 3m - 6$$
$$2m + 4 = -6$$
$$2m + 4 - 4 = -6 - 4$$
$$2m = -10$$
$$\frac{2m}{2} = \frac{-10}{2}$$
$$m = -5$$

-5 checks as the solution.

The number is -5.

You Try It 2

Translate "seven less than a number is equal to five more than three times the number" into an equation and solve.

Your Solution

Example 3

The sum of two numbers is 9. Eight times the smaller number is five less than three times the larger number. Find the numbers.

Solution

the smaller number: p
the larger number: $9 - p$

Eight times the smaller number	is	five less than three times the larger number

$$8p = 3(9 - p) - 5$$
$$8p = 27 - 3p - 5$$
$$8p = 22 - 3p$$
$$8p + 3p = 22 - 3p + 3p$$
$$11p = 22$$
$$\frac{11p}{11} = \frac{22}{11}$$
$$p = 2$$

$9 - p = 9 - 2 = 7$

These numbers check as solutions.

The smaller number is 2.
The larger number is 7.

You Try It 3

The sum of two numbers is 14. One more than three times the smaller equals the sum of the larger number and three. Find the two numbers.

Your Solution

Solutions on p. A21

OBJECTIVE B
Applications

Example 4
A new laser printer prints at a rate of 16 ppm (pages per minute). This is four times the rate of a slower laser printer. Find the rate of the slower printer.

Strategy
To find the rate of the slower printer, write and solve an equation using r to represent the rate of the slower printer.

Solution

| 16 | is | four times the rate of the slower printer |

$$16 = 4r$$
$$\frac{16}{4} = \frac{4r}{4}$$
$$4 = r$$

The rate of the slower printer is 4 ppm.

You Try It 4
The selling price of a large-screen television in 1989 was $4,200. This is $1,700 more than the price today. Find the price of the large-screen television today.

Your Strategy

Your Solution

Example 5
A wallpaper hanger charges a fee of $25 plus $12 for each roll of wallpaper used in a room. If the total charge for hanging wallpaper is $97, how many rolls of wallpaper were used?

Strategy
To find the number of rolls of wallpaper used, write and solve an equation using n to represent the number of rolls of wallpaper.

Solution

| $25 plus $12 for each roll of wallpaper | is | $97 |

$$25 + 12n = 97$$
$$25 - 25 + 12n = 97 - 25$$
$$12n = 72$$
$$\frac{12n}{12} = \frac{72}{12}$$
$$n = 6$$

The wallpaper hanger used 6 rolls.

You Try It 5
The fee charged by a ticketing agency for a concert is $3.50 plus $17.50 for each ticket purchased. If your total charge for tickets is $161, how many tickets are you purchasing?

Your Strategy

Your Solution

Solutions on p. A21

Example 6

A bank charges a check service fee of $5.00 per month plus $.08 for each check that is written. For the month of July, a customer paid a check service fee of $5.96. How many checks did the customer write?

Strategy

To find the number of checks written in July, write and solve an equation using n for the number of checks.

Solution

| $5.00 plus $.08 per check | is | $5.96 |

$$5 + 0.08n = 5.96$$
$$5 - 5 + 0.08n = 5.96 - 5$$
$$0.08n = 0.96$$
$$\frac{0.08n}{0.08} = \frac{0.96}{0.08}$$
$$n = 12$$

The customer wrote 12 checks in July.

Example 7

A wire 22 in. long is cut into two pieces. The longer piece is 4 in. more than twice the shorter piece. Find the length of the shorter piece.

Strategy

To find the length, write and solve an equation using x to represent the length of the shorter piece and $22 - x$ to represent the length of the longer piece.

Solution

| The longer piece | is | 4 in. more than twice the shorter piece |

$$22 - x = 2x + 4$$
$$22 - x - 2x = 2x - 2x + 4$$
$$22 - 3x = 4$$
$$22 - 22 - 3x = 4 - 22$$
$$-3x = -18$$
$$\frac{-3x}{-3} = \frac{-18}{-3}$$
$$x = 6$$

The shorter piece is 6 in.

You Try It 6

A computer bulletin board service charges a monthly fee of $9.95 plus $.13 per minute of access time. If a customer was charged $24.77, how many minutes of access time did the customer use?

Your Strategy

Your Solution

You Try It 7

A board 18 ft long is cut into two pieces. One foot more than twice the shorter piece is 2 ft less than the longer piece. Find the length of each piece.

Your Strategy

Your Solution

6.4 EXERCISES

OBJECTIVE A

Translate into an equation and solve.

1. The sum of a number and twelve is twenty. Find the number.

2. The difference between nine and a number is seven. Find the number.

3. Three-fifths of a number is negative thirty. Find the number.

4. The quotient of a number and six is twelve. Find the number.

5. Four more than three times a number is thirteen. Find the number.

6. The sum of twice a number and five is fifteen. Find the number.

7. The difference between nine times a number and six is twelve. Find the number.

8. Six less than four times a number is twenty-two. Find the number.

9. Eight less than the product of eleven and a number is negative nineteen. Find the number.

10. Seven more than the product of six and a number is eight less than the product of three and the number. Find the number.

11. Fifteen less than the product of four and a number is the product of the number minus eleven and six. Find the number.

12. Five less than the product of four and a number is the product of three and the sum of the number and seven. Find the number.

13. Six more than twice the sum of three times a number and eight is negative two. Find the number.

14. Three times the difference between four times a number and seven is fifteen. Find the number.

15. The sum of two numbers is twenty-one. Twice the smaller number is three more than the larger number. Find the two numbers.

16. The sum of two numbers is thirty. Three times the smaller number is twice the larger number. Find the two numbers.

17. The sum of two numbers is twenty-three. The larger number is five more than twice the smaller number. Find the two numbers.

18. The sum of two numbers is twenty-five. The larger number is five less than four times the smaller number. Find the two numbers.

OBJECTIVE B

Write an equation and solve.

19. *Consumerism* Due to depreciation, the value of a car one year after it was purchased is $13,500. This is four-fifths of its original value. Find the original value of the car.

20. *The Arts* An oil painting was purchased at an auction for $250,000. This is two and one-half times the value of the painting five years ago. What was the value of the painting five years ago?

21. *Computers* A high-density computer disk can store approximately 1,400,000 bytes of data. This is four times as much as a low-density disk. How many bytes can be stored on a low-density disk?

22. *Computers* The height of a computer monitor screen is 15 in. This is three-fourths of the length of the screen. What is the length of the computer monitor screen?

Write an equation and solve.

23. *The Arts*   The value of a silk Turkish rug is four times the value of a Chinese wool rug. The total value of the two rugs is $4,500. What is the value of the Turkish rug?

24. *Labor Unions*   A union charges monthly dues of $4.00 plus $.15 for each hour worked during the month. A union member's dues for March were $29.20. How many hours did the union member work during the month of March?

25. *Business*   The advertising dollars spent by three companies during the 1996 Olympic Games is shown in the figure at the right. The amount spent by General Motors was ten million dollars less than twice the amount spent by Visa. How many advertising dollars were spent by Visa during the Olympics?

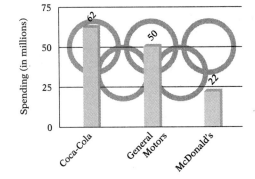

1996 Olympic Advertisers
Source: **Competitive Media Reporting**

26. *Consumerism*   A technical information hotline charges a customer $9 plus $.50 per minute to answer questions about software. How many minutes did a customer who received a bill for $14.50 use this service?

27. *Consumerism*   The total cost to paint the inside of a house was $1,346. This cost included $125 for materials and $33 per hour for labor. How many hours of labor were required?

28. *Consumerism*   The cellular phone service for a business executive is $35 per month plus $.40 per minute of phone use. In a month when the executive's cellular phone bill was $51.80, how many minutes did the executive use the phone?

29. *Carpentry*   A 12-foot board is cut into two pieces. Twice the length of the shorter piece is three feet less than the longer piece. Find the length of each piece.

30. *Sports*   A 14-yard fishing line is cut into two pieces. Three times the length of the longer piece is four times the length of the shorter piece. Find the length of each piece.

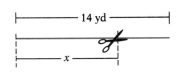

Write an equation and solve.

31. *Financial Aid* Seven thousand dollars is divided into two scholarships. Twice the amount of the smaller scholarship is $1,000 less than the larger scholarship. What is the amount of the larger scholarship?

32. *Investments* An investment of $10,000 is divided into two accounts, one for stocks and one for mutual funds. The value of the stock account is $2,000 less than twice the value of the mutual fund account. Find the amount in each account.

33. *Food Mixtures* A 10-pound blend of coffee contains Colombian coffee, French Roast, and Java. There is one pound more of French Roast than of Colombian and two pounds more of Java than of French Roast. How many pounds of each are in the mixture?

34. *Agriculture* A 60-pound soil supplement contains nitrogen, iron, and potassium. There is twice as much potassium as iron and three times as much nitrogen as iron. How many pounds of each are in the soil supplement?

CRITICAL THINKING

An equation that is never true is called a *contradiction*. For example, the equation $x = x + 1$ is a contradiction. There is no value of x that will make the equation true. An equation that is true for all real numbers is called an *identity*. The equation $x + x = 2x$ is an identity. This equation is true for any real number. A *conditional equation* is one that is true for some real numbers and false for some real numbers. The equation $2x = 4$ is a conditional equation. This equation is true when x is 2 and false for any other real number. Determine whether each equation below is a contradiction, identity, or conditional equation. If it is a conditional equation, find the solution.

35. $6x + 2 = 5 + 3(2x - 1)$

36. $3 - 2(4x + 1) = 5 + 8(1 - x)$

37. $3t - 5(t + 1) = 2(2 - t) - 9$

38. $6 + 4(2y + 1) = 5 - 8y$

39. $3v - 2 = 5v - 2(2 + v)$

40. $9z = 15z$

41. It is always important to check the answer to an application problem to be sure the answer makes sense. Consider the following problem. A 4-quart mixture of fruit juices is made from apple juice and cranberry juice. There are 6 more quarts of apple juice than of cranberry juice. Write and solve an equation for the number of quarts of each juice used. Does the answer to this question make sense? Explain.

42. A formula is an equation that relates variables in a known way. Find two examples of formulas that are used in your college major. Explain what each of the variables represents.

SECTION 6.5   The Rectangular Coordinate System

OBJECTIVE A

The rectangular coordinate system

Before the fifteenth century, geometry and algebra were considered separate branches of mathematics. That all changed when René Descartes, a French mathematician who lived from 1596 to 1650, founded **analytic geometry.** In this geometry, a *coordinate system* is used to study relationships between variables.

A **rectangular coordinate system** is formed by two number lines, one horizontal and one vertical, that intersect at the zero point of each line. The point of intersection is called the **origin.** The two lines are called **coordinate axes,** or simply **axes.**

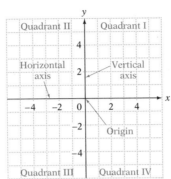

The axes determine a **plane,** which can be thought of as a large, flat sheet of paper. The two axes divide the plane into four regions called **quadrants,** which are numbered counterclockwise from I to IV.

POINT OF INTEREST

Although Descartes is given credit for introducing analytic geometry, others were working on the same concept, notably Pierre Fermat. Nowhere in Descartes's work is there a coordinate system as we draw it with two axes. Descartes did not use the word *coordinate* in his work. This word was introduced by Gottfried Leibnitz, who also first used the words *abscissa* and *ordinate.*

Each point in the plane can be identified by a pair of numbers called an **ordered pair.** The first number of the pair measures a horizontal distance and is called the **abscissa,** or *x*-coordinate. The second number of the pair measures a vertical distance and is called the **ordinate,** or *y*-coordinate. The ordered pair (a, b) associated with a point is also called the **coordinates** of the point.

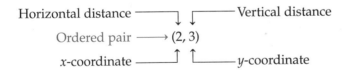

To **graph** or **plot** a point in the plane, place a dot at the location given by the ordered pair. The **graph of an ordered pair** is the dot drawn at the coordinates of the point in the plane. The points whose coordinates are $(3, 4)$ and $(-2.5, -3)$ are graphed in the figures below.

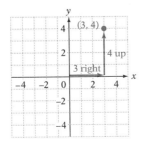

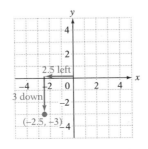

Example 1 Graph the ordered pairs (−2, −3), (3, −2), (1, 3), and (4, 1).

Solution

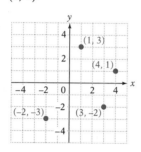

You Try It 1 Graph the ordered pairs (−1, 3), (1, 4), (−4, 0), and (−2, −1).

Your Solution

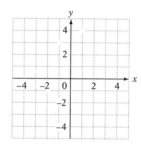

Example 2 Find the coordinates of each point.

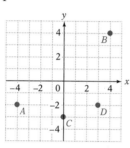

Solution $A(−4, −2)$ $C(0, −3)$
            $B(4, 4)$ $D(3, −2)$

You Try It 2 Find the coordinates of each point.

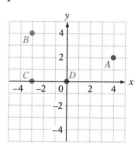

Your Solution

Solutions on p. A22

OBJECTIVE B

Scatter diagrams

Discovering a relationship between two variables is an important task in the study of mathematics. These relationships occur in many forms and in a wide variety of applications. Here are some examples.

A botanist wants to know the relationship between the number of bushels of wheat yielded per acre and the amount of watering per acre.

An environmental scientist wants to know the relationship between the incidence of skin cancer and the amount of ozone in the atmosphere.

A business analyst wants to know the relationship between the price of a product and the number of products that are sold at that price.

A researcher may investigate the relationship between two variables by means of *regression analysis*, which is a branch of statistics. The study of the relationship between two variables may begin with a **scatter diagram,** which is a graph of the ordered pairs of the known data.

The following table gives data collected by a university registrar comparing the grade point averages of graduating high school seniors and their scores on a national test.

GPA, x	3.25	3.00	3.00	3.50	3.50	2.75	2.50	2.50	2.00	2.00	1.50
Test, y	1200	1200	1000	1500	1100	1000	1000	900	800	900	700

The scatter diagram for these data is shown below. Each ordered pair represents the GPA and test score for a student. For example, the ordered pair (2.75, 1000) indicates a student with a GPA of 2.75 who had a test score of 1000.

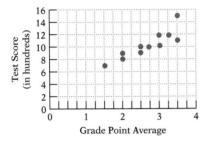

Example 3

A nutritionist collected data on the number of grams of sugar and grams of fiber in one-ounce servings of six popular brands of cereal. The data are recorded in the following table. Graph the scatter diagram for the data.

Sugar, x	6	8	6	5	7	5
Fiber, y	2	1	4	4	2	3

Strategy

Graph the ordered pairs on a rectangular coordinate system where the horizontal axis represents the grams of sugar and the vertical axis represents the grams of fiber.

Solution

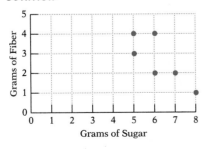

You Try It 3

A sports statistician collected data on the total number of yards gained by a college football team and the number of points scored by the team. The data are recorded in the following table. Graph the scatter diagram for the data.

Yards, x	300	400	350	400	300	450
Points, y	18	24	14	21	21	30

Your Strategy

Your Solution

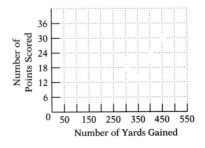

Solution on p. A22

OBJECTIVE C

Solutions of linear equations in two variables

A **solution of an equation in two variables** is an ordered pair (x, y) whose coordinates make the equation a true statement.

Is $(-3, 7)$ a solution of $y = -2x + 1$?

Replace x by -3; replace y by 7. Simplify. Compare the results. If the results are equal, the given ordered pair is a solution. If the results are not equal, the given ordered pair is not a solution.

$$y = -2x + 1$$

7	$-2(-3) + 1$
7	$6 + 1$
$7 = 7$	

Yes, $(-3, 7)$ is a solution of the equation $y = -2x + 1$.

Besides $(-3, 7)$, there are many other ordered-pair solutions of $y = -2x + 1$. For example, $(0, 1)$, $\left(-\frac{3}{2}, 4\right)$, and $(4, -7)$ are also solutions. In general, an equation in two variables has an infinite number of solutions. By choosing any value of x and substituting that value into the equation, we can calculate the corresponding value of y.

Find the ordered pair solution of $y = \frac{2}{3}x - 3$ that corresponds to $x = 6$.

$$y = \frac{2}{3}x - 3$$

Replace x by 6.

$$y = \frac{2}{3}(6) - 3$$

Solve for y.

$$y = 4 - 3$$

$$y = 1$$

The ordered-pair solution is $(6, 1)$.

Example 4

Is $(-3, 2)$ a solution of $y = 2x + 2$?

Solution

$$y = 2x + 2$$

2	$2(-3) + 2$
	$-6 + 2$
$2 \neq -4$	

No, $(-3, 2)$ is not a solution of $y = 2x + 2$.

You Try It 4

Is $(2, -4)$ a solution of $y = -\frac{1}{2}x - 3$?

Your Solution

Solution on p. A22

OBJECTIVE D

Equations of the form $y = mx + b$

The **graph of an equation in two variables** is a graph of the ordered-pair solutions of the equation.

Consider $y = 2x + 1$. Choosing $x = -2, -1, 0, 1,$ and 2 and determining the corresponding values of y produces some of the ordered pairs of the equation. These are recorded in the table at the right. See the graph of the ordered pairs in Figure 1.

x	$y = 2x + 1$	y	(x, y)
-2	$2(-2) + 1$	-3	$(-2, -3)$
-1	$2(-1) + 1$	-1	$(-1, -1)$
0	$2(0) + 1$	1	$(0, 1)$
1	$2(1) + 1$	3	$(1, 3)$
2	$2(2) + 1$	5	$(2, 5)$

Choosing values of x that are not integers produces more ordered pairs to graph, such as $\left(-\dfrac{5}{2}, -4\right)$ and $\left(\dfrac{3}{2}, 4\right)$, as shown in Figure 2. Choosing still other values of x would result in more and more ordered pairs being graphed. The result would be so many dots that the graph would appear as the straight line shown in Figure 3, which is the graph of $y = 2x + 1$.

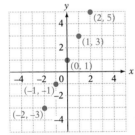

Figure 1

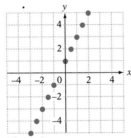

Figure 2

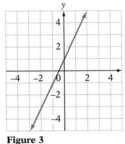

Figure 3

Equations in two variables have characteristic graphs. The equation $y = 2x + 1$ is an example of a *linear equation* because its graph is a straight line.

Linear Equation in Two Variables

Any equation of the form $y = mx + b$, where m and b are constants, is a **linear equation in two variables.** The graph of a linear equation in two variables is a straight line.

Examples of linear equations are shown at the right.

$$y = 2x + 1 \qquad (m = 2, b = 1)$$

$$y = -2x - 5 \quad (m = -2, b = -5)$$

$$y = -\frac{3}{4}x \qquad \left(m = -\frac{3}{4}, b = 0\right)$$

The equation $y = x^2 + 4x + 3$ is not a linear equation in two variables because there is a term with a variable squared. The equation $y = \dfrac{3}{x-4}$ is not a linear equation because a variable occurs in the denominator of the fraction.

To graph a linear equation, choose some values of x and then find the corresponding values of y. Because a straight line is determined by two points, it is sufficient to find only two ordered-pair solutions. However, it is recommended that at least three ordered-pair solutions be used to ensure accuracy.

TAKE NOTE

If the three points you graph do not lie on a straight line, you have made an arithmetic error in calculating a point or you have plotted a point incorrectly.

Graph $y = -\dfrac{3}{2}x + 2$.

This is a linear equation with $m = -\dfrac{3}{2}$ and $b = 2$. Find at least three solutions. Because m is a fraction, choose values of x that will simplify the calculations. We have chosen -2, 0, and 4 for x. (Any values of x could have been selected.)

x	$y = -\dfrac{3}{2}x + 2$	y	(x, y)
-2	$-\dfrac{3}{2}(-2) + 2$	5	$(-2, 5)$
0	$-\dfrac{3}{2}(0) + 2$	2	$(0, 2)$
4	$-\dfrac{3}{2}(4) + 2$	-4	$(4, -4)$

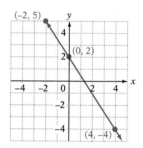

Remember that a graph is a drawing of the ordered-pair solutions of the equation. Therefore, every point on the graph is a solution of the equation and every solution of the equation is a point on the graph.

The graph at the right is the graph of $y = x + 2$. Note that $(-4, -2)$ and $(1, 3)$ are points on the graph and that these points are solutions of $y = x + 2$. The point whose coordinates are $(4, 1)$ is not a point on the graph and is not a solution of the equation.

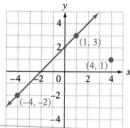

Example 5 Graph $y = 3x - 2$.

Solution

x	y
0	-2
-1	-5
2	4

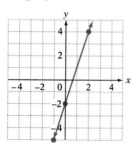

You Try It 5 Graph $y = 3x + 1$.

Your Solution

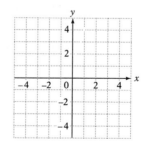

Solution on p. A22

6.5 EXERCISES

OBJECTIVE A

1. Graph the ordered pairs $(-2, 1)$, $(3, -5)$, $(-2, 4)$, and $(0, 3)$.

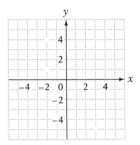

2. Graph the ordered pairs $(5, -1)$, $(-3, -3)$, $(-1, 0)$, and $(1, -1)$.

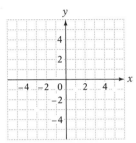

3. Graph the ordered pairs $(0, 0)$, $(0, -5)$, $(-3, 0)$, and $(0, 2)$.

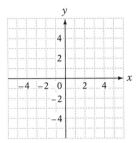

4. Graph the ordered pairs $(-4, 5)$, $(-3, 1)$, $(3, -4)$, and $(5, 0)$.

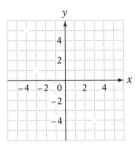

5. Graph the ordered pairs $(-1, 4)$, $(-2, -3)$, $(0, 2)$, and $(4, 0)$.

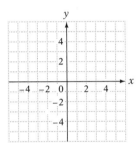

6. Graph the ordered pairs $(5, 2)$, $(-4, -1)$, $(0, 0)$ and $(0, 3)$.

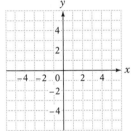

7. Find the coordinates of each point.

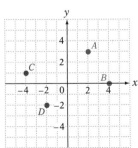

8. Find the coordinates of each point.

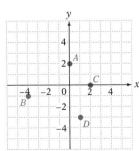

9. Find the coordinates of each point.

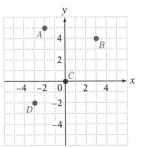

10. Find the coordinates of each point.

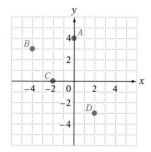

11.
a. Name the abscissas of points *A* and *C*.
b. Name the ordinates of points *B* and *D*.

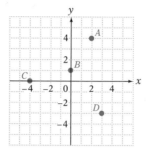

12.
a. Name the abscissas of points *A* and *C*.
b. Name the ordinates of points *B* and *D*.

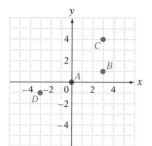

OBJECTIVE B

13. *Employment* The number of years of previous work experience and the salary of a person who completes a Master of Business Administration degree are recorded in the following table. Graph the scatter diagram of these data.

Years experience, *x*	2	0	5	2	3	1
Salary (in hundreds), *y*	30	25	45	35	30	35

14. *Business* The number of miles, in thousands, a rental car is driven and the cost to service that vehicle were recorded by the manager of the rental agency. The data are recorded in the following table. Graph the scatter diagram for the data.

Miles (in thousands), *x*	10	10	5	20	15	5
Cost of service, *y*	100	250	250	500	300	150

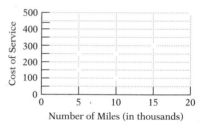

15. *Criminology* Sherlock Holmes solved a crime by recognizing a relationship between the length, in inches, of a person's stride and the height of that person. The data for six people are recorded in the table below. Graph the scatter diagram of these data.

Length of stride, *x*	15	25	20	25	15	30
Height, *y*	60	70	65	65	65	75

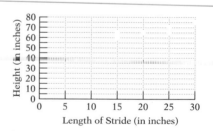

16. *Physiology* An exercise physiologist measured the time, in minutes, a person spent on a treadmill at a fast walk and the heart rate of that person. The results are recorded in the following table. Draw a scatter diagram of these data.

Time on treadmill, *x*	2	10	5	8	6	5
Heart rate, *y*	75	90	80	90	85	85

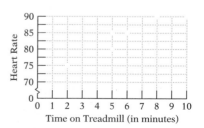

OBJECTIVE C

17. Is $(3, 4)$ a solution of
$y = -x + 7$?

18. Is $(2, -3)$ a solution of
$y = x + 5$?

19. Is $(-1, 2)$ a solution of
$y = \frac{1}{2}x - 1$?

20. Is $(1, -3)$ a solution of
$y = -2x - 1$?

21. Is $(4, 1)$ a solution of
$y = \frac{1}{4}x + 1$?

22. Is $(-5, 3)$ a solution of
$y = -\frac{2}{5}x + 1$?

23. Is $(0, 4)$ a solution of
$y = \frac{3}{4}x + 4$?

24. Is $(-2, 0)$ a solution of
$y = -\frac{1}{2}x - 1$?

25. Is $(0, 0)$ a solution of
$y = 3x + 2$?

26. Is $(0, 0)$ a solution of
$y = -\frac{3}{4}x$?

27. Find the ordered-pair
solution of $y = 3x - 2$
corresponding to $x = 3$.

28. Find the ordered-pair
solution of $y = 4x + 1$
corresponding to $x = -1$.

29. Find the ordered-pair
solution of $y = \frac{2}{3}x - 1$
corresponding to $x = 6$.

30. Find the ordered-pair
solution of $y = \frac{3}{4}x - 2$
corresponding to $x = 4$.

31. Find the ordered-pair
solution of $y = -3x + 1$
corresponding to $x = 0$.

32. Find the ordered-pair
solution of $y = \frac{2}{5}x - 5$
corresponding to $x = 0$.

33. Find the ordered-pair
solution of $y = \frac{2}{5}x + 2$
corresponding to $x = -5$.

34. Find the ordered-pair
solution of $y = -\frac{1}{6}x - 2$
corresponding to $x = 12$.

OBJECTIVE D

Graph.

35. $y = 2x - 4$

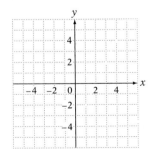

36. $y = 2x + 1$

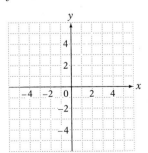

37. $y = -x + 2$

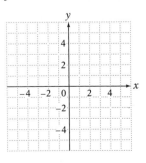

Graph.

38. $y = x + 3$

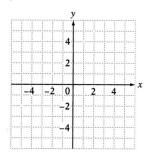

39. $y = x - 3$

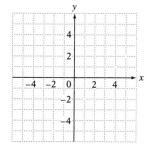

40. $y = x + 2$

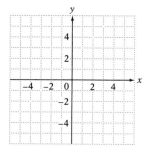

41. $y = -2x + 3$

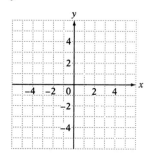

42. $y = -4x + 1$

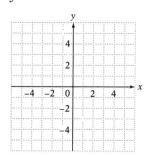

43. $y = -3x + 4$

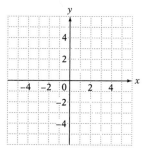

44. $y = 4x - 5$

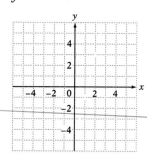

45. $y = 2x - 1$

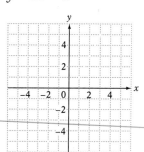

46. $y = 2x$

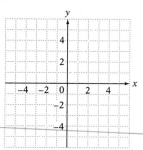

47. $y = 3x$

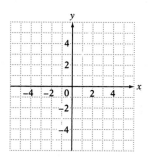

48. $y = \dfrac{3}{2}x$

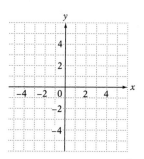

49. $y = \dfrac{1}{3}x$

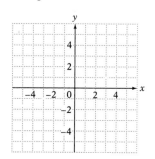

Graph.

50. $y = -\dfrac{5}{2}x$

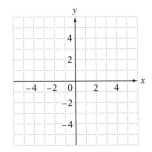

51. $y = -\dfrac{4}{3}x$

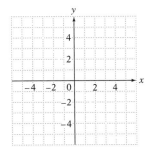

52. $y = \dfrac{2}{3}x + 1$

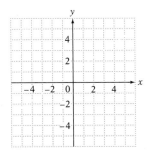

53. $y = \dfrac{3}{2}x - 1$

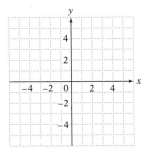

54. $y = \dfrac{1}{4}x + 2$

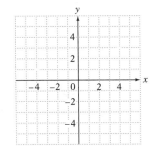

55. $y = \dfrac{2}{5}x - 1$

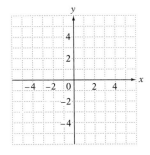

56. $y = -\dfrac{1}{2}x + 3$

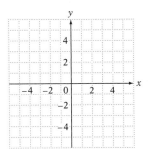

57. $y = -\dfrac{2}{3}x + 1$

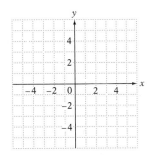

58. $y = -\dfrac{3}{4}x - 3$

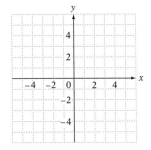

59. $y = -\dfrac{5}{3}x - 2$

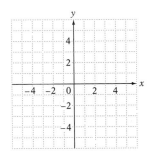

60. $y = \dfrac{1}{2}x - 1$

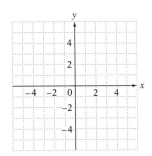

61. $y = \dfrac{5}{2}x - 1$

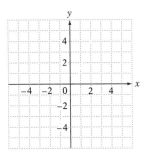

Graph.

62. $y = -\dfrac{1}{4}x + 1$

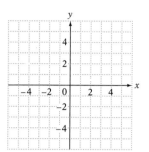

63. $y = x$

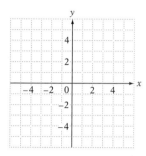

64. $y = -x$

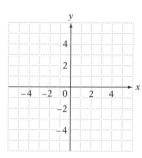

CRITICAL THINKING

65. A computer screen has a coordinate system that is different from the xy-coordinate system we have discussed. In one mode, the origin of the coordinate system is the top left point of the screen, as shown at the right. Plot the points whose coordinates are (200, 400), (0, 100), and (300, 0).

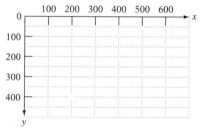

66. (a) What are the coordinates of the point at which the graph of $y = 2x + 1$ crosses the y-axis? (b) What are the coordinates of the point at which the graph of $y = 3x - 6$ crosses the x-axis?

67. Select the correct word and fill in the blank: (a) If $y = 3x - 4$ and the value of x changes from 3 to 4, then the value of y increases/decreases by ? . (b) If $y = -2x + 1$ and the value of x changes from 3 to 4, then the value of y increases/decreases by ? .

68. Graph $y = 2x + b$ for $b = -1, 0, 1,$ and 2. From the graphs, what observations can you make about the graphs as the value of b changes?

69. Graph $y = mx + 1$ for $m = \dfrac{1}{2}, 1, \dfrac{3}{2}, 2,$ and $\dfrac{5}{2}$. From the graphs, what observations can you make about the graphs as the value of m increases?

70. Graph $y = mx + 1$ for $m = -2, -1, 1,$ and 2. From the graphs, what observations can you make about the graphs of the lines when m is negative and when m is positive?

71. There is a coordinate system on Earth that consists of *longitude* and *latitude*. Write a report on how location is determined on the surface of Earth. Include in your report the longitude and latitude coordinates of your school.

PROJECTS IN MATHEMATICS

Modular Arithmetic

Sun	Mon	Tue	Wed	Thu	Fri	Sat
					1	2
3	4	5	6	7	8	9
10	11	12	13	14	15	16
17	18	19	20	21	22	23
24	25	26	27	28	29	30

For the calendar at the left, note that if each of the dates under Friday were divided by 7, the remainder would be the same. For example,

$$15 \div 7 = 2 \text{ with remainder } 1 \qquad 29 \div 7 = 4 \text{ with a remainder } 1$$

Dividing each of the dates under Tuesday by 7 results in a remainder of 5.

$$19 \div 7 = 2 \text{ with remainder } 5$$

The same idea can be applied to each of the seven days of the week. Numbers that have the same remainder when divided by a given number n are said to be **congruent modulo n**. For example, 5, 12, 19, and 26 are congruent modulo 7.

The reason the remainders are the same is that there are 7 days in one week. For the given calendar, Tuesday is on the 5th, 12th (5 + 7), 19th (12 + 7), and 26th (19 + 7).

The notation $a \equiv b \pmod{n}$ is used to denote that a and b have the same remainder when divided by n. For example, $19 \equiv 26 \pmod 7$ because 19 and 26 have the same remainder when divided by 7. The remainder is 5.

For each of the problems below, mark the statement true or false.

1. $34 \equiv 9 \pmod 5$ **2.** $78 \equiv 23 \pmod 9$

3. $16 \equiv 52 \pmod{12}$ **4.** $20 \equiv 0 \pmod{10}$

There are many applications of congruence. The Universal Product Code (UPC) that is used by grocery stores is one application. The UPC identification number consists of 12 digits.

To be a valid UPC number, the following modular equation must be true.

$$3a_1 + a_2 + 3a_3 + a_4 + 3a_5 + a_6 + 3a_7 + a_8 + 3a_9 + a_{10} + 3a_{11} + a_{12} \equiv 0 \pmod{10}$$

Each a in this equation is one of the numbers of the UPC identification number. The first 11 numbers identify the country in which the product was made, the manufacturer, and type of product. The twelfth digit, a_{12}, is called the **check digit** and is chosen so that the equation is true.

ISBN 0-395-75524-7

0 46442 77018 7
/
7 is the
check digit

For example, the first 11 numbers of the UPC shown at the left identify *The 1996 Information Please Almanac* published by Houghton Mifflin Company. Substituting the first 11 numbers into the equation gives

$$3(0) + 4 + 3(6) + 4 + 3(4) + 2 + 3(7) + 7 + 3(0) + 1 + 3(8) + a_{12} \equiv 0 \pmod{10}$$
$$0 + 4 + 18 + 4 + 12 + 2 + 21 + 7 + 0 + 1 + 24 + a_{12} \equiv 0 \pmod{10}$$
$$93 + a_{12} \equiv 0 \pmod{10}$$

To have a valid UPC number, a_{12} is chosen so that the result is congruent to $0 \pmod{10}$. For $93 + a_{12} \equiv 0 \pmod{10}$, $93 + a_{12}$ must be divisible by 10. The single digit that can be added to 93 so that the sum is divisible by 10 is 7. Therefore, $a_{12} = 7$, which is the check digit shown in the UPC number.

If a bookstore ordered this book and incorrectly wrote the number 046443770187, a computer processing the order would be able to determine that there had been a mistake because the number is not congruent to 0 (mod 10) and therefore does not belong to any product.

Another number shown above the bar coding is 0-395-75524-7, which is the International Standard Book Number (ISBN). The first number identifies the book as being published in an English-speaking country. The next group of numbers is the specific publisher, the next group of five digits identifies the particular book, and the last digit is the check digit. In this case, a certain sum must be congruent to 0 (mod 11). The formula for an ISBN is

$$10a_1 + 9a_2 + 8a_3 + 7a_4 + 6a_5 + 5a_6 + 4a_7 + 3a_8 + 2a_9 + a_{10} \equiv 0 \text{ (mod 11)}$$

Use this formula to verify the ISBN for *The 1996 Information Please Almanac.*

Data Analysis

Decide on two quantities that may be related. Here are some examples: height and weight, time studying for a test and the test grade, age of a car and its cost. Collect at least 10 pairs of values and then draw a scatter diagram for the data. Is there any trend? That is, as the values on the horizontal axis increase, do the values on the vertical axis increase or decrease?

CHAPTER SUMMARY

Key Words

An *equation* expresses the equality of two mathematical expressions. A *solution* of an equation is a number that, when substituted for the variable, results in a true equation.

To *solve* an equation means to find a solution of the equation. The goal is to rewrite the equation in the form *variable = constant*.

Some of the words and phrases that translate to equal are: *equals, is, is the same as, amounts to, totals,* and *represents*.

A *rectangular coordinate system* is formed by two numbers lines, one horizontal and one vertical, that intersect at the zero point of each line. The point of intersection is called the *origin*. The number lines that make up a rectangular coordinate system are called *coordinate axes*.

A rectangular coordinate system divides the plane into four regions called *quadrants*.

An *ordered pair* (a, b) is used to locate a point in a rectangular coordinate system. The *coordinates* of a point are the numbers in the ordered pair associated with the point.

Essential Rules

Addition Property of Equations The same number or variable expression can be added to each side of an equation without changing the solution of the equation.

Multiplication Property of Equations Each side of an equation can be multiplied by the same nonzero number without changing the solution of the equation.

An equation of the form $y = mx + b$ is a *linear equation in two variables*.

CHAPTER REVIEW EXERCISES

1. Solve: $z + 5 = 2$

2. Solve: $-8x + 4x = -12$

3. Solve: $7 = 8a - 5$

4. Solve: $7 + a = 0$

5. Solve: $40 = -\dfrac{5}{3}y$

6. Solve: $-\dfrac{3}{8} = \dfrac{4}{5}z$

7. Solve: $9 - 5y = -1$

8. Solve: $-4(2 - x) = x + 9$

9. Solve: $3a + 8 = 12 - 5a$

10. Solve: $12p - 7 = 5p - 21$

11. Solve: $3(2n - 3) = 2n + 3$

12. Solve: $3m = -12$

13. Solve: $4 - 3(2p + 1) = 3p + 11$

14. Solve: $1 + 4(2c - 3) = 3(3c - 5)$

15. Solve: $\dfrac{3x}{4} + 10 = 7$

16. Is $(-10, 0)$ a solution of $y = \dfrac{1}{5}x + 2$?

17. Graph the points whose coordinates are $(-2, 3)$, $(4, 5)$, $(0, -2)$, and $(-4, 0)$.

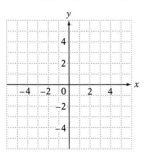

18. Graph $y = 3x - 5$.

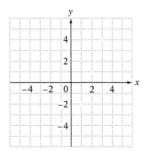

19. Graph $y = -\dfrac{1}{2}x + 3$.

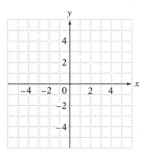

20. Find the ordered pair solution of $y = 4x - 9$ that corresponds to $x = 2$.

Solve.

21. Translate "the difference between seven and the product of five and a number is thirty-seven" into an equation and solve.

22. *Music* A piano wire 24 in. long is cut into two pieces. Twice the length of the shorter piece is equal to the length of the longer piece. Find the length of the longer piece.

23. *Business* The consulting fee for a security specialist was $1,300. This included $250 for supplies and $150 for each hour of consultation. Find the number of hours of consultation.

24. *Landmarks* The height of the Eiffel Tower is 302 m. This is 28 m less than six times the height of the leaning tower of Pisa. Find the height of the leaning tower of Pisa.

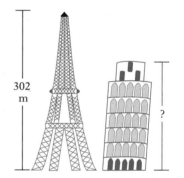

25. *Education* The math midterm scores and the final exam scores for six students are given in the following table. Graph the scatter diagram for these data.

Midterm score, x	90	85	75	80	85	70
Final exam score, y	95	75	80	75	90	70

26. *Physics* A lever is 18 ft long. A force of 25 lb is applied at a distance of 6 ft from the fulcrum. How large a force must be applied to the other end of the lever so that the system will balance? Use the lever system equation $F_1 x = F_2(d - x)$.

27. *Business* A business analyst has determined that the cost per unit for a stereo amplifier is $127 and that the fixed costs per month are $20,000. Find the number of amplifiers produced during a month in which the total cost was $38,669. Use the equation $T = U \cdot N + F$, where T is the total cost, U is the cost per unit, N is the number of units produced, and F is the fixed cost.

CUMULATIVE REVIEW EXERCISES

1. Evaluate $-3ab$ when $a = -2$ and $b = 3$.

2. Simplify: $-3(4p - 7)$

3. Simplify: $\left(\dfrac{2}{3}\right)\left(-\dfrac{9}{8}\right) + \dfrac{3}{4}$

4. Solve: $-\dfrac{2}{3}y = 12$

5. Evaluate $(-b)^3$ when $b = -2$.

6. Evaluate $4xy^2 - 2xy$ when $x = -2$ and $y = 3$.

7. Simplify: $\sqrt{121}$

8. Simplify: $\sqrt{48}$

9. Simplify: $4(3v - 2) - 5(2v - 3)$

10. Simplify: $-4(-3m)$

11. Is -9 a solution of the equation $-5d = -45$?

12. Solve: $5 - 7a = 3 - 5a$

13. Simplify: $6 - 2(7z - 3) + 4z$

14. Evaluate $\dfrac{a^2 + b^2}{2ab}$ when $a = -2$ and $b = -1$.

15. Solve: $8z - 9 = 3$

16. Simplify: $(2m^2n^5)^5$

17. Multiply: $-3a^3(2a^2 + 3ab - 4b^2)$

18. Multiply: $(2x - 3)(3x + 1)$

19. Simplify: 2^{-4}

20. Simplify: $\dfrac{x^8}{x^2}$

21. Simplify: $(-5x^3y)(-3x^5y^2)$

22. Solve: $5 - 3(2x - 8) = -2(1 - x)$

23. Graph $y = \dfrac{5}{3}x + 1$.

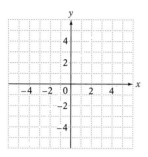

24. Graph $y = -\dfrac{2}{5}x$.

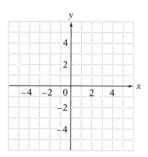

25. Write 3.5×10^{-8} in decimal notation.

Solve.

26. Translate "the product of five and the sum of a number and two" into a variable expression. Then simplify the variable expression.

27. *Physics*   Find the time it takes a falling object to increase its speed from 50 ft/s to 98 ft/s. Use the equation $v = v_0 + 32t$, where v is the final velocity, v_0 is the initial velocity, and t is the time it takes for the object to fall.

28. *Aviation*   A turboprop plane can fly at one-half the speed of sound. Express the speed of the turboprop plane in terms of the speed of sound.

29. *Sports*   The figure at the right shows the top NBA stars' salaries in 1995–1996. Find the total paid in salaries to all five players listed.

30. *Finances*   A homeowner's mortgage payment for one month for principal and interest was $949. The principal payment was $204 less than the interest payment. Find the amount of the interest payment.

31. *Geography*   The Aleutian Trench in the Pacific Ocean is 8,100 m deep. Each story of an average skyscraper is 4.2 m tall. How many stories, to the nearest whole number, would a skyscraper as tall as the Aleutian Trench have?

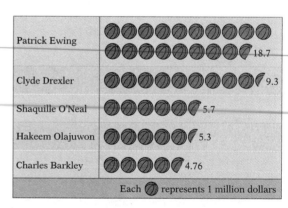

Top NBA Stars' Salaries (in millions), 1995–1996
Source: Forbes Magazine

32. *Charities*   A donation of $12,000 is given to two charities. One charity received twice the amount of the other charity. How much did each charity receive?

7

Ratio and Proportion

FOCUS On Problem Solving

Problems in mathematics or real life involve a question, or a need, and information or circumstances relating to that need. Solving problems in the sciences usually involves a question, observations, and measurements of some kind.

One of the challenges of problem solving in the sciences is to separate the relevant information about a problem from other information. Following is an example from the physical sciences in which some relevant information was omitted.

Hooke's Law states that the distance that a weight will stretch a spring is directly proportional to the weight on the spring. That is, $d = kF$, where d is the distance the spring is stretched and F is the force. In an experiment to verify this law, some physics students were continually getting the wrong results. Finally, the instructor discovered that the heat produced when the lights were turned on was affecting the experiment. In this case, some relevant information affecting the experiment was being omitted.

A lawyer drove 8 mi to the train station. After a 35-minute ride of 18 mi, the lawyer walked 10 min to the office. Find the total time it took the lawyer to get to work.

From this situation, answer the following before reading on.

a. What is asked for?
b. Is there enough information to answer the question?
c. Is information given that is not necessary?

Here are the answers.

a. We want the total time for the lawyer to get to work.

b. No. We do not know the time it takes to get to the train station.
c. Yes. The distance to the train station and the distance of the train ride are not necessary to answer the question.

In the following problems:

a. What is asked for?
b. Is there enough information to answer the question?
c. Is there information that is not needed?

1. A customer bought 6 boxes of strawberries and paid with a $20 bill. What was the change?

2. A board is cut into two pieces. One piece is 3 ft longer than the other piece. What is the length of the original board?

3. A family rented a car for their vacation and drove 680 mi. The cost of the rental car was $21 per day with 150 free miles per day and $.15 for each mile over 150. How many miles did the family drive per day?

4. An investor bought 8 acres of land for $80,000. One and one-half acres were set aside for a park, and the remainder was developed into one-half-acre lots. How many lots were available for sale?

5. You wrote checks of $43.67, $122.88, and $432.22 after making a deposit of $768.55. How much do you have left in your checking account?

SECTION 7.1   Ratios and Rates

OBJECTIVE **A**

Ratios and rates

In previous work, we have used quantities with units, such as 12 ft, 3 h, 2 shirts, and 15 acres. A **ratio** is the quotient or comparison of two quantities with the *same* unit.

We can compare the measure of 3 ft to the measure of 8 ft by writing a quotient.

$$\frac{3 \text{ ft}}{8 \text{ ft}} = \frac{3}{8} \qquad 3 \text{ ft is } \frac{3}{8} \text{ of 8 ft.}$$

This quotient or comparison of two numbers can be written in three ways:

1. As a fraction: $\frac{3}{8}$

2. As two numbers separated by a colon: $3 : 8$

3. Or as two numbers separated by the word TO: 3 TO 8

The ratio of 15 mi to 45 mi is written as

$$\frac{15 \text{ mi}}{45 \text{ mi}} = \frac{15}{45} = \frac{1}{3}$$

A ratio is in **simplest form** when the two numbers do not have a common factor. The units are not written in a ratio.

The comparison of two quantities with *different* units is called a **rate**. A rate is written as a fraction. The rate of 200 mi to 6 h is written as

$$\frac{200 \text{ mi}}{6 \text{ h}} = \frac{100 \text{ mi}}{3 \text{ h}}$$

A rate is in **simplest form** when the numbers have no common factors. The units are written as part of the rate.

A **unit rate** is a rate in which the denominator is 1. **To find a unit rate, divide the number in the numerator of the rate by the number in the denominator of the rate.** A unit rate is often written in decimal form.

> A student received $39 for working 6 h in the bookstore. Find the wage for one hour (the unit rate).
>
> The rate is written with a fraction bar as $\frac{\$39}{6 \text{ h}}$.
>
> Divide the number in the numerator of the rate
> (39) by the number in the denominator (6). $39 \div 6 = 6.5$
>
> The unit rate is written as $\frac{\$6.50}{1 \text{ h}} = \$6.50/\text{h}$, which is read "$6.50 per hour."

> ◆◆
> **POINT OF INTEREST**
>
> It is believed that billiards was invented in France during the reign of Louis XI (1423–1483). In the United States, the standard billiard table is 4 ft 6 in. by 9 ft. This is a ratio of 1 : 2. The same ratio holds for carom and snooker tables, which are 5 ft by 10 ft.

Example 1 Write the comparison of 12 to 8 as a ratio in simplest form using a colon and the word TO.

Solution $12:8 = 3:2$
$12 \text{ TO } 8 = 3 \text{ TO } 2$

You Try It 1 Write the comparison of 12 to 20 as a ratio in simplest form using a colon and the word TO.

Your Solution

Example 2 Write "12 hits in 26 times at bat" as a rate in simplest form.

Solution $\dfrac{12 \text{ hits}}{26 \text{ at-bats}} = \dfrac{6 \text{ hits}}{13 \text{ at-bats}}$

You Try It 2 Write "20 bags of grass seed for 12 acres" as a rate in simplest form.

Your Solution

Example 3 Write "285 mi in 5 h" as a unit rate in simplest form.

Solution $\dfrac{285 \text{ mi}}{5 \text{ h}} = 57 \text{ mph}$

You Try It 3 Write "$4.48 for 3.5 lb" as a unit rate in simplest form.

Your Solution

Solutions on p. A22

OBJECTIVE B
Dimensional analysis

In solving application problems, it may be useful to include the units in order to organize the problem so that the answer is in the proper units. Using units to organize and check the correctness of an application is called **dimensional analysis**. We will use conversion of units, as well as the operations of multiplying units and dividing units, in applying dimensional analysis to application problems.

The following example converts the unit miles to feet. The equivalent measures $1 \text{ mi} = 5{,}280 \text{ ft}$ are used to form the following rates, which are called conversion factors: $\dfrac{1 \text{ mi}}{5{,}280 \text{ ft}}$ or $\dfrac{5{,}280 \text{ ft}}{1 \text{ mi}}$. Because $1 \text{ mi} = 5{,}280 \text{ ft}$, both of the conversion factors $\dfrac{1 \text{ mi}}{5{,}280 \text{ ft}}$ and $\dfrac{5{,}280 \text{ ft}}{1 \text{ mi}}$ are equal to 1.

To convert 3 mi to feet, multiply 3 mi by the conversion factor $\dfrac{5{,}280 \text{ ft}}{1 \text{ mi}}$.

$$3 \text{ mi} = 3 \text{ mi} \cdot \mathbf{1} = \dfrac{3 \text{ mi}}{1} \cdot \dfrac{5{,}280 \text{ ft}}{1 \text{ mi}} = \dfrac{3 \text{ mi} \cdot 5{,}280 \text{ ft}}{1 \text{ mi}} = 3 \cdot 5{,}280 \text{ ft} = 15{,}840 \text{ ft}$$

There are two important points in the above illustration. First, you can think of dividing the numerator and denominator by the common unit mile just as you would divide the numerator and denominator of a fraction by a common factor. Second, the conversion factor $\dfrac{5{,}280 \text{ ft}}{1 \text{ mi}}$ is equal to one, and multiplying an expression by one does not change the value of the expression.

In the example below, we convert 1 h to seconds. Because $60 \text{ min} = 1 \text{ h}$ and $60 \text{ s} = 1 \text{ min}$, we use the conversion factors $\frac{60 \text{ min}}{1 \text{ h}}$ and $\frac{60 \text{ s}}{1 \text{ min}}$.

$$1 \text{ h} = \frac{1 \cancel{h}}{1} \cdot \frac{60 \cancel{min}}{1 \cancel{h}} \cdot \frac{60 \text{ s}}{1 \cancel{min}} = 60 \cdot 60 \text{ s} = 3{,}600 \text{ s}$$

The conversion factor $\frac{60 \text{ min}}{1 \text{ h}}$ was used rather than $\frac{1 \text{ h}}{60 \text{ min}}$ so that we could divide by the unit "hours." The conversion factor $\frac{60 \text{ s}}{1 \text{ min}}$ was used rather than $\frac{1 \text{ min}}{60 \text{ s}}$ so that we could divide by the unit "minutes."

Since there are 3,600 s in 1 h, we can write the conversion factor $\frac{3{,}600 \text{ s}}{1 \text{ h}}$. This conversion factor will be used in this section.

Other conversion factors that we will use in this section can be found from the equivalent measures listed in the table on page 406.

TAKE NOTE

If you are unfamiliar with the metric system of measurement, you will find it helpful to study Appendix A on page A1 at the back of this textbook.

Convert 800 yd to miles.

The unit "yards" must divide.
Use the conversion factor $\frac{3 \text{ ft}}{1 \text{ yd}}$.
The unit "feet" must divide.
Use the conversion factor $\frac{1 \text{ mi}}{5{,}280 \text{ ft}}$.

800 yd

$$= \frac{800 \cancel{yd}}{1} \cdot \frac{3 \cancel{ft}}{1 \cancel{yd}} \cdot \frac{1 \text{ mi}}{5{,}280 \cancel{ft}}$$

$$= \frac{800 \cdot 3 \text{ mi}}{5{,}280} = 0.\overline{45} \text{ mi}$$

Convert 88 ft/s to miles per hour.

This example requires two conversions. We must convert feet to miles and seconds to hours.

The unit "feet" must divide.
Use the conversion factor $\frac{1 \text{ mi}}{5{,}280 \text{ ft}}$.

$$88 \frac{\text{ft}}{\text{s}} = \frac{88 \cancel{ft}}{1 \text{ s}} \cdot \frac{1 \text{ mi}}{5{,}280 \cancel{ft}} = \frac{88 \text{ mi}}{5{,}280 \text{ s}}$$

To convert seconds to hours, use the conversion factor $\frac{3{,}600 \text{ s}}{1 \text{ h}}$.

$$= \frac{88 \text{ mi}}{5{,}280 \cancel{s}} \cdot \frac{3{,}600 \cancel{s}}{1 \text{ h}}$$

$$= \frac{88 \text{ mi} \cdot 3{,}600}{5{,}280 \text{ h}} = \frac{60 \text{ mi}}{\text{h}}$$

The complete conversion would look like this:

$$88 \frac{\text{ft}}{\text{s}} = \frac{88 \cancel{ft}}{1 \cancel{s}} \cdot \frac{1 \text{ mi}}{5{,}280 \cancel{ft}} \cdot \frac{3{,}600 \cancel{s}}{1 \text{ h}} = \frac{88 \cdot 3{,}600 \text{ mi}}{5{,}280 \text{ h}} = \frac{60 \text{ mi}}{\text{h}} = 60 \text{ mph}$$

Table of Equivalent Measures

	U.S. System	Metric System	Conversion Between U.S. and Metric Systems
Length	1 ft = 12 in.	1 m = 1,000 mm	1 in. ≈ 2.54 cm
	1 yd = 36 in.	1 m = 100 cm	1 m ≈ 3.28 ft
	1 yd = 3 ft	1 km = 1,000 m	1 m ≈ 1.09 yd
	1 mi = 5,280 ft		1 mi ≈ 1.61 km
Weight or Mass	1 lb = 16 oz	1 g = 1,000 mg	1 oz ≈ 28.35 g
	1 ton = 2,000 lb	1 kg = 1,000 g	1 lb ≈ 454 g
			1 kg ≈ 2.2 lb
Area	1 ft^2 = 144 in^2		1 in^2 ≈ 6.45 cm^2
	1 yd^2 = 9 ft^2		1 m^2 ≈ 1.196 yd^2
	1 acre = 43,560 ft^2		
	1 mi^2 = 640 acres		
Liquid Measure or Volume	1 cup = 8 fl oz	1 ml = 1 cm^3	1 in^3 ≈ 16.39 ml
	1 pt = 2 c	1 L = 1,000 ml	1 L ≈ 1.06 qt
	1 qt = 2 pt	1 kl = 1,000 L	1 gal ≈ 3.79 L
	1 gal = 4 qt		
	1 gal = 231 in^3		
	1 yd^3 = 27 ft^3		

Example 4

Convert 0.5 acre to square feet.

Solution

Use the conversion factor $\dfrac{43{,}560 \text{ ft}^2}{1 \text{ acre}}$.

$$0.5 \text{ acre} = \frac{0.5 \text{ acre}}{1} \cdot \frac{43{,}560 \text{ ft}^2}{1 \text{ acre}}$$

$$= 0.5 \cdot 43{,}560 \text{ ft}^2 = 21{,}780 \text{ ft}^2$$

You Try It 4

Convert 1.5 mi^2 to acres.

Your Solution

Example 5

Convert 150 lb to kilograms. Round to the nearest tenth.

Solution

Use the conversion factor $\dfrac{1 \text{ kg}}{2.2 \text{ lb}}$.

$$150 \text{ lb} = \frac{150 \text{ lb}}{1} \cdot \frac{1 \text{ kg}}{2.2 \text{ lb}}$$

$$= \frac{150 \text{ kg}}{2.2} \approx 68.2 \text{ lb}$$

You Try It 5

Convert 10 L to gallons. Round to the nearest tenth.

Your Solution

Example 6

Convert 8 m/s to kilometers per hour.

Solution

Use the conversion factors $\dfrac{1 \text{ km}}{1{,}000 \text{ m}}$ and $\dfrac{3{,}600 \text{ s}}{1 \text{ h}}$.

$$8 \text{ m/s} = \frac{8 \text{ m}}{1 \text{ s}} = \frac{8 \text{ m}}{1 \text{ s}} \cdot \frac{1 \text{ km}}{1{,}000 \text{ m}} \cdot \frac{3{,}600 \text{ s}}{1 \text{ h}}$$

$$= \frac{8 \cdot 3{,}600 \text{ km}}{1{,}000 \text{ h}} = 28.8 \text{ km/h}$$

You Try It 6

Convert 90 km/h to meters per second.

Your Solution

Solutions on p. A22

Objective C
Applications

In solving these application problems, we will keep the units throughout the solution as we work through the arithmetic. Note that, just as in Objective B above, conversion factors are used to set up the units before the arithmetic is performed.

In 1980, a horse named Fiddle Isle ran a 1.5-mile race in 2 min 23 s. Find Fiddle Isle's average speed for that race in miles per hour. Round to the nearest tenth.

Strategy   To find the average speed:

▶ Convert 2 min 23 s to decimal minutes.

▶ Use the formula $r = \frac{d}{t}$, where r is the speed, d is the distance, and t is the time. Use the conversion factor $\frac{60 \text{ min}}{1 \text{ h}}$.

Solution   2 min 23 s $= 2\frac{23}{60}$ min ≈ 2.383 min

$$r = \frac{d}{t} = \frac{1.5 \text{ mi}}{2.383 \text{ min}} = \frac{1.5 \text{ mi}}{2.383 \text{ min}} \cdot \frac{60 \text{ min}}{1 \text{ h}}$$

$$= \frac{90 \text{ mi}}{2.383 \text{ h}} \approx 37.8 \text{ mph}$$

Fiddle Isle's average speed was 37.8 mph.

Example 7

A carpet is to be placed in a room that is 20 ft wide and 30 ft long. At $18.50 per square yard, how much will it cost to carpet the area? Use the formula $A = LW$, where A is the area, L is the length, and W is the width, to find the area of the room.

Strategy

To find the cost of the carpet:

▶ Use the formula $A = LW$ to find the area.

▶ Use the conversion factor $\frac{1 \text{ yd}^2}{9 \text{ ft}^2}$ to find the area in square yards.

▶ Multiply by $\frac{\$18.50}{\text{yd}^2}$ to find the cost.

Solution

$A = LW = 30 \text{ ft} \cdot 20 \text{ ft} = 600 \text{ ft}^2$ ▶ ft · ft = ft²

$600 \text{ ft}^2 = \frac{600 \text{ ft}^2}{1} \cdot \frac{1 \text{ yd}^2}{9 \text{ ft}^2} = \frac{600 \text{ yd}^2}{9} = \frac{200 \text{ yd}^2}{3}$

$\text{Cost} = \frac{200 \text{ yd}^2}{3} \cdot \frac{\$18.50}{\text{yd}^2} \approx \$1,233.33$

The cost of the carpet is $1,233.33.

You Try It 7

Find the number of gallons of water in a fish tank that is 36 in. long and 24 in. wide and is filled to a depth of 16 in. Use the formula $V = LWH$, where V is the volume, L is the length, W is the width, and H is the depth of the water. Round to the nearest tenth.

Your Strategy

Your Solution

Solution on p. A22

7.1 EXERCISES

OBJECTIVE A

Write the comparison as a ratio in simplest form using a fraction, a colon, and the word TO.

1. 16 in. to 24 in.

2. 8 lb to 60 lb

3. 9 h to 24 h

4. $55 to $150

5. 9 ft to 2 ft

6. 50 min to 6 min

The figure at the right shows the lung capacity of inactive vs. athletic 45-year-olds. Use this graph for Exercise 7.

7. *Physiology* Write the comparison of the lung capacity of an inactive male to that of an athletic male as a ratio in simplest form using a fraction, a colon, and the word TO.

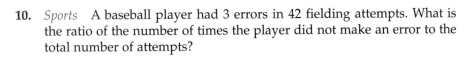

Lung Capacity (in milliliters of oxygen per kilogram of body weight per minute)

Write as a ratio in simplest form using a fraction.

8. *Construction* The cost of building a patio cover was $600 for labor and $1,600 for materials. Find the ratio of the cost of materials to the cost of labor.

9. *Mechanics* Find the ratio of two meshed gears if one gear has 24 teeth and the other gear has 36 teeth.

10. *Sports* A baseball player had 3 errors in 42 fielding attempts. What is the ratio of the number of times the player did not make an error to the total number of attempts?

11. *Sports* A basketball team won 18 games and lost 8 games during the season. What is the ratio of the number of games won to the total number of games?

Write as a rate in simplest form.

12. $65 for 3 shirts

13. 150 mi in 6 h

14. $76 for 8 h work

15. $3.28 for 6 candy bars

16. 252 avocado trees on 6 acres

17. 9 children in 4 families

Write as a unit rate.

18. $460 earned for 40 h of work

19. $38,700 earned in 12 months

20. 387.8 mi in 7 h

21. 364.8 mi on 9.5 gal of gas

22. $9.54 for 4.5 lb

23. $3.36 for 15 oz

Solve.

24. *Travel* An airplane flew 1,155 mi in 2.5 h. Find the rate of travel. Use the formula $r = \frac{d}{t}$, where r is the speed, d is the distance, and t is the time.

25. *Travel* A ship sailed 513.6 mi in 24 h. Find the rate of travel. Use the formula $r = \frac{d}{t}$, where r is the speed, d is the distance, and t is the time.

26. *Sports* A professional basketball team scored 8,432 points in an 82-game season. To the nearest tenth, find the average number of points scored per game.

27. *Sports* A softball player had 67 hits in 202 times at bat. Find the player's batting average. Batting average is the ratio of the number of hits to the total number of times at bat. Round to the nearest thousandth.

28. *Investments* An investor purchased 100 shares of stock for $2,500. One year later the investor sold the stock for $3,200. What was the investor's profit per share?

OBJECTIVE B

Perform the following conversions. Round to the nearest hundredth.

29. Convert 122 oz to pounds.

30. Convert 3,500 lb to tons.

31. Convert 32 c to gallons.

32. Convert 3.4 gal to quarts.

33. Convert 3.2 mi to feet.

34. Convert 12,000 ft to miles.

35. Convert 0.25 acre to square feet.

36. Convert the 100-yd dash to meters.

Perform the following conversions. Round to the nearest tenth.

37. Find the weight in pounds of a 98-kilogram person.

38. Find the weight in kilograms of a 135-pound person.

39. Find the width of 35-mm film in inches.

40. Find the distance of the 1,500-meter race in feet.

41. Convert 66 mph to feet per second.

42. Convert 100 ft/s to miles per hour.

43. Convert 65 mph to kilometers per hour.

44. Convert 100 yd in 9.6 s to miles per hour.

45. The cost of gasoline is $1.47/gal. Find the cost per liter.

46. Gasoline costs 38.5¢/L. Find the cost per gallon.

47. Express 80 km/h in miles per hour.

48. Express 30 m/s in feet per second.

OBJECTIVE C

Solve. Round to the nearest tenth.

49. *Catering*   Two cups of coffee are prepared for each invited guest at a reception. How many gallons of coffee would be prepared for a guest list of 200?

50. *Catering*   Twelve ounces of punch are prepared for each guest invited to a reception. How many quarts of punch would be prepared for a guest list of 40?

51. *Interior Decorating*   A carpet is to be placed in a meeting hall that is 36 ft wide and 80 ft long. At $17.25 per square yard, how much will it cost to carpet the meeting hall? Use the formula $A = LW$, where A is the area, L is the length, and W is the width.

52. *Real Estate*   A one-fourth acre commercial lot is on sale for $2.15 per square foot. Find the sale price of the commercial lot.

53. *Real Estate*   A 0.75-acre industrial parcel was sold for $98,010. Find the parcel's price per square foot.

54. *Construction*   A new driveway required 900 ft³ of concrete. Concrete is ordered by the cubic yard. How much concrete is needed for the driveway? If concrete can be ordered only in increments of 10 yd³, how many cubic yards must be ordered?

Solve. Round to the nearest tenth.

55. *Astronomy* (a) The distance around Earth is 41,000 km. What is the distance, in miles, around Earth? (b) The distance from Earth to the sun is 93,000,000 mi. What is the distance, in kilometers, from Earth to the sun?

56. *Astronomy* Solar wind, made up of hot gases, travels from the sun toward Earth at a speed of 1 million mph. Find the speed, in kilometers per hour, at which solar winds approach Earth.

57. *Sports* The largest Lake Trout ever caught in the state of Utah weighed 51 lb 8 oz. Find the trout's weight in kilograms.

58. *Aquariums* Find the number of gallons of water in a fish tank that is 24 in. long and 18 in. wide and is filled to a depth of 12 in. Use the formula $V = LWH$, where V is the volume, L is the length, W is the width, and H is the height.

59. *Race Car Driving* A piston-engined dragster traveled 440 yd in 4.936 s at Ennis, Texas, on October 9, 1988. Find the average speed of the dragster in miles per hour. Use the formula $r = \frac{d}{t}$, where r is the rate, d is the distance, and t is the time.

60. *Sonar* The Marianas Trench in the Pacific Ocean is the deepest part of the ocean. The depth is 6.85 mi. The speed of sound under water is 4,700 ft/s. Find the time it takes a sound signal to travel from the surface to the bottom of the Marianas Trench and back. Use the formula $r = \frac{d}{t}$, where r is the rate, d is the distance, and t is the time.

CRITICAL THINKING

61. A bank uses the ratio of a borrower's total monthly debts to the borrower's total monthly income to determine the maximum monthly payment for a potential homeowner. This ratio is called the debt–equity ratio. Compute the debt–equity ratio for the potential homeowner whose debts and income are shown at the right. Write the ratio as a decimal to the nearest hundredth.

Income	Debts
$2,700	$1,125
350	320
500	60
	135

62. a. Is the square of the ratio $\frac{a}{b}$ equal to the original ratio?

    b. Is the sum of a ratio and itself equal to the original ratio?

63. Write a paragraph explaining the concept of scaling. The description on a box containing a model airplane states that the scale of the model is 1 : 50 of the actual plane. Discuss the meaning of 1 : 50. The measurement of the propeller of the model is 1 in. What is the measurement of a propeller of the actual plane?

64. Write a paragraph explaining the use of ratios by Eratosthenes to measure the circumference of Earth.

SECTION 7.2 Proportion

OBJECTIVE A
Proportion

A **proportion** is the equality of two ratios or rates.

The equality $\dfrac{250 \text{ mi}}{5 \text{ h}} = \dfrac{50 \text{ mi}}{1 \text{ h}}$ is a proportion.

> ### Definition of Proportion
>
> If $\dfrac{a}{b}$ and $\dfrac{c}{d}$ are equal ratios or rates, then $\dfrac{a}{b} = \dfrac{c}{d}$ is a proportion.

POINT OF INTEREST

Proportions were studied by the earliest mathematicians. Clay tablets uncovered by archeologists show evidence of proportions in Egyptian and Babylonian cultures dating from 1800 B.C.

Each of the four numbers in a proportion is called a **term**. Each term is numbered according to the following diagram.

$$\begin{array}{lcl} \text{first term} \longleftarrow & \dfrac{a}{b} = \dfrac{c}{d} & \longrightarrow \text{third term} \\ \text{second term} \longleftarrow & & \longrightarrow \text{fourth term} \end{array}$$

The first and fourth terms of the proportion are called the **extremes** and the second and third terms are called the **means**.

If we multiply the proportion by the least common multiple of the denominators, we obtain the following result:

$$\frac{a}{b} = \frac{c}{d}$$

$$bd\left(\frac{a}{b}\right) = bd\left(\frac{c}{d}\right)$$

$$ad = bc \qquad ad \text{ is the product of the extremes.}$$
$$bc \text{ is the product of the means.}$$

In any true proportion, **the product of the means equals the product of the extremes.** This is sometimes phrased as "the cross products are equal."

In the true proportion $\dfrac{3}{4} = \dfrac{9}{12}$, the cross products are equal.

$$\dfrac{3}{4} \times \dfrac{9}{12} \longrightarrow \begin{array}{l} 4 \cdot 9 = 36 \longleftarrow \text{ Product of the means} \\ 3 \cdot 12 = 36 \longleftarrow \text{ Product of the extremes} \end{array}$$

Determine whether the proportion $\dfrac{47 \text{ mi}}{2 \text{ gal}} = \dfrac{304 \text{ mi}}{13 \text{ gal}}$ is a true proportion.

The product of the means: The product of the extremes:

$$2 \cdot 304 = 608 \qquad\qquad 47 \cdot 13 = 611$$

The proportion is not true because $608 \neq 611$.

When three terms of a proportion are given, the fourth term can be found.

Solve: $\frac{n}{5} = \frac{9}{16}$

Find the number (n) that will make the proportion true.

$$\frac{n}{5} = \frac{9}{16}$$

The product of the means equals the product of the extremes. Solve for n.

$$5 \cdot 9 = n \cdot 16$$

$$45 = 16n$$

$$\frac{45}{16} = \frac{16n}{16}$$

$$2.8125 = n$$

Example 1 Determine whether $\frac{15}{3} = \frac{90}{18}$ is a true proportion.

Solution $\frac{15}{3} \diagdown\diagup \frac{90}{18} \longrightarrow 3 \cdot 90 = 270$
$\phantom{\frac{15}{3}} \longrightarrow 15 \cdot 18 = 270$

The product of the means equals the product of the extremes.

The proportion is true.

You Try It 1 Determine whether $\frac{50 \text{ mi}}{3 \text{ gal}} = \frac{250 \text{ mi}}{12 \text{ gal}}$ is a true proportion.

Your Solution

Example 2 Solve: $\frac{5}{9} = \frac{x}{45}$

Solution $9 \cdot x = 5 \cdot 45$

$$9x = 225$$

$$\frac{9x}{9} = \frac{225}{9}$$

$$x = 25$$

You Try It 2 Solve: $\frac{7}{12} = \frac{42}{x}$

Your Solution

Example 3 Solve: $\frac{6}{n} = \frac{45}{124}$. Round to the nearest tenth.

Solution $n \cdot 45 = 6 \cdot 124$

$$45n = 744$$

$$\frac{45n}{45} = \frac{744}{45}$$

$$n \approx 16.5$$

You Try It 3 Solve: $\frac{6}{n} = \frac{3}{321}$.

Your Solution

Solutions on pp. A22–A23

Example 4   Solve: $\dfrac{x+2}{3} = \dfrac{7}{8}$

You Try It 4   Solve: $\dfrac{4}{5} = \dfrac{3}{x-3}$

Solution $\dfrac{x+2}{3} = \dfrac{7}{8}$

Your Solution

$$3 \cdot 7 = (x+2)8$$

$$21 = 8x + 16$$

$$5 = 8x$$

$$\dfrac{5}{8} = x$$

Solution on p. A23 ||||||

OBJECTIVE B
Applications

Proportions are useful in many types of application problems. In recipes, proportions are used when a larger batch of ingredients is used than the recipe calls for. In mixing cement, the amounts of cement, sand, and rock are mixed in the same ratio. A map is drawn on a proportional basis, such as one inch representing 50 mi.

In setting up a proportion, keep the same units in the numerators and the same units in the denominators. For example, if feet is in the numerator on one side of the proportion, then feet must be in the numerator on the other side of the proportion.

A customer sees an ad in a newspaper advertising 2 tires for $82.50. The customer wants to buy 5 tires and use one for the spare. How much will the 5 tires cost?

Use a proportion with the cost of the 5 tires as the unknown.

$$\frac{2 \text{ tires}}{\$82.50} = \frac{5 \text{ tires}}{\text{cost}}$$

$$\$82.50 \cdot 5 \text{ tires} = 2 \text{ tires} \cdot \text{cost}$$

Note that the unit "tires" will cancel.

$$\frac{\$82.50 \cdot 5 \text{ tires}}{2 \text{ tires}} = \text{cost}$$

$$\$206.25 = \text{cost}$$

The 5 tires will cost $206.25.

TAKE NOTE

It is also correct to write the proportion with the costs in the numerators and the number of tires in the denominators. The solution will be the same.

Example 5

During a Friday, the ratio of stocks declining in price to those advancing was 5 to 3. If 450,000 shares advanced, how many shares declined on that day?

Strategy

To find the number of shares declining in price, write and solve a proportion using n to represent the number of shares declining in price.

Solution

$$\frac{5 \text{ (declining)}}{3 \text{ (advancing)}} = \frac{n \text{ shares declining}}{450{,}000 \text{ shares advancing}}$$

$$3n = 5 \cdot 450{,}000$$

$$3n = 2{,}250{,}000$$

$$\frac{3n}{3} = \frac{2{,}250{,}000}{3}$$

$$n = 750{,}000$$

750,000 shares declined in price.

Example 6

From previous experience, a manufacturer knows that in an average production of 5,000 calculators, 40 will be defective. What number of defective calculators can be expected from a run of 45,000 calculators?

Strategy

To find the number of defective calculators, write and solve a proportion using n to represent the number of defective calculators.

Solution

$$\frac{40 \text{ defective calculators}}{5{,}000 \text{ calculators}} = \frac{n \text{ defective calculators}}{45{,}000 \text{ calculators}}$$

$$5{,}000 \cdot n = 40 \cdot 45{,}000$$

$$5{,}000n = 1{,}800{,}000$$

$$\frac{5{,}000n}{5{,}000} = \frac{1{,}800{,}000}{5000}$$

$$n = 360$$

The manufacturer can expect 360 defective calculators.

You Try It 5

An automobile can travel 396 mi on 11 gal of gas. At the same rate, how many gallons of gas would be necessary to travel 832 mi? Round to the nearest tenth.

Your Strategy

Your Solution

You Try It 6

An automobile recall was based on tests that showed 15 transmission defects in 1,200 cars. At this rate, how many defective transmissions will be found in 120,000 cars?

Your Strategy

Your Solution

Solutions on p. A23

7.2 EXERCISES

OBJECTIVE A

Determine whether the proportion is true or not true.

1. $\dfrac{27}{8} = \dfrac{9}{4}$

2. $\dfrac{3}{18} = \dfrac{4}{19}$

3. $\dfrac{45}{135} = \dfrac{3}{9}$

4. $\dfrac{3}{4} = \dfrac{54}{72}$

5. $\dfrac{16}{3} = \dfrac{48}{9}$

6. $\dfrac{15}{5} = \dfrac{3}{1}$

7. $\dfrac{6 \text{ min}}{5 \text{ cents}} = \dfrac{30 \text{ min}}{25 \text{ cents}}$

8. $\dfrac{7 \text{ tiles}}{4 \text{ ft}} = \dfrac{42 \text{ tiles}}{20 \text{ ft}}$

9. $\dfrac{15 \text{ ft}}{3 \text{ yd}} = \dfrac{90 \text{ ft}}{18 \text{ yd}}$

10. $\dfrac{\$65}{5 \text{ days}} = \dfrac{\$26}{2 \text{ days}}$

11. $\dfrac{1 \text{ gal}}{4 \text{ qt}} = \dfrac{7 \text{ gal}}{28 \text{ qt}}$

12. $\dfrac{300 \text{ ft}}{4 \text{ rolls}} = \dfrac{450 \text{ ft}}{7 \text{ rolls}}$

Solve. Round to the nearest hundredth.

13. $\dfrac{2}{3} = \dfrac{n}{15}$

14. $\dfrac{7}{15} = \dfrac{n}{15}$

15. $\dfrac{n}{5} = \dfrac{12}{25}$

16. $\dfrac{n}{8} = \dfrac{7}{8}$

17. $\dfrac{3}{8} = \dfrac{n}{12}$

18. $\dfrac{5}{8} = \dfrac{40}{n}$

19. $\dfrac{3}{n} = \dfrac{7}{40}$

20. $\dfrac{7}{12} = \dfrac{25}{n}$

21. $\dfrac{16}{n} = \dfrac{25}{40}$

22. $\dfrac{15}{45} = \dfrac{72}{n}$

23. $\dfrac{120}{n} = \dfrac{144}{25}$

24. $\dfrac{65}{20} = \dfrac{14}{n}$

25. $\dfrac{0.5}{2.3} = \dfrac{n}{20}$

26. $\dfrac{1.2}{2.8} = \dfrac{n}{32}$

27. $\dfrac{0.7}{1.2} = \dfrac{6.4}{n}$

28. $\dfrac{2.5}{0.6} = \dfrac{165}{n}$

29. $\dfrac{x}{6.25} = \dfrac{16}{87}$

30. $\dfrac{x}{2.54} = \dfrac{132}{640}$

31. $\dfrac{1.2}{0.44} = \dfrac{y}{14.2}$

32. $\dfrac{12.5}{y} = \dfrac{102}{55}$

33. $\dfrac{n+2}{5} = \dfrac{1}{2}$

34. $\dfrac{5+n}{8} = \dfrac{3}{4}$

35. $\dfrac{4}{3} = \dfrac{n-2}{6}$

36. $\dfrac{3}{5} = \dfrac{n-7}{8}$

Solve. Round to the nearest hundredth.

37. $\dfrac{2}{n+3} = \dfrac{7}{12}$

38. $\dfrac{5}{n+1} = \dfrac{7}{3}$

39. $\dfrac{7}{10} = \dfrac{3+n}{2}$

40. $\dfrac{3}{2} = \dfrac{5+n}{4}$

41. $\dfrac{x-4}{3} = \dfrac{3}{4}$

42. $\dfrac{x-1}{8} = \dfrac{5}{2}$

43. $\dfrac{6}{1} = \dfrac{x-2}{5}$

44. $\dfrac{7}{3} = \dfrac{x-4}{8}$

45. $\dfrac{5}{8} = \dfrac{2}{x-3}$

46. $\dfrac{5}{2} = \dfrac{1}{x-6}$

47. $\dfrac{3}{x-4} = \dfrac{5}{3}$

48. $\dfrac{8}{x-6} = \dfrac{5}{4}$

OBJECTIVE B

Solve.

49. *Biology* In a drawing, the length of an amoeba is 2.6 in. The scale of the drawing is 1 in. on the drawing equals 0.002 in. on the amoeba. Find the actual length of the amoeba.

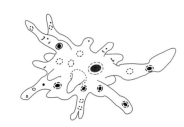

50. *Insurance* A life insurance policy costs $5.22 for every $1,000 of insurance. At this rate, what is the cost of $75,000 of insurance?

51. *Sewing* Six children's robes can be made from 6.5 yd of material. How many robes can be made from 26 yd of material?

52. *Computers* A computer manufacturer finds that an average of 3 defective hard disks are found in every 100 disks manufactured. How many defective disks are expected to be found in the production of 1,200 hard disks?

53. *Taxes* The property tax on a $90,000 home is $4,320. At this rate, what is the property tax on a home appraised at $140,000?

54. *Medicine* The dosage of a certain medication is 2 mg for every 80 lb of body weight. How many milligrams of this medication are required for a person who weighs 220 lb?

55. *Travel* An automobile was driven 84 mi and used 3 gal of gasoline. At the same rate of consumption, how far would the car travel on 14.5 gal of gasoline?

Solve. Round to the nearest hundredth.

56. *Nutrition* If a 56-gram serving of pasta contains 7 g of protein, how many grams of protein are in a 454-gram box of the pasta?

57. *Consumerism* If 4 grapefruit sell for 52¢, how much do 14 grapefruit cost?

58. *Sports* A halfback on a college football team has rushed for 435 yd in 5 games. At this rate, how many rushing yards will the halfback have in 12 games?

59. *Construction* A building contractor estimates that five overhead lights are needed for every 400 ft² of office space. Using this estimate, how many light fixtures are necessary for an office building of 35,000 ft²?

60. *Sports* A softball player has hit 9 home runs in 32 games. At the same rate, how many home runs will the player hit in a 160-game schedule?

61. *Health* A dieter has lost 3 lb in 5 weeks. At this rate, how long will it take the dieter to lose 36 lb?

62. *Consumerism* Steak costs $12.60 for 3 lb. At this rate, how much does 8 lb of steak cost?

63. *Business* An automobile recall was based on engineering tests that showed 22 defects in 1,000 cars. At this rate, how many defects would be found in 125,000 cars?

64. *Health* Walking 5 mi in 2 h will use 650 calories. Walking at the same rate, how many miles would a person need to walk to lose one pound? (The burning of 3,500 calories is equivalent to the loss of one pound.)

65. *Travel* An account executive bought a new car and drove 22,000 mi in the first four months. At the same rate, how many miles will the account executive drive in three years?

66. *Investments* An investment of $1,500 earns $120 each year. At the same rate, how much additional money must be invested to earn $300 each year?

$1,500	$1,500 + x
earns	earns
$120	$300

67. *Investments* A stock investment of $3,500 earns a dividend of $280. At the same rate, how much additional money would have to be invested so that the total dividend is $400?

Solve. Round to the nearest hundredth.

68. *Cartography* The scale on a map is one-half inch equals 8 mi. What is the actual distance between two points that are $1\frac{1}{4}$ in. apart on the map?

69. *Energy* A slow-burning candle will burn 1.5 in. in 40 min. How many inches of the candle will burn in 4 h?

70. *Mixtures* A saltwater solution is made by dissolving $\frac{2}{3}$ lb of salt in 5 gal of water. At this rate, how many pounds of salt are required for 12 gal of water?

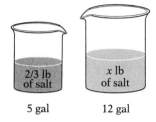

2/3 lb of salt   x lb of salt

5 gal   12 gal

71. *Business* A management consulting firm recommends that the ratio of mid-management salaries to junior management salaries be 7:5. Using this recommendation, what is the yearly mid-management salary when the junior management salary is $45,000?

CRITICAL THINKING

72. Determine whether the statement is true or false.

  a. A quotient $(a \div b)$ is a ratio.

  b. If $\frac{a}{b} = \frac{c}{d}$, then $\frac{b}{a} = \frac{d}{c}$.

  c. If $\frac{a}{b} = \frac{c}{d}$, then $\frac{a}{c} = \frac{b}{d}$.

  d. If $\frac{a}{b} = \frac{c}{d}$, then $\frac{a}{d} = \frac{c}{b}$.

73. If $\frac{a}{b} = \frac{c}{d}$, does $\frac{a}{b} = \frac{a+c}{b+d}$? Explain your answer.

74. If $\frac{a}{b} = \frac{c}{d}$, show that $\frac{a+b}{b} = \frac{c+d}{d}$.

75. A survey of voters in a city claimed that 2 people of every 5 who voted cast a ballot in favor of city amendment A, and 3 people of every 4 who voted cast a ballot against amendment A. Is this possible? Explain your answer.

76. Let the circle at the right represent $10,000. The circle is divided into 8 equal parts. (a) Use a proportion to find the amount of money that 1 part of the circle would represent. (b) Use a proportion to find the amount of money that 5 parts of the circle would represent.

$10,000

77. ✏️ Write a paragraph describing how proportional representation is used to select the members in the U.S. House of Representatives.

78. ✏️ Write a paragraph explaining the term *harmonic mean*.

SECTION 7.3   Direct and Inverse Variation

OBJECTIVE A
Direct variation

An equation of the form $y = kx$ describes many important relationships in business, science, and engineering. The equation $y = kx$, where k is a constant, is an example of a **direct variation**. The constant k is called the **constant of variation** or the **constant of proportionality**. The equation $y = kx$ is read "y varies directly as x."

For example, the distance traveled by a car traveling at a constant rate of 55 miles per hour is represented by $y = 55x$, where x is the number of hours and y is the total distance traveled. The number 55 is called the constant of proportionality.

> The distance (d) sound travels varies directly as the time (t) it travels. If sound travels 8,920 ft in 8 s, find the distance that sound travels in 3 s.
>
> This is a direct variation. k is the constant of proportionality.
>
> $$d = kt$$
>
> Substitute 8,920 for d and 8 for t.
>
> $$8,920 = k \cdot 8$$
>
> Solve for k.
>
> $$\frac{8,920}{8} = k$$
> $$1,115 = k$$
>
> Write the direct variation equation for d, substituting 1,115 for k.
>
> $$d = 1,115t$$
>
> Find d when $t = 3$.
>
> $$d = 1,115 \cdot 3$$
> $$d = 3,345$$
>
> Sound travels 3,345 ft in 3 s.

A direct variation equation can be written in the form $y = kx^n$, where n is a positive integer. For example, the equation $y = kx^2$ is read "y varies directly as the square of x."

> The load (L) that a horizontal beam can safely support is directly proportional to the square of the depth (d) of the beam. A beam with a depth of 8 in. can support 800 lb. Find the load that a beam with a depth of 6 in. can support.
>
> This is a direct variation. k is the constant of proportionality.
>
> $$L = kd^2$$
>
> Substitute 800 for L and 8 for d.
>
> $$800 = k \cdot 8^2$$
> $$800 = k \cdot 64$$
>
> Solve for k.
>
> $$\frac{800}{64} = k$$
> $$12.5 = k$$
>
> Write the direct variation equation for L, substituting 12.5 for k.
>
> $$L = 12.5d^2$$
>
> Find L when $d = 6$.
>
> $$L = 12.5 \cdot 6^2$$
> $$L = 450$$
>
> The beam can support a load of 450 lb.

Example 1

Find the constant of variation if y varies directly as x, and $y = 5$ when $x = 35$.

Strategy

To find the constant of variation, substitute 5 for y and 35 for x in the direct variation equation $y = kx$ and solve for k.

Solution

$$y = kx$$
$$5 = k \cdot 35$$
$$\frac{5}{35} = k$$
$$\frac{1}{7} = k$$

The constant of variation is $\frac{1}{7}$.

You Try It 1

Find the constant of variation if y varies directly as x, and $y = 120$ when $x = 8$.

Your Strategy

Your Solution

Example 2

Given that L varies directly as P, and $L = 24$ when $P = 16$, find P when $L = 80$. Round to the nearest tenth.

Strategy

To find P when $L = 80$:

▶ Write the basic direct variation equation, replace the variables by the given values, and solve for k.
▶ Write the direct variation equation, replacing k by its value. Substitute 80 for L and solve for P.

Solution

$$L = kP$$
$$24 = k \cdot 16$$
$$\frac{24}{16} = k$$
$$1.5 = k$$

$$L = 1.5P$$
$$80 = 1.5P$$
$$\frac{80}{1.5} = P$$
$$53.3 \approx P$$

The value of P is approximately 53.3 when $L = 80$.

You Try It 2

Given that S varies directly as R, and $S = 8$ when $R = 30$, find S when $R = 200$. Round to the nearest tenth.

Your Strategy

Your Solution

Example 3

The distance (*d*) required for a car to stop varies directly as the square of the velocity (*v*) of the car. If a car traveling 40 mph requires 130 ft to stop, find the stopping distance for a car traveling 60 mph.

Strategy

To find the stopping distance:

▶ Write the basic direct variation equation, replace the variables by the given values, and solve for *k*.
▶ Write the direct variation equation, replacing *k* by its value. Substitute 60 for *v* and solve for *d*.

Solution

$$d = kv^2$$
$$130 = k \cdot 40^2$$
$$130 = k \cdot 1{,}600$$
$$\frac{13}{160} = k$$
$$d = \frac{13}{160} \cdot v^2 = \frac{13}{160} \cdot 60^2 = \frac{13}{160} \cdot 3{,}600 = 292.5$$

The stopping distance is 292.5 ft.

You Try It 3

The distance (*d*) a body falls from rest varies directly as the square of the time (*t*) of the fall. An object falls 64 ft in 2 s. How far will the object fall in 9 s?

Your Strategy

Your Solution

Solution on p. A23 | | | | |

OBJECTIVE **B**

Inverse variation

The equation $y = \frac{k}{x}$, where *k* is a constant, is an example of an **inverse variation**. The equation $y = \frac{k}{x}$ is read "*y* varies inversely as *x*" or "*y* is inversely proportional to *x*." In general, an inverse variation equation can be written $y = \frac{k}{x^n}$, where *n* is a positive integer. For example, the equation $y = \frac{k}{x^2}$ is read "*y* varies inversely as the square of *x*."

The volume (*V*) of a gas varies inversely as the pressure (*P*). The inverse variation equation would be written as

$$V = \frac{k}{P}$$

The gravitational force (*F*) between two planets is inversely proportional to the square of the distance (*d*) between the planets. This inverse variation would be written as

$$F = \frac{k}{d^2}$$

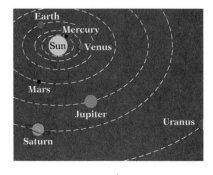

Given that y varies inversely as the square of x, and $y = 5$ when $x = 2$, find y when $x = 40$.

Write the basic inverse variation equation where y varies inversely as the square of x.	$y = \dfrac{k}{x^2}$
Replace x and y by the given values.	$5 = \dfrac{k}{2^2}$
Solve for the constant of variation.	$20 = k$
Write the inverse variation equation by substituting the value of k into the basic inverse variation equation.	$y = \dfrac{20}{x^2}$
To find y when $x = 40$, substitute 40 for x in the equation and solve for y.	$y = \dfrac{20}{40^2}$
	$y = 0.0125$

Example 4

A company that produces personal computers has determined that the number of computers it can sell (S) is inversely proportional to the price (P) of the computer. Two thousand computers can be sold when the price is $2,500. How many computers can be sold if the price of a computer is $2,000?

Strategy

To find the number of computers:

▶ Write the basic inverse variation equation, replace the variables by the given values, and solve for k.

▶ Write the inverse variation equation, replacing k by its value. Substitute 2,000 for the price and solve for the number sold.

Solution

$$S = \frac{k}{P}$$

$$2{,}000 = \frac{k}{2{,}500}$$

$$5{,}000{,}000 = k$$

$$S = \frac{5{,}000{,}000}{P} = \frac{5{,}000{,}000}{2{,}000} = 2{,}500$$

If the price is $2,000, then 2,500 computers can be sold.

You Try It 4

The resistance (R) to the flow of electric current in a wire of fixed length is inversely proportional to the square of the diameter (d) of the wire. If a wire of diameter 0.01 cm has a resistance of 0.5 ohm, what is the resistance in a wire that is 0.02 cm in diameter?

Your Strategy

Your Solution

Solution on p. A23

7.3 EXERCISES

OBJECTIVE A

Solve.

1. Find the constant of variation when y varies directly as x, and $y = 15$ when $x = 2$.

2. Find the constant of variation when t varies directly as s, and $t = 24$ when $s = 120$.

3. Find the constant of variation when n varies directly as the square of m, and $n = 64$ when $m = 2$.

4. Find the constant of variation when y varies directly as the square of x, and $y = 30$ when $x = 3$.

5. Given that P varies directly as R, and $P = 20$ when $R = 5$, find P when $R = 6$.

6. Given that T varies directly as S, and $T = 36$ when $S = 9$, find T when $S = 2$.

7. Given that M is directly proportional to P, and $M = 15$ when $P = 30$, find M when $P = 20$.

8. Given that A is directly proportional to B, and $A = 6$ when $B = 18$, find A when $B = 21$.

9. Given that y is directly proportional to the square of x, and $y = 10$ when $x = 2$, find y when $x = 0.5$.

10. Given that W is directly proportional to the square of V, and $W = 50$ when $V = 5$, find W when $V = 12$.

11. If A varies directly as the square of r, and $A = 3.14$ when $r = 1$, find A when $r = 2$.

12. If A varies directly as the square of r, and $A = \frac{22}{7}$ when $r = 1$, find A when $r = 7$.

Solve.

13. *Compensation* A worker's wage (*w*) is directly proportional to the number of hours (*h*) worked. If $82 is earned for working 8 h, how much is earned for working 30 h?

14. *Mechanics* The distance (*d*) a spring will stretch varies directly as the force (*F*) applied to the spring. If a force of 12 lb is required to stretch a spring 3 in., what force is required to stretch the spring 5 in.?

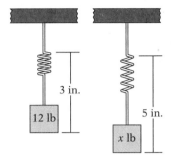

15. *Sports* The pressure (*P*) on a diver in the water varies directly as the depth (*d*). If the pressure is 2.25 lb/in² when the depth is 5 ft, what is the pressure when the depth is 12 ft?

16. *Computers* The number of words typed (*w*) is directly proportional to the time (*t*) spent typing. A typist can type 260 words in 4 min. Find the number of words typed in 15 min.

17. *Travel* The stopping distance (*s*) of a car varies directly as the square of its speed (*v*). If a car traveling 50 mph requires 170 ft to stop, find the stopping distance for a car traveling 65 mph.

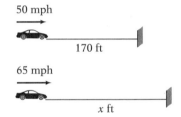

18. *Physics* The distance (*d*) an object falls is directly proportional to the square of the time (*t*) of the fall. If an object falls a distance of 8 ft in 0.5 s, how far will the object fall in 5 s?

19. *Energy* The current (*I*) varies directly as the voltage (*V*) in an electric circuit. If the current is 4 amps when the voltage is 100 volts, find the current when the voltage is 75 volts.

20. *Travel* The distance traveled (*d*) varies directly as the time (*t*) of travel, assuming that the speed is constant. If it takes 45 min to travel 50 mi, how many hours would it take to travel 180 mi?

OBJECTIVE B

Solve.

21. Find the constant of variation when *y* varies inversely as *x* and *y* = 10 when *x* = 5.

22. Find the constant of proportionality when *T* varies inversely as *S* and *T* = 0.2 when *S* = 8.

Solve.

23. Find the constant of variation when p varies inversely as the square of q and $p = 4$ when $q = 5$.

24. Find the constant of variation when W varies inversely as the square of V and $W = 5$ when $V = 0.5$.

25. If y varies inversely as x and $y = 500$ when $x = 4$, find y when $x = 10$.

26. If W varies inversely as L and $W = 20$ when $L = 12$, find L when $W = 90$.

27. If y varies inversely as the square of x and $y = 40$ when $x = 4$, find y when $x = 10$.

28. If L varies inversely as the square of d and $L = 25$ when $d = 2$, find L when $d = 5$.

29. *Geometry*   The length (L) of a rectangle of fixed area varies inversely as the width (W). If the length of the rectangle is 8 ft when the width is 5 ft, find the length of the rectangle when the width is 4 ft.

5 ft

8 ft

4 ft

x ft

30. *Travel*   The time (t) of travel of an automobile trip varies inversely as the speed (v). Traveling at an average speed of 65 mph, a trip took 4 h. The return trip took 5 h. Find the average speed of the return trip.

31. *Energy*   The current (I) in an electric circuit is inversely proportional to the resistance (R). If the current is 0.25 amp when the resistance is 8 ohms, find the resistance when the current is 1.2 amps.

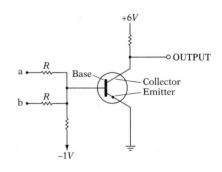

32. *Physics*   The volume (V) of a gas varies inversely as the pressure (P) on the gas. If the volume of the gas is 12 ft^3 when the pressure is 15 lb/ft^2, find the volume of the gas when the pressure is 4 lb/ft^2.

33. *Business*   A computer company that produces personal computers has determined that the number of computers it can sell (S) is inversely proportional to the price (P) of the computer. Eighteen hundred computers can be sold if the price is $1,800. How many computers can be sold if the price is $1,500?

34. *Mechanics*   The speed (s) of a gear varies inversely as the number of teeth (t). If a gear that has 40 teeth makes 15 revolutions per minute, how many revolutions per minute will a gear that has 32 teeth make?

Solve.

35. *Energy* The intensity (I) of a light source is inversely proportional to the square of the distance (d) from the source. If the intensity is 20 lumens at a distance of 8 ft, what is the intensity when the distance is 5 ft?

36. *Magnetism* The repulsive force (f) between the north poles of two magnets is inversely proportional to the square of the distance (d) between them. If the repulsive force is 18 lb when the distance is 3 in., find the repulsive force when the distance is 1.2 in.

37. *Sound* The loudness (L) measured in decibels of a stereo speaker is inversely proportional to the square of the distance (d) from the speaker. The loudness is 20 decibels at a distance of 10 ft. What is the loudness at a distance of 6 ft from the speaker?

38. *Consumerism* The number of items (N) that can be purchased for a given amount of money is inversely proportional to the cost (C) of an item. If 390 items can be purchased when the cost per item is $.50, how many items can be purchased when the cost per item is $.20?

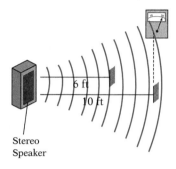

Stereo
Speaker

CRITICAL THINKING

39. Determine whether the statement is true or false.
    a. In the direct variation equation $f = kx$, if x increases, then f increases.
    b. In the inverse variation equation $x = \dfrac{k}{y}$, if x increases, then y increases.
    c. In the direct variation equation $T = ks^2$, if s is doubled, then T doubles.

40. If y varies directly as x, and $x = 10$ when $y = 4$, find y when $x = 15$. Show that this variation can be solved by using a proportion.

41. (a) The variable y varies directly as the cube of x. If x is doubled, by what factor is y increased? (b) The variable y varies inversely as the cube of x. If x is doubled, by what factor is y decreased?

42. Discuss how variation is used in prescribing medication for individuals.

43. Explain how proportions may be used in pricing large quantities of a purchase as compared to small quantities of a purchase. Is the unit price of a large purchase always smaller than the unit price of a small purchase?

44. Write a short history of pi, one of the universal constants of nature.

45. Explain the relationship between direct variation and proportion.

PROJECTS IN MATHEMATICS

Earned Run Average

One measure of a pitcher's success is the earned run average. The **earned run average (ERA)** is the number of earned runs a pitcher gives up for every 9 innings pitched. The definition of an earned run is somewhat complicated, but basically an earned run is a run that is scored as a result of hits and base running that involve no errors on the part of the pitcher's team. If the opposing team scores a run on an error (for example, a fly ball that should have been caught in the outfield was fumbled), then that is not an earned run.

A proportion is used to calculate a pitcher's ERA. Remember that the statistic involves the number of earned runs per *9 innings*. The answer is always rounded to the nearest hundredth.

During the 1995 baseball season, Randy Johnson gave up 59 earned runs and pitched 214.3 innings for the Seattle Mariners. Randy Johnson's ERA is calculated as shown below:

Strategy

To find Johnson's ERA, let $x =$ the number of earned runs for every 9 innings pitched. Then set up a proportion. Solve the proportion for x.

Solution

$$\frac{59 \text{ earned runs}}{214.3 \text{ innings}} = \frac{x}{9 \text{ innings}}$$

$$214.3x = 59(9)$$

$$214.3x = 531$$

$$\frac{214.3x}{214.3} = \frac{531}{214.3}$$

$$x \approx 2.48$$

Randy Johnson's ERA for 1995 was 2.48.

1. Roger Clemens' first year with the Boston Red Sox was 1984. During that season, he pitched 133.3 innings and gave up 64 earned runs. Calculate Clemens' ERA for 1984.

2. In 1987, Nolan Ryan had the lowest ERA of any pitcher in the Major Leagues. He gave up 65 earned runs and pitched 211.7 innings for the Astros. Calculate Ryan's ERA for 1987.

3. During the 1995 baseball season, Andy Ashby of the San Diego Padres pitched 192.2 innings and gave up 63 earned runs. During the same season, Ismael Valdes of the Los Angeles Dodgers gave up 67 earned runs and pitched 197.7 innings. Who had the lower ERA for that season? How much lower?

4. Find the necessary statistics for the pitcher of your "home team" and calculate that pitcher's ERA.

*Joint and
Combined Variation*

A variation may involve more than two variables. If a quantity varies directly as the product of two or more variables, it is known as a **joint variation**.

The weight of a rectangular metal box is directly proportional to the volume of the box, given by length · width · height.

Thus, weight = $kLWH$.

The weight of a box with $L = 24$ in., $W = 12$ in., and $H = 12$ in. is 72 lb. Find the weight of another box with $L = 18$ in., $W = 9$ in., and $H = 18$ in.

$$\text{weight} = kLWH$$

$$72 = k(24)(12)(12)$$

$$\frac{72}{(24)(12)(12)} = k$$

$$\frac{1}{48} = k$$

$$\text{weight} = kLWH$$

$$\text{weight} = \frac{1}{48}(18)(9)(18) \qquad \text{Substitute } \frac{1}{48} \text{ for } k.$$

$$\text{weight} = 60.75 \qquad\qquad \begin{array}{l}\text{Substitute the dimensions of the}\\ \text{other box into the equation.}\end{array}$$

The weight of the other box is 60.75 lb.

Direct and inverse variation can occur in the same problem. When this occurs, it is called a **combined variation**.

The electrical resistance of a wire is directly proportional to the length and inversely proportional to the square of the diameter of the wire. This is written as

$$R = \frac{kL}{d^2}$$

1. The weight that a horizontal beam with a rectangular cross section can safely support varies jointly as the width and square of the depth of the cross section and inversely as the length of the beam.

   a. Write the joint variation.
   b. If a 2-inch by 4-inch beam 8 ft long can safely support a load of 300 lb, what load can be safely supported by a beam made of the same material with a width of 4 in., a depth of 6 in., and a length of 12 ft?

2. The force on a flat surface that is perpendicular to a wind is directly proportional to the product of the area of the surface and the square of the speed of the wind.

   a. Write the joint variation.
   b. What effect does doubling the area have on the force of the wind?
   c. What effect does doubling the speed of the wind have on the force of the wind?

3. The force of attraction between two magnets is directly proportional to the product of the strengths of the two magnets and inversely proportional to the square of the distance between the magnets.

   a. Write the combined variation.
   b. What effect does doubling the strength of one magnet have on the force between the magnets?
   c. What effect does doubling the distance between the magnets have on the force between the magnets?

Travelers' Safety   According to a recent study, the crime rate against travelers in the United States is lower than that against the general population. The article reporting this study included a bar graph like the one in Figure 7.1.

1. Why are the figures reported based on crime victims per 1,000 adults per year?

2. Use the given figures for personal crimes and property crimes to set up a proportion. Is the proportion true?

3. Why might the crime rate against travelers be lower than that against the general population?

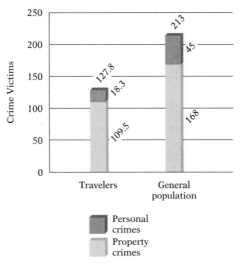

Figure 7.1   Crime Victims per 1,000 Adults
per Year
Source: Travel Industry Association of America

Price–Earnings Ratio

The price–earnings ratio of a company's stock is one measure a stock market analyst uses to determine the financial well-being of the company.

1. Explain the meaning of the price–earnings ratio.

2. If the price–earnings ratio for a company's stock is 7.3, express the price of the stock in terms of the stock's earnings per share.

3. Find the price–earnings ratio for the stocks that make up the Dow Jones Industrial Average. Name the five companies with the highest price–earnings ratios.

4. As a small group activity, have each member of the group find the price–earnings ratios for several stocks within the same industry, for example, airline companies, public utilities, retail sales companies. Are the price–earnings ratios within each industry close? Describe and discuss your findings.

CHAPTER SUMMARY

Key Words

A *ratio* is the comparison of two quantities with the same unit. A *rate* is the comparison of two quantities with different units. A ratio or rate is in simplest form when the two quantities do not have a common factor. A *unit rate* is a rate in which the denominator is one.

Dimensional analysis is a process of using units to organize and check the correctness of an application.

A *proportion* is the equality of two ratios or rates. The first and fourth terms of the proportion are called the *extremes*, and the second and third terms of the proportion are called the *means*.

The equation $y = kx^n$, where k is a constant, is an example of a *direct variation*. The equation $y = \dfrac{k}{x^n}$, where k is a constant, is an example of an *inverse variation*. The constant k in a direct variation or in an inverse variation is called the *constant of variation* or the *constant of proportionality*.

Essential Rules

To find a unit rate, divide the number in the numerator of the rate by the number in the denominator of the rate.

To set up a proportion, keep the same units in the numerator and the same units in the denominator.

To solve a proportion, use the fact that the product of the means equals the product of the extremes. For the proportion $\dfrac{a}{b} = \dfrac{c}{d}$, $ad = bc$.

CHAPTER REVIEW EXERCISES

1. Write 100 lb to 100 lb as a ratio in simplest form.

2. Write 6 roof supports for every 9 ft as a rate in simplest form.

3. Convert 25 mi to kilometers.

4. Solve: $\dfrac{n}{3} = \dfrac{8}{15}$

5. Write 8 h to 15 h as a ratio in simplest form.

6. Write $274 earned in 40 h as a unit rate.

7. Solve: $\dfrac{6}{n} = \dfrac{15}{32}$

8. Convert 50,000 ft^2 to acres. Round to the nearest tenth.

9. Write 171 mi driven in 3 h as a unit rate.

10. Solve: $\dfrac{5}{8} = \dfrac{22}{n}$

11. Convert 10,500 ft to miles. Round to the nearest tenth.

12. Solve $\dfrac{2}{3.5} = \dfrac{n}{12}$. Round to the nearest hundredth.

13. Write 3 yd to 24 yd as a ratio in simplest form.

14. Write 15 lb of fertilizer for 12 trees as a rate in simplest form.

15. Solve: $\dfrac{n}{4.5} = \dfrac{32}{45}$

16. Convert 200 ft^3 to cubic yards. Round to the nearest tenth.

17. Write 20 oz to 5 oz as a ratio in simplest form.

18. Solve: $\dfrac{0.05}{0.08} = \dfrac{234}{n}$

Solve.

19. Find the constant of variation when y varies directly as x and $x = 30$ when $y = 10$.

20. Find the constant of proportionality when y varies inversely as x and $y = 10$ when $x = 2$.

Solve.

21. Given that R varies directly as P and $P = 20$ when $R = 4$, find P when $R = 15$.

22. Given that T varies directly as the square of S, and $T = 50$ when $S = 5$, find T when $S = 120$.

23. Given that y varies inversely as x and $y = 0.2$ when $x = 5$, find y when $x = 25$.

24. Given that U varies inversely as the square of V, and $V = 4$ when $U = 20$, find U when $V = 2$.

25. *Technology* In 3 years the price of a graphing calculator went from $125 to $75. What is the ratio of the decrease in price to the original price?

26. *Investments* An investment of $8,000 earns $520 in dividends. At the same rate, how much money must be invested to earn $780 in dividends?

27. *Taxes* The sales tax on a $95 purchase is $7.60. Find the sales tax on a car costing $11,600.

28. *Lawn Care* The directions on a bag of plant food recommend one-half pound for every 50 ft² of lawn. How many pounds of plant food should be used on a lawn of 275 ft²?

29. *Elections* A pre-election survey showed that 3 out of 4 registered voters would vote in a county election. At this rate, how many registered voters would vote in a county with 325,000 registered voters?

30. *Physics* Hooke's Law states that the distance (d) a spring will stretch is directly proportional to the weight (w) on the spring. A weight of 5 lb will stretch a spring 2 in. How far will a weight of 28 lb stretch a spring?

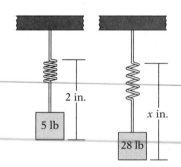

31. *Measurement* Find the speed in feet per second of a baseball pitched at 87 mph.

32. *Physics* Boyle's Law states that the volume (V) of a gas varies inversely as the pressure (P), assuming the temperature remains constant. The pressure of the gas in a balloon is 6 lb/in² when the volume is 2.5 ft³. Find the volume of the balloon if the pressure increases to 12 lb/in².

33. *Travel* The stopping distance (d) of a car varies directly as the square of the speed (v) of the car. For a car traveling at 50 mph, the stopping distance is 170 ft. Find the stopping distance of a car that is traveling at 30 mph.

CUMULATIVE REVIEW EXERCISES

1. Simplify: $18 \div \dfrac{6-3}{9} - (-3)$

2. Convert 1.2 gal to quarts.

3. Subtract: $7\dfrac{5}{12} - 3\dfrac{5}{9}$

4. Simplify: $\dfrac{4}{5} \div \dfrac{4}{5} + \dfrac{2}{3}$

5. Find the quotient of 342 and -3.

6. Evaluate $2a - 3ab$ when $a = 2$ and $b = -3$.

7. Solve: $5x - 20 = 0$

8. Solve: $3(x - 4) + 2x = 3$

9. Graph -3.5 on the number line.

10. Graph $x < -3$.

11. Simplify: $(-5)^2 - (-8) \div (7 - 5)^2 \cdot 2 - 8$

12. Simplify: $\left(-\dfrac{2}{3}\right)\left(-\dfrac{3}{4}\right)^2$

13. Simplify: $\sqrt{169}$

14. Simplify: $5 - 2(1 - 3a) + 2(a - 3)$

15. Multiply: $(4a^3b)(-5a^2b^3)$

16. Simplify: $-3y^2 + 3y - y^2 - 6y$

17. Find the ordered pair solution of $y = 3x - 2$ that corresponds to $x = -1$.

18. Write 30 cents to one dollar as a ratio in simplest form.

19. Write \$9,425 in 5 months as a unit rate.

20. Convert 45 mph to feet per second.

21. Solve: $\dfrac{2}{3} = \dfrac{n}{48}$

22. Simplify: $\dfrac{\dfrac{1}{2} + \dfrac{3}{4}}{2 - \dfrac{5}{8}}$

23. Evaluate $-2\sqrt{x^2 - 3y}$ when $x = 4$ and $y = -3$.

24. Solve: $3x + 3(x + 4) = 4(x + 2)$

25. *Government* Nine states in the United States employ full-time state legislators. The figure at the right shows the annual compensation by state of those legislators. What is the monthly salary of a state legislator in Pennsylvania? Round to the nearest cent.

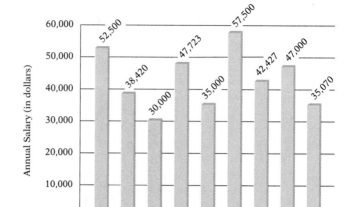

Annual Compensation of Full-Time State Legislators
Source: National Conference of State Legislators

Solve.

26. Five less than two-thirds of a number is three. Find the number.

27. Translate "the difference between four times a number and three times the sum of the number and two" into a variable expression. Then simplify.

28. *Travel* Your odometer reads 18,325 mi before embarking on a 125-mile trip. After driving one and one-half hours, the odometer reads 18,386 mi. How many miles are left to drive?

29. *Banking* You had a balance of $422.89 in your checking account. You then made a deposit of $122.35 and wrote a check for $279.76. Find the new balance in your checking account.

30. *Computers* A data processor finished $\frac{2}{5}$ of a job on the first day and $\frac{1}{3}$ on the second day. What part of the job is to be finished on the third day?

31. *Elections* In a recent city election, $\frac{2}{3}$ of the registered voters voted. How many votes were cast if the city had 31,281 registered voters?

32. *Travel* A car is driven 402.5 mi on 11.5 gal of gas. Find the number of miles traveled per gallon of gas.

33. *Mechanics* At a certain speed, the engine rpm (revolutions per minute) of a car in fourth gear is 2,500. This is two-thirds of the rpm of the engine in third gear. Find the rpm of the engine in third gear.

Percent

FOCUS On Problem Solving

A calculator is an important tool of problem solving. It can be used as an aid to guessing or estimating a solution of a problem. Here are a few problems to solve with a calculator. You may need to research some of the questions to find information you do not know.

1. Choose any single-digit positive number. Multiply the number by 1,507. Now multiply the result by 7,519. What is the answer? Choose another positive single-digit number and again multiply by 1,507 and 7,519. What is the answer? What pattern do you see? Why does this work?

2. Are there enough people in the United States so that if they held hands in a line, they would stretch around the world at the equator? To answer this question, begin by asking yourself what information you need. What assumptions must you make?

3. The gross domestic product in 1996 was about $7,200,000,000,000. Is this more or less than the amount of money that would be placed on the last square of a standard checkerboard if 1¢ is placed on the first square, 2¢ is placed on the second square, 4¢ is placed on the third square, 8¢ is placed on the fourth square, and so on until the 64th square is reached?

4. Which of the reciprocals of the first 16 natural numbers have a terminating decimal representation, and which have a repeating decimal representation?

5. What is the largest natural number n for which $4^n > 1 \cdot 2 \cdot 3 \cdots n$?

6. If $1,000 bills were stacked one on top of another, is the height of one billion dollars less than or more than the height of the Washington Monument?

7. What is the value of $1 + \cfrac{1}{1 + \cfrac{1}{1 + \cfrac{1}{1 + \cfrac{1}{1 + 1}}}}$?

8. Calculate 15^2, 35^2, 65^2, and 85^2. Study the results. Make a conjecture about a relationship between a number ending in 5 and its square. Use your conjecture to find 75^2 and 95^2. Does your conjecture work for 125^2?

9. Find the sum of the first 1,000 natural numbers. (*Hint:* You could just start adding $1 + 2 + 3 + 4 + \cdots$, but even if you performed one operation each second, it would take over 15 minutes to find the sum. Instead, try pairing the numbers and then adding the numbers in each pair. Pair 1 and 1000, 2 and 999, 3 and 998, and so on. What is the sum of each pair? How many pairs are there? Use this information to answer the original question.)

10. To qualify for a home loan, a bank requires that the monthly mortgage payment be less than 25% of a borrower's monthly take-home income. A laboratory technician has deductions for taxes, insurance, and retirement that amount to 25% of the technician's monthly gross income. What minimum gross monthly income must this person earn to receive a bank loan that has a $1,200 per month mortgage payment?

SECTION 8.1   Percent

OBJECTIVE A
Percents as decimals or fractions

Percent means "parts of 100." The figure at the right has 100 parts. Because 19 of the 100 parts are shaded, 19% of the figure is shaded.

19 parts to 100 parts can be expressed as the ratio $\frac{19}{100}$. One percent can be expressed as 1 part to 100, or $\frac{1}{100}$. Thus 1% is $\frac{1}{100}$ or 0.01.

"A population growth rate of 5%," "a manufacturer's discount of 40%," and "an 8% increase in pay" are typical examples of the many ways in which percent is used in applied problems. When solving problems involving a percent, it is usually necessary either to rewrite the percent as a fraction or a decimal, or to rewrite a fraction or a decimal as a percent.

To write a percent as a fraction, remove the percent sign and multiply by $\frac{1}{100}$.

Write 67% as a fraction.
Remove the percent sign and multiply by $\frac{1}{100}$. $67\% = 67\left(\frac{1}{100}\right) = \frac{67}{100}$

To write a percent as a decimal, remove the percent sign and multiply by 0.01.

Write 19% as a decimal.

Remove the percent sign and multiply by 0.01. This is the same as moving the decimal point two places to the left.

$19\% \quad = \quad 19(0.01) \quad = \quad 0.19$

Move the decimal point two places to the left. Then remove the percent sign.

Example 1   Write 150% as a fraction and as a decimal.

Solution $150\% = 150\left(\frac{1}{100}\right) = \frac{150}{100} = 1\frac{1}{2}$

$150\% = 150(0.01) = 1.50$

You Try It 1   Write 110% as a fraction and as a decimal.

Your Solution

Example 2   Write $66\frac{2}{3}\%$ as a fraction.

Solution $66\frac{2}{3}\% = 66\frac{2}{3}\left(\frac{1}{100}\right)$

$= \frac{200}{3}\left(\frac{1}{100}\right) = \frac{2}{3}$

You Try It 2   Write $16\frac{3}{8}\%$ as a fraction.

Your Solution

Solutions on pp. A23–A24

Example 3 Write 0.35% as a decimal.

Solution $0.35\% = 0.35(0.01) = 0.0035$

You Try It 3 Write 0.8% as a decimal.

Your Solution

Solution on p. A24

OBJECTIVE B
Fractions and decimals as percents

A fraction or decimal can be written as a percent by multiplying by 100%. Since 100% is $\frac{100}{100} = 1$, **multiplying by 100% is the same as multiplying by 1.**

Write $\frac{7}{8}$ as a percent.

Multiply $\frac{7}{8}$ by 100%.

$$\frac{7}{8} = \frac{7}{8}(100\%) = \frac{700}{8}\% = 87.5\%$$

Write 0.64 as a percent.

Multiply by 100%. This is the same as moving the decimal point two places to the right.

$$0.64 = 0.64(100\%) = 64\%$$

Move the decimal point two places to the right. Then write the percent sign.

Example 4 Write 1.78 as a percent.

Solution $1.78 = 1.78(100\%) = 178\%$

You Try It 4 Write 0.038 as a percent.

Your Solution

Example 5 Write $\frac{3}{11}$ as a percent. Write the remainder in fractional form.

Solution
$$\frac{3}{11} = \frac{3}{11}(100\%) = \frac{300}{11}\%$$
$$= 27\frac{3}{11}\%$$

You Try It 5 Write $\frac{9}{7}$ as a percent. Write the remainder in fractional form.

Your Solution

Example 6 Write $1\frac{1}{7}$ as a percent. Round to the nearest tenth of a percent.

Solution
$$1\frac{1}{7} = \frac{8}{7} = \frac{8}{7}(100\%)$$
$$= \frac{800}{7}\% \approx 114.3\%$$

You Try It 6 Write $1\frac{5}{9}$ as a percent. Round to the nearest tenth of a percent.

Your Solution

Solutions on p. A24

8.1 EXERCISES

OBJECTIVE A

Write as a fraction and as a decimal.

1. 5%
2. 60%
3. 30%
4. 90%

5. 250%
6. 140%
7. 28%
8. 66%

9. 35%
10. 85%
11. 6%
12. 8%

13. 122%
14. 166%
15. 29%
16. 83%

Write as a fraction.

17. $11\frac{1}{9}\%$
18. $12\frac{1}{2}\%$
19. $37\frac{1}{2}\%$
20. $31\frac{1}{4}\%$

21. $3\frac{1}{8}\%$
22. $66\frac{2}{3}\%$
23. $45\frac{5}{11}\%$
24. $6\frac{2}{3}\%$

25. $\frac{3}{8}\%$
26. $\frac{1}{4}\%$
27. $\frac{1}{2}\%$
28. $68\frac{3}{4}\%$

29. $83\frac{1}{3}\%$
30. $6\frac{1}{4}\%$
31. $87\frac{1}{2}\%$
32. $3\frac{1}{3}\%$

Write as a decimal.

33. 7.3%
34. 9.1%
35. 15.8%
36. 16.7%

37. 0.3%
38. 0.9%
39. 121.2%
40. 18.23%

41. 62.14%
42. 0.15%
43. 8.25%
44. 5.05%

Solve.

45. ✓ *Pets* The figure at the right shows some ways in which owners pamper their dogs. What fraction of the owners surveyed would buy a house or a car with their dog in mind?

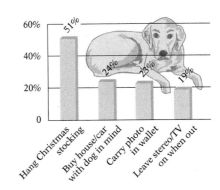

How Owners Pamper Their Dogs
Source: Purina Survey

OBJECTIVE B

Write as a percent.

46. 0.15   **47.** 0.37   **48.** 0.05   **49.** 0.02

50. 0.175   **51.** 0.125   **52.** 1.15   **53.** 1.36

54. 0.62   **55.** 0.96   **56.** 2.09   **57.** 0.07

Write as a percent. Round to the nearest tenth of a percent.

58. $\dfrac{27}{50}$ **59.** $\dfrac{83}{100}$ **60.** $\dfrac{37}{200}$ **61.** $\dfrac{1}{3}$

62. $\dfrac{3}{8}$ **63.** $\dfrac{5}{11}$ **64.** $\dfrac{4}{9}$ **65.** $\dfrac{7}{8}$

66. $\dfrac{9}{20}$ **67.** $1\dfrac{2}{3}$ **68.** $2\dfrac{1}{2}$ **69.** $1\dfrac{2}{7}$

70. $1\dfrac{11}{12}$ **71.** $\dfrac{2}{5}$ **72.** $\dfrac{1}{8}$ **73.** $\dfrac{1}{6}$

Write as a percent. Write the remainder in fractional form.

74. $\dfrac{225}{200}$ **75.** $\dfrac{17}{50}$ **76.** $\dfrac{17}{25}$ **77.** $\dfrac{3}{8}$

78. $\dfrac{9}{16}$ **79.** $\dfrac{5}{14}$ **80.** $\dfrac{3}{16}$ **81.** $\dfrac{4}{7}$

82. $1\dfrac{1}{4}$ **83.** $2\dfrac{5}{8}$ **84.** $1\dfrac{5}{9}$ **85.** $2\dfrac{5}{6}$

86. $\dfrac{12}{25}$ **87.** $\dfrac{7}{30}$ **88.** $\dfrac{3}{7}$ **89.** $\dfrac{2}{9}$

CRITICAL THINKING

90. Determine whether the statement is true or false. If the statement is false, give an example to show that it is false.
   a. Multiplying a number by a percent always decreases the number.
   b. Dividing by a percent always increases the number.
   c. The word *percent* means "per hundred."
   d. A percent is always less than one.

91. Explain in your own words how to change an improper fraction to a percent.

92. Write an essay on the history of the symbol for percent.

SECTION 8.2   The Basic Percent Equation

OBJECTIVE A
The basic percent equation

A real estate broker receives a payment that is 6% of a $175,000 sale. To find the amount the broker receives requires answering the question, "6% of $175,000 is what?" This sentence can be written using mathematical symbols and then solved for the unknown number. Recall that **of** is written as · (times), **is** is written as = (equals), and **what** is written as n (the unknown number).

$$
\begin{array}{ccccc}
6\% & \text{of} & \$175{,}000 & \text{is} & \text{what?} \\
\downarrow & \downarrow & \downarrow & \downarrow & \downarrow \\
\text{percent} & \cdot & \text{base} & = & \text{amount} \\
6\% & & \$175{,}000 & & n
\end{array}
$$

$$0.06 \cdot \$175{,}000 = n$$
$$\$10{,}500 = n$$

The broker receives a payment of $10,500.

The solution was found by solving the basic percent equation for amount.

> ### The Basic Percent Equation
> Percent · base = amount

Find 2.5% of 800.

Use the basic percent equation.
Percent = 2.5% = 0.025,
base = 800, amount = n

Percent · base = amount
$$0.025 \cdot 800 = n$$
$$20 = n$$

2.5% of 800 is 20.

A recent promotional game at a grocery store listed the probability of winning a prize as "1 chance in 2." A percent can be used to describe the chance of winning. This requires answering the question, "What percent of 2 is 1?"

The chance of winning can be found by solving the basic percent equation for percent.

$$
\begin{array}{ccccc}
\text{What} & \text{percent} & \text{of} & 2 & \text{is} & 1? \\
& \downarrow & \downarrow & \downarrow & \downarrow & \downarrow \\
& \text{percent} & \cdot & \text{base} & = & \text{amount} \\
& n & & 2 & & 1
\end{array}
$$

$$n \cdot 2 = 1$$
$$n = \frac{1}{2}$$

Write the fraction as a percent. $$n = \frac{1}{2}(100\%) = 50\%$$

There is a 50% chance of winning a prize.

32 is what percent of 20?

Use the basic percent equation.
Percent = n, base = 20,
amount = 32

$$\text{Percent} \cdot \text{base} = \text{amount}$$
$$n \cdot 20 = 32$$
$$\frac{20n}{20} = \frac{32}{20}$$
$$n = 1.6$$
$$n = 160\%$$

Write 1.6 as a percent.

32 is 160% of 20.

Each year an investor receives a payment that equals 8% of the value of an investment. This year that payment amounted to $640. To find the value of the investment this year, we must answer the question, "8% of what value is $640?"

The value of the investment can be found by solving the basic percent equation for the base.

$$
\begin{array}{ccccc}
8\% & \text{of} & \text{what} & \text{is} & \$640? \\
\downarrow & \downarrow & \downarrow & \downarrow & \downarrow \\
\text{Percent} & \cdot & \text{base} & = & \text{amount} \\
8\% & & n & & 640
\end{array}
$$

$$0.08 \cdot n = 640$$
$$\frac{0.08n}{0.08} = \frac{640}{0.08}$$
$$n = 8,000$$

This year the investment is worth $8,000.

62% of what is 800? Round to the nearest tenth.

Use the basic percent equation.
Percent = 62% = 0.62, base − n,
amount = 800

$$\text{Percent} \cdot \text{base} = \text{amount}$$
$$0.62 \cdot n = 800$$
$$\frac{0.62n}{0.62} = \frac{800}{0.62}$$
$$n \approx 1,290.3$$

62% of 1,290.3 is approximately 800.

Note from the previous three problems that, if any two parts of the basic percent equation are given, the third part can be found.

Example 1 Find 9.4% of 240.

Strategy To find the amount, solve the basic percent equation.
Percent = 9.4% = 0.094, base = 240, amount = n

Solution Percent · base = amount
$$0.094 \cdot 240 = n$$
$$22.56 = n$$

22.56 is 9.4% of 240.

You Try It 1 Find $33\frac{1}{3}\%$ of 45.

Your Strategy

Your Solution

Solution on p. A24

Example 2   What percent of 30 is 12?

    Strategy   To find the percent, solve the
basic percent equation.
Percent = n, base = 30,
amount = 12

    Solution   Percent · base = amount

$$n \cdot 30 = 12$$

$$\frac{30n}{30} = \frac{12}{30} = 0.4$$

$$n = 40\%$$

12 is 40% of 30.

You Try It 2   25 is what percent of 40?

    Your Strategy

    Your Solution

Example 3   60 is 2.5% of what?

    Strategy   To find the base, solve the basic
percent equation.
Percent = 2.5% = 0.025,
base = n, amount = 60

    Solution   Percent · base = amount

$$0.025 \cdot n = 60$$

$$\frac{0.025n}{0.025} = \frac{60}{0.025}$$

$$n = 2{,}400$$

60 is 2.5% of 2,400.

You Try It 3 $16\frac{2}{3}\%$ of what is 15?

    Your Strategy

    Your Solution

Solutions on p. A24

OBJECTIVE B
Percent problems using proportions

Percent problems can also be solved by using proportions. The proportion
method is based on writing two ratios with quantities that can be found in the
basic percent equation. One ratio is the percent ratio, written as $\frac{\text{percent}}{100}$. The
second ratio is the amount-to-base ratio, written as $\frac{\text{amount}}{\text{base}}$. These two ratios
form the proportion

$$\frac{\textbf{percent}}{\textbf{100}} = \frac{\textbf{amount}}{\textbf{base}}$$

The proportion method can be illustrated by a diagram. The rectangle at the
right is divided into two parts. The whole rectangle is represented by 100 and
the part by percent. On the other side, the whole rectangle is represented by
the base and the part by amount. The ratio of the percent to 100 is equal to the
ratio of the *amount* to the *base*.

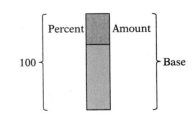

What is 32% of 85?

Sketch a diagram.

Percent = 32,
base = 85,
amount = n

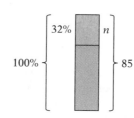

32% of 85 is 27.2.

$$\frac{\text{percent}}{100} = \frac{\text{amount}}{\text{base}}$$

$$\frac{32}{100} = \frac{n}{85}$$

$$100 \cdot n = 32 \cdot 85$$
$$100n = 2{,}720$$
$$\frac{100n}{100} = \frac{2{,}720}{100}$$
$$n = 27.2$$

Example 4 24% of what is 16? Round to the nearest hundredth.

Solution Percent = 24, base = n, amount = 16

$$\frac{24}{100} = \frac{16}{n}$$
$$24 \cdot n = 100 \cdot 16$$
$$24n = 1{,}600$$
$$n = \frac{1{,}600}{24} \approx 66.67$$

16 is approximately 24% of 66.67.

You Try It 4 What percent of 182 is 56? Round to the nearest hundredth.

Your Solution

Example 5 Find 1.2% of 42.

Solution Percent = 1.2, base = 42, amount = n

$$\frac{1.2}{100} = \frac{n}{42}$$
$$1.2 \cdot 42 = 100 \cdot n$$
$$50.4 = 100n$$
$$\frac{50.4}{100} = \frac{100n}{100}$$
$$0.504 = n$$

1.2% of 42 is 0.504.

You Try It 5 Find 0.74% of 1,200.

Your Solution

Example 6 What percent of 52 is 13?

Solution Percent = n, base = 52, amount = 13

$$\frac{n}{100} = \frac{13}{52}$$
$$n \cdot 52 = 100 \cdot 13$$
$$52n = 1{,}300$$
$$\frac{52n}{52} = \frac{1{,}300}{52}$$
$$n = 25$$

25% of 52 is 13.

You Try It 6 8 is 25% of what?

Your Solution

Solutions on p. A24

Objective C
Applications

A computer programmer receives a weekly wage of $650, and $110.50 is deducted for income tax. Find the percent of the computer programmer's salary deducted for income tax.

Use the basic percent equation.
Percent = n, base = 650,
amount = 110.50

$$\text{Percent} \cdot \text{base} = \text{amount}$$
$$n \cdot 650 = 110.50$$
$$\frac{650n}{650} = \frac{110.50}{650}$$
$$n = 0.17$$

17% of the computer programmer's salary is deducted for income tax.

A number of U.S. companies will sell their stock directly to an individual through dividend reinvestment plans, or DRIPs. The circle graph in Figure 8.1 shows what Charles Carlson, editor of *DRIP Investor*, recommended for a start-up selection of stocks in 1996. What percent of the total is invested in Exxon? (*USA Today*, 8/27/96)

Find the total amount invested.

$$100 + 50 + 250 = 400$$

To find what percent of the total (400) is invested in Exxon, use the basic percent equation.   Percent = n,   base = 400, amount = 250.

$$\text{Percent} \cdot \text{base} = \text{amount}$$
$$n \cdot 400 = 250$$
$$\frac{400n}{400} = \frac{250}{400}$$
$$n = 0.625$$

62.5% of the total is invested in Exxon.

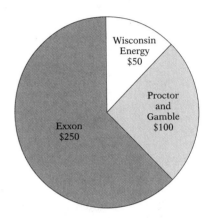

Figure 8.1   Recommended Start-Up Selection of Stocks

Example 7

Twelve percent of a company's $60,000 budget is used for advertising. Find the amount of the company's budget spent for advertising.

Strategy

To find the amount, use the basic percent equation.
Percent = 12% = 0.12, base = 60,000, amount = n

Solution

$$\text{Percent} \cdot \text{base} = \text{amount}$$
$$0.12 \cdot 60,000 = n$$
$$7,200 = n$$

The company spent $7,200 for advertising.

You Try It 7

An instructor receives a monthly salary of $2,165, and $324.75 is deducted for income tax. Find the percent of the instructor's salary deducted for income tax.

Your Strategy

Your Solution

Solution on p. A24

Example 8

A taxpayer pays a tax rate of 35% for state and federal taxes. The taxpayer has an income of $37,500. Find the amount of state and federal taxes paid by the taxpayer.

Strategy

To find the amount, solve the basic percent equation.
Percent = 35% = 0.35, base = 37,500, amount = n

Solution

Percent · base = amount
$\quad 0.35 \cdot 37{,}500 = n$
$\quad\quad\quad 13{,}125 = n$

The amount of taxes paid is $13,125.

You Try It 8

Seventy percent of the people polled in a public opinion survey approved of the way the mayor was performing the duties of government. If 210 persons approved, how many were polled?

Your Strategy

Your Solution

Example 9

A department store has a blue blazer on sale for $114, which is 60% of the original price. What is the difference between the original price and the sale price?

Strategy

To find the difference between the original price and the sale price:

▶ Find the original price. Solve the basic percent equation.
Percent = 60% = 0.60, amount = 114, base = n
▶ Subtract the sale price from the original price.

Solution

Percent · base = amount
$\quad 0.60 \cdot n = 114$
$\quad \dfrac{0.60n}{0.60} = \dfrac{114}{0.60}$
$\quad\quad\quad n = 190$

$190 - 114 = 76$

The difference in price is $76.

You Try It 9

An electrician's wage this year is $20.01 per hour, which is 115% of last year's wage. What was the increase in the hourly wage over last year?

Your Strategy

Your Solution

Solutions on pp. A24–A25

8.2 EXERCISES

OBJECTIVE A

Solve. Use the basic percent equation.

1. 8% of 100 is what?

2. 16% of 50 is what?

3. 0.05% of 150 is what?

4. 0.075% of 625 is what?

5. 15 is what percent of 90?

6. 24 is what percent of 60?

7. What percent of 16 is 6?

8. What percent of 24 is 18?

9. 10 is 10% of what?

10. 37 is 37% of what?

11. 2.5% of what is 30?

12. 10.4% of what is 52?

13. Find 10.7% of 485.

14. Find 12.8% of 625.

15. 80% of 16.25 is what?

16. 26% of 19.5 is what?

17. 54 is what percent of 2,000?

18. 8 is what percent of 2,500?

19. 16.4 is what percent of 4.1?

20. 5.3 is what percent of 50?

21. 18 is 240% of what?

22. 24 is 320% of what?

OBJECTIVE B

Solve. Use the proportion method.

23. 26% of 250 is what?

24. Find 18% of 150.

25. 37 is what percent of 148?

26. What percent of 150 is 33?

Solve. Use the proportion method.

27. 68% of what is 51?

28. 126 is 84% of what?

29. What percent of 344 is 43?

30. 750 is what percent of 50?

31. 82 is 20.5% of what?

32. 2.4% of what is 21?

33. What is 6.5% of 300?

34. Find 96% of 75.

35. 7.4 is what percent of 50?

36. What percent of 1,500 is 693?

37. Find 50.5% of 124.

38. What is 87.4% of 225?

39. 120% of what is 6?

40. 14 is 175% of what?

41. What is 250% of 18?

42. 325% of 4.4 is what?

43. 87 is what percent of 29?

44. What percent of 38 is 95?

OBJECTIVE C

Solve.

45. *Automotive Technology*   A mechanic estimates that the brakes of an RV still have 6,000 mi of wear. This amount is 12% of the estimated safe-life use of the brakes. What is the estimated safe-life use of the brakes?

46. *Business*   A charity organization spent $2,940 for administrative expenses. This amount is 12% of the money it collected. What is the total amount the charity organization collected?

47. *Taxes*   A sales clerk receives a salary of $2,240 per month, and 18% of this amount is deducted for income tax. Find the amount deducted for income tax.

48. *Astronomy*   The aphelion of Earth is its distance when it is farthest from the sun. The perihelion is when it is nearest the sun, as shown in the figure at the right. What percent of the aphelion is the perihelion? Round to the nearest tenth of a percent.

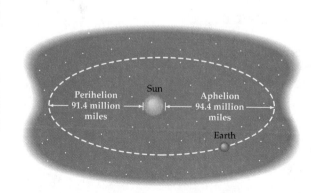

Solve.

49. *Fire Science*   A fire department received 24 false alarms out of a total of 200 alarms received. What percent of the alarms received were false alarms?

50. *Demographics*   The table at the right shows the predicted increase in population from 1996 to 1997 for each of four cities. What percent of the 1996 population of Denver, Colorado, is the increase in population? Round to the nearest tenth of a percent.

51. *Demographics*   The table at the right shows the predicted increase in population from 1996 to 1997 for each of four cities. What percent of the 1996 population of Austin, Texas, is the increase in population? Round to the nearest tenth of a percent.

City	1996 Population	Projected Increase
Austin, TX	623,000	31,000
Fresno, CA	488,000	24,000
Atlanta, GA	398,000	1,000
Denver, CO	558,000	19,000

52. *Business*   An antique shop owner expects to receive $16\frac{2}{3}\%$ of the shop's sales as profit. What is the expected profit in a month when the total sales are $24,000?

53. *Manufacturing*   In 1996, General Mills sold approximately 599 million pounds of cold cereal. This was 102.5% of 1995 sales. How many pounds, to the nearest million, did General Mills sell in 1995?

54. *Depreciation*   A used mobile home was purchased for $18,000. This amount was 64% of the cost of the mobile home when it was new. What was the new mobile home cost?

55. *Agriculture*   A farmer is given an income tax credit of 15% of the cost of some farm machinery. What tax credit would the farmer receive on farm equipment that cost $85,000?

56. *Financing*   A car is sold for $8,900. The buyer of the car makes a down payment of $1,780. What percent of the selling price is the down payment?

57. *Medicine*   The active ingredient in a prescription skin cream is clobetasol propionate. It is 0.05% of the total ingredients. How many grams of clobetasol propionate are in a 30-gram tube of this cream?

58. *Telecommunications*   According to Sprint Paging, approximately 30 million people in the United States had pagers in 1996. The figure at the right shows the number of hours people leave their pagers on during the day. How many people left their pagers on 24 h a day?

59. *Astronomy*   The diameter of Earth is approximately 8,000 mi, and the diameter of the sun is approximately 870,000 mi. What percent of Earth's diameter is the sun's diameter?

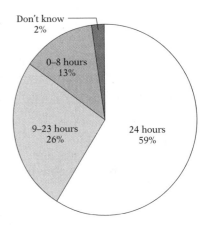

Number of Hours per Day Pagers Are Left On
Source: Sprint Paging

Solve.

60. *Business* In 1996, the price of a Power Mac 7200 at CompuAmerica was $1,699, while at MacBase™ the same computer was selling for $1,799. What percent of the MacBase™ price is the price at CompuAmerica? Round to the nearest tenth of a percent.

61. *Sports* The table at the right shows the teams with the best opening day records in the National Football League through 1997. Is the Bears opening day winning percent better or worse than that of the Giants?

62. *Sports* The table at the right shows the teams with the best opening day records in the National Football League through 1997. Is the Broncos opening day winning percent better or worse than that of the Bears?

Team	Win	Loss	Tie
Cowboys	27	9	1
Broncos	23	13	1
Giants	35	25	4
Chiefs	22	15	0
Bears	38	25	1

Source: NFL

63. *Manufacturing* During a quality control test, a manufacturer of computer boards found that 56 boards were defective. This was 0.7% of the total number of computer boards tested. How many of the tested computer boards were not defective?

64. *Agriculture* During 1996, Texas suffered through one of its longest droughts in history. Of the $5 billion in losses caused by the drought, $1.1 billion of that was direct losses to ranchers. What percent of the total losses was direct losses to ranchers?

────────────

CRITICAL THINKING

65. Find 10% of a number and subtract it from the original number. Now take 10% of the new number and subtract it from the new number. Is this the same as taking 20% of the original number? Explain.

66. Increase a number by 10%. Now decrease the new number by 10%. Is the result the original number? Explain.

67. Your employer agrees to give you a 5% raise after one year on the job, a 6% raise the next year, and a 7% raise the following year. Is your salary after the third year greater than, less than, or the same as it would be if you had received a 6% raise each year?

68. Visit a savings and loan institution or credit union to research and write a report on the meaning of *points* as it relates to a loan.

69. Find five different uses of percents and explain why percent was used in those instances.

SECTION 8.3 Percent Increase and Percent Decrease

OBJECTIVE A
Percent increase

Percent increase is used to show how much a quantity has increased over its original value. The statements "sales volume increased by 11% over last year's sales volume" and "employees received an 8% pay increase" are illustrations of the use of percent increase.

There were approximately 51.7 million students in public elementary and secondary schools in 1996. This number is expected to increase as shown in Figure 8.2. To the nearest tenth of a percent, what percent increase is expected from 1996 to 2006?

Find the increase in the number of students from 1996 to 2006.

54.7 million − 51.7 million = 3 million

The increase is 3 million students.

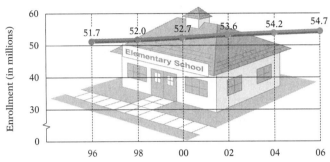

Figure 8.2 Elementary and Secondary School Enrollments
Source: U.S. Department of Education

Find the percent increase. Use the basic percent equation. Percent = n, base = 51.7, amount = 3.

There will be a 5.8% increase in enrollment.

$$\text{Percent} \cdot \text{base} = \text{amount}$$
$$n \cdot 51.7 = 3$$
$$\frac{51.7n}{51.7} = \frac{3}{51.7}$$
$$n \approx 0.058$$

Example 1

A sales clerk was earning $5.60 per hour before an 8% increase in pay. What is the new hourly wage? Round to the nearest cent.

Strategy

To find the new hourly wage:

▶ Use the basic percent equation to find the increase in pay.
Percent = 8% = 0.08, base = 5.60, amount = n
▶ Add the amount of increase to the original wage.

Solution

$$\text{Percent} \cdot \text{base} = \text{amount}$$
$$0.08 \cdot 5.60 = n$$
$$0.45 \approx n$$

$5.60 + $.45 = $6.05

The new hourly wage is $6.05.

You Try It 1

An automobile manufacturer increased the average mileage on a car from 17.5 mi/gal to 18.2 mi/gal. Find the percent increase in mileage.

Your Strategy

Your Solution

Solution on p. A25

Objective B
Percent decrease

Percent decrease is used to show how much a quantity has decreased from its original value. The statements "the president's approval rating has decreased 9% over last month" and "there has been a 15% decrease in the number of industrial accidents" are illustrations of the use of percent decrease.

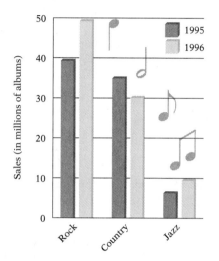

Figure 8.3 Sales of Albums
Source: USA Today, 8/29/96

Figure 8.3 shows the sales of albums for the first six months of 1995 and 1996. The sales of country music declined from 32.9 million albums to 29.4 million albums. What percent decrease does this represent?

Find the decrease in sales.

$$32.9 \text{ million} - 29.4 \text{ million} = 3.5 \text{ million}$$

The decrease in sales is 3.5 million albums.

Find the percent decrease. Use the basic percent equation. Percent = n, base = 32.9, amount = 3.5

The sales of country music albums decreased 10.6%.

$$\text{Percent} \cdot \text{base} = \text{amount}$$
$$n \cdot 32.9 = 3.5$$
$$\frac{32.9n}{32.9} = \frac{3.5}{32.9}$$
$$n \approx 0.106$$

Example 2

Violent crime in a small city decreased from 27 per 1,000 people to 24 per 1,000 people. Find the percent decrease in violent crime. Round to the nearest tenth of a percent.

Strategy

To find the percent decrease in crime:

▶ Find the decrease in the number of crimes.
▶ Use the basic percent equation to find the percent decrease in crime.
   Percent = n, base = 27,
   amount = decrease in the number of crimes

Solution

$$27 - 24 = 3$$

$$\text{Percent} \cdot \text{base} = \text{amount}$$
$$n \cdot 27 = 3$$
$$n = \frac{3}{27} \approx 0.111$$

Violent crime decreased by approximately 11.1% during the year.

You Try It 2

The market value of a luxury car decreased 24% during the year. Find the value of a luxury car that cost $47,000 last year. Round to the nearest dollar.

Your Strategy

Your Solution

8.3 EXERCISES

OBJECTIVE A

Solve. Round percents to the nearest tenth of a percent.

1. *Demographics*   A metropolitan statistical area (MSA) is a designation given to a city that has a population of at least 50,000 people. The 1996 population of Bowling Green, Kentucky, was 46,000 people. What is the smallest percent increase in population necessary for Bowling Green to become an MSA?

2. *Sports*   In 1995, the average salary of the top five professional baseball players was $4.76 million. In 1996, the average salary of the top five professional baseball players was $5.14 million. What was the percent increase in the average salaries between the two years?

3. *Forestry*   The number of acres of forest that burned in 1995 was approximately 1.7 million. In 1996, forest fires consumed approximately 5.1 million acres. What percent increase does this represent?

4. *Business*   Between the first six months of 1995 and the first six months of 1996, Johnson and Johnson saw its net earnings increase from $1.3 billion to $1.5 billion. What percent increase does this represent?

5. *Computer Science*   In 1988, the speed of the central processing unit (CPU) of a personal computer was approximately 8 megahertz. In 1996, the speed of a personal computer's CPU was 200 megahertz. What percent increase in speed does this represent?

6. *Federal Deficit*   The figure at the right shows the increase in the federal deficit between 1980 and 1996. What was the percent increase in the federal deficit between 1990 and 1995?

7. *Air Transportation*   Beginning in September 1996, there was a 10% increase in the price of an airline ticket due to a "ticket tax" that was added to the price of all airline tickets. If the total price of a ticket (including tax) was $1,100, what amount was the ticket tax?

8. *Air Transportation*   In July 1997, Boeing Company increased production of its 777 jetliner from 5 per month to 7 per month. What percent increase did this represent?

9. *Finances*   The annual dividend per share at Pepsico was raised from $.36 per share in 1994 to $.40 per share in 1995. What percent increase does this represent?

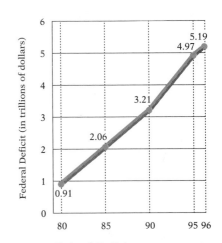

Federal Deficit
Source: Treasury Department

OBJECTIVE B

Solve. Round percents to the nearest tenth of a percent.

10. *Consumerism*   A family reduced its normal monthly food budget of $320 by $50. What percent decrease does this represent?

Solve. Round percents to the nearest tenth of a percent.

11. *Air Transportation* One configuration of the Boeing 777-300 has a seating capacity of 394. This is 26 more seats than the corresponding 777-200 airplane. What is the percent decrease in capacity from the 777-300 to the 777-200 model?

12. *Finances* The initial selling price of CompuServe's stock was $30 in April 1996. In August 1996, the share price was $13\frac{3}{8}$. What percent decrease does this represent?

13. *Medicine* The figure at the right shows the average salary of physicians for the years shown. What percent decrease in average salary did physicians experience between 1993 and 1994?

14. *Automotive* The price of an ES-300 Lexus sedan in 1997 was $29,000, a 7.7% decrease from the previous year. What was the price, to the nearest hundred dollars, of an ES-300 in 1996?

15. *Depreciation* An estimate for the decreasing value of a new car purchase is 25% the first year of ownership and 15% the second year. Using these estimates, what is the value of a car two years after it was purchased for $11,500?

16. *Production Technology* A new production method reduced the time needed to clean a piece of metal from 8 min to 5 min. What percent decrease does this represent?

17. *Business* A sales manager's average monthly expense for gasoline was $92. After joining a car pool, the manager was able to decrease gasoline expenses by 22%. What is the average monthly gasoline bill now?

18. *Consumer Science* As a result of an increased number of service lines at a grocery store, the average amount of time a customer waits in a line has decreased from 3.8 min to 2.5 min. Find the percent decrease.

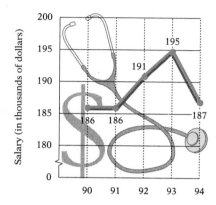

Average Physicians' Salary
Source: American Medical Association

CRITICAL THINKING

19. A department store gives you three discount coupons: a 10% discount, a 20% discount, and a 30% discount off any item. All three coupons can be used to purchase just one item. You have decided to purchase a desk chair costing $225 and use all three coupons. Is there a particular order in which you should ask to have the discount coupons applied to your purchase so that your purchase price is as small as possible? Explain.

20. A wide-screen TV costing $3,000 was on sale for 30% off. An additional 10% off the sale price was offered to customers who paid by check. Calculate the sales price after the two discounts. Is this the same as a discount of 40%? Find the equivalent discount of the successive discounts.

21. Define "per millage." Explain its relation to percent.

SECTION 8.4   Business and Consumer Applications

Objective A
Markup

Cost is the price a merchandising business or retailer pays for a product. **Selling price**, or **retail price**, is the price for which a merchandising business or retailer sells a product to a customer. The difference between selling price and cost is called **markup**. Markup is added to cost to cover the expenses of operating a business and provide a profit to the owners.

Markup can be expressed as a percent of the cost, or it can be expressed as a percent of the selling price. The percent markup is called the **markup rate**, and it is expressed as the markup based on the cost or the markup based on the selling price.

A diagram is useful when expressing the markup equations. In the diagram at the right, the total length is the selling price. One part of the diagram is the cost, and the other part is the markup.

The Markup Equations When the Markup Is Based on Cost

$M = S - C$ $M = $ markup
$M = r \cdot C$ $S = $ selling price
$S = C + M$ $C = $ cost
$S = C + rC = (1 + r)C$ $r = $ markup rate

The manager of a clothing store buys a suit for \$80 and sells the suit for \$116. Find the markup rate on the cost.

Find the markup by solving the formula $M = S - C$ for M.
$S = 116, C = 80$

$$M = S - C$$
$$= 116 - 80$$
$$= 36$$

Find the markup rate by solving the formula $M = r \cdot C$ for r.
$M = 36, C = 80$

$$M = r \cdot C$$
$$36 = r \cdot 80$$
$$\frac{36}{80} = r$$
$$0.45 = r$$

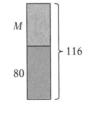

The markup rate on the cost is 45%.

The Markup Equations When the Markup Is Based on the Selling Price

$M = S - C$ $M = $ markup
$M = r \cdot S$ $S = $ selling price
$C = S - M$ $C = $ cost
$C = S - rS = (1 - r)S$ $r = $ markup rate

A retailer buys a shipment of shirts for $13.80 each. The retailer uses a markup rate of 40%, based on the selling price. Find the selling price of each shirt.

To find the selling price, solve the formula
$C = S(1 - r)$ for S.
$C = 13.80$, $r = 40\% = 0.40$

$$C = (1 - r)S$$
$$13.80 = (1 - 0.40)S$$
$$13.80 = (0.60) \cdot S$$
$$\frac{13.80}{0.60} = S$$
$$23 = S$$

Markup

13.80

S

The selling price is $23.

The difference in the markup rate when it is based on cost and when it is based on selling price can be illustrated by the following problem.

A camera cost a department store $225. The selling price of the camera is $281.25. Find the markup rate on the cost and the markup rate on the selling price.

To find the markup, solve the formula
$M = S - C$ for M.
$S = 281.25$, $C = 225$

$$M = S - C$$
$$M = 281.25 - 225$$
$$M = 56.25$$

M

225

281.25

The markup is $56.25.

To find the markup rate on the cost, solve the formula $M = r \cdot C$ for r.
$M = 56.25$, $C = 225$

$$M = r \cdot C$$
$$56.25 = r \cdot 225$$
$$\frac{56.25}{225} = r$$
$$0.25 = r$$

The markup rate on the cost is 25%.

To find the markup rate on the selling price, solve the formula
$M = r \cdot S$ for r.
$M = 56.25$, $S = 281.25$

$$M = r \cdot S$$
$$56.25 = r \cdot 281.25$$
$$\frac{56.25}{281.25} = r$$
$$0.20 = r$$

The markup rate on the selling price is 20%.

Example 1

A graphing calculator costing $45 is sold for $80. Find the markup rate based on the cost. Round to the nearest tenth of a percent.

Strategy

To find the markup rate:

▶ Solve the formula $M = S - C$ for M.
$S = 80, C = 45$
▶ Solve the formula $M = r \cdot C$ for r.

Solution

$$M = S - C \qquad\qquad M = r \cdot C$$
$$= 80 - 45 \qquad\qquad 35 = r \cdot 45$$
$$= 35 \qquad\qquad\qquad \frac{35}{45} = r$$
$$0.778 \approx r$$

The markup rate on the cost is 77.8%.

You Try It 1

A laser printer costing $950 is sold for $1,450. Find the markup rate based on the cost. Round to the nearest tenth of a percent.

Your Strategy

Your Solution

Example 2

A refrigerator costing $453 is sold for $755. Find the markup rate based on the selling price.

Strategy

To find the markup rate:

▶ Solve the formula $M = S - C$ for M.
$S = 755, C = 453$
▶ Solve the formula $M = r \cdot S$ for r.

Solution

$$M = S - C \qquad\qquad M = r \cdot S$$
$$= 755 - 453 \qquad\quad 302 = r \cdot 755$$
$$= 302 \qquad\qquad\qquad \frac{302}{755} = r$$
$$0.40 = r$$

The markup rate on the selling price is 40%.

You Try It 2

An outboard motor costing $517 is sold for $940. Find the markup rate based on the selling price.

Your Strategy

Your Solution

Solutions on p. A25

OBJECTIVE B

Discount

A retailer may reduce the regular price of a product for a promotional sale because the goods are damaged, odd sizes or colors, or discontinued items. The **discount**, or **markdown**, is the amount by which a retailer reduces the regular price of a product. The percent discount is called the **discount rate** and is usually expressed as a percent of the original selling price (the regular price).

The Discount Equations

$M = R - S$

$M = r \cdot R$

$S = R - M$

$S = R - r \cdot R = (1 - r)R$

M = discount or markdown

S = sale price

R = regular price

r = discount rate

A portable computer that regularly sells for $1,850 is on sale for $1,480. Find the discount rate.

To find the discount, solve the formula $M = R - S$ for M.
$R = 1,850$, $S = 1,480$

$M = R - S$
$= 1,850 - 1,480$
$= 370$

To find the discount rate, solve the formula $M = r \cdot R$ for r.
$M = 370$, $R = 1,850$

$M = r \cdot R$
$370 = r \cdot 1,850$
$\dfrac{370}{1,850} = r$
$0.20 = r$

The discount rate on the portable computer is 20%.

Example 3

A necklace that regularly sells for $450 is on sale for 32% off the regular price. Find the sale price.

Strategy

To find the sale price, solve the formula
$S = (1 - r)R$ for S.
$r = 0.32$, $R = 450$

Solution

$S = (1 - r)R$
$S = (1 - 0.32) \cdot 450$
$S = 0.68 \cdot 450$
$S = 306$

The sale price is $306.

You Try It 3

A garage door opener with a regular price of $212 is on sale for 25% off the regular price. Find the sale price.

Your Strategy

Your Solution

Solution on p. A25

||||||

Example 4

A video game that sold for $160 is on sale for $104. Find the discount rate.

Strategy

To find the discount rate:

▶ Solve the formula $M = R - S$ for M.
  $R = 160$, $S = 104$
▶ Solve the formula $M = r \cdot R$ for r.

Solution

$M = R - S$
$M = 160 - 104$
$M = 56$

$M = r \cdot R$
$56 = r \cdot 160$
$\dfrac{56}{160} = r$
$0.35 = r$

The discount rate is 35%.

You Try It 4

A large-screen TV that sold for $2,300 is on sale for $1,495. Find the discount rate.

Your Strategy

Your Solution

Solution on p. A25 ||||||

OBJECTIVE C
Simple interest

When money is deposited in a bank account, the bank pays the depositor for the privilege of using that money. When money is borrowed from a bank, the borrower pays the bank for the privilege of using that money. The original amount deposited or borrowed is called the **principal**. The amount paid for the privilege of using the money is called **interest**. The amount of interest to be paid is usually computed as a percent of the principal. The percent used to determine the amount of interest to be paid is the **interest rate**.

Interest computed on the original principal is called **simple interest**. Simple interest is given by the formula

$I = Prt$ I = simple interest, P = principal,
              r = annual interest rate, and t = time in years

The time (t) in the formula above is in years. A time period given in days or months must be converted to a fraction of a year before being substituted in the formula for t.

We will use the formula $t = \dfrac{\text{number of days}}{360}$ to convert days to a part of a year.

This method is known as the "banker's method," as contrasted with the exact method, which uses 365 days to compute t.

POINT OF INTEREST

A person receiving a mortgage loan of $125,000 for 30 years at a fixed interest rate of 7% has a monthly payment of $649.18. If the loan is kept for 30 years, the total amount of interest paid is $108,704.80.

Find the simple interest on a 90-day loan of $5,000 at an annual interest rate of 7.5%. Round to the nearest cent.

Use the simple interest formula. $\qquad$ $I = Prt$
$P = 5,000, r = 7.5\% = 0.075$

Convert days to a fraction of a year. $t = \dfrac{90}{360}$ $\qquad$ $I = 5,000(0.075)\left(\dfrac{90}{360}\right)$

$$I = 93.75$$

The interest is $93.75.

Example 5

Find the simple interest on a 9-month loan of $8,000 at an annual interest rate of 9.5%.

Strategy

To find the simple interest, solve the formula $I = Prt$ for I.

$P = 8,000, r = 0.095, t = \dfrac{9}{12}$

Solution

$I = Prt$

$I = 8,000(0.095)\left(\dfrac{9}{12}\right)$

$I = 570$

The interest on the loan is $570.

You Try It 5

Find the simple interest on a 60-day loan of $4,000 at an annual interest rate of 8.6%.

Your Strategy

Your Solution

Example 6

The simple interest on a 45-day loan of $12,000 is $168. Find the simple interest rate.

Strategy

To find the interest rate, solve the simple interest formula for r.

$I = 168, P = 12,000, t = \dfrac{45}{360}$

Solution

$I = Prt$

$168 = 12,000(r)\left(\dfrac{45}{360}\right)$

$168 = 1,500r$

$\dfrac{168}{1,500} = r$

$0.112 = r$

The simple interest rate is 11.2%.

You Try It 6

The simple interest on a 180-day loan of $5,000 is $225. Find the simple interest rate.

Your Strategy

Your Solution

Solutions on pp. A25–A26

8.4 Exercises

Objective A

Solve.

1. A watch costing $98 is sold for $156.80. Find the markup rate based on the cost.

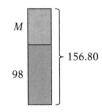

2. A set of golf clubs costing $360 is sold for $630. Find the markup rate based on the cost.

3. A pair of shoes costing $61.20 is sold for $85. Find the markup rate based on the selling price.

4. A lawnmower costing $189.75 is sold for $345. Find the markup rate based on the selling price.

5. A bicycle costing $110 has a markup rate of 55% of the cost. Find the markup.

6. A television set costing $315 has a markup rate of 30% of the cost. Find the markup.

7. A computer costing $1,750 has a markup rate of 25% of the cost. Find the selling price.

8. A flat of strawberries costing $7.60 has a markup rate of 125% of the cost. Find the selling price.

9. A basketball with a selling price of $32.50 has a markup rate of 40% of the selling price. Find the cost of the basketball.

10. A radio with a selling price of $67.80 has a markup rate of 50% of the selling price. Find the cost of the radio.

11. A calculator with a selling price of $80 has a markup rate of 32% of the selling price. Find the markup.

12. A fishing reel with a selling price of $62 has a markup rate of 22% of the selling price. Find the markup.

Solve.

13. A freezer costing $360 is sold for $520. Find the markup rate on the cost. Round to the nearest tenth of a percent.

14. A sofa costing $320 is sold for $479. Find the markup rate on the cost. Round to the nearest tenth of a percent.

15. A dress costing $42 is sold for $117. Find the markup rate on the selling price. Round to the nearest tenth of a percent.

16. A gold chain costing $560 is sold for $1,450. Find the markup rate on the selling price. Round to the nearest tenth of a percent.

OBJECTIVE B

Solve.

17. An exercise bicycle that regularly sells for $460 is on sale for $350. Find the markdown.

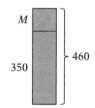

18. A suit with a regular price of $179 is on sale for $119. Find the markdown.

19. An oak bedroom set with a regular price of $1,295 is on sale for $995. Find the markdown rate. Round to the nearest tenth of a percent.

20. A stereo set with a regular price of $495 is on sale for $380. Find the markdown rate. Round to the nearest tenth of a percent.

21. A computer with a regular price of $1,995 is on sale for 30% off the regular price. Find the sale price.

22. A painting with a regular price of $1,600 is on sale for 45% off the regular price. Find the sale price.

23. A soccer ball with a regular price of $42 is on sale for 40% off the regular price. Find the sale price.

24. A gold ring with a regular price of $415 is on sale for 55% off the regular price. Find the sale price.

25. A mechanic's tool set is on sale for $180 after a markdown of 40% off the regular price. Find the regular price.

Solve.

26. A battery with a discount price of $65 is on sale for 22% off the regular price. Find the regular price. Round to the nearest cent.

27. A wristwatch is on sale for $80 after a markdown of 35% off the regular price. Find the regular price. Round to the nearest cent.

28. A compact disk player with a regular price of $325 is on sale for $201.50. Find the discount rate.

29. A luggage set with a regular price of $178 is on sale for $103.24. Find the discount rate.

30. A telescope is on sale for $165 after a markdown of 40% off the regular price. Find the regular price.

31. During a recent promotion, Nestlé offered a 25¢-off coupon on a bag of Toll House Morsels. If the grocery store offers "double coupons" for this item (the customer receives a discount that is double the value of a manufacturer's coupon), what is the percent discount in price on a 12-ounce bag costing $3.89? Round to the nearest tenth of a percent.

32. Some grocery stores offer "triple coupons" for an item (the customer receives a discount that is triple the value of a manufacturer's coupon). What is the percent discount in price on a six-pack of Snapple costing $3.29 if the customer has a 50¢-off coupon? Round to the nearest tenth of a percent.

Save 25¢
on one bag of Nestlé
Toll House Morsels
any variety
(except 6 oz. size)

OBJECTIVE C

Solve.

33. Find the simple interest on a 90-day loan of $15,000 at an annual interest rate of 7.4%.

34. Find the simple interest on a 75-day loan of $7,500 at an annual interest rate of 9.6%.

35. Find the simple interest on an 8-month loan of $4,500 at an annual interest rate of 10.2%.

36. Find the simple interest on a 9-month loan of $20,000 at an annual interest rate of 8.8%.

37. Find the simple interest on a 2-year loan of $8,000 at an annual interest rate of 9%.

Solve.

38. Find the simple interest on a one-and-one-half-year loan of $1,500 at an annual interest rate of 7.5%.

39. A $12,000 investment earned $462 in interest in 6 months. Find the annual simple interest rate.

40. A $3,000 investment earned $168.75 in 9 months. Find the annual simple interest rate.

41. A $50,000 investment earned $937.50 in 75 days. Find the annual simple interest rate.

42. A car dealer borrowed $150,000 at a 9.5% annual simple interest rate for 4 years. What was the simple interest due on the loan?

43. A corporate executive was offered a $25,000 loan at an 8.2% annual simple interest rate for 4 years. Find the simple interest due on the loan.

44. In a grocery store, 25 new cash registers were installed for a total of $32,000. The entire amount was financed for 2 years at an annual simple interest rate of 7.8%. Find the monthly payment. Round to the nearest cent. (*Hint:* Add the simple interest to the amount financed. Divide the sum by the number of months.)

45. A hobby store owner borrowed $58,000 to increase the size of the store. This amount was financed for 3 years at an annual simple interest rate of 8%. Find the monthly payment. Round to the nearest cent. (*Hint:* Add the simple interest to the amount financed. Divide the sum by the number of months.)

CRITICAL THINKING

46. A used car is on sale for a discount of 20% off the regular price of $5,500. An additional 10% discount on the sale price was offered. Is the result a 30% discount? What is the single discount that would give the same sale price?

47. Explain why a store owner might use a markup rate based on the selling price instead of the cost.

48. Explain the fundamental difference between simple interest and compound interest.

49. Visit a savings and loan office to collect information about the different kinds of home loans. Write a short essay describing the different kinds of loans available.

PROJECTS IN MATHEMATICS

Buying a Car   Besides the initial expense of buying a car, there are continuing expenses involved in owning a car. These ongoing expenses include car insurance, gas and oil, general maintenance, and monthly car payments.

A student has an after-school job to earn money to buy and maintain a car. How many hours per week must the student work to support a car? Assume that the student is earning $5.50 per hour. Here are some assumptions about the monthly cost in several categories.

Monthly Payment:

Assume that the car cost $2,500 with a down payment of $300. The remainder is financed for 3 years at an annual simple interest rate of 11%.

Amount financed = 2,500 − 300 = 2,200
The amount of interest = $2,200 · 0.11 · 3 = $726
The total amount to be repaid = $2,200 + $726 = $2,926
Monthly payment = $\dfrac{2,926}{36} \approx$ $81.28

Insurance: Assume $1,020 per year = $\dfrac{\$1,020}{12 \text{ months}}$ = $85 per month

Gas: Assume that the student travels 600 mi per month. At 25 mi/gal of gas, the student uses 24 gal of gas per month. The gas costs $1.20 per gallon.

24 · $1.20 = $28.80 per month for gas

Miscellaneous: Assume $.10 per mile for upkeep.

600 · $.10 = $60 per month for upkeep

Total monthly expenses = $81.28 + $85 + $28.80 + $60 = $255.08

To find the number of hours per month that the student must work to finance the car, divide the total monthly expenses by the hourly rate.

Number of hours per month = $\dfrac{\$255.08}{\$5.50} \approx 46.4$ h

Number of hours per week = $\dfrac{46.4 \text{ h}}{1 \text{ month}} \cdot \dfrac{1 \text{ month}}{4 \text{ weeks}} = 11.6$ h/week

The student has to work approximately 12 h per week to pay the monthly car expenses.

If you own a car, make out your own expense record. If you do not own a car, make assumptions on the kind of car that you would want to purchase, and calculate the total monthly expenses that you would have. An insurance company will give you rates on different kinds of insurance. An automobile club can give you approximations of miscellaneous expenses.

Compound Interest

In this chapter, we discussed simple interest. If the interest received on the principal of an investment is added to the principal and the new amount is reinvested at the same interest rate, the interest is called **compound interest**. Compound interest is usually compounded annually (once a year), semiannually (twice a year), quarterly (four times a year), monthly (twelve times a year), or daily (once a day). The time period of each addition of interest to principal is the **conversion period** or **compounding period**.

The discussion below shows how compound interest differs from simple interest.

Suppose on January 1 an investor places $100 in an account that earns 8% annual interest compounded quarterly. The value of the investor's deposit after one year is calculated in the following manner.

▶ January 1–March 31 (1st quarter, 3 months) $I = Prt$

$P = 100, r = 0.08, t = \dfrac{3}{12} = \dfrac{1}{4}$ $= 100(0.08)\left(\dfrac{1}{4}\right) = 2$

New principal = $100 + $2 = $102

▶ April 1–June 30 (2nd quarter, 3 months) $I = Prt$

$P = 102, r = 0.08, t = \dfrac{3}{12} = \dfrac{1}{4}$ $= 102(0.08)\left(\dfrac{1}{4}\right) = 2.04$

New principal = $102 + $2.04 = $104.04

▶ July 1–September 30 (3rd quarter, 3 months) $I = Prt$

$P = 104.04, r = 0.08, t = \dfrac{3}{12} = \dfrac{1}{4}$ $= 104.04(0.08)\left(\dfrac{1}{4}\right) = 2.08$

New principal = $104.04 + $2.08 = $106.12

▶ October 1–December 31 (4th quarter, 3 months) $I = Prt$

$P = 106.12, r = 0.08, t = \dfrac{3}{12} = \dfrac{1}{4}$ $= 106.12(0.08)\left(\dfrac{1}{4}\right) = 2.12$

New principal = $106.12 + $2.12 = $108.24

The value of the investor's deposit after one year is $108.24.

To determine the compound interest earned on the investment, subtract the original investment ($100) from the value of the investment after one year ($108.24). $108.24 - 100 = 8.24$

The compound interest earned is $8.24.

The *simple interest* earned on the same $100 investment after one year is calculated at the right. $I = Prt$
$= 100(0.08)(1) = 8$

The simple interest earned is $8.

The compound interest earned is $.24 more than the simple interest earned on the investment.

If it were necessary to compute the value of this $100 investment after 10 years, we would have to continue the above calculations another 9 times. Fortunately, there is a formula for computing compound interest.

> **Compound Interest Formula**
>
> The value, A, of an investment, P, at an annual interest rate of $r\%$ is given by
>
> $$A = P(1 + i)^n$$
>
> where n = compounding periods per year × number of years and
>
> $$i = \frac{r}{\text{compounding periods per year}}.$$

For example, if $5,000 is deposited for 3 years in an account earning 9% annual interest compounded monthly, then the value of P (the investment) is 5,000. The investment is compounded 12 times per year for 3 years: $n = 12 \cdot 3$. The interest rate r is 9% = 0.09; $i = \frac{0.09}{12}$.

CALCULATOR NOTE

This problem can be completed with a calculator using the following keystrokes.

2500 $\times$ ((1 + 0.08 / 365))
y^x (365 $\times$ 5)) =

On some calculators, the y^x key may be shown as $\wedge$ or x^y.

An investment of $2,500 is placed in an account that earns 8% annual interest compounded daily. Find the value of the account in 5 years.

Substitute the given values into the compound interest formula.

$P = 2,500$, $n = 365 \cdot 5 = 1,825$, $i = \frac{0.08}{365}$

$A = 2,500\left(1 + \frac{0.08}{365}\right)^{1,825}$

$\approx 2,500(1.491759)$

$\approx 3,729.40$

The value of the investment will be $3,729.40.

1. An investment of $10,000 is placed in an account that earns 9% annual interest compounded monthly. Find the value of the investment in 10 years.

2. You invest $2,000 in an account that earns 10% interest compounded quarterly. Find the value of the investment after 6 years.

3. If $1,000 is deposited in an account earning 8% annual interest, what is the difference between the amount of interest earned from daily compounding and the amount of interest earned from monthly compounding?

4. Visit several banks, savings and loans, or credit unions and determine the interest rates on certificates of deposit (CDs). Using this information, compare the final value of a $1,000 investment after 10 years in each of these institutions.

5. Research the meaning of *effective interest rate* and explain how simple interest and compound interest relate to that term.

6. What is the effective interest rate for an annual interest rate of 6% compounded quarterly? daily?

CHAPTER SUMMARY

Key Words

Percent means "parts of 100." *Percent increase* is used to show how much a quantity has increased over its original value. *Percent decrease* is used to show how much a quantity has decreased from its original value.

Cost is the price a business pays for a product. *Selling price* is the price for which a business sells a product to a customer.

Markup is the difference between selling price and cost. The percent markup is called the *markup rate* and is expressed as the markup rate based on cost or the markup rate based on selling price.

Discount or *markdown* is the difference between the regular price and the discount price. The discount is frequently stated as a percent, called the *discount rate*.

Principal is the amount of money originally deposited or borrowed. *Interest* is the amount paid for the privilege of using someone else's money. The percent used to determine the amount of interest is the *interest rate*. Interest computed on the original amount is called *simple interest*.

Essential Rules

To write a percent as a fraction, drop the percent sign and multiply by $\frac{1}{100}$.

To write a percent as a decimal, drop the percent sign and multiply by 0.01.

To write a fraction as a percent, multiply by 100%.

To write a decimal as a percent, multiply by 100%.

The Basic Percent Equation

Percent · base = amount

Markup Equations:

Based on cost:
$$M = S - C$$
$$M = r \cdot C$$
$$S = (1 + r)C$$

Based on selling price:
$$M = S - C$$
$$M = r \cdot S$$
$$C = (1 - r)S$$

Discount Equations:
$$M = R - S$$
$$M = r \cdot R$$
$$S = (1 - r)R$$

Simple Interest Equation

$$I = Prt$$

CHAPTER REVIEW EXERCISES

1. Write 32% as a fraction.

2. Write 22% as a decimal.

3. Write 25% as a fraction and as a decimal.

4. Write $3\frac{2}{5}$% as a fraction.

5. Write $\frac{7}{40}$ as a percent.

6. Write $1\frac{2}{7}$ as a percent. Round to the nearest tenth of a percent.

7. Write 2.8 as a percent.

8. 42% of 50 is what?

9. What percent of 3 is 15?

10. 12 is what percent of 18? Round to the nearest tenth of a percent.

11. 150% of 20 is what number?

12. Find 18% of 85.

13. 32% of what number is 180?

14. 4.5 is what percent of 80?

15. Find 0.58% of 2.54.

16. 0.0048 is 0.05% of what number?

Solve.

17. _Mining_ The table at the right shows the world's top five gold-producing countries and the amount each produced in 1995. What percent of the gold produced by these countries was produced in South Africa? Round to the nearest tenth of a percent.

Country	Gold Produced (in millions of troy ounces)
Australia	8.2
Canada	4.6
Russia	4.4
South Africa	16.8
U.S.	10.6

Source: The Gold Institute

18. _Business_ A company spent 7% of its $120,000 budget for advertising. How much of the budget was spent for advertising?

19. _Manufacturing_ A quality control inspector found that 1.2% of 4,000 cellular telephones were defective. How many of the phones were not defective?

20. _Consumerism_ An auto manufacturer offered a customer a rebate of $1,000 on each car sold by a dealership. A customer bought a car from the dealership for $16,500. What percent of the cost is the rebate? Round to the nearest tenth of a percent.

Solve.

21. *Business* A resort lodge expects to make a profit of 22% of total income. What is the expected profit on $750,000 of income?

22. *Sports* A basketball auditorium increased its 9,000 seating capacity by 18%. How many seats were added to the auditorium?

23. *Travel* An airline knowingly overbooks flights by selling 12% more tickets than there are seats available. How many tickets would this airline sell for an airplane that has 175 seats?

24. *Elections* In a recent city election, 25,400 out of 112,000 registered voters voted. What percent of the registered voters voted in the election? Round to the nearest tenth of a percent.

25. *Compensation* A clerk typist was earning $10.50 an hour before an 8% increase in pay. What is the typist's new hourly wage?

26. *Computers* A computer system that sold for $2,400 one year ago can now be bought for $1,800. What percent decrease does this represent?

27. *Business* A car dealer advertises a 6% markup rate on the cost of a car. Find the selling price of a car that costs the dealer $14,500.

28. *Business* A suit with a selling price of $299 has a markup of 30% of the selling price. Find the cost of the suit.

29. *Business* A tennis racket that regularly sells for $80 is on sale for 30% off the regular price. Find the sale price.

30. *Travel* An airline is offering a 40% discount on round-trip air fares. Find the sale price of a round-trip ticket that normally sells for $650.

31. *Finances* Find the simple interest on a 45-day loan of $3,000 at an annual simple interest rate of 8.6%.

32. *Finances* Find the simple interest on an 8-month loan of $4,500 at an annual simple interest rate of 7.8%.

33. *Finances* A realtor borrowed $80,000 to restore an apartment building. This amount was financed for 3 years at an annual simple interest rate of 9%. Find the monthly payment. Round to the nearest cent.

CUMULATIVE REVIEW EXERCISES

1. Evaluate $a - b$ when $a = 102.5$ and $b = 77.546$.

2. Evaluate 5^4.

3. Find the product of 4.67 and 3.007.

4. Multiply: $(2x - 3)(2x - 5)$

5. Divide: $3\dfrac{5}{8} \div 2\dfrac{7}{12}$

6. Multiply: $-2a^2b(-3ab^2 + 4a^2b^3 - ab^3)$

7. 120% of 35 is what?

8. Solve: $x - 2 = -5$

9. Find the product of 1.005 and 10^5.

10. Simplify: $-\dfrac{5}{8} - \left(-\dfrac{3}{4}\right) + \dfrac{5}{6}$

11. Simplify: $\dfrac{3 - \dfrac{7}{8}}{\dfrac{11}{12} + \dfrac{1}{4}}$

12. Multiply: $(-3a^2b)(4a^5b^4)$

13. Graph $y = -2x + 5$.

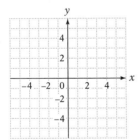

14. Graph $y = \dfrac{5}{3}x - 2$.

15. Find the quotient of $\dfrac{7}{8}$ and $\dfrac{5}{16}$.

16. Simplify: $4 - (-3) + 5 - 8$

17. Solve: $\dfrac{3}{4}x = -9$

18. Solve: $6x - 9 = -3x + 36$

19. Write 322.4 mi in 5 h as a unit rate.

20. Solve: $\dfrac{32}{n} = \dfrac{5}{7}$

21. 2.5 is what percent of 30? Round to the nearest tenth of a percent.

22. Find 42% of 160.

23. Simplify: $44 - (-6)^2 \div (-3) + 2$

24. Solve: $3(x - 2) + 2 = 11$

Solve.

25. *Health* According to the table at the right, what fraction of the population aged 75–84 are affected by Alzheimer's Disease?

Age Group	Percent Affected by Alzheimer's Disease
65 - 74	4%
75 - 84	10%
85 +	17%

Source: Mayo Clinic Family Health Book, Encyclopedia Americana, Associated Press

26. *Business* A suit that regularly sells for $202.50 is on sale for 36% off the regular price. Find the sale price.

27. *Business* A graphing calculator with a selling price of $67.20 has a markup of 60% of the cost. Find the cost of the graphing calculator.

28. *Sports* A baseball team has won 13 out of the first 18 games played. At this rate, how many games will the team win in a 162-game season?

29. *Sports* A wrestler needs to lose 8 lb in three days in order to make the proper weight class. The wrestler loses $3\frac{1}{2}$ lb the first day and $2\frac{1}{4}$ lb the second day. How many pounds must the wrestler lose the third day in order to make the weight class?

30. *Physics* The speed of a falling object is given by the formula $v = \sqrt{64d}$, where v is the speed of the falling object in feet per second and d is the distance in feet that the object has fallen. Find the speed of an object that has fallen 81 ft.

31. *Sports* On March 7, 1987, Ben Johnson ran the 60-meter dash in 6.41 s. Convert this speed to kilometers per hour. Round to the nearest hundredth.

32. *Compensation* A plumber charged $1,632 for work done on a medical building. This charge included $192 for materials and $40 per hour for labor. Find the number of hours the plumber worked on the medical building.

33. *Physics* The current (I) in an electric circuit is inversely proportional to the resistance (R). If the current is 2 amperes when the resistance is 20 ohms, find the resistance when the current is 8 amperes.

CHAPTER 9

Geometry

FOCUS On Problem Solving

Some problems in mathematics are solved by using **trial and error**. The trial-and-error method of arriving at a solution to a problem involves repeated tests or experiments until a satisfactory conclusion is reached.

Many of the Critical Thinking exercises in this text require a trial-and-error method of solution. For example, a Critical Thinking exercise on page 530 of this chapter reads:

Explain how you could cut through a cube so that the face of the resulting solid is (a) a square, (b) an equilateral triangle, (c) a trapezoid, (d) a hexagon.

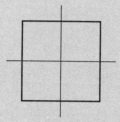

There is no formula to apply to this problem; there is no computation to perform. This problem requires picturing a cube and the results after cutting through it at different places on its surface and at different angles. For part (a), cutting perpendicular to the top and bottom of the cube and parallel to two of its sides will result in a square. The other shapes may prove more difficult.

When solving problems of this type, keep an open mind. Sometimes, when using the trial-and-error method, we are hampered by narrowness of vision; we cannot expand our thinking to include other possibilities. Then when we see someone else's solution, it appears so obvious to us! For example, for the Critical Thinking question above, it is necessary to conceive of cutting through the cube at places other than the top surface; we need to be open to the idea of beginning the cut at one of the corner points of the cube.

A topic of the Projects in Mathematics feature at the end of this chapter is symmetry. Here again, trial and error is used to determine the lines of symmetry inherent in an object. For example, in determining lines of symmetry for a square, begin by drawing a square. The horizontal line of symmetry and the vertical line of symmetry may be immediately obvious to you.

But there are two others. Do you see that a line drawn through opposite corners of the square is also a line of symmetry?

Many of the questions in this text that require an answer of "always true, sometimes true, or never true" are best solved by the trial-and-error method. For example, consider the statement presented in Section 2 of this chapter.

If two rectangles have the same area, then they have the same perimeter.

Try some numbers. Each of two rectangles, one measuring 6 units by 2 units and another measuring 4 units by 3 units, has an area of 12 square units, but the perimeter of the first is 16 units and the perimeter of the second is 14 units. So the answer "always true" has been eliminated. We still need to determine whether there is a case for which it is true. After experimenting with a lot of numbers, you may come to realize that we are trying to determine whether it is possible for two different pairs of factors of a number to have the same sum. Is it?

Don't be afraid to make many experiments, and remember that <u>errors</u>, or tests that "don't work," are a part of the trial-and-<u>error</u> process.

   **Introduction to Geometry**

Objective A

Problems involving lines and angles

The word *geometry* comes from the Greek words for *earth* and *measure*. The original purpose of geometry was to measure land. Today geometry is used in many fields, such as physics, medicine, and geology. Geometry is used in applied fields such as mechanical drawing and astronomy. Geometric forms are used in art and design.

Three basic concepts of geometry are point, line, and plane. A **point** is symbolized by drawing a dot. A **line** is determined by two distinct points and extends indefinitely in both directions, as the arrows on the line shown at the right indicate. This line contains points A and B and is represented by $\overleftrightarrow{AB}$. A line can also be represented by a single letter, such as ℓ.

A **ray** starts at a point and extends indefinitely in *one* direction. The point at which a ray starts is called the **endpoint** of the ray. The ray shown at the right is denoted by $\overrightarrow{AB}$. Point A is the endpoint of the ray.

A **line segment** is part of a line and has two endpoints. The line segment shown at the right is denoted by $\overline{AB}$.

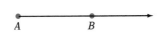

The distance between the endpoints of $\overline{AC}$ is denoted by AC. If B is a point on $\overline{AC}$, then AC (the distance from A to C) is the sum of AB (the distance from A to B) and BC (the distance from B to C).

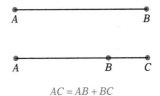

$AC = AB + BC$

Given $AB = 22$ cm and $AC = 31$ cm, find BC.

Write an equation for the distances between points on the line segment.	$AC = AB + BC$
Substitute the given distances for AB and AC into the equation.	$31 = 22 + BC$
Solve for BC.	$9 = BC$

$BC = 9$ cm

In this section we will be discussing figures that lie in a plane. A **plane** is a flat surface and can be pictured as a table top or blackboard that extends in all directions. Figures that lie in a plane are called **plane figures**.

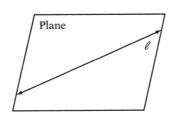

Lines in a plane can be intersecting or parallel. **Intersecting lines** cross at a point in the plane. **Parallel lines** never meet. The distance between them is always the same.

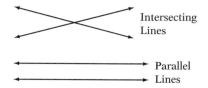

Intersecting Lines

Parallel Lines

The symbol ∥ means "is parallel to." In the figure at the right, $j \parallel k$ and $\overline{AB} \parallel \overline{CD}$. Note that j contains $\overline{AB}$ and k contains $\overline{CD}$. Parallel lines contain parallel line segments.

An **angle** is formed by two rays with the same endpoint. The **vertex** of the angle is the point at which the two rays meet. The rays are called the **sides** of the angle.

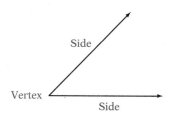

If A and C are points on rays r_1 and r_2, and B is the vertex, then the angle is called $\angle B$ or $\angle ABC$, where $\angle$ is the symbol for angle. Note that the angle is named by the vertex, or the vertex is the second point listed when the angle is named by giving three points. $\angle ABC$ could also be called $\angle CBA$.

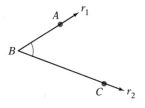

An angle can also be named by a variable written between the rays close to the vertex. In the figure at the right, $\angle x = \angle QRS$ and $\angle y = \angle SRT$. Note that in this figure, more than two rays meet at R. In this case, the vertex cannot be used to name an angle.

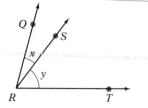

POINT OF INTEREST

The Babylonians knew that Earth was in approximately the same position in the sky every 365 days. Historians suggest that the reason one complete revolution of a circle is 360° is that 360 is the closest number to 365 that is divisible by many numbers.

An angle is measured in **degrees**. The symbol for degrees is a small raised circle, °. Probably because early Babylonians believed that Earth revolves around the sun in approximately 360 days, the angle formed by a circle has a measure of 360° (360 degrees).

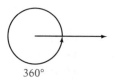

360°

A **protractor** is used to measure an angle. Place the center of the protractor at the vertex of the angle with the edge of the protractor along a side of the angle. The angle shown in the figure below measures 58°.

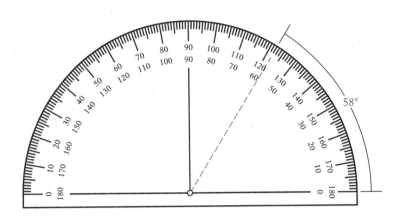

A 90° angle is called a **right angle**. The symbol ∟ represents a right angle.

Perpendicular lines are intersecting lines that form right angles.

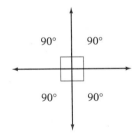

The symbol ⊥ means "is perpendicular to." In the figure at the right, $p \perp q$ and $\overline{AB} \perp \overline{CD}$. Note that line p contains $\overline{AB}$ and line q contains $\overline{CD}$. Perpendicular lines contain perpendicular line segments.

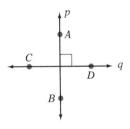

Complementary angles are two angles whose measures have the sum 90°.

$$\angle A + \angle B = 70° + 20° = 90°$$

$\angle A$ and $\angle B$ are complementary angles.

A 180° angle is called a **straight angle.**

∠*AOB* is a straight angle.

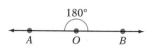

Supplementary angles are two angles whose measures have the sum 180°.

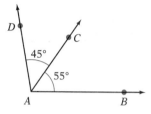

$$\angle A + \angle B = 130° + 50° = 180°$$

∠*A* and ∠*B* are supplementary angles.

An **acute angle** is an angle whose measure is between 0° and 90°. ∠*B* above is an acute angle. An **obtuse angle** is an angle whose measure is between 90° and 180°. ∠*A* above is an obtuse angle.

Two angles that share a common side are **adjacent angles.** In the figure at the right, ∠*DAC* and ∠*CAB* are adjacent angles. ∠*DAC* = 45° and ∠*CAB* = 55°.

$$\angle DAB = \angle DAC + \angle CAB$$
$$= 45° + 55° = 100°$$

In the figure at the right, ∠*EDG* = 80°. ∠*FDG* is three times the measure of ∠*EDF*. Find the measure of ∠*EDF*.

Let x = the measure of ∠*EDF*. Then $3x$ = the measure of ∠*FDG*. Write an equation and solve for x, the measure of ∠*EDF*.

$$\angle EDF + \angle FDG = \angle EDG$$
$$x + 3x = 80$$
$$4x = 80$$
$$x = 20$$

∠*EDF* = 20°

Example 1

Given *MN* = 15 mm, *NO* = 18 mm, and *MP* = 48 mm, find *OP*.

Solution

$$MN + NO + OP = MP$$
$$15 + 18 + OP = 48$$
$$33 + OP = 48$$
$$OP = 15$$

OP = 15 mm

You Try It 1

Given *QR* = 24 cm, *ST* = 17 cm, and *QT* = 62 cm, find *RS*.

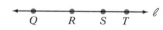

Your Solution

Solution on p. A26

Example 2

Given $XY = 9$ m and YZ is twice XY, find XZ.

Solution

$XZ = XY + YZ$
$XZ = XY + 2(XY)$
$XZ = 9 + 2(9)$
$XZ = 9 + 18$
$XZ = 27$

$XZ = 27$ m

Example 3

Find the complement of a 38° angle.

Strategy

Complementary angles are two angles whose sum is 90°. To find the complement, let x represent the complement of a 38° angle. Write an equation and solve for x.

Solution

$x + 38° = 90°$
$\quad x = 52°$

The complement of a 38° angle is a 52° angle.

Example 4

Find the measure of $\angle x$.

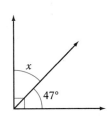

Strategy

To find the measure of $\angle x$, write an equation using the fact that the sum of the measure of $\angle x$ and 47° is 90°. Solve for $\angle x$.

Solution

$\angle x + 47° = 90°$
$\quad \angle x = 43°$

The measure of $\angle x$ is 43°.

You Try It 2

Given $BC = 16$ ft and $AB = \frac{1}{4}(BC)$, find AC.

Your Solution

You Try It 3

Find the supplement of a 129° angle.

Your Strategy

Your Solution

You Try It 4

Find the measure of $\angle a$.

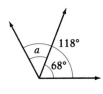

Your Strategy

Your Solution

Solutions on p. A26

OBJECTIVE B
Problems involving angles formed by intersecting lines

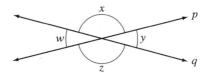

Four angles are formed by the intersection of two lines. If the two lines are perpendicular, each of the four angles is a right angle. If the two lines are not perpendicular, then two of the angles formed are acute angles and two of the angles are obtuse angles. The two acute angles are always opposite each other, and the two obtuse angles are always opposite each other.

In the figure at the right, $\angle w$ and $\angle y$ are acute angles. $\angle x$ and $\angle z$ are obtuse angles.

Two angles that are on opposite sides of the intersection of two lines are called **vertical angles**. Vertical angles have the same measure. $\angle w$ and $\angle y$ are vertical angles. $\angle x$ and $\angle z$ are vertical angles.

Vertical angles have the same measure.

$$\angle w = \angle y$$
$$\angle x = \angle z$$

Two angles that share a common side are called **adjacent angles**. For the figure shown above, $\angle x$ and $\angle y$ are adjacent angles, as are $\angle y$ and $\angle z$, $\angle z$ and $\angle w$, and $\angle w$ and $\angle x$. Adjacent angles of intersecting lines are supplementary angles.

Adjacent angles of intersecting lines are supplementary angles.

$$\angle x + \angle y = 180°$$
$$\angle y + \angle z = 180°$$
$$\angle z + \angle w = 180°$$
$$\angle w + \angle x = 180°$$

Given that $\angle c = 65°$, find the measures of angles a, b, and d.

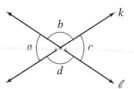

$\angle a = \angle c$ because $\angle a$ and $\angle c$ are vertical angles.

$\angle a = 65°$

$\angle b$ is supplementary to $\angle c$ because $\angle b$ and $\angle c$ are adjacent angles of intersecting lines.

$\angle b + \angle c = 180°$
$\angle b + 65° = 180°$
$\angle b = 115°$

$\angle d = \angle b$ because $\angle d$ and $\angle b$ are vertical angles.

$\angle d = 115°$

A line that intersects two other lines at different points is called a **transversal**.

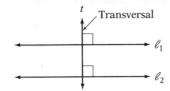

If the lines cut by a transversal t are parallel lines and the transversal is perpendicular to the parallel lines, all eight angles formed are right angles.

If the lines cut by a transversal t are parallel lines and the transversal is not perpendicular to the parallel lines, all four acute angles have the same measure and all four obtuse angles have the same measure. For the figure at the right:

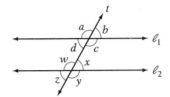

$$\angle b = \angle d = \angle x = \angle z$$

$$\angle a = \angle c = \angle w = \angle y$$

Alternate interior angles are two angles that are on opposite sides of the transversal and between the lines. In the figure above, $\angle c$ and $\angle w$ are alternate interior angles; $\angle d$ and $\angle x$ are alternate interior angles. Alternate interior angles have the same measure.

Alternate interior angles have the same measure.

$$\angle c = \angle w$$
$$\angle d = \angle x$$

Alternate exterior angles are two angles that are on opposite sides of the transversal and outside the parallel lines. In the figure above, $\angle a$ and $\angle y$ are alternate exterior angles; $\angle b$ and $\angle z$ are alternate exterior angles. Alternate exterior angles have the same measure.

Alternate exterior angles have the same measure.

$$\angle a = \angle y$$
$$\angle b = \angle z$$

Corresponding angles are two angles that are on the same side of the transversal and are both acute angles or are both obtuse angles. For the figure above, the following pairs of angles are corresponding angles: $\angle a$ and $\angle w$, $\angle d$ and $\angle z$, $\angle b$ and $\angle x$, $\angle c$ and $\angle y$. Corresponding angles have the same measure.

Corresponding angles have the same measure.

$$\angle a = \angle w$$
$$\angle d = \angle z$$
$$\angle b = \angle x$$
$$\angle c = \angle y$$

Given that $\ell_1 \parallel \ell_2$ and $\angle c = 58°$, find the measures of $\angle f$, $\angle h$, and $\angle g$.

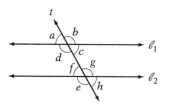

$\angle c$ and $\angle f$ are alternate interior angles.

$\angle c$ and $\angle h$ are corresponding angles.

$\angle g$ is supplementary to $\angle h$.

$\angle f = \angle c = 58°$

$\angle h = \angle c = 58°$

$\angle g + \angle h = 180°$
$\angle g + 58° = 180°$
$\angle g = 122°$

Example 5

Find x.

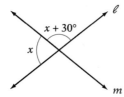

Strategy

The angles labeled are adjacent angles of intersecting lines and are therefore supplementary angles. To find x, write an equation and solve for x.

Solution

$x + (x + 30°) = 180°$
$2x + 30° = 180°$
$2x = 150°$
$x = 75°$

You Try It 5

Find x.

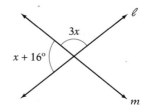

Your Strategy

Your Solution

Example 6

Given $\ell_1 \parallel \ell_2$, find x.

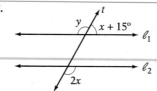

Strategy

$2x = y$ because alternate exterior angles have the same measure. $(x + 15°) + y = 180°$ because adjacent angles of intersecting lines are supplementary angles. Substitute $2x$ for y and solve for x.

Solution

$(x + 15°) + 2x = 180°$
$3x + 15° = 180°$
$3x = 165°$
$x = 55°$

You Try It 6

Given $\ell_1 \parallel \ell_2$, find x.

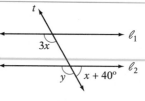

Your Strategy

Your Solution

Solutions on p. A26 ||||||

OBJECTIVE C
Problems involving the angles of a triangle

If the lines cut by a transversal are not parallel lines, the three lines will intersect at three points. In the figure at the right, the transversal t intersects lines p and q. The three lines intersect at points A, B, and C. These three points define three line segments, $\overline{AB}$, $\overline{BC}$, and $\overline{AC}$. The plane figure formed by these three line segments is called a **triangle**.

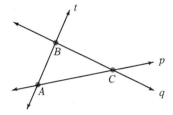

Each of the three points of intersection is the vertex of four angles. The angles within the region enclosed by the triangle are called **interior angles**. In the figure at the right, angles a, b, and c are interior angles. The sum of the measures of the interior angles of a triangle is 180°.

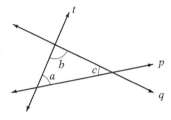

$$\angle a + \angle b + \angle c = 180°$$

The Sum of the Measures of the Interior Angles of a Triangle

The sum of the measures of the interior angles of a triangle is 180°.

An angle adjacent to an interior angle is an **exterior angle**. In the figure at the right, angles m and n are exterior angles for angle a. The sum of the measures of an interior and an exterior angle is 180°.

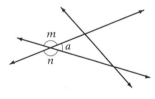

$$\angle a + \angle m = 180°$$
$$\angle a + \angle n = 180°$$

Given that $\angle c = 40°$ and $\angle d = 100°$, find the measure of $\angle e$.

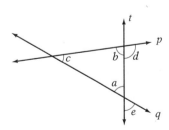

$\angle d$ and $\angle b$ are supplementary angles.

$$\angle d + \angle b = 180°$$
$$100° + \angle b = 180°$$
$$\angle b = 80°$$

The sum of the interior angles is 180°.

$$\angle c + \angle b + \angle a = 180°$$
$$40° + 80° + \angle a = 180°$$
$$120° + \angle a = 180°$$
$$\angle a = 60°$$

$\angle a$ and $\angle e$ are vertical angles.

$$\angle e = \angle a = 60°$$

Example 7

Given that $\angle y = 55°$, find the measures of angles a, b, and d.

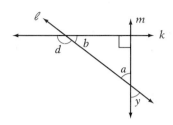

Strategy

▶ To find the measure of angle a, use the fact that $\angle a$ and $\angle y$ are vertical angles.
▶ To find the measure of angle b, use the fact that the sum of the measures of the interior angles of a triangle is 180°.
▶ To find the measure of angle d, use the fact that the sum of an interior and an exterior angle is 180°.

Solution

$\angle a = \angle y = 55°$

$\angle a + \angle b + 90° = 180°$
$55° + \angle b + 90° = 180°$
$\angle b + 145° = 180°$
$\angle b = 35°$

$\angle d + \angle b = 180°$
$\angle d + 35° = 180°$
$\angle d = 145°$

You Try It 7

Given that $\angle a = 45°$ and $\angle x = 100°$, find the measures of angles b, c, and y.

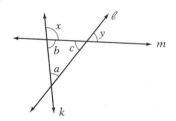

Your Strategy

Your Solution

Example 8

Two angles of a triangle measure 53° and 78°. Find the measure of the third angle.

Strategy

To find the measure of the third angle, use the fact that the sum of the measures of the interior angles of a triangle is 180°. Write an equation using x to represent the measure of the third angle. Solve the equation for x.

Solution

$x + 53° + 78° = 180°$
$x + 131° = 180°$
$x = 49°$

The measure of the third angle is 49°.

You Try It 8

One angle in a triangle is a right angle, and one angle measures 34°. Find the measure of the third angle.

Your Strategy

Your Solution

9.1 EXERCISES

OBJECTIVE A

Use a protractor to measure the angle. State whether the angle is acute, obtuse, or right.

1.

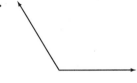

2.

3.

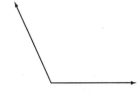

4.

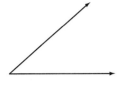

5.

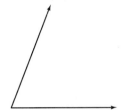

6.

Solve.

7. Find the complement of a 62° angle.

8. Find the complement of a 31° angle.

9. Find the supplement of a 162° angle.

10. Find the supplement of a 72° angle.

11. Given $AB = 12$ cm, $CD = 9$ cm, and $AD = 35$ cm, find the length of BC.

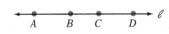

12. Given $AB = 21$ mm, $BC = 14$ mm, and $AD = 54$ mm, find the length of CD.

13. Given $QR = 7$ ft and RS is three times the length of QR, find the length of QS.

14. Given $QR = 15$ in. and RS is twice the length of QR, find the length of QS.

15. Given $EF = 20$ m and FG is $\frac{1}{2}$ the length of EF, find the length of EG.

Solve.

16. Given $EF = 18$ cm and FG is $\frac{1}{3}$ the length of EF, find the length of EG.

17. Given $\angle LOM = 53°$ and $\angle LON = 139°$, find the measure of $\angle MON$.

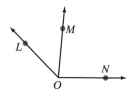

18. Given $\angle MON = 38°$ and $\angle LON = 85°$, find the measure of $\angle LOM$.

Find the measure of $\angle x$.

19.

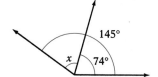

20.

Given that $\angle LON$ is a right angle, find the measure of $\angle x$.

21.

22.

23.

24.

Find the measure of $\angle a$.

25.

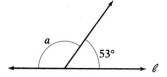

26.

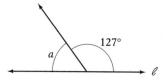

27.

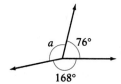

28.

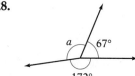

Find *x*.

29.

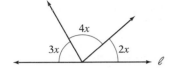

30.

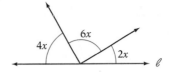

31.

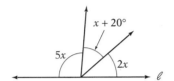

32.

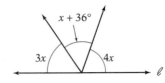

33.

34.

Solve.

35. Given $\angle a = 51°$, find the measure of $\angle b$.

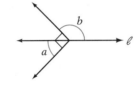

36. Given $\angle a = 38°$, find the measure of $\angle b$.

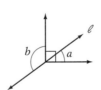

OBJECTIVE B

Find the measure of $\angle x$.

37.

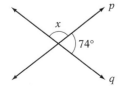

38.

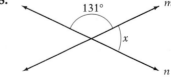

Find x.

39.

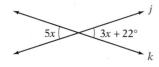

$5x$ $3x + 22°$

40.

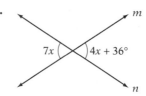

$7x$ $4x + 36°$

Given that $\ell_1 \parallel \ell_2$, find the measures of angles a and b.

41.

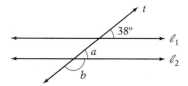

$38°$ a b

42.

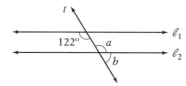

$122°$ a b

43.

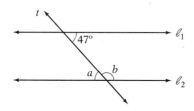

$47°$ a b

44.

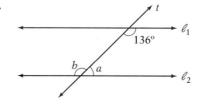

$136°$ b a

Given that $\ell_1 \parallel \ell_2$, find x.

45.

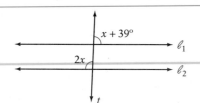

$5x$ $4x$

46.

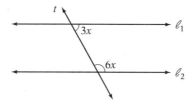

$3x$ $6x$

47.

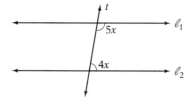

$x + 39°$ $2x$

48.

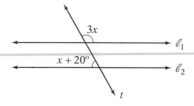

$3x$ $x + 20°$

OBJECTIVE C

Solve.

49. Given that $\angle a = 95°$ and $\angle b = 70°$, find the measures of angles x and y.

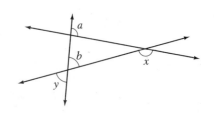

a b x y

Solve.

50. Given that ∠ a = 35° and ∠b = 55°, find the measures of angles x and y.

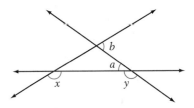

51. Given that ∠y = 45°, find the measures of angles a and b.

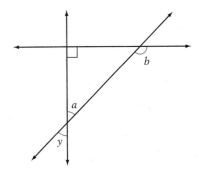

52. Given that ∠y = 130°, find the measures of angles a and b.

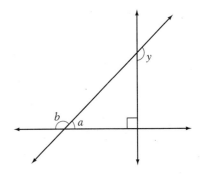

53. Given that $\overline{AO} \perp \overline{OB}$, express in terms of x the number of degrees in ∠BOC.

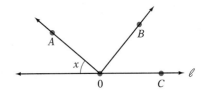

54. Given that $\overline{AO} \perp \overline{OB}$, express in terms of x the number of degrees in ∠AOC.

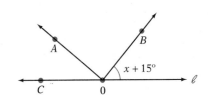

55. One angle in a triangle is a right angle, and one angle is equal to 30°. What is the measure of the third angle?

Solve.

56. A triangle has a 45° angle and a right angle. Find the measure of the third angle.

57. Two angles of a triangle measure 42° and 103°. Find the measure of the third angle.

58. Two angles of a triangle measure 62° and 45°. Find the measure of the third angle.

59. A triangle has a 13° angle and a 65° angle. What is the measure of the third angle?

CRITICAL THINKING

60. (a) What is the smallest possible whole number of degrees in an angle of a triangle? (b) What is the largest possible whole number of degrees in an angle of a triangle?

61. Cut out a triangle and then tear off two of the angles, as shown at the right. Position the pieces you tore off so that angle a is adjacent to angle b and angle c is adjacent to angle b. Describe what you observe. What does this demonstrate?

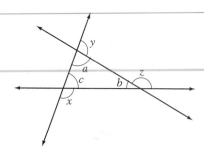

62. Construct a triangle with the given angle measures.
   a. 45°, 45°, and 90°   **b.** 30°, 60°, and 90°   **c.** 40°, 40°, and 100°

63. Determine whether the statement is always true, sometimes true, or never true.
   a. Two lines that are parallel to a third line are parallel to each other.
   b. A triangle contains two acute angles.
   c. Vertical angles are complementary angles.

64. For the figure at the right, find the sum of the measures of angles x, y, and z.

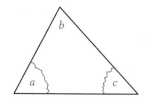

65. For the figure at the right, explain why $\angle a + \angle b = \angle x$. Write a rule that describes the relationship between an exterior angle of a triangle and the opposite interior angles. Use the rule to write an equation involving angles a, c, and z.

66. If $\overline{AB}$ and $\overline{CD}$ intersect at point O, and $\angle AOC = \angle BOC$, explain why $\overline{AB} \perp \overline{CD}$.

67. Do some research on the principle of reflection. Explain how this principle applies to the operation of a periscope and to the game of billiards.

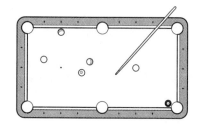

SECTION 9.2   Plane Geometric Figures

OBJECTIVE A
Perimeter of a plane geometric figure

A **polygon** is a closed figure determined by three or more line segments that lie in a plane. The line segments that form the polygon are called its **sides**. The figures below are examples of polygons.

 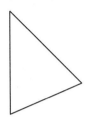 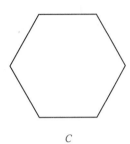 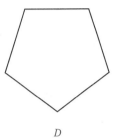 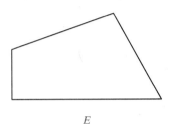

A	*B*	*C*	*D*	*E*

A **regular polygon** is one in which each side has the same length and each angle has the same measure. The polygons in Figures *A*, *C*, and *D* above are regular polygons.

The name of a polygon is based on the number of its sides. The table below lists the names of polygons that have from 3 to 10 sides.

Number of Sides	Name of the Polygon
3	Triangle
4	Quadrilateral
5	Pentagon
6	Hexagon
7	Heptagon
8	Octagon
9	Nonagon
10	Decagon

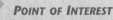

POINT OF INTEREST

Although a polygon is defined in terms of its sides, the word actually comes from the Latin word *polygonum*, which means having many *angles*. This is certainly the case for a polygon.

Triangles and quadrilaterals are two of the most common types of polygons. Triangles are distinguished by the number of equal sides and also by the measures of their angles.

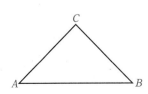   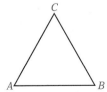   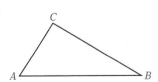

An **isosceles triangle** has two sides of equal length. The angles opposite the equal sides are of equal measure.
$AC = BC$
$\angle A = \angle B$

The three sides of an **equilateral triangle** are of equal length. The three angles are of equal measure.
$AB = BC = AC$
$\angle A = \angle B = \angle C$

A **scalene triangle** has no two sides of equal length. No two angles are of equal measure.

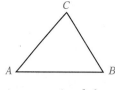

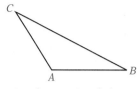

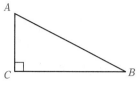

An **acute triangle** has three acute angles.

An **obtuse triangle** has an obtuse angle.

A **right triangle** has a right angle.

Quadrilaterals are also distinguished by their sides and angles, as shown below. Note that a rectangle, a square, and a rhombus are different forms of a parallelogram.

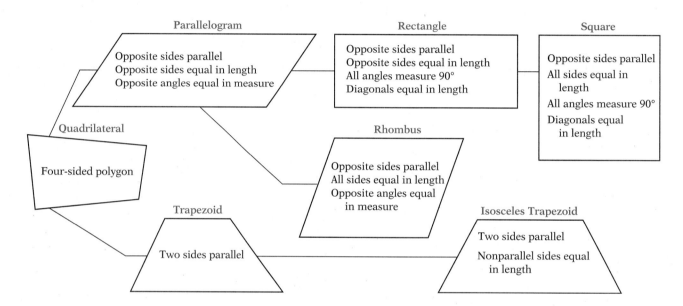

The **perimeter** of a plane geometric figure is a measure of the distance around the figure. Perimeter is used in buying fencing for a lawn or determining how much baseboard is needed for a room.

The perimeter of a triangle is the sum of the lengths of the three sides.

Perimeter of a Triangle

Let a, b, and c be the lengths of the sides of a triangle. The perimeter, P, of the triangle is given by $P = a + b + c$.

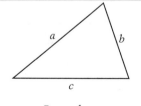

$$P = a + b + c$$

Find the perimeter of the triangle shown at the right.

$P = 5 + 7 + 10 = 22$

The perimeter is 22 ft.

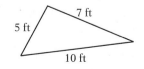

The perimeter of a quadrilateral is the sum of the lengths of its four sides.

A rectangle has four right angles and opposite sides of equal length. Usually the length, L, of a rectangle refers to the length of one of the longer sides of the rectangle, and the width, W, refers to the length of one of the shorter sides. The perimeter can then be represented $P = L + W + L + W$.

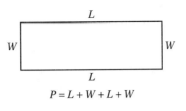

$P = L + W + L + W$

The formula for the perimeter of a rectangle is derived by combining like terms.

$P = 2L + 2W$

Perimeter of a Rectangle

Let L represent the length and W the width of a rectangle. The perimeter, P, of the rectangle is given by $P = 2L + 2W$.

Find the perimeter of the rectangle shown at the right.

The length is 5 m. Substitute 5 for L.
The width is 2 m. Substitute 2 for W.
Solve for P.

$P = 2L + 2W$
$P = 2(5) + 2(2)$
$P = 10 + 4$
$P = 14$

5 m

2 m

The perimeter is 14 m.

A square is a rectangle in which each side has the same length. Letting s represent the length of each side of a square, the perimeter of a square can be represented $P = s + s + s + s$.

$P = s + s + s + s$

The formula for the perimeter of a square is derived by combining like terms.

$P = 4s$

Perimeter of a Square

Let s represent the length of a side of a square. The perimeter, P, of the square is given by $P = 4s$.

Find the perimeter of the square shown at the right.

$P = 4s = 4(8) = 32$

8 in.

The perimeter is 32 in.

A **circle** is a plane figure in which all points are the same distance from point O, called the **center** of the circle.

The **diameter** of a circle is a line segment across the circle through point O. AB is a diameter of the circle at the right. The variable d is used to designate the diameter of a circle.

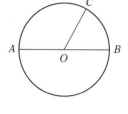

The **radius** of a circle is a line segment from the center of the circle to a point on the circle. OC is a radius of the circle at the right. The variable r is used to designate a radius of a circle.

The length of the diameter is twice the length of the radius.

$$d = 2r \text{ or } r = \frac{1}{2}d$$

The distance around a circle is called the **circumference**. The circumference, C, of a circle is equal to the product of π (pi) and the diameter.

$$C = \pi d$$

Because $d = 2r$, the formula for the circumference can be written in terms of r.

$$C = 2\pi r$$

POINT OF INTEREST

Archimedes (c. 287–212 B.C.) is the person who calculated that $\pi \approx 3\frac{1}{7}$. He actually showed that $3\frac{10}{71} < \pi < 3\frac{1}{7}$. The approximation $3\frac{10}{71}$ is more accurate but more difficult to use.

The Circumference of a Circle

The circumference, C, of a circle with diameter d and radius r is given by $C = \pi d$ or $C = 2\pi r$.

The formula for circumference uses the number π, which is an irrational number. The value of π can be approximated by a fraction or by a decimal.

$$\pi \approx \frac{22}{7} \text{ or } \pi \approx 3.14$$

The π key on a scientific calculator gives a closer approximation of π than 3.14. Use a scientific calculator to find approximate values in calculations involving π.

Find the circumference of a circle with a diameter of 6 in.

The diameter of the circle is given. Use the circumference formula that involves the diameter. $d = 6$.

$$C = \pi d$$
$$C = \pi(6)$$

The exact circumference of the circle is 6π in.

$$C = 6\pi$$

An approximate measure is found by using the π key on a calculator.

$$C \approx 18.85$$

The approximate circumference is 18.85 in.

CALCULATOR NOTE

The π key on your calculator can be used to find decimal approximations to formulas that contain π. To perform the calculation at the right, enter

6 $\boxed{\times}$ $\boxed{\pi}$ $\boxed{=}$.

Example 1

A carpenter is designing a square patio with a perimeter of 44 ft. What is the length of each side?

Strategy

To find the length of each side, use the formula for the perimeter of a square. Substitute 44 for P and solve for s.

Solution

$P = 4s$
$44 = 4s$
$11 = s$

The length of each side of the patio is 11 ft.

You Try It 1

The infield for a softball field is a square with each side of length 60 ft. Find the perimeter of the infield.

Your Strategy

Your Solution

Example 2

The dimensions of a triangular sail are 18 ft, 11 ft, and 15 ft. What is the perimeter of the sail?

Strategy

To find the perimeter, use the formula for the perimeter of a triangle. Substitute 18 for a, 11 for b, and 15 for c. Solve for P.

Solution

$P = a + b + c$
$P = 18 + 11 + 15$
$P - 44$

The perimeter of the sail is 44 ft.

You Try It 2

What is the perimeter of a standard piece of typing paper that measures $8\frac{1}{2}$ in. by 11 in.?

Your Strategy

Your Solution

Example 3

Find the circumference of a circle with a radius of 15 cm. Round to the nearest hundredth.

Strategy

To find the circumference, use the circumference formula that involves the radius. An approximation is asked for; use the π key on a calculator. $r = 15$.

Solution

$C = 2\pi r = 2\pi(15) = 30\pi \approx 94.25$

The circumference is 94.25 cm.

You Try It 3

Find the circumference of a circle with a diameter of 9 in. Give the exact measure.

Your Strategy

Your Solution

Solutions on pp. A26–A27

OBJECTIVE B
Area of a plane geometric figure

Area is the amount of surface in a region. Area can be used to describe the size of a rug, a parking lot, a farm, or a national park. Area is measured in square units.

A square that measures 1 in. on each side has an area of 1 square inch, written 1 in².

A square that measures 1 cm on each side has an area of 1 square centimeter, written 1 cm².

Larger areas can be measured in square feet (ft²), square meters (m²), square miles (mi²), acres (43,560 ft²), or any other square unit.

The area of a geometric figure is the number of squares that are necessary to cover the figure. In the figures below, two rectangles have been drawn and covered with squares. In the figure on the left, 12 squares, each of area 1 cm², were used to cover the rectangle. The area of the rectangle is 12 cm². In the figure on the right, 6 squares, each of area 1 in², were used to cover the rectangle. The area of the rectangle is 6 in².

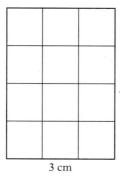

3 cm

The area of the rectangle is 12 cm².

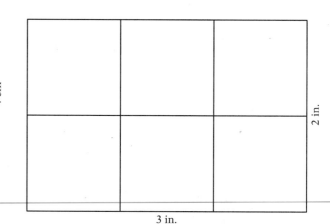

3 in.

The area of the rectangle is 6 in².

Note from the above figures that the area of a rectangle can be found by multiplying the length of the rectangle by its width.

Area of a Rectangle

Let *L* represent the length and *W* the width of a rectangle. The area, *A*, of the rectangle is given by $A = LW$.

Find the area of the rectangle shown at the right.

$A = LW = 11(7) = 77$

The area is 77 m².

A square is a rectangle in which all sides are the same length. Therefore, both the length and the width of a square can be represented by s, and $A - LW = s \cdot s = s^2$

Area of a Square

Let s represent the length of a side of a square. The area, A, of the square is given by $A = s^2$.

$A = s \cdot s = s^2$

Find the area of the square shown at the right.

$A = s^2 = 9^2 = 81$

The area is 81 mi².

9 mi

Figure $ABCD$ is a parallelogram. BC is the **base**, b, of the parallelogram. AE, perpendicular to the base, is the **height**, h, of the parallelogram.

Any side of a parallelogram can be designated as the base. The corresponding height is found by drawing a line segment perpendicular to the base from the opposite side.

A rectangle can be formed from a parallelogram by cutting a right triangle from one end of the parallelogram and attaching it to the other end. The area of the resulting rectangle will equal the area of the original parallelogram.

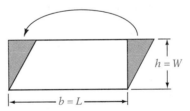

Area of a Parallelogram

Let b represent the length of the base and h the height of a parallelogram. The area, A, of the parallelogram is given by $A = bh$.

Find the area of the parallelogram shown at the right.

$A = bh = 12 \cdot 6 = 72$

The area is 72 m².

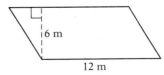

Figure *ABC* is a triangle. *AB* is the **base**, *b*, of the triangle. *CD*, perpendicular to the base, is the **height**, *h*, of the triangle.

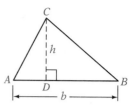

Any side of a triangle can be designated as the base. The corresponding height is found by drawing a line segment perpendicular to the base from the vertex opposite the base.

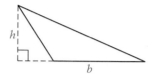

Consider the triangle with base *b* and height *h* shown at the right. By extending a line from *C* parallel to the base *AB* and equal in length to the base, a parallelogram is formed. The area of the parallelogram is *bh* and is twice the area of the triangle. Therefore, the area of the triangle is one-half the area of the parallelogram, or $\frac{1}{2}bh$.

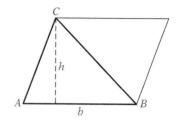

Area of a Triangle

Let *b* represent the length of the base and *h* the height of a triangle. The area, *A*, of the triangle is given by $A = \frac{1}{2}bh$.

Find the area of a triangle with a base of 18 cm and a height of 6 cm.

$A = \frac{1}{2}bh = \frac{1}{2} \cdot 18 \cdot 6 = 54$

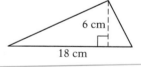

The area is 54 cm².

Figure *ABCD* is a trapezoid. *AB* is one **base**, b_1, of the trapezoid, and *CD* is the other base, b_2. *AE*, perpendicular to the two bases, is the **height**, *h*.

In the trapezoid at the right, the line segment *BD* divides the trapezoid into two triangles, *ABD* and *BCD*. In triangle *ABD*, b_1 is the base and *h* is the height. In triangle *BCD*, b_2 is the base and *h* is the height. The area of the trapezoid is the sum of the areas of the two triangles.

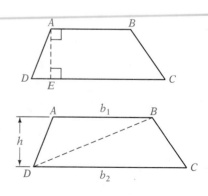

Area of trapezoid *ABCD* = Area of triangle *ABD* + Area of triangle *BCD*

$$= \frac{1}{2}b_1 h + \frac{1}{2}b_2 h = \frac{1}{2}h(b_1 + b_2)$$

Area of a Trapezoid

Let b_1 and b_2 represent the lengths of the bases and h the height of a trapezoid. The area, A, of the trapezoid is given by

$A = \frac{1}{2}h(b_1 + b_2).$

Find the area of a trapezoid that has bases measuring 15 in. and 5 in. and a height of 8 in.

$A = \frac{1}{2}h(b_1 + b_2)$

$= \frac{1}{2} \cdot 8(15 + 5) = 4(20) = 80$

The area is 80 in^2.

The area of a circle is equal to the product of π and the square of the radius.

$A = \pi r^2$

POINT OF INTEREST

The *New Yorker* published an article about two brothers in New York City, David and Gregory Chudnovsky, who have built their own supercomputer out of mail-order electronic hardware. As of 1992, the brothers had used the computer to calculate π to 2.25 billion places. The Chudnovskys are hoping that with enough numbers, some pattern will emerge in the decimal representation of this irrational number.

The Area of a Circle

The area, A, of a circle with radius r is given by $A = \pi r^2$.

Find the area of a circle that has a radius of 6 cm.

Use the formula for the area of a circle. $r = 6$.

$A = \pi r^2$
$A = \pi(6)^2$
$A = \pi(36)$

The exact area of the circle is 36π cm^2.

$A = 36\pi$

An approximate measure is found by using the π key on a calculator.

$A \approx 113.10$

The approximate area of the circle is 113.10 cm^2.

CALCULATOR NOTE

To approximate 36π on your calculator, enter 36 $\boxed{\times}$ $\boxed{\pi}$ $\boxed{=}$.

For your reference, all of the formulas for the perimeter and the area of the geometric figures presented in this section are listed in the Chapter Summary on page 544.

Example 4

The Parks and Recreation Department of a city plans to plant grass seed in a playground that has the shape of a trapezoid, as shown below. Each bag of grass seed will seed 1,500 ft². How many bags of grass seed should the department purchase?

You Try It 4

An interior designer decides to wallpaper two walls of a room. Each roll of wallpaper will cover 30 ft². Each wall measures 8 ft by 12 ft. How many rolls of wallpaper should be purchased?

Strategy

To find the number of bags to be purchased:

▶ Use the formula for the area of a trapezoid to find the area of the playground.
▶ Divide the area of the playground by the area one bag will seed (1,500).

Your Strategy

Solution

$A = \dfrac{1}{2}h(b_1 + b_2)$

$A = \dfrac{1}{2} \cdot 64(80 + 115)$

$A = 6,240$ The area of the playground is 6,240 ft².

$6,240 \div 1,500 = 4.16$

Because a portion of a fifth bag is needed, 5 bags of grass seed should be purchased.

Your Solution

Example 5

Find the area of a circle with a diameter of 5 ft. Give the exact measure.

You Try It 5

Find the area of a circle with a radius of 11 cm. Round to the nearest hundredth.

Strategy

To find the area:

▶ Find the radius of the circle.
▶ Use the formula for the area of a circle. Leave the answer in terms of π.

Your Strategy

Solution

$r = \dfrac{1}{2}d = \dfrac{1}{2}(5) = 2.5$

$A = \pi r^2 = \pi(2.5)^2 = \pi(6.25) = 6.25\pi$

The area of the circle is 6.25π ft².

Your Solution

Solutions on p. A27

9.2 EXERCISES

OBJECTIVE A

Name each polygon.

1.
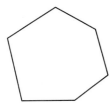

2.
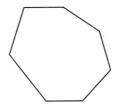

3.
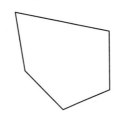

4.
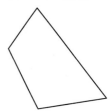

Classify the triangle as isosceles, equilateral, or scalene.

5.

6.
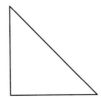

7.
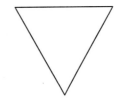

8.
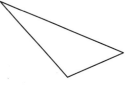

Classify the triangle as acute, obtuse, or right.

9.
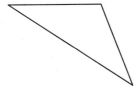

10.
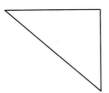

11.

12.
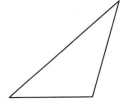

Find the perimeter of the figure.

13.
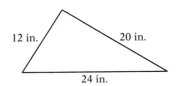
12 in.   20 in.   24 in.

14.
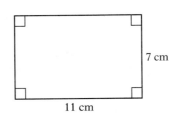
7 cm   11 cm

15.
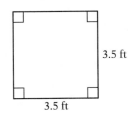
3.5 ft   3.5 ft

16.
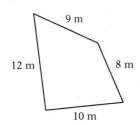
9 m   12 m   8 m   10 m

17.
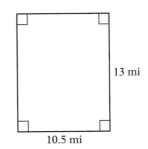
13 mi   10.5 mi

18.
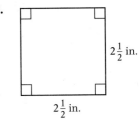
$2\frac{1}{2}$ in. $2\frac{1}{2}$ in.

Find the circumference of the figure.
Give both the exact value and an approximation to the nearest hundredth.

19.

4 cm

20.

12 m

21.

5.5 mi

22.

18 in.

23.

17 ft

24.

6.6 km

Solve.

25. The lengths of the three sides of a triangle are 3.8 cm, 5.2 cm, and 8.4 cm. Find the perimeter of the triangle.

26. The lengths of the three sides of a triangle are 7.5 m, 6.1 m, and 4.9 m. Find the perimeter of the triangle.

27. The length of each of two sides of an isosceles triangle is $2\frac{1}{2}$ cm. The third side measures 3 cm. Find the perimeter of the triangle.

28. The length of each side of an equilateral triangle is $4\frac{1}{2}$ in. Find the perimeter of the triangle.

29. A rectangle has a length of 8.5 m and a width of 3.5 m. Find the perimeter of the rectangle.

30. Find the perimeter of a rectangle that has a length of $5\frac{1}{2}$ ft and a width of 4 ft.

31. The length of each side of a square is 12.2 cm. Find the perimeter of the square.

32. Find the perimeter of a square that is 0.5 m on each side.

33. Find the perimeter of a regular pentagon that measures 3.5 in. on each side.

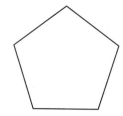

Solve.

34. What is the perimeter of a regular hexagon that measures 8.5 cm on each side?

35. Find the circumference of a circle that has a diameter of 1.5 in. Give the exact value.

36. The diameter of a circle is 4.2 ft. Find the circumference of the circle. Round to the nearest hundredth.

37. The radius of a circle is 36 cm. Find the circumference of the circle. Round to the nearest hundredth.

38. Find the circumference of a circle that has a radius of 2.5 m. Give the exact value.

39. How many feet of fencing should be purchased for a rectangular garden that is 18 ft long and 12 ft wide?

40. How many meters of binding are required to bind the edge of a rectangular quilt that measures 3.5 m by 8.5 m?

41. Wall-to-wall carpeting is installed in a room that is 12 ft long and 10 ft wide. The edges of the carpet are nailed to the floor. Along how many feet must the carpet be nailed down?

42. The length of a rectangular park is 55 yd. The width is 47 yd. How many yards of fencing are needed to surround the park?

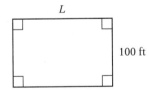

43. The perimeter of a rectangular playground is 440 ft. If the width is 100 ft, what is the length of the playground?

44. A rectangular vegetable garden has a perimeter of 64 ft. The length of the garden is 20 ft. What is the width of the garden?

45. Each of two sides of a triangular banner measures 18 in. If the perimeter of the banner is 46 in., what is the length of the third side of the banner?

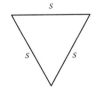

46. The perimeter of an equilateral triangle is 13.2 cm. What is the length of each side of the triangle?

47. The perimeter of a square picture frame is 48 in. Find the length of each side of the frame.

Solve.

48. A square rug has a perimeter of 32 ft. Find the length of each edge of the rug.

49. The circumference of a circle is 8 cm. Find the length of a diameter of the circle. Round to the nearest hundredth.

50. The circumference of a circle is 15 in. Find the length of a radius of the circle. Round to the nearest hundredth.

51. Find the length of molding needed to put around a circular table that is 4.2 ft in diameter. Round to the nearest hundredth.

52. How much binding is needed to bind the edge of a circular rug that is 3 m in diameter? Round to the nearest hundredth.

53. A bicycle tire has a diameter of 24 in. How many feet does the bicycle travel when the wheel makes eight revolutions? Round to the nearest hundredth.

54. A tricycle tire has a diameter of 12 in. How many feet does the tricycle travel when the wheel makes twelve revolutions? Round to the nearest hundredth.

55. The distance from the surface of Earth to its center is 6,356 km. What is the circumference of Earth? Round to the nearest hundredth.

56. Bias binding is to be sewed around the edge of a rectangular tablecloth measuring 72 in. by 45 in. If the bias binding comes in packages containing 15 ft of binding, how many packages of bias binding are needed for the tablecloth?

OBJECTIVE B

Find the area of the figure.

57.

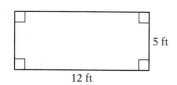

5 ft

12 ft

58.

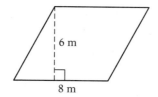

6 m

8 m

59.

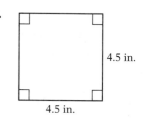

4.5 in.

4.5 in.

Find the area of the figure.

60.

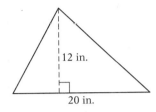

12 in.

20 in.

61.

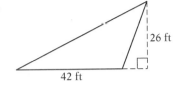

26 ft

42 ft

62.

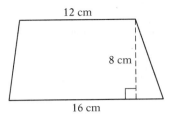

12 cm

8 cm

16 cm

Find the area of the figure.
Give both the exact value and an approximation to the nearest hundredth.

63.

4 cm

64.

12 m

65.

5.5 mi

66.

18 in.

67.

17 ft

68.

6.6 km

Solve.

69. The length of a side of a square is 12.5 cm. Find the area of the square.

70. Each side of a square measures $3\frac{1}{2}$ in. Find the area of the square.

71. The length of a rectangle is 38 in., and the width is 15 in. Find the area of the rectangle.

72. Find the area of a rectangle that has a length of 6.5 m and a width of 3.8 m.

73. The length of the base of a parallelogram is 16 in., and the height is 12 in. Find the area of the parallelogram.

74. The height of a parallelogram is 3.4 m, and the length of the base is 5.2 m. Find the area of the parallelogram.

Solve.

75. The length of the base of a triangle is 6 ft. The height is 4.5 ft. Find the area of the triangle.

76. The height of a triangle is 4.2 cm. The length of the base is 5 cm. Find the area of the triangle.

77. The length of one base of a trapezoid is 35 cm, and the length of the other base is 20 cm. If the height is 12 cm, what is the area of the trapezoid?

78. The height of a trapezoid is 5 in. The bases measure 16 in. and 18 in. Find the area of the trapezoid.

79. The radius of a circle is 5 in. Find the area of the circle. Give the exact value.

80. The diameter of a circle is 6.5 m. Find the area of the circle. Give the exact value.

81. The Hale telescope at Mount Palomar, California, has a diameter of 200 in. Find its area. Give the exact value.

82. An irrigation system waters a circular field that has a 50-foot radius. Find the area watered by the irrigation system. Give the exact value.

83. Find the area of a rectangular flower garden that measures 14 ft by 9 ft.

84. What is the area of a square patio that measures 8.5 m on each side?

85. Artificial turf is being used to cover a playing field. If the field is rectangular with a length of 100 yd and a width of 75 yd, how much artificial turf must be purchased to cover the field?

86. A fabric wall hanging is to fill a space that measures 5 m by 3.5 m. Allowing for 0.1 m of the fabric to be folded back along each edge, how much fabric must be purchased for the wall hanging?

87. The area of a rectangle is 300 in². If the length of the rectangle is 30 in., what is the width?

30 in.

W

88. The width of a rectangle is 12 ft. If the area is 312 ft², what is the length of the rectangle?

Solve.

89. The height of a triangle is 5 m. The area of the triangle is 50 m². Find the length of the base of the triangle.

90. The area of a parallelogram is 42 m². If the height of the parallelogram is 7 m, what is the length of the base?

91. You plan to stain the wooden deck attached to your house. The deck measures 10 ft by 8 ft. If a quart of stain will cover 50 ft², how many quarts of stain should you buy?

92. You want to tile your kitchen floor. The floor measures 12 ft by 9 ft. How many tiles, each a square with side $1\frac{1}{2}$ ft, should you purchase for the job?

93. You are wallpapering two walls of a child's room, one measuring 9 ft by 8 ft and the other measuring 11 ft by 8 ft. The wallpaper costs $18.50 per roll, and each roll of the wallpaper will cover 40 ft². What is the cost to wallpaper the two walls?

94. An urban renewal project involves reseeding a park that is in the shape of a square, 60 ft on each side. Each bag of grass seed costs $5.75 and will seed 1,200 ft². How much money should be budgeted for buying grass seed for the park?

95. A circle has a radius of 8 in. Find the increase in area when the radius is increased by 2 in. Round to the nearest hundredth.

96. A circle has a radius of 6 cm. Find the increase in area when the radius is doubled. Round to the nearest hundredth.

97. You want to install wall-to-wall carpeting in your living room, which measures 15 ft by 24 ft. If the cost of the carpet you would like to purchase is $15.95 per square yard, what is the cost of the carpeting for your living room? (*Hint:* 9 ft² = 1 yd²)

98. You want to paint the walls of your bedroom. Two walls measure 15 ft by 9 ft, and the other two walls measure 12 ft by 9 ft. The paint you wish to purchase costs $12.98 per gallon, and each gallon will cover 400 ft² of wall. Find the total amount you will spend on paint.

99. A walkway 2 m wide surrounds a rectangular plot of grass. The plot is 30 m long and 20 m wide. What is the area of the walkway?

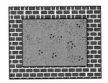

Solve.

100. Pleated draperies for a window must be twice as wide as the width of the window. Draperies are being made for four windows, each 2 ft wide and 4 ft high. Since the drapes will fall slightly below the window sill and extra fabric will be needed for hemming the drapes, 1 ft must be added to the height of the window. How much material must be purchased to make the drapes?

CRITICAL THINKING

101. Find the ratio of the areas of two squares if the ratio of the lengths of their sides is 2:3.

102. If both the length and the width of a rectangle are doubled, how many times larger is the area of the resulting rectangle?

103. If the formula $C = \pi d$ is solved for π, the resulting equation is $\pi = \dfrac{C}{d}$.

Therefore, π is the ratio of the circumference of a circle to the length of its diameter. Use several circular objects, such as coins, plates, tin cans, and wheels, to show that the ratio of the circumference of each object to its diameter is approximately equal to 3.14.

104. Derive a formula for the area of a circle in terms of the diameter of the circle.

105. Determine whether the statement is always true, sometimes true, or never true.
   a. Two triangles that have the same perimeter have the same area.
   b. Two rectangles that have the same area have the same perimeter.
   c. If two squares have the same area, then the sides of the squares have the same length.
   d. An equilateral triangle is also an isosceles triangle.
   e. All the radii (plural of radius) of a circle are equal.
   f. All the diameters of a circle are equal.

106. Suppose a circle is cut into 16 equal pieces, which are then arranged as shown at the right. The figure formed resembles a parallelogram. What variable expression could describe the base of the parallelogram? What variable could describe its height? Explain how the formula for the area of a circle is derived from this approach.

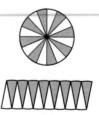

107. Prepare a report on the history of quilts in America. Find examples of quilt patterns that incorporate regular polygons. Use pieces of cardboard to create the shapes needed for one block of one of the quilt patterns you learned about.

108. The **apothem** of a regular polygon is the distance from the center of the polygon to a side. Explain how to derive a formula for the area of a regular polygon using the apothem. (*Hint:* Use the formula for the area of a triangle.)

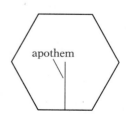

apothem

SECTION 9.3   Triangles

OBJECTIVE A
The Pythagorean Theorem

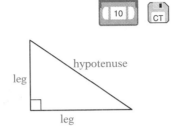

A **right triangle** contains one right angle. The side opposite the right angle is called the **hypotenuse**. The other two sides are called **legs**.

The angles in a right triangle are usually labeled with the capital letters A, B, and C, with C reserved for the right angle. The side opposite angle A is side a, the side opposite angle B is side b, and c is the hypotenuse.

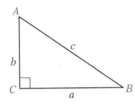

The Greek mathematician Pythagoras is generally credited with the discovery that the square of the hypotenuse of a right triangle is equal to the sum of the squares of the two legs. This is called the **Pythagorean Theorem**.

The figure at the right is a right triangle with legs measuring 3 units and 4 units and a hypotenuse measuring 5 units. Each side of the triangle is also the side of a square. The number of square units in the area of the largest square is equal to the sum of the numbers of square units in the areas of the smaller squares.

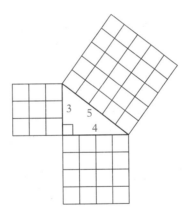

POINT OF INTEREST

The first known proof of the Pythagorean Theorem is in a Chinese textbook that dates from 150 B.C. The book is called the *Nine Chapters on the Mathematical Art*. The diagram below is from that book and was used in the proof of the theorem.

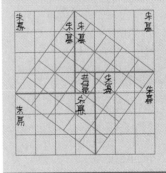

Square of the hypotenuse	=	sum of the squares of the two legs

$$5^2 = 3^2 + 4^2$$
$$25 = 9 + 16$$
$$25 = 25$$

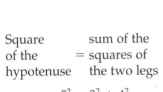

Pythagorean Theorem

If a and b are the lengths of the legs of a right triangle and c is the length of the hypotenuse, then $c^2 = a^2 + b^2$.

If the lengths of two sides of a right triangle are known, the Pythagorean Theorem can be used to find the length of the third side.

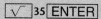

5 cm / 12 cm / c

Consider a right triangle with legs that measure 5 cm and 12 cm. Use the Pythagorean Theorem, with $a = 5$ and $b = 12$, to find the length of the hypotenuse. (If you let $a = 12$ and $b = 5$, the result is the same.)

$$c^2 = a^2 + b^2$$
$$c^2 = 5^2 + 12^2$$
$$c^2 = 25 + 144$$
$$c^2 = 169$$

This equation states that the square of c is 169. Since $13^2 = 169$, $c = 13$, and the length of the hypotenuse is 13 cm. We can find c by taking the square root of 169: $\sqrt{169} = 13$. This suggests the following property.

CALCULATOR NOTE

The way in which you evaluate the square root of a number will depend on the type of calculator you have. Here are two possible calculator keystrokes to find $\sqrt{35}$.

Method 1:

35 √

Method 2:

√ 35 ENTER

Method 1 is used on many scientific calculators; Method 2 is used on many graphing calculators.

The Principal Square Root Property

If $r^2 = s$, then $r = \sqrt{s}$, and r is called the square root of s.

The Principal Square Root Property and its application can be illustrated as follows: Because $5^2 = 25$, $5 = \sqrt{25}$. Therefore, if $c^2 = 25$, $c = \sqrt{25} = 5$.

Recall that numbers whose square roots are integers, such as 25, are perfect squares. If a number is not a perfect square, a calculator can be used to find an approximate square root when a decimal approximation is required.

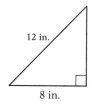

12 in. / 8 in.

The length of one leg of a right triangle is 8 in. The hypotenuse is 12 in. Find the length of the other leg. Round to the nearest hundredth.

Use the Pythagorean Theorem.
$a = 8$, $c = 12$
Solve for b^2.
(If you let $b = 8$ and solve for a^2, the result is the same.)

$$a^2 + b^2 = c^2$$
$$8^2 + b^2 = 12^2$$
$$64 + b^2 = 144$$
$$b^2 = 80$$

Use the Principal Square Root Property.

Since $b^2 = 80$, b is the square root of 80. $b = \sqrt{80}$

Use a calculator to approximate $\sqrt{80}$. $b \approx 8.94$

The length of the other leg is approximately 8.94 in.

Example 1

The two legs of a right triangle measure 12 ft and 9 ft. Find the hypotenuse of the right triangle.

Strategy

To find the hypotenuse, use the Pythagorean Theorem. $a = 12$, $b = 9$

Solution

$$c^2 = a^2 + b^2$$
$$c^2 = 12^2 + 9^2$$
$$c^2 = 144 + 81$$
$$c^2 = 225$$
$$c = \sqrt{225}$$
$$c = 15$$

The length of the hypotenuse is 15 ft.

You Try It 1

The hypotenuse of a right triangle measures 6 m, and one leg measures 2 m. Find the measure of the other leg. Round to the nearest hundredth.

Your Strategy

Your Solution

Solution on p. A27

OBJECTIVE B
Similar triangles

Similar objects have the same shape but not necessarily the same size. A tennis ball is similar to a basketball. A model ship is similar to an actual ship.

Similar objects have corresponding parts; for example, the rudder on the model ship corresponds to the rudder on the actual ship. The relationship between the sizes of each of the corresponding parts can be written as a ratio, and each ratio will be the same. If the rudder on the model ship is $\frac{1}{100}$ the size of the rudder on the actual ship, then the model wheelhouse is $\frac{1}{100}$ the size of the actual wheelhouse, the width of the model is $\frac{1}{100}$ the width of the actual ship, and so on.

The two triangles ABC and DEF shown at the right are similar. Side AB corresponds to side DE, side BC corresponds to side EF, and side AC corresponds to side DF. The ratios of corresponding sides are equal.

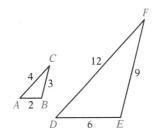

$$\frac{AB}{DE} = \frac{2}{6} = \frac{1}{3}, \frac{BC}{EF} = \frac{3}{9} = \frac{1}{3}, \text{ and } \frac{AC}{DF} = \frac{4}{12} = \frac{1}{3}.$$

Since the ratios of corresponding sides are equal, three proportions can be formed.

$$\frac{AB}{DE} = \frac{BC}{EF}, \frac{AB}{DE} = \frac{AC}{DF}, \text{ and } \frac{BC}{EF} = \frac{AC}{DF}.$$

The corresponding angles in similar triangles are equal. Therefore,

$$\angle A = \angle D, \angle B = \angle E, \text{ and } \angle C = \angle F.$$

Triangles ABC and DEF at the right are similar triangles. AH and DK are the heights of the triangles. The ratio of heights of similar triangles equals the ratio of corresponding sides.

Ratio of corresponding sides $= \dfrac{1.5}{6} = \dfrac{1}{4}$

Ratio of heights $= \dfrac{1}{4}$

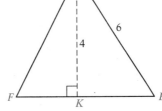

Properties of Similar Triangles

For similar triangles, the ratios of corresponding sides are equal. The ratio of corresponding heights is equal to the ratio of corresponding sides.

The two triangles at the right are similar triangles. Find the length of side *EF*. Round to the nearest tenth.

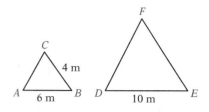

The triangles are similar, so the ratios of corresponding sides are equal.

$$\frac{EF}{BC} = \frac{DE}{AB}$$

$$\frac{EF}{4} = \frac{10}{6}$$

$$6(EF) = 4(10)$$
$$6(EF) = 40$$
$$EF \approx 6.7$$

The length of side *EF* is approximately 6.7 m.

Example 2

Triangles *ABC* and *DEF* are similar. Find *FG*, the height of triangle *DEF*.

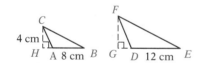

Strategy

To find *FG*, write a proportion using the fact that, in similar triangles, the ratio of corresponding sides equals the ratio of corresponding heights. Solve the proportion for *FG*.

Solution

$$\frac{AB}{DE} = \frac{CH}{FG}$$

$$\frac{8}{12} = \frac{4}{FG}$$

$$8(FG) = 12(4)$$
$$8(FG) = 48$$
$$FG = 6$$

The height *FG* of triangle *DEF* is 6 cm.

You Try It 2

Triangles *ABC* and *DEF* are similar. Find *FG*, the height of triangle *DEF*.

C 7 m   15 m   *F*
10 m
A   *H*   *B*   *D*   *G*   *E*

Your Strategy

Your Solution

OBJECTIVE C
Congruent triangles

Congruent objects have the same shape *and* the same size.

The two triangles at the
right are congruent. They
have the same size.

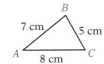 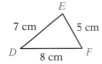

Congruent and similar triangles differ in that congruent means that the corresponding sides and angles of the triangle must be equal; for similar triangles, corresponding angles are equal, but corresponding sides are not
necessarily the same length.

The three major rules used to determine whether two triangles are congruent
are given below.

Side-Side-Side Rule (SSS)

Two triangles are congruent if the three sides of one triangle
equal the corresponding three sides of a second triangle.

In the triangles at the right,
$AC = DE$, $AB = EF$, and
$BC = DF$. The corresponding sides of triangles ABC
and DEF are equal. The triangles are congruent by the
SSS Rule.

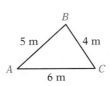

Side-Angle-Side Rule (SAS)

If two sides and the included angle of one triangle equal two
sides and the included angle of a second triangle, the two
triangles are congruent.

In the two triangles at
the right, $AB = EF$,
$AC = DE$, and
$\angle BAC = \angle DEF$.   The
triangles are congru-
ent by the SAS Rule.

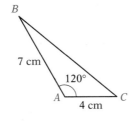 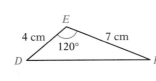

Angle-Side-Angle Rule (ASA)

If two angles and the included side of one triangle equal two angles and the included side of a second triangle, the two triangles are congruent.

For triangles *ABC* and *DEF* at the right, ∠*A* = ∠*F*, ∠*C* = ∠*E*, and *AC* = *EF*. The triangles are congruent by the ASA Rule.

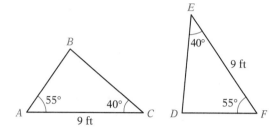

Given triangle *PQR* and triangle *MNO*, do the conditions ∠*P* = ∠*O*, ∠*Q* = ∠*M*, and *PQ* = *MO* guarantee that triangle *PQR* is congruent to triangle *MNO*?

Draw a sketch of the two triangles and determine whether one of the rules for congruence is satisfied.

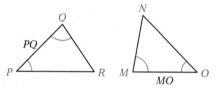

Because two angles and the included side of one triangle equal two angles and the included side of the second triangle, the triangles are congruent by the ASA Rule.

Example 3

In the figure below, is triangle *ABC* congruent to triangle *DEF*?

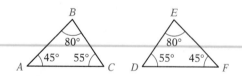

Strategy

To determine whether the triangles are congruent, determine whether one of the rules for congruence is satisfied.

Solution

The triangles do not satisfy the SSS Rule, the SAS Rule, or the ASA Rule. The triangles are not necessarily congruent.

You Try It 3

In the figure below, is triangle *PQR* congruent to triangle *MNO*?

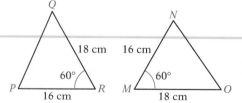

Your Strategy

Your Solution

Solution on p. A27

9.3 EXERCISES

OBJECTIVE A

Find the unknown side of the triangle. Round to the nearest tenth.

1.
3 in.
4 in.

2.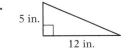
5 in.
12 in.

3.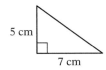
5 cm
7 cm

4.
7 cm
9 cm

5.
15 ft
10 ft

6.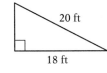
20 ft
18 ft

7.
4 cm   6 cm

8.
9 m   12 m

9.
9 yd
9 yd

Solve. Round to the nearest tenth.

10. A ladder 8 m long is leaning against a building. How high on the building will the ladder reach when the bottom of the ladder is 3 m from the building?

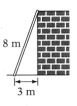
8 m
3 m

11. Find the distance between the centers of the holes in the metal plate.

12. If you travel 18 mi east and then 12 mi north, how far are you from your starting point?

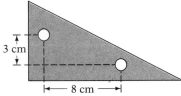
3 cm
8 cm

13. Find the perimeter of a right triangle with legs that measure 5 cm and 9 cm.

14. Find the perimeter of a right triangle with legs that measure 6 in. and 8 in.

OBJECTIVE B

Find the ratio of corresponding sides for the similar triangles.

15.

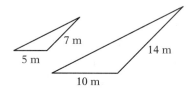

16.

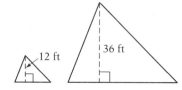

17.

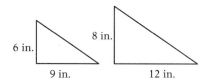

18.

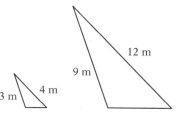

In Exercises 19–26, triangles *ABC* and *DEF* are similar triangles. Solve and round to the nearest tenth.

19. Find side *DE*.

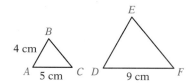

20. Find side *DE*.

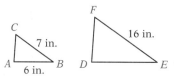

21. Find the height of triangle *DEF*.

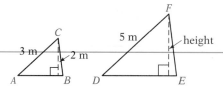

22. Find the height of triangle *ABC*.

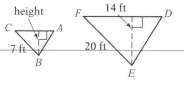

23. Find the perimeter of triangle *ABC*.

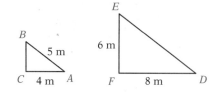

24. Find the perimeter of triangle *DEF*.

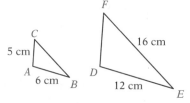

25. Find the perimeter of triangle *ABC*.

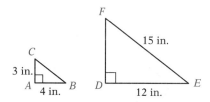

26. Find the area of triangle *DEF*.

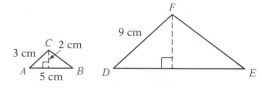

In Exercises 27 and 28, triangles *ABC* and *DEF* are similar triangles. Solve and round to the nearest tenth.

27. Find the area of triangle *ABC*.

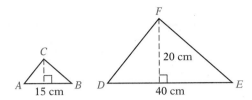

28. Find the area of triangle *DEF*.

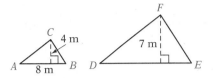

The sun's rays, objects on Earth, and the shadows cast by them form similar triangles. Use this fact to solve Exercises 29–32.

29. Find the height of the flagpole.

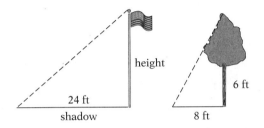

30. Find the height of the flagpole.

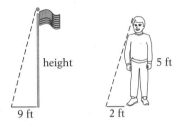

31. Find the height of the building.

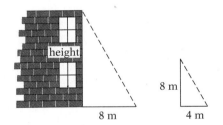

32. Find the height of the building.

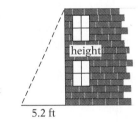

OBJECTIVE C

Determine whether the two triangles are congruent. If they are congruent, state by what rule they are congruent.

33.

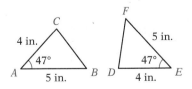

34.

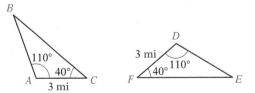

35.

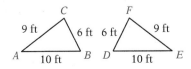

36.

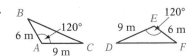

Determine whether the two triangles are congruent. If they are congruent, state by what rule they are congruent.

37.

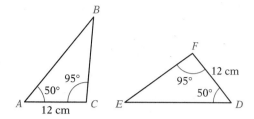

38.

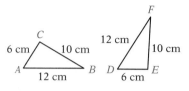

Solve.

39. Given triangle *ABC* and triangle *DEF*, do the conditions $\angle C = \angle E$, $AC = EF$, and $BC = DE$ guarantee that triangle *ABC* is congruent to triangle *DEF*? If they are congruent, by what rule are they congruent?

40. Given triangle *PQR* and triangle *MNO*, do the conditions $PR = NO$, $PQ = MO$, and $QR = MN$ guarantee that triangle *PQR* is congruent to triangle *MNO*? If they are congruent, by what rule are they congruent?

41. Given triangle *LMN* and triangle *QRS*, do the conditions $\angle M = \angle S$, $\angle N = \angle Q$, and $\angle L = \angle R$ guarantee that triangle *LMN* is congruent to triangle *QRS*? If they are congruent, by what rule are they congruent?

42. Given triangle *DEF* and triangle *JKL*, do the conditions $\angle D = \angle K$, $\angle E = \angle L$, and $DE = KL$ guarantee that triangle *DEF* is congruent to triangle *JKL*? If they are congruent, by what rule are they congruent?

43. Given triangle *ABC* and triangle *PQR*, do the conditions $\angle B = \angle P$, $BC = PQ$, and $AC = QR$ guarantee that triangle *ABC* is congruent to triangle *PQR*? If they are congruent, by what rule are they congruent?

CRITICAL THINKING

44. Congruent triangles were a topic of this section. Use the concept of congruent triangles to derive the formula for the area of a parallelogram given that the area of a rectangle is $A = LW$.

45. Determine whether the statement is always true, sometimes true, or never true.
   a. If two angles of one triangle are equal to two angles of a second triangle, then the triangles are similar triangles.
   b. Two isosceles triangles are similar triangles.
   c. Two equilateral triangles are similar triangles.

46. What is a Pythagorean triple? Provide at least three examples of Pythagorean triples.

47. Provide definitions of the words *property*, *principle*, *axiom*, and *theorem*.

SECTION 9.4   Solids

OBJECTIVE A
Volume of a solid

Geometric solids are figures in space. Five common geometric solids are the rectangular solid, the sphere, the cylinder, the cone, and the pyramid.

A **rectangular solid** is one in which all six sides, called **faces**, are rectangles. The variable L is used to represent the length of a rectangular solid, W its width, and H its height.

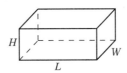

A **sphere** is a solid in which all points are the same distance from point O, called the **center** of the sphere. The **diameter**, d, of a sphere is a line across the sphere going through point O. The **radius**, r, is a line from the center to a point on the sphere. AB is a diameter and OC is a radius of the sphere shown at the right.

$$d = 2r \quad \text{or} \quad r = \frac{1}{2}d$$

The most common cylinder, called a **right circular cylinder**, is one in which the bases are circles and are perpendicular to the height of the cylinder. The variable r is used to represent the radius of a base of a cylinder, and h represents the height. In this text, only right circular cylinders are discussed.

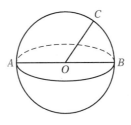

A **right circular cone** is obtained when one base of a right circular cylinder is shrunk to a point, called the **vertex**, V. The variable r is used to represent the radius of the base of the cone, and h represents the height. The variable l is used to represent the **slant height**, which is the distance from a point on the circumference of the base to the vertex. In this text, only right circular cones are discussed.

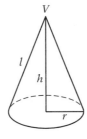

The base of a **regular pyramid** is a regular polygon, and the sides are isosceles triangles. The height, h, is the distance from the vertex, V, to the base and is perpendicular to the base. The variable l is used to represent the **slant height**, which is the height of one of the isosceles triangles on the face of the pyramid. The regular square pyramid at the right has a square base. This is the only type of pyramid discussed in this text.

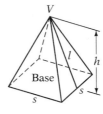

A **cube** is a special type of rectangular solid. Each of the six faces of a cube is a square. The variable *s* is used to represent the length of one side of a cube.

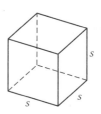

Volume is a measure of the amount of space inside a figure in space. Volume can be used to describe the amount of heating gas used for cooking, the amount of concrete delivered for the foundation of a house, or the amount of water in storage for a city's water supply.

A cube that is 1 ft on each side has a volume of 1 cubic foot, which is written 1 ft^3. A cube that measures 1 cm on each side has a volume of 1 cubic centimeter, written 1 cm^3.

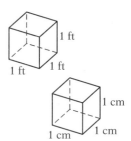

The volume of a solid is the number of cubes that are necessary to exactly fill the solid. The volume of the rectangular solid at the right is 24 cm^3 because it will hold exactly 24 cubes, each 1 cm on a side. Note that the volume can be found by multiplying the length times the width times the height.

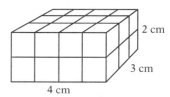

$4 \cdot 3 \cdot 2 = 24$

The formulas for the volumes of the geometric solids described above are given below.

Volumes of Geometric Solids

The volume, *V*, of a **rectangular solid** with length *L*, width *W*, and height *H* is given by $V = LWH$.

The volume, *V*, of a **cube** with side *s* is given by $V = s^3$.

The volume, *V*, of a **sphere** with radius *r* is given by $V = \frac{4}{3}\pi r^3$.

The volume, *V*, of a **right circular cylinder** is given by $V = \pi r^2 h$, where *r* is the radius of the base and *h* is the height.

The volume, *V*, of a **right circular cone** is given by $V = \frac{1}{3}\pi r^2 h$, where *r* is the radius of the circular base and *h* is the height.

The volume, *V*, of a **regular square pyramid** is given by $V = \frac{1}{3}s^2 h$, where *s* is the length of a side of the base and *h* is the height.

Find the volume of a sphere with a diameter of 6 in.

First find the radius of the sphere.

$r = \dfrac{1}{2}d = \dfrac{1}{2}(6) = 3$

Use the formula for the volume of a sphere.

$V = \dfrac{4}{3}\pi r^3$

$V = \dfrac{4}{3}\pi(3)^3$

$V = \dfrac{4}{3}\pi(27)$

The exact volume of the sphere is 36π in^3.

$V = 36\pi$

An approximate measure can be found by using the π key on a calculator.

$V \approx 113.10$

The approximate volume is 113.10 in^3.

CALCULATOR NOTE

To approximate 36π on your calculator, enter 36 $\boxed{\times}$ $\boxed{\pi}$ $\boxed{=}$.

Example 1

The length of a rectangular solid is 5 m, the width is 3.2 m, and the height is 4 m. Find the volume of the solid.

Strategy

To find the volume, use the formula for the volume of a rectangular solid. $L = 5$, $W = 3.2$, $H = 4$

Solution

$V = LWH = 5(3.2)(4) = 64$

The volume of the rectangular solid is 64 m^3.

You Try It 1

Find the volume of a cube that measures 2.5 m on a side.

Your Strategy

Your Solution

Example 2

The radius of the base of a cone is 8 cm. The height is 12 cm. Find the volume of the cone. Round to the nearest hundredth.

Strategy

To find the volume, use the formula for the volume of a cone. An approximation is asked for; use the π key on a calculator. $r = 8$, $h = 12$

Solution

$V = \dfrac{1}{3}\pi r^2 h$

$V = \dfrac{1}{3}\pi(8)^2(12) = \dfrac{1}{3}\pi(64)(12) = 256\pi \approx 804.25$

The volume is approximately 804.25 cm^3.

You Try It 2

The diameter of the base of a cylinder is 8 ft. The height of the cylinder is 22 ft. Find the exact volume of the cylinder.

Your Strategy

Your Solution

Solutions on p. A27

OBJECTIVE B
Surface area of a solid

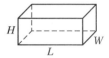

The **surface area** of a solid is the total area on the surface of the solid.

When a rectangular solid is cut open and flattened out, each face is a rectangle. The surface area, SA, of the rectangular solid is the sum of the areas of the six rectangles,

$$SA = LW + LH + WH + LW + WH + LH$$

which simplifies to

$$SA = 2LW + 2LH + 2WH$$

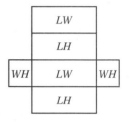

The surface area of a cube is the sum of the areas of the six faces of the cube. The area of each face is s^2. Therefore, the surface area, SA, of a cube is given by the formula $SA = 6s^2$.

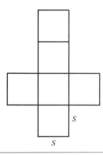

When a cylinder is cut open and flattened out, the top and bottom of the cylinder are circles. The side of the cylinder flattens out to a rectangle. The length of the rectangle is the circumference of the base, which is $2\pi r$; the width is h, the height of the cylinder. Therefore, the area of the rectangle is $2\pi rh$. The surface area, SA, of the cylinder is

$$SA = \pi r^2 + 2\pi rh + \pi r^2$$

which simplifies to

$$SA = 2\pi r^2 + 2\pi rh$$

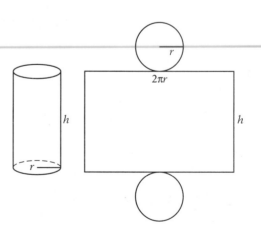

The surface area of a pyramid is the area of the base plus the area of the four isosceles triangles. A side of the square base is s; therefore, the area of the base is s^2. The slant height, l, is the height of each triangle, and s is the base of each triangle. The surface area, SA, of a pyramid is

$$SA = s^2 + 4\left(\frac{1}{2}sl\right)$$

which simplifies to

$$SA = s^2 + 2sl$$

Formulas for the surface areas of geometric solids are given below.

Surface Areas of Geometric Solids

The surface area, SA, of a **rectangular solid** with length L, width W, and height H is given by $SA = 2LW + 2LH + 2WH$.

The surface area, SA, of a **cube** with side s is given by $SA = 6s^2$.

The surface area, SA, of a **sphere** with radius r is given by $SA = 4\pi r^2$.

The surface area, SA, of a **right circular cylinder** is given by $SA = 2\pi r^2 + 2\pi rh$, where r is the radius of the base and h is the height.

The surface area, SA, of a **right circular cone** is given by $SA = \pi r^2 + \pi rl$, where r is the radius of the circular base and l is the slant height.

The surface area, SA, of a **regular pyramid** is given by $SA = s^2 + 2sl$, where s is the length of a side of the base and l is the slant height.

Find the surface area of a sphere with a diameter of 18 cm.

First find the radius of the sphere. $r = \dfrac{1}{2}d = \dfrac{1}{2}(18) = 9$

Use the formula for the surface area of a sphere.

$SA = 4\pi r^2$
$SA = 4\pi(9)^2$
$SA = 4\pi(81)$
$SA = 324\pi$

The exact surface area of the sphere is 324π cm².

An approximate measure can be found by using the π key on a calculator. $SA \approx 1{,}017.88$

The approximate surface area is 1,017.88 cm².

CALCULATOR NOTE

To approximate 324π on your calculator, enter

324 ☐×☐ ☐π☐ ☐=☐.

Example 3

The diameter of the base of a cone is 5 m, and the slant height is 4 m. Find the surface area of the cone. Give the exact measure.

Strategy

To find the surface area of the cone:

▶ Find the radius of the base of the cone.
▶ Use the formula for the surface area of a cone. Leave the answer in terms of π.

Solution

$$r = \frac{1}{2}d = \frac{1}{2}(5) = 2.5$$

$$SA = \pi r^2 + \pi r l$$
$$SA = \pi(2.5)^2 + \pi(2.5)(4)$$
$$SA = \pi(6.25) + \pi(2.5)(4)$$
$$SA = 6.25\pi + 10\pi$$
$$SA = 16.25\pi$$

The surface area of the cone is 16.25π m^2.

You Try It 3

The diameter of the base of a cylinder is 6 ft, and the height is 8 ft. Find the surface area of the cylinder. Round to the nearest hundredth.

Your Strategy

Your Solution

Example 4

Find the area of a label used to cover a soup can that has a radius of 4 cm and a height of 12 cm. Round to the nearest hundredth.

Strategy

To find the area of the label, use the fact that the surface area of the sides of a cylinder is given by $2\pi rh$. An approximation is asked for; use the π key on a calculator. $r = 4$, $h = 12$

Solution

Area of the label $= 2\pi rh$
Area of the label $= 2\pi(4)(12) = 96\pi \approx 301.59$

The area is approximately 301.59 cm^2.

You Try It 4

Which has a larger surface area, a cube with a side measuring 10 cm or a sphere with a diameter measuring 8 cm?

Your Strategy

Your Solution

9.4 EXERCISES

OBJECTIVE A

Find the volume of the figure. For calculations involving π, give both the exact value and an approximation to the nearest hundredth.

1.
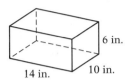
6 in.
14 in.   10 in.

2.
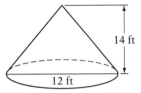
14 ft
12 ft

3.
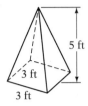
5 ft
3 ft
3 ft

4.
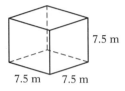
7.5 m
7.5 m   7.5 m

5.

3 cm

6.
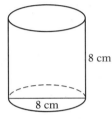
8 cm
8 cm

Solve.

7. A rectangular solid has a length of 6.8 m, a width of 2.5 m, and a height of 2 m. Find the volume of the solid.

8. Find the volume of a rectangular solid that has a length of 4.5 ft, a width of 3 ft, and a height of 1.5 ft.

9. Find the volume of a cube whose side measures 2.5 in.

10. The length of a side of a cube is 7 cm. Find the volume of the cube.

11. The diameter of a sphere is 6 ft. Find the volume of the sphere. Give the exact measure.

12. Find the volume of a sphere that has a radius of 1.2 m. Round to the nearest tenth.

13. The diameter of the base of a cylinder is 24 cm. The height of the cylinder is 18 cm. Find the volume of the cylinder. Round to the nearest hundredth.

14. The height of a cylinder is 7.2 m. The radius of the base is 4 m. Find the volume of the cylinder. Give the exact measure.

15. The radius of the base of a cone is 5 in. The height of the cone is 9 in. Find the volume of the cone. Give the exact measure.

Solve.

16. The height of a cone is 15 cm. The diameter of the cone is 10 cm. Find the volume of the cone. Round to the nearest hundredth.

17. The length of a side of the base of a pyramid is 6 in., and the height is 10 in. Find the volume of the pyramid.

18. The height of a pyramid is 8 m, and the length of a side of the base is 9 m. What is the volume of the pyramid?

19. The volume of a freezer with a length of 7 ft and a height of 3 ft is 52.5 ft³. Find the width of the freezer.

20. The length of an aquarium is 18 in., and the width is 12 in. If the volume of the aquarium is 1,836 in³, what is the height of the aquarium?

21. The volume of a cylinder with a height of 10 in. is 502.4 in³. Find the radius of the base of the cylinder. Round to the nearest hundredth.

22. The diameter of the base of a cylinder is 14 cm. If the volume of the cylinder is 2,310 cm³, find the height of the cylinder. Round to the nearest hundredth.

23. A rectangular solid has a square base and a height of 5 in. If the volume of the solid is 125 in³, find the length and the width.

24. The volume of a rectangular solid is 864 m³. The rectangular solid has a square base and a height of 6 m. Find the dimensions of the solid.

25. An oil storage tank, which is in the shape of a cylinder, is 4 m high and has a diameter of 6 m. The oil tank is two-thirds full. Find the number of cubic meters of oil in the tank. Round to the nearest hundredth.

26. A silo, which is in the shape of a cylinder, is 16 ft in diameter and has a height of 30 ft. The silo is three-fourths full. Find the volume of the portion of the silo that is not being used for storage. Round to the nearest hundredth.

OBJECTIVE B

Find the surface area of the figure.

27.
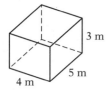
3 m
5 m
4 m

28.
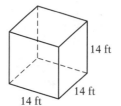
14 ft
14 ft
14 ft

29.
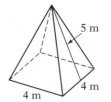
5 m
4 m
4 m

Find the surface area of the figure. Give both the exact value and an approximation to the nearest hundredth.

30.

2 cm

31.

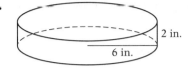

2 in.

6 in.

32.

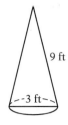

9 ft

-3 ft-

Solve.

33. The height of a rectangular solid is 5 ft. The length is 8 ft, and the width is 4 ft. Find the surface area of the solid.

34. The width of a rectangular solid is 32 cm. The length is 60 cm, and the height is 14 cm. What is the surface area of the solid?

35. The side of a cube measures 3.4 m. Find the surface area of the cube.

36. Find the surface area of a cube that has a side measuring 1.5 in.

37. Find the surface area of a sphere with a diameter of 15 cm. Give the exact value.

38. The radius of a sphere is 2 in. Find the surface area of the sphere. Round to the nearest hundredth.

39. The radius of the base of a cylinder is 4 in. The height of the cylinder is 12 in. Find the surface area of the cylinder. Round to the nearest hundredth.

40. The diameter of the base of a cylinder is 1.8 m. The height of the cylinder is 0.7 m. Find the surface area of the cylinder. Give the exact value.

41. The slant height of a cone is 2.5 ft. The radius of the base is 1.5 ft. Find the surface area of the cone. Give the exact value.

42. The diameter of the base of a cone is 21 in. The slant height is 16 in. What is the surface area of the cone? Round to the nearest hundredth.

43. The length of a side of the base of a pyramid is 9 in., and the slant height is 12 in. Find the surface area of the pyramid.

44. The slant height of a pyramid is 18 m, and the length of a side of the base is 16 m. What is the surface area of the pyramid?

Solve.

45. The surface area of a rectangular solid is 108 cm². The height of the solid is 4 cm, and the length is 6 cm. Find the width of the rectangular solid.

46. The length of a rectangular solid is 12 ft. The width is 3 ft. If the surface area is 162 ft², find the height of the rectangular solid.

47. A can of paint will cover 300 ft². How many cans of paint should be purchased in order to paint a cylinder that has a height of 30 ft and a radius of 12 ft?

48. A hot air balloon is in the shape of a sphere. Approximately how much fabric was used to construct the balloon if its diameter is 32 ft? Round to the nearest whole number.

49. How much glass is needed to make a fish tank that is 12 in. long, 8 in. wide, and 9 in. high? The fish tank is open at the top.

50. Find the area of a label used to cover a can of juice that has a diameter of 16.5 cm and a height of 17 cm. Round to the nearest hundredth.

51. The length of a side of the base of a pyramid is 5 cm, and the slant height is 8 cm. How much larger is the surface area of this pyramid than the surface area of a cone with a diameter of 5 cm and a slant height of 8 cm? Round to the nearest hundredth.

CRITICAL THINKING

52. Half of a sphere is called a **hemisphere**. Derive formulas for the volume and surface area of a hemisphere.

53. Determine whether the statement is always true, sometimes true, or never true.
    a. The slant height of a regular pyramid is longer than the height.
    b. The slant height of a cone is shorter than the height.
    c. The four triangular faces of a regular pyramid are equilateral triangles.

54. a. What is the effect on the surface area of a rectangular solid if the width and height are doubled?
    b. What is the effect on the volume of a rectangular solid if both the length and the width are doubled?
    c. What is the effect on the volume of a cube if the length of each side of the cube is doubled?
    d. What is the effect on the surface area of a cylinder if the radius and height are doubled?

55. Explain how you could cut through a cube so that the face of the resulting solid is (a) a square, (b) an equilateral triangle, (c) a trapezoid, (d) a hexagon.

OBJECTIVE A
Perimeter of a composite plane figure

Composite geometric figures are made from two or more geometric figures. The composite figure below is made from parts of a rectangle and a circle.

Composite figure = 3 sides of a rectangle $+ \dfrac{1}{2}$ the circumference of a circle

Perimeter   = $2L + W$ + $\dfrac{1}{2}\pi d$

Find the perimeter of the composite figure shown above if the width of the rectangle is 4 m and the length of the rectangle is 8 m. Round to the nearest hundredth.

Use the equation given above. $L = 8, W = 4$. The diameter of the circle equals the width of the rectangle, 4.

$$P = 2L + W + \dfrac{1}{2}\pi d$$

$$P = 2(8) + 4 + \dfrac{1}{2}\pi(4)$$

Use the π key on a calculator to approximate the perimeter.

$$P = 20 + 2\pi$$
$$P \approx 26.28$$

To the nearest hundredth, the perimeter of the figure is 26.28 m.

> **CALCULATOR NOTE**
> To evaluate $20 + 2\pi$ on your calculator, enter 2 $\boxed{\times}$ $\boxed{\pi}$ $\boxed{+}$ 20 $\boxed{=}$. Round the number in the display to two decimal places.

Example 1
Find the perimeter of the figure. Round to the nearest hundredth.

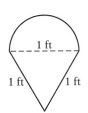

You Try It 1
The circumference of the circle in the figure is 6π cm. Find the perimeter of square *ABCD*.

Strategy

The perimeter is equal to 2 sides of a triangle plus $\dfrac{1}{2}$ the circumference of a circle. An approximation is asked for; use the π key on a calculator.

Your Strategy

Solution

$$P = a + b + \dfrac{1}{2}\pi d$$

$$P = 1 + 1 + \dfrac{1}{2}\pi(1) = 2 + 0.5\pi \approx 3.57$$

The perimeter is approximately 3.57 ft.

Your Solution

Solution on p. A28

25 | CT

OBJECTIVE **B**

Area of a composite plane figure

The area of the composite figure shown below is found by calculating the area of the rectangle and then subtracting the area of the triangle.

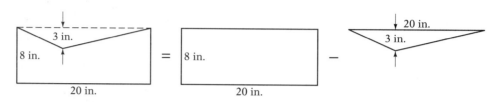

Area of the composite figure = area of the rectangle − area of the triangle

$$= \quad LW \quad - \quad \frac{1}{2}bh$$

$$= \quad 20(8) \quad - \quad \frac{1}{2}(20)(3)$$

$$= \quad 160 \quad - \quad 30$$

$$= \quad 130$$

The area of the composite figure is 130 in².

Example 2

Find the area of the shaded portion of the figure. Round to the nearest hundredth.

8 m

8 m

Strategy

The area is equal to the area of the square minus the area of the circle. The radius of the circle is one-half the length of a side of the square (8). An approximation is asked for; use the π key on a calculator.

Solution

$$r = \frac{1}{2}s = \frac{1}{2}(8) = 4$$

$$A = s^2 - \pi r^2$$
$$A = (8)^2 - \pi(4)^2 = 64 - 16\pi \approx 13.73$$

The area is approximately 13.73 m².

You Try It 2

Find the area of the composite figure.

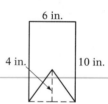

6 in.

4 in.

10 in.

Your Strategy

Your Solution

OBJECTIVE C
Volume of a composite solid

Composite geometric solids are solids made from two or more geometric solids. The following solid is made from a cylinder and one-half of a sphere.

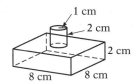

Composite solid   = a cylinder + one-half of a sphere

Volume of the composite solid = $\pi r^2 h$ + $\dfrac{1}{2} \cdot \dfrac{4}{3} \pi r^3$

Find the volume of the solid shown above if the radius of the base of the cylinder is 3 in. and the height of the cylinder is 10 in. Give the exact measure.

Use the equation given above. $r = 3$, $h = 10$. The radius of the sphere equals the radius of the base of the cylinder, 3.

$V = \pi r^2 h + \dfrac{1}{2} \cdot \dfrac{4}{3} \pi r^3$

$V = \pi(3)^2(10) + \dfrac{1}{2} \cdot \dfrac{4}{3} \pi(3)^3$

$V = \pi(9)(10) + \dfrac{2}{3}\pi(27)$

$V = 90\pi + 18\pi = 108\pi$

The volume of the solid is 108π in³.

Example 3

Find the volume of the solid. Round to the nearest hundredth.

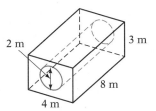

You Try It 3

Find the volume of the solid. Give the exact measure.

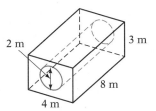

Strategy

The volume is equal to the volume of the rectangular solid minus the volume of the cylinder. The radius of the circle is one-half the diameter of the circle. An approximation is asked for; use the π key on a calculator.

Your Strategy

Solution

$r = \dfrac{1}{2}d = \dfrac{1}{2}(2) = 1$

$V = LWH - \pi r^2 h$
$V = 8(4)(3) - \pi(1)^2(8) = 96 - 8\pi \approx 70.87$

The volume is approximately 70.87 m³.

Your Solution

Solution on p. A28

OBJECTIVE D
Surface area of a composite solid

The composite solid shown below is made from a cone, a cylinder, and one-half of a sphere.

Surface area of the solid = the surface area of a cone minus the base +
the surface area of the sides of a cylinder +
one-half of the surface area of a sphere

$$= \pi r l \quad + \quad 2\pi r h \quad + \quad \frac{1}{2}(4\pi r^2)$$

Find the surface area of the solid shown above. The radius of the base of the cylinder is 4 m and the height is 5 m. The slant height of the cone is 6 m. Give the exact measure.

Use the equation given above. $r = 4$, $h = 5$, $l = 6$. The radius of the base of the cone and the radius of the sphere equal the radius of the base of the cylinder, 4.

$$SA = \pi r l + 2\pi r h + \frac{1}{2}(4\pi r^2)$$

$$SA = \pi(4)(6) + 2\pi(4)(5) + \frac{1}{2}[4\pi(4)^2]$$

$$SA = \pi(24) + 40\pi + 2\pi(16)$$

$$SA = 24\pi + 40\pi + 32\pi = 96\pi$$

The surface area of the solid is 96π m².

Example 4

Find the surface area of the solid. Round to the nearest hundredth.

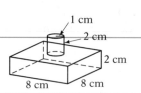

You Try It 4

Find the surface area of the solid. Round to the nearest hundredth.

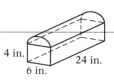

Strategy

The total surface area equals the surface area of the rectangular solid, minus the bottom of the cylinder, plus the surface area of the cylinder, minus the bottom of the cylinder.

Your Strategy

Solution

$$SA = 2LW + 2LH + 2HW - \pi r^2 + 2\pi r^2 + 2\pi r h - \pi r^2$$
$$SA = 2LW + 2LH + 2HW + 2\pi r h$$
$$SA = 2(8)(8) + 2(8)(2) + 2(2)(8) + 2\pi(1)(2)$$
$$SA = 128 + 32 + 32 + 4\pi$$
$$SA = 192 + 4\pi \approx 204.57$$

The surface area is approximately 204.57 cm².

Your Solution

Solution on p. A28

9.5 EXERCISES

OBJECTIVE A

Find the perimeter of the composite figure. For calculations involving π, give both the exact value and an approximation to the nearest hundredth.

1.
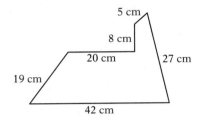

2.
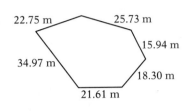

3.
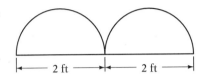

4.
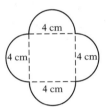

5.
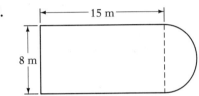

6.
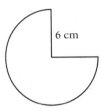

7.
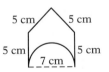

8.
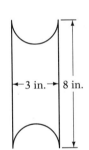

9.
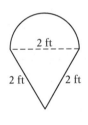

10.
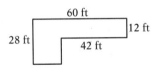

11.
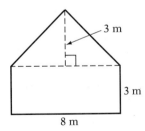

12.
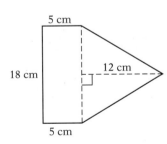

Solve.

13. Find the length of weather stripping installed around the arched door shown in the figure at the right. Round to the nearest hundredth.

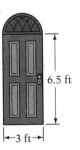

Solve.

14. Find the perimeter of the roller rink shown in the figure at the right. Round to the nearest hundredth.

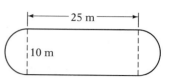

25 m

10 m

15. The rectangular lot shown in the figure at the right is being fenced. The fencing along the road will cost $2.70 per foot. The rest of the fencing will cost $2.10 per foot. Find the total cost to fence the lot.

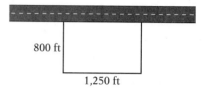

800 ft

1,250 ft

16. A rain gutter is being installed on a home that has the dimensions shown in the figure at the right. At a cost of $11.30 per meter, how much will it cost to install the rain gutter?

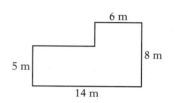

6 m

8 m

5 m

14 m

────

OBJECTIVE B

Find the area of the composite figure. For calculations involving π, give both the exact value and an approximation to the nearest hundredth.

17.

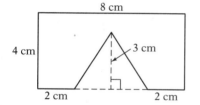

8 cm

4 cm

3 cm

2 cm 2 cm

18.

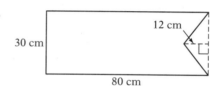

30 cm

12 cm

80 cm

19.

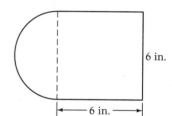

6 in.

6 in.

20.

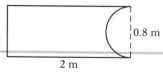

0.8 m

2 m

21.

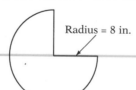

Radius = 8 in.

22.

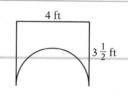

4 ft

$3\frac{1}{2}$ ft

23.

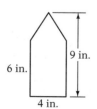

9 in.

6 in.

4 in.

24.

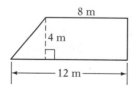

8 m

4 m

12 m

25.

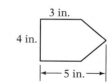

3 in.

4 in.

5 in.

Find the area of the composite figure. For calculations involving π, give both the exact value and an approximation to the nearest hundredth.

26.

22 cm

22 cm

27.

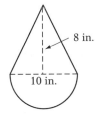

8 in.

10 in.

28.

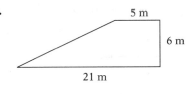

5 m

6 m

21 m

Solve.

29. A carpet is to be installed in one room and a hallway, as shown in the diagram at the right. At a cost of $18.50 per square meter, how much will it cost to carpet the area?

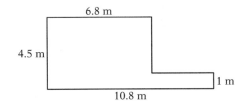

6.8 m

4.5 m

10.8 m

1 m

30. Find the area of the 2-meter boundary around the swimming pool shown in the figure at the right.

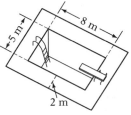

5 m

8 m

2 m

31. How much hardwood floor is needed to cover the roller rink shown in the figure at the right? Round to the nearest hundredth.

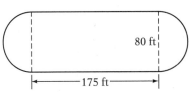

80 ft

175 ft

32. Find the total area of a national park with the dimensions shown in the figure at the right. Round to the nearest hundredth.

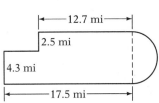

12.7 mi

2.5 mi

4.3 mi

17.5 mi

OBJECTIVE C

Find the volume of the composite figure. For calculations involving π, give both the exact value and an approximation to the nearest hundredth.

33.

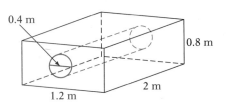

0.4 m

0.8 m

2 m

1.2 m

34.

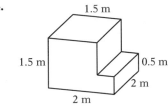

1.5 m

1.5 m

0.5 m

2 m

2 m

35.

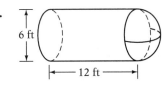

6 ft

12 ft

Find the volume of the composite figure. For calculations involving π, give both the exact value and an approximation to the nearest hundredth.

36.
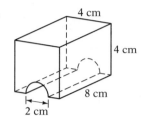
4 cm
4 cm
8 cm
2 cm

37.
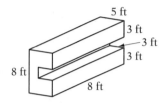
5 ft
3 ft
3 ft
3 ft
8 ft
8 ft

38.
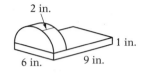
2 in.
1 in.
6 in.
9 in.

39.
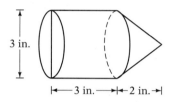
3 in.
3 in. 2 in.

40.
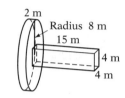
2 m
Radius 8 m
15 m
4 m
4 m

41.
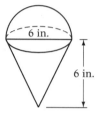
6 in.
6 in.

42.
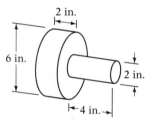
2 in.
6 in.
2 in.
4 in.

43.
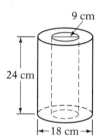
9 cm
24 cm
18 cm

44.
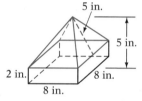
5 in.
5 in.
2 in.
8 in.
8 in.

Solve.

45. Find the volume of the bushing shown in the figure at the right. Round to the nearest hundredth.

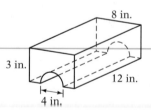
8 in.
3 in.
12 in.
4 in.

46. A truck is carrying an oil tank, as shown in the figure at the right. If the tank is half full, how many cubic feet of oil is the truck carrying? Round to the nearest hundredth.

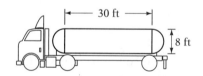
30 ft
8 ft

47. The concrete floor of a building is shown in the figure at the right. At a cost of $3.15 per cubic foot, find the cost of having the floor poured. Round to the nearest cent.

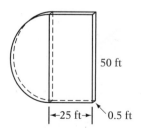
50 ft
25 ft
0.5 ft

Solve.

48. How many liters of water are needed to fill the swimming pool shown at the right? (1 m³ contains 1,000 L)

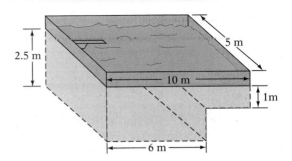

OBJECTIVE D

Find the surface area of the composite figure. For calculations involving π, give both the exact value and an approximation to the nearest hundredth.

49.

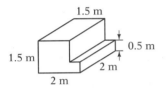

50.

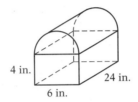

51.

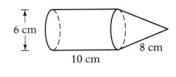

52.

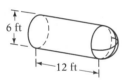

53.

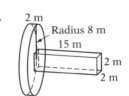

54.

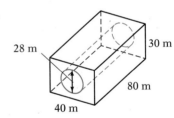

55.

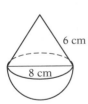

56.

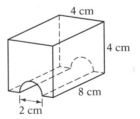

57.

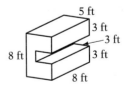

58.

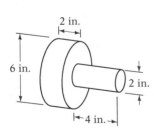

59.

60.

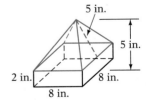

Solve.

61. A can of paint will cover 250 ft². Find the number of cans that should be purchased in order to paint the exterior of the auditorium shown in the figure at the right.

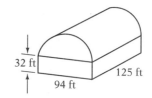

62. A piece of sheet metal is cut and formed into the shape shown at the right. Given that there are 0.24 g in 1 cm² of the metal, find the total number of grams of metal used. Round to the nearest hundredth.

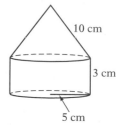

63. The walls of a room that is 25.5 ft long, 22 ft wide, and 8 ft high are being plastered. There are two doors in the room, each 2.5 ft by 7 ft. Each of the six windows in the room measures 2.5 ft by 4 ft. At a cost of $.75 per square foot, find the cost of plastering the walls of the room.

CRITICAL THINKING

64. You plan on painting the bookcase shown at the right. The bookcase is 10 in. deep, 36 in. high, and 36 in. long. The wood is 1 in. thick. You do not plan to paint the back side or the bottom. Find the surface area of the wood that needs to be painted.

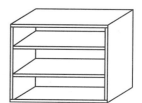

65. Bottles of apple juice are being packaged six to a carton for shipping. The diameter of the base of the bottles is 4 in. The height of the bottles is 8 in. The cartons are made of corrugated cardboard that is $\frac{1}{8}$ in. thick. Pieces of cardboard, each $\frac{1}{16}$ in. thick, are placed between bottles. Find the dimensions of the shipping carton.

66. A sphere fits inside a cylinder as shown at the right. The height of the cylinder equals the diameter of the sphere. Show that the surface area of the sphere equals the surface area of the sides of the cylinder.

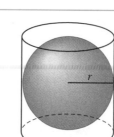

67. ✐ Explain the meaning of the "vanishing point" in a drawing. Find examples of its use.

68. ✐ Prepare a report on the use of geometric forms in architecture. Include examples of both plane geometric figures and geometric solids.

69. ✐ Write a paper on the artist M.C. Escher. Explain how he used mathematics and geometry in his works.

PROJECTS IN MATHEMATICS

Lines of Symmetry

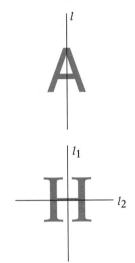

Look at the letter A printed at the left. If the letter were folded along line l, the two sides of the letter would match exactly. This letter has **symmetry** with respect to line l. Line l is called the **axis of symmetry**.

Now consider the letter H printed below at the left. Both lines l_1 and l_2 are axes of symmetry for this letter; the letter could be folded along either line and the two sides would match exactly.

Does the letter A have more than one axis of symmetry? Find axes of symmetry for other capital letters of the alphabet. Which lower-case letters have one axis of symmetry? Do any of the lower-case letters have more than one axis of symmetry?

Find the number of axes of symmetry for each of the plane geometric figures presented in this chapter.

There are other types of symmetry. Look up the meaning of point symmetry and rotational symmetry. Which plane geometric figures provide examples of these types of symmetry?

Find examples of symmetry in nature, art, and architecture.

Preparing a Circle Graph

In Section 9.1, a protractor was used to measure angles. Preparing a circle graph requires the ability to use a protractor to draw angles.

To draw an angle of 142°, first draw a line. Place the dot at the beginning of the line. This dot will be the vertex of the angle.

Place the protractor on the line and locate 142°. Place a dot next to the number.

Remove the protractor and draw a line from the beginning of the first line to the dot that marks 142°.

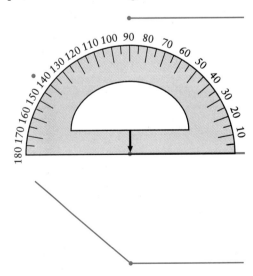

An example of preparing a circle graph is given on the next page.

The revenues (in thousands of dollars) from four segments of a car dealership for the first quarter of 1997 were:

New Car Sales:	$2,100	Used Car/Truck Sales:	$1,500
New Truck Sales:	$1,200	Parts/Service:	$700

To draw a circle graph to represent the percent that each segment contributed to the total revenue from all four segments:

Find the total revenue from all four segments.

$$2,100 + 1,200 + 1,500 + 700 = 5,500$$

Find what percent each segment is of the total revenue of $5,500.

New car sales: $\dfrac{2,100}{5,500} \approx 38.2\%$

New truck sales: $\dfrac{1,200}{5,500} \approx 21.8\%$

Used car/truck sales: $\dfrac{1,500}{5,500} \approx 27.3\%$

Parts/service: $\dfrac{700}{5,500} \approx 12.7\%$

Each percent represents the part of the circle for that sector. Because the circle contains 360°, multiply each percent by 360° to find the measure of the angle for each sector. Round to the nearest whole number.

New car sales:

$$0.382 \times 360° \approx 138°$$

New truck sales:

$$0.218 \times 360° \approx 78°$$

Used car/truck sales:

$$0.273 \times 360° \approx 98°$$

Parts/service:

$$0.127 \times 360° \approx 46°$$

Draw a circle and use a protractor to draw the sectors representing the percents that each segment contributed to the total revenue from all four segments.

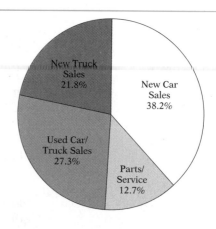

Collect data appropriate for display in a circle graph. [Some possibilities are: last year's sales for the top three car manufacturers in the United States, votes cast in the last election for your state governor, the majors of the students in your math class, the number of students enrolled in each class (senior, junior, etc.) at your college.] Then prepare the circle graph.

CHAPTER SUMMARY

Key Words   A *line* is determined by two distinct points and extends indefinitely in both directions. A *line segment* is part of a line that has two endpoints.

Parallel lines never meet; the distance between them is always the same. *Perpendicular lines* are intersecting lines that form right angles.

A *ray* starts at a point and extends indefinitely in one direction. The point at which a ray starts is the *endpoint* of the ray.

An *angle* is formed by two rays with the same endpoint. The *vertex* of an angle is the point at which the two rays meet.

An angle is measured in *degrees*. A 90° angle is a *right angle*. A 180° angle is a *straight angle*. An *acute angle* is an angle whose measure is between 0° and 90°. An *obtuse angle* is an angle whose measure is between 90° and 180°.

Complementary angles are two angles whose measures have the sum 90°. *Supplementary angles* are two angles whose measures have the sum 180°.

Two angles that are on opposite sides of the intersection of two lines are *vertical angles*; vertical angles have the same measure. Two angles that share a common side are *adjacent angles*; adjacent angles of intersecting lines are supplementary angles.

A line that intersects two other lines at two different points is a *transversal*. If the lines cut by a transversal are parallel lines, equal angles are formed: *alternate interior angles*, *alternate exterior angles*, and *corresponding angles*.

A *polygon* is a closed figure determined by three or more line segments. The line segments that form the polygon are its *sides*. A *regular polygon* is one in which each side has the same length and each angle has the same measure. Polygons are classified by the number of sides.

A *triangle* is a plane figure formed by three line segments. An *isosceles triangle* has two sides of equal length. The three sides of an *equilateral triangle* are of equal length. A *scalene triangle* has no two sides of equal length. An *acute triangle* has three acute angles. An *obtuse triangle* has one obtuse angle. A *right triangle* has a right angle.

Similar triangles have the same shape but not necessarily the same size. *Congruent triangles* have the same shape and the same size.

A *quadrilateral* is a four-sided polygon. A parallelogram, a rectangle, a square, a rhombus, and a trapezoid are all quadrilaterals.

A *circle* is a plane figure in which all points are the same distance from the center of the circle. A *diameter* of a circle is a line segment across the circle through the center. A *radius* of a circle is a line segment from the center of the circle to a point on the circle.

The *perimeter* of a plane geometric figure is a measure of the distance around the figure. The distance around a circle is called the *circumference*.

Area is the amount of surface in a region. *Volume* is a measure of the amount of space inside a figure in space. The *surface area* of a solid is the total area on the surface of the solid.

Essential Rules

Triangles
Sum of the measures of the interior angles = 180°
Sum of an interior and corresponding exterior angle = 180°
Rules to determine congruence: SSS rule, SAS rule, ASA rule

Perimeter

Triangle:	$P = a + b + c$	
Rectangle:	$P = 2L + 2W$	
Square:	$P = 4s$	
Circle:	$C = \pi d$ or $C = 2\pi r$	

Area

Triangle:	$A = \dfrac{1}{2}bh$
Rectangle:	$A = LW$
Square:	$A = s^2$
Circle:	$A = \pi r^2$
Parallelogram:	$A = bh$
Trapezoid:	$A = \dfrac{1}{2}h(b_1 + b_2)$

Volume

Rectangular solid:	$V = LWH$
Cube:	$V = s^3$
Sphere:	$V = \dfrac{4}{3}\pi r^3$
Right circular cylinder:	$V = \pi r^2 h$
Right circular cone:	$V = \dfrac{1}{3}\pi r^2 h$
Regular pyramid:	$V = \dfrac{1}{3}s^2 h$

Surface Area

Rectangular solid:	$SA = 2LW + 2LH + 2WH$
Cube:	$SA = 6s^2$
Sphere:	$SA = 4\pi r^2$
Right circular cylinder:	$SA = 2\pi r^2 + 2\pi rh$
Right circular cone:	$SA = \pi r^2 + \pi rl$
Regular pyramid:	$SA = s^2 + 2sl$

Principal Square Root Property If $r^2 = s$, then $r = \sqrt{s}$, and r is called the square root of s.

Pythagorean Theorem If a and b are the legs of a right triangle and c is the length of the hypotenuse, then $c^2 = a^2 + b^2$.

Similar Triangles The ratios of corresponding sides are equal. The ratio of corresponding heights is equal to the ratio of corresponding sides.

CHAPTER REVIEW EXERCISES

1. Given that $\angle a = 74°$ and $\angle b = 52°$, find the measures of angles x and y.

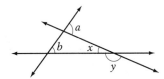

2. Triangles ABC and DEF are similar. Find the perimeter of triangle ABC.

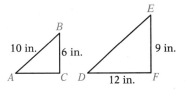

3. Find the volume of the composite figure.

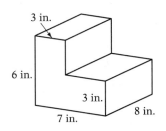

4. Find the measure of $\angle x$.

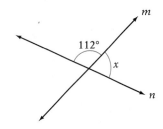

5. Determine whether the two triangles are congruent. If they are congruent, state by what rule they are congruent.

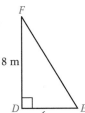

6. Find the surface area of the composite figure. Round to the nearest hundredth.

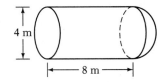

7. Given that $BC = 11$ cm and AB is three times the length of BC, find the length of AC.

8. Find x.

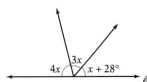

9. Find the area of the composite figure. Round to the nearest hundredth.

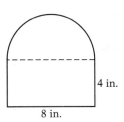

10. Find the volume of the figure.

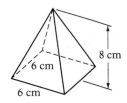

11. Find the perimeter of the composite figure. Round to the nearest hundredth.

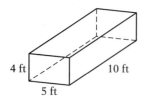

12. Given that $\ell_1 \parallel \ell_2$, find the measures of angles a and b.

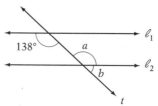

13. Find the surface area of the figure.

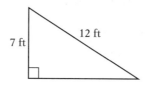

14. Find the unknown side of the triangle. Round to the nearest hundredth.

Solve.

15. Find the supplement of a 32° angle.

16. Find the volume of a rectangular solid with a length of 6.5 ft, a width of 2 ft, and a height of 3 ft.

17. Two angles of a triangle measure 37° and 48°. Find the measure of the third angle.

18. The height of a triangle is 7 cm. The area of the triangle is 28 cm². Find the length of the base of the triangle.

19. Find the volume of a sphere that has a diameter of 12 mm. Give the exact value.

20. The perimeter of a square picture frame is 86 cm. Find the length of each side of the frame.

21. A can of paint will cover 200 ft². How many cans of paint should be purchased in order to paint a cylinder that has a height of 15 ft and a radius of 6 ft?

22. The length of a rectangular park is 56 yd. The width is 48 yd. How many yards of fencing are needed to surround the park?

23. What is the area of a square patio that measures 9.5 m on each side?

24. A walkway 2 m wide surrounds a rectangular plot of grass. The plot is 40 m long and 25 m wide. What is the area of the walkway?

CUMULATIVE REVIEW EXERCISES

1. Find 8.5% of 2,400.

2. Find all the factors of 78.

3. Divide: $4\frac{2}{3} \div 5\frac{3}{5}$

4. Add: $(3x^2 + 5x - 2) + (4x^2 - x + 7)$

5. Divide and round to the nearest tenth:
   $82.93 \div 6.5$

6. Write 0.000029 in scientific notation.

7. Find the measure of $\angle x$.

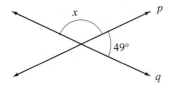

8. Find the unknown side of the triangle.

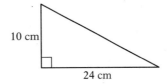

9. Find the area of the composite figure.

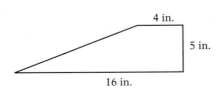

10. Find the volume of the composite figure. Round to the nearest hundredth.

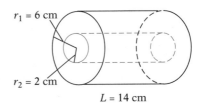

11. Multiply: $(4x^2y^2)(-3x^3y)$

12. Solve: $3(2x + 5) = 18$

13. Find the perimeter of the figure. Round to the nearest hundredth.

14. Graph $x > -3$.

$$\xleftarrow{\quad\quad} \underset{-6\ -5\ -4\ -3\ -2\ -1\ \ 0\ \ 1\ \ 2\ \ 3\ \ 4\ \ 5\ \ 6}{+\ +\ +\ +\ +\ +\ +\ +\ +\ +\ +\ +\ +} \xrightarrow{\quad\quad}$$

15. Simplify: $5(2x + 4) - (3x + 2)$

16. Evaluate $2x + 3y^2z$ when $x = 5$, $y = -1$, and $z = -4$.

17. Solve: $4x + 2 = 6x - 8$

18. Graph $y = -\dfrac{3}{2}x + 3$.

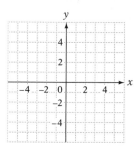

19. Evaluate $x^2y - 2z$ when $x = \dfrac{1}{2}$, $y = \dfrac{4}{5}$, and $z = -\dfrac{3}{10}$.

Solve.

20. *Catering* Two hundred fifty people are expected to attend a reception. Assuming that each person drinks 12 oz of coffee, how many gallons of coffee should be prepared? Round to the nearest whole number.

21. *Business* The charge for cellular phone service for a business executive is $42 per month plus $.35 per minute of phone use. In a month when the executive's phone bill was $72.45, how many minutes did the executive use the cellular phone?

22. *Taxes* If the sales tax on a $12.50 purchase is $.75, what is the sales tax on a $75 purchase?

23. *Business* The figure at the right shows sales, in millions of liters, of champagne and sparkling wine. Find the percent decrease in sales from 1984 to 1994. Round to the nearest tenth of a percent.

24. *Geometry* The volume of a box is 144 ft³. The length of the box is 12 ft, and the width is 4 ft. Find the height of the box.

25. *Sports* The pressure, P, in pounds per square inch, at a certain depth in the ocean can be approximated by the equation $P = 15 + \dfrac{1}{2}D$, where D is the depth in feet. Use this equation to find the depth when the pressure is 35 lb/in².

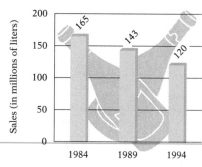

Champagne and Sparkling Wine Sales
Source: Jobson's Wine Handbook, *Wine Investor*

26. *Astronautics* The weight of an object is related to the distance the object is above the surface of Earth. A formula for this relationship is $d = 4{,}000\sqrt{\dfrac{E}{S}} - 4{,}000$, where E is the object's weight on the surface of Earth and S is the object's weight at a distance of d miles above Earth's surface. A space explorer who weighs 196 lb on the surface of Earth weighs 49 lb in space. How far above Earth's surface is the space explorer?

CHAPTER 10

Statistics and Probability

FOCUS On Problem Solving

Problem solving in the previous chapters concentrates on solving a certain problem. After the problem is solved, there is an important question to be asked: "Does the solution to this problem apply to other types of problems?"

To illustrate this extension to problem solving, we will consider *triangular numbers*, which were studied by ancient Greek mathematicians. The numbers 1, 3, 6, 10, 15, 21 are the first six triangular numbers. What is the next triangular number?

To answer this question, note in the diagram below that a triangle can be formed using the number of dots that correspond to a triangular number.

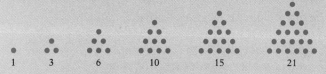

Observe that the number of dots in a row is one more than the number in the row above. The total number of dots can be found by addition.

$$1 = 1 \qquad 1 + 2 = 3 \qquad 1 + 2 + 3 = 6$$
$$1 + 2 + 3 + 4 = 10 \qquad 1 + 2 + 3 + 4 + 5 = 15$$
$$1 + 2 + 3 + 4 + 5 + 6 = 21$$

The pattern suggests that the next triangular number (the 7th one) is the sum of the first 7 natural numbers.

$$1 + 2 + 3 + 4 + 5 + 6 + 7 = 28$$

The 7th triangular number is 28.

The diagram at the right shows the 7th triangular number.

Using the pattern for triangular numbers, the 10th triangular number is

$$1 + 2 + 3 + 4 + 5 + 6 + 7 + 8 + 9 + 10 = 55$$

Now consider a situation that may seem to be totally unrelated to triangular numbers. Suppose you are in charge of scheduling softball games for a league. There are seven teams in the league, and each team must play every other team once. How many games must be scheduled?

We label the teams A, B, C, D, E, F, and G. (See the figure at the right).

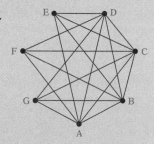

A line between two teams indicates that the two teams play each other. Beginning with A, there are 6 lines for the 6 teams that A must play.

Now consider B. There are 6 teams that B must play, but the line between A and B has already been drawn, so there are only 5 remaining games to schedule for B. Now move on to C. The lines between C and A and C and B have already been drawn, so there are 4 additional lines to be drawn to represent the teams C will play. Moving on to D, the lines between D and A, D and B, and D and C have already been drawn, so there are 3 more lines to be drawn to represent the teams D will play.

Note that as we move from team to team, one fewer line needs to be drawn. When we reach F, there is only one line to be drawn, the one between F and G. The total number of lines drawn is $6 + 5 + 4 + 3 + 2 + 1 = 21$, the sixth triangular number. For a league with 7 teams, the number of games that must be scheduled so that each team plays every other team once is the 6th triangular number. If there were 10 teams in the league, the number of games that must be scheduled would be the 9th triangular number, which is 45.

SECTION 10.1   Organizing Data

OBJECTIVE A
Frequency distributions

Statistics is the study of collecting, organizing, and interpreting data. Data are collected from a **population**, which is the set of all observations of interest. Here are some examples of populations.

> An auto insurance company wants to determine information about the size of claims for auto accidents. The population for the insurance company is the dollar amount of each claim.

> A medical researcher wants to determine the effectiveness of a new drug to control blood pressure. The population for the researcher is the amount of change in blood pressure for each patient receiving the medication.

> The quality control inspector of a precision instrument company wants to determine the diameters of ball bearings. The population for the inspector is the measure of the diameter of each ball bearing.

A **frequency distribution** is one method of organizing the data collected from a population. A frequency distribution is constructed by dividing the data gathered from the population into **classes**. Here is an example.

A ski association surveys 40 of its members, asking them to report the percent of their ski terrain that is rated expert. The results of the survey follow.

Percent of Expert Terrain at 40 Ski Resorts

14	24	8	31	27	9	12	32	24	27
12	21	24	23	12	31	30	31	26	34
13	18	29	33	34	21	28	23	11	10
25	20	14	18	15	11	17	29	21	25

To organize these data into a frequency distribution:

1. Find the smallest number (8) and the largest number (34) in the table. The difference between these two numbers is the **range** of the data.

    Range = 34 − 8 = 26

2. Decide how many classes the frequency distribution will contain. Usually frequency distributions have from 6 to 12 classes. The frequency distribution for this example will contain 6 classes.

3. Divide the range by the number of classes. If necessary, round the quotient to a whole number. This number is called the **class width**.

    $\frac{26}{6} \approx 4$. The class width is 4.

4. Form the classes of the frequency distribution.

Classes	
8–12	Add 4 to the smallest number.
13–17	Add 4 again.
18–22	
23–27	Continue until a class contains
28–32	the largest number in the set
33–37	of data.

14	12	13	25
24	21	18	20
8	24	29	14
31	23	33	18
27	12	34	15
9	31	21	11
12	30	28	17
32	31	23	29
24	26	11	21
27	34	10	25

5. Complete the table by tabulating the data for each class. For each number from the data, place a slash next to the class that contains the number. Count the number of tallies in each class. This is the **class frequency**.

Frequency Distribution for Ski Resort Data

Classes	Tally	Frequency
8–12	/////////	8
13–17	/////	5
18–22	//////	6
23–27	//////////	10
28–32	////////	8
33–37	///	3

By organizing data into a frequency distribution, statements about the data can be made. For example, twenty-seven $(6 + 10 + 8 + 3)$ of the ski resorts reported that 18% or more of their terrain was rated expert.

An insurance adjuster had tabulated the dollar amount of 50 auto accident claims. The results are given in the following table. Use these data for Example 1 and You Try It 1.

Dollar Amount of 50 Auto Insurance Claims

475	224	722	721	815	351	596	625	981	748
993	881	361	560	574	742	703	998	435	873
882	278	455	803	985	305	522	900	638	810
677	688	410	505	890	186	829	631	882	991
484	339	950	579	539	422	326	793	453	118

Example 1

For the table of Auto Insurance Claims, make a frequency distribution that has 6 classes.

Strategy

To make the frequency distribution:
▶ Find the range.
▶ Divide the range by 6, the number of classes. Round the quotient to the nearest whole number. This is the class width.
▶ Tabulate the data for each class.

Solution

Range = 998 − 118 = 880

Class width = $\dfrac{880}{6} \approx 147$

Dollar Amount of Insurance Claims

Classes	Tally	Frequency
118–265	///	3
266–413	////////	7
414–561	//////////	10
562–709	/////////	9
710–857	/////////	9
858–1,005	////////////	12

You Try It 1

For the table of Auto Insurance Claims, make a frequency distribution that has 8 classes.

Your Strategy

Your Solution

Solution on pp. A28–A29

OBJECTIVE B
Histograms

A **histogram** is a bar graph that represents the data in a frequency distribution. The width of a bar represents each class, and the height of the bar corresponds to the frequency of the class.

A survey of 105 households is conducted, and the number of kilowatt-hours (kWh) of electricity that are used by each in a one-month period is recorded in the frequency distribution, shown at the left below. The histogram for the frequency distribution is shown in Figure 10.1.

Classes (kWh)	Frequency
850–900	9
900–950	14
950–1,000	17
1,000–1,050	25
1,050–1,100	16
1,100–1,150	14
1,150–1,200	10

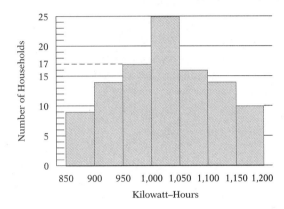

Figure 10.1

From the frequency distribution or the histogram, we can see that 17 households used between 950 kWh and 1,000 kWh during the one-month period.

Example 2

Use the histogram in Figure 10.1 to find the number of households that used 950 kWh of electricity or less during the month.

Strategy

To find the number:

▶ Read the histogram to find the number of households whose use was between 850 and 900 kWh and the number whose use was between 900 and 950 kWh.
▶ Add the two numbers.

Solution

Number between 850 and 900 kWh: 9
Number between 900 and 950 kWh: 14

9 + 14 = 23

23 households used 950 kWh of electricity or less during the month.

You Try It 2

Use the histogram in Figure 10.1 to find the number of households that used 1,100 kWh of electricity or more during the month.

Your Strategy

Your Solution

Solution on p. A29

Objective C
Frequency polygons

A **frequency polygon** is a graph that displays information in a manner similar to a histogram. A dot is placed above the center of each class interval at a height corresponding to that class's frequency. The dots are then connected to form a broken-line graph. The center of a class interval is called the **class midpoint**.

The per capita incomes in a recent year for the 50 states are recorded in the frequency polygon in Figure 10.2. The number of states with a per capita income between $21,000 and $24,000 is 14.

The percent of states for which the per capita income is between $21,000 and $24,000 can be determined by solving the basic percent equation. The base is 50 and the amount is 14.

$$pB = A$$
$$p(50) = 14$$
$$p = \frac{14}{50}$$
$$p = 0.28$$

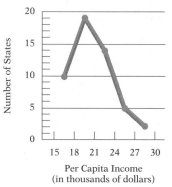

Figure 10.2
Source: U.S. Department of Commerce, Bureau of Economic Analysis, Survey of Current Business

28% of the states had a per capita income between $21,000 and $24,000.

Example 3

Use Figure 10.2 to find the number of states for which the per capita income was $24,000 or above.

You Try It 3

Use Figure 10.2 to find the ratio of the number of states with a per capita income between $15,000 and $18,000 to the number with a per capita income between $24,000 and $27,000.

Strategy

To find the number of states:

▶ Read the frequency polygon to find the number of states with a per capita income between $24,000 and $27,000 and the number of states with a per capita income between $27,000 and $30,000.
▶ Add the numbers.

Your Strategy

Solution

Number with per capita income between $24,000 and $27,000: 5

Number with per capita income between $27,000 and $30,000: 2

$$5 + 2 = 7$$

The per capita income was $24,000 or above in 7 states.

Your Solution

Solution on p. A29

10.1 EXERCISES

OBJECTIVE A

 Education   Use the table below for Exercises 1 10.

Annual Tuition at 40 Universities (hundreds of dollars)

75	77	85	38	31	81	78	82
61	64	53	41	60	77	74	85
62	84	51	42	78	39	45	50
67	43	79	81	35	86	39	48
73	26	29	22	26	49	85	57

1. What is the range of data in the Annual Tuition table?

2. Make a frequency distribution for the Annual Tuition table. Use 8 classes.

3. Which class has the greatest frequency?

4. How many universities charge a tuition that is between $6,700 and $7,500?

5. How many universities charge a tuition that is between $4,000 and $4,800?

6. How many universities charge a tuition that is less than or equal to $5,700?

7. What percent of the universities charge a tuition that is between $8,500 and $9,300?

8. What percent of the universities charge a tuition that is between $2,200 and $3,000?

9. What percent of the universities charge a tuition that is greater than or equal to $5,800?

10. What percent of the universities charge a tuition that is less than or equal to $6,600?

The Hotel Industry   Use the table below for Exercises 11–20.

Corporate Room Rate for 50 Hotels

40	67	57	97	94	62	71	45	49	43
86	51	54	66	86	58	81	80	87	89
37	86	83	80	75	48	79	92	87	57
44	48	79	92	87	56	96	80	62	66
61	78	72	58	75	69	71	82	95	107

11.   Make a frequency distribution for the hotel room rates. Use 7 classes.

12.   How many hotels charge a corporate room rate that is between $59 and $69 per night?

13.   How many hotels charge a corporate room rate that is between $37 and $47 per night?

14.   How many hotels charge a corporate room rate that is between $92 and $113 per night?

15.   How many hotels charge a corporate room rate that is less than or equal to $80?

16.   What percent of the hotels charge a corporate room rate that is between $81 and $91 per night?

17.   What percent of the hotels charge a corporate room rate that is between $70 and $80 per night?

18.   What percent of the hotels charge a corporate room rate that is greater than or equal to $81 per night?

19.   What percent of the hotels charge a corporate room rate that is less than or equal to $58 per night?

20.   What is the ratio of the number of hotels whose room rates are between $59 and $69 to those whose room rates are between $70 and $80?

OBJECTIVE B

  Consumer Credit   A total of 50 monthly credit account balances are recorded in the figure at the right.

21. How many account balances were between $750 and $1,000?

22. How many account balances were less than $1,000?

23. What percent of the account balances were greater than $750?

24. What percent of the account balances were between $1,000 and $1,250?

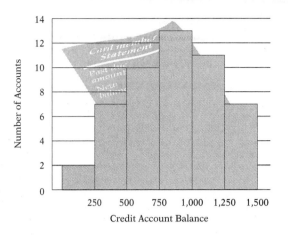

  Marathons   The times (in minutes) for 100 runners in a marathon are recorded in the figure at the right.

25. What is the ratio of the number of runners with times that were between 150 min and 155 min to those with times between 175 min and 180 min?

26. What is the ratio of the number of runners with times that were between 165 min and 170 min to those with times between 155 min and 160 min?

27. What percent of the runners had times greater than 165 min?

28. What percent of the runners had times less than 170 min?

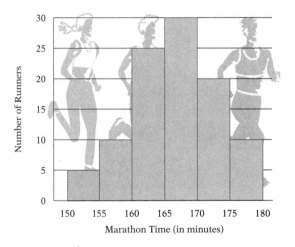

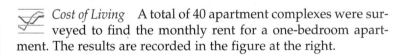 *Cost of Living*   A total of 40 apartment complexes were surveyed to find the monthly rent for a one-bedroom apartment. The results are recorded in the figure at the right.

29. What percent of the apartments had rents between $550 and $650?

30. What percent of the apartments had rents between $250 and $350?

31. What percent of the apartments had rents greater than $450?

32. What percent of the apartments had rents less than $550?

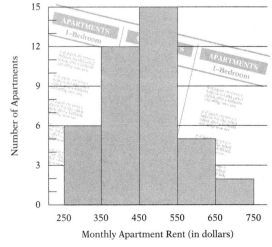

OBJECTIVE C

Education   The scores of 50 nurses taking a state board exam are given in the figure at the right.

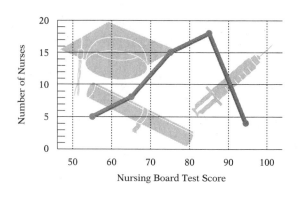

33. How many nurses had scores that were greater than 80?

34. How many nurses had scores that were less than 70?

35. What percent of the nurses had scores between 70 and 90?

36. What percent of the nurses had scores greater than 70?

Emergency Calls   The response times for 75 emergency 911 calls for a city are recorded in the figure at the right.

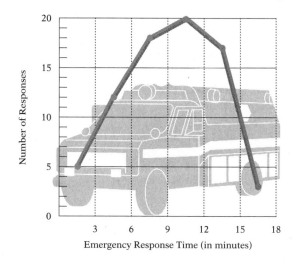

37. What is the ratio of the number of response times between 6 min and 9 min to the number of response times between 15 min and 18 min?

38. What is the ratio of the number of response times of 0 min to 3 min to the total number of recorded response times?

39. What percent of the response times are greater than 9 min? Round to the nearest tenth of a percent.

40. What percent of the response times are less than 12 min? Round to the nearest tenth of a percent.

CRITICAL THINKING

41. Toss two dice 100 times. Record the sum of the dots on the upward faces. Make a histogram showing the number of times the sum was 2, 3, . . . , 12.

42. If each frequency in a frequency table is divided by the total number of observations, a relative frequency table is formed. A relative frequency histogram is a histogram of the relative frequencies. Draw a relative frequency histogram for the data in Exercise 41.

43. In your own words, describe a frequency table.

44. How are a frequency table and a histogram alike? How are they different?

45. The frequency table at the right contains data from a survey of the type of vehicle a prospective buyer would consider. Explain why these data could be shown in a bar graph but not in a histogram.

Body Style	Frequency
Sedan	25
Convertible	16
Minivan	9
Sports car	20
Truck	23

SECTION **10.2** **Statistical Measures**

OBJECTIVE A
The mean, median, and mode of a distribution

The average score on an exam was 73. The Dow Jones Industrial Average was 6798. The average rainfall on a tropical island is 350 in. per year. Each of these statements uses one number to describe an entire collection of numbers. Such a number is called an *average*. In statistics, there are various ways to calculate an average. Three of the most common—*mean, median,* and *mode*—are discussed here.

An automotive engineer tests the miles-per-gallon ratings of 15 cars and records the results as follows:

Miles-per-Gallon Ratings of 15 Cars

25   22   21   27   25   35   29   31   25   26   21   39   34   32   28

The **mean** of the data is the sum of the measurements divided by the number of measurements. The symbol for the mean is $\bar{x}$.

> **Formula for the Mean**
>
> $$\bar{x} = \frac{\text{Sum of all data values}}{\text{Number of data values}}$$

To find the mean for the data above, add the numbers and then divide by 15.

$$\bar{x} = \frac{25 + 22 + 21 + 27 + 25 + 35 + 29 + 31 + 25 + 26 + 21 + 39 + 34 + 32 + 28}{15}$$

$$= \frac{420}{15} = 28$$

The mean number of miles per gallon for the 15 cars tested was 28 mi/gal.

The mean is one of the most frequently computed averages. It is the one that is commonly used to calculate a student's performance in a class.

> The scores for a history student on 5 tests were 78, 82, 91, 87, and 93. What was the mean score for this student?
>
> To find the mean, add the numbers. Then divide by 5.
> $$\bar{x} = \frac{78 + 82 + 91 + 87 + 93}{5}$$
> $$= \frac{431}{5} = 86.2$$
>
> The mean score for the history student was 86.2.

The **median** of data is the number that separates the data into two equal parts when the numbers are arranged from smallest to largest (or largest to smallest). There are always an equal number of values above the median and below the median.

To find the median of a set of numbers, first arrange the numbers from smallest to largest. The median is the number in the middle. The result of arranging the data, given on the previous page, for the miles per gallon ratings from smallest to largest is given below.

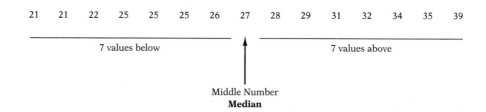

The median is 27.

If the data contain an *even* number of values, the median is the sum of the two middle numbers divided by 2.

The selling prices of the last six homes sold by a real estate agent were $175,000, $150,000 $250,000, $130,000, $245,000, and $190,000. Find the median selling price of these homes.

Arrange the numbers from smallest to largest. Because there are an even number of values, the median is the sum of the two middle numbers divided by two.

130,000   150,000   175,000   190,000   245,000   250,000

$$\text{middle 2 numbers}$$

$$\text{Median} = \frac{175,000 + 190,000}{2} = 182,500$$

The median selling price of a home was $182,500.

The **mode** of a set of numbers is the value that occurs most frequently. If a set of numbers has no number that occurs more than once, then the data have no mode.

Here again are the data for the gasoline mileage ratings of cars.

Miles per Gallon Ratings of 15 Cars

25   22   21   27   25   35   29   31   25   26   21   39   34   32   28

25 is the number that occurs most frequently. The mode is 25.

The gasoline mileage rating data show that the mean, median, and mode of a set of numbers do not have to be the same value. For the data on rating the miles per gallon for 15 cars,

Mean = 28   Median = 27   Mode = 25

Although any of the averages can be used when the data collected consist of numbers, the mean and median are not appropriate for *qualitative* data. Examples of qualitative data are recording a person's favorite color or recording a person's preference from among classical music, hard rock, jazz, rap, and country western. It does not make sense to say that the *average* favorite color is red or the *average* musical choice is jazz. The mode is used to indicate the most frequently chosen color or musical category. The **modal response** is the category that receives the greatest number of responses.

A survey asked people to state whether they strongly disagree, disagree, have no opinion, agree, or strongly agree with the position a state's governor has taken on increasing taxes for health care. What was the modal response for these data?

Strongly disagree   57
Disagree   68
No opinion   12
Agree   45
Strongly agree   58

Because a response of "disagree" was recorded most frequently, the modal response was disagree.

Example 1

Twenty students were asked the number of units in which they were currently enrolled. The responses were

| 15 | 12 | 13 | 15 | 17 | 18 | 13 | 20 | 9 | 16 |
| 14 | 10 | 15 | 12 | 17 | 16 | 6 | 14 | 15 | 12 |

Find the mean and median number of units taken by these students.

Strategy

To find the mean number of units taken by the 20 students:
▶ Determine the sum of the numbers.
▶ Divide the sum by 20.

To find the median number of units taken by the 20 students:
▶ Arrange the numbers from smallest to largest.
▶ Because there is an even number of values, the median is the sum of the two middle numbers divided by 2.

Solution

The sum of the numbers is 279.

$$\bar{x} = \frac{279}{20} = 13.95$$

The mean is 13.95 units.

| 6 | 9 | 10 | 12 | 12 | 12 | 13 | 13 | 14 | 14 |
| 15 | 15 | 15 | 15 | 16 | 16 | 17 | 17 | 18 | 20 |

$$\text{Median} = \frac{14 + 15}{2} = 14.5$$

The median is 14.5 units.

You Try It 1

The amounts spent by the last 10 customers at a fast-food restaurant were

| 2.32 | 4.21 | 3.45 | 3.90 | 3.58 | 2.45 | 3.05 | 4.00 | 1.59 | 2.75 |

Find the mean and median amount spent on lunch by these customers.

Your Strategy

Your Solution

Example 2

A bowler has scores of 165, 172, 168, and 185 for four games. What score must the bowler have on the next game so that the mean for the five games is 174?

Strategy

To find the score, use the formula for the mean, letting n be the score on the fifth game.

Solution

$$174 = \frac{165 + 172 + 168 + 185 + n}{5}$$

$$174 = \frac{690 + n}{5}$$

$870 = 690 + n$ Multiply each side by 5.
$180 = n$

The score on the fifth game must be 180.

You Try It 2

You have scores of 82, 91, 79, and 83 on four exams. What score must you receive on the fifth exam to have a mean of 84 for the five exams?

Your Strategy

Your Solution

Solution on p. A29 |||||||

OBJECTIVE **B**

Box-and-whiskers plots

The purpose of calculating a mean or median is to obtain one number that describes some measurements. That one number alone, however, may not adequately represent the data. A **box-and-whiskers plot** is a graph that gives a more complete picture of the data. A box-and-whiskers plot shows five numbers: the smallest value, the *first quartile*, the median, the *third quartile*, and the largest value. The **first quartile**, symbolized by Q_1, is the number below which one-quarter of the data lie. The **third quartile**, symbolized by Q_3, is the number above which one-quarter of the data lie.

Find the first quartile Q_1 and the third quartile Q_3 for the prices of 15 half-gallon cartons of ice cream.

| 3.26 | 4.71 | 4.18 | 4.45 | 5.49 | 3.18 | 3.86 | 3.58 | 4.29 | 5.44 | 4.83 | 4.56 | 4.36 | 2.39 | 2.66 |

To find the quartiles, first arrange the data from the smallest value to the largest value. Then find the median.

| 2.39 | 2.66 | 3.18 | 3.26 | 3.58 | 3.86 | 4.18 | 4.29 | 4.36 | 4.45 | 4.56 | 4.71 | 4.83 | 5.44 | 5.49 |

The median is 4.29.

Now separate the data into two groups: those values below the median and those values above the median.

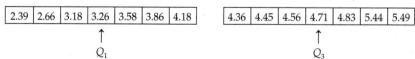

The first quartile Q_1 is the median of the lower half of the data: $Q_1 = 3.26$.
The third quartile Q_3 is the median of the upper half of the data: $Q_3 = 4.71$.

The **interquartile range** is the difference between Q_3 and Q_1.

Interquartile range = $Q_3 - Q_1 = 4.71 - 3.26 = 1.45$

A box-and-whiskers plot shows the data in the interquartile range as a box. The box-and-whiskers plot for the data on the cost of ice cream is shown below.

Note that the box-and-whiskers plot labels five values: the smallest, 2.39; the first quartile Q_1, 3.26; the median, 4.29; the third quartile Q_3, 4.71; and the largest value, 5.49.

For Example 3 and You Try It 3, use the data in the table at the right, which gives the number of people registered for a software training program.

Participants in Software Training

30	45	54	24	48	38	43
38	46	53	62	64	40	35

||||||

Example 3

Find Q_1 and Q_3 for the data in the software training table.

Strategy

To find Q_1 and Q_3, arrange the data from smallest to largest. Find the median. Then find Q_1, the median of the lower half of the data, and Q_3, the median of the upper half of the data.

Solution

24	30	35	38	38	40	43
45	46	48	53	54	62	64

Median = $\dfrac{43 + 45}{2} = 44$

$Q_1 = 38$ The median of the top row of data.
$Q_3 = 53$ The median of the bottom row of data.

You Try It 3

Draw the box-and-whiskers plot for the data in the software training table.

Your Strategy

Your Solution

Solution on p. A29 ||||||

Objective C

The standard deviation of a distribution

[CT]

Consider two students, each of whom has taken five exams.

Student A

84	86	83	85	87

$\bar{x} = \dfrac{84 + 86 + 83 + 85 + 87}{5} = \dfrac{425}{5} = 85$

The mean for Student A is 85.

Student B

90	75	94	68	98

$\bar{x} = \dfrac{90 + 75 + 94 + 68 + 98}{5} = \dfrac{425}{5} = 85$

The mean for Student B is 85.

For each of these students, the mean (average) for the 5 tests is 85. However, Student A has a more consistent record of scores than Student B. One way to measure the consistency or "clustering" of data near the mean is the **standard deviation**.

To calculate the standard deviation:

1. Sum the squares of the differences between each value of data and the mean.

2. Divide the result in Step 1 by the number of items in the set of data.

3. Take the square root of the result in Step 2.

Here is the calculation for Student A. The symbol for standard deviation is the Greek letter *sigma*, denoted by σ.

Step 1	x	$(x - \bar{x})$	$(x - \bar{x})^2$
	84	$(84 - 85)$	$(-1)^2 = 1$
	86	$(86 - 85)$	$1^2 = 1$
	83	$(83 - 85)$	$(-2)^2 = 4$
	85	$(85 - 85)$	$0^2 = 0$
	87	$(87 - 85)$	$2^2 = 4$
			Total = 10

Step 2 $\dfrac{10}{5} = 2$

Step 3 $\sigma = \sqrt{2} \approx 1.414$

The standard deviation for Student A's scores is approximately 1.414.

Following a similar procedure for Student B, the standard deviation for Student B's scores is approximately 11.524. Since the standard deviation of Student B's scores is greater than that of Student A's ($11.524 > 1.414$), Student B's scores are not as consistent as those of Student A.

Example 4

The weights in pounds of the five-man front line of a college football team are 210, 245, 220, 230, and 225. Find the standard deviation of the weights.

Strategy

To calculate the standard deviation:

▶ Find the mean of the weights.
▶ Use the procedure for calculating standard deviation.

Solution

$$\bar{x} = \frac{210 + 245 + 220 + 230 + 225}{5} = 226$$

Step 1

x	$(x - \bar{x})^2$
210	$(210 - 226)^2 = 256$
245	$(245 - 226)^2 = 361$
220	$(220 - 226)^2 = 36$
230	$(230 - 226)^2 = 16$
225	$(225 - 226)^2 = 1$
	Total = 670

Step 2

$\dfrac{670}{5} = 134$

Step 3

$\sigma = \sqrt{134}$
≈ 11.576

The standard deviation of the weights is approximately 11.576 lb.

You Try It 4

The number of miles a runner recorded for the last six days of running were 5, 7, 3, 6, 9, and 6. Find the standard deviation of the miles run.

Your Strategy

Your Solution

Solution on pp. A29–A30

10.2 EXERCISES

OBJECTIVE A

Solve.

1. *Business*   The number of big-screen televisions sold each month for one year was recorded by an electronics store. For 1997, the results were 15, 12, 20, 20, 19, 17, 22, 24, 17, 20, 15, and 27. Calculate the mean and the median number of televisions sold per month.

2. *The Airline Industry*   The number of seats occupied on a jet for 16 transatlantic flights was recorded. The numbers were 309, 422, 389, 412, 401, 352, 367, 319, 410, 391, 330, 408, 399, 387, 411, and 398. Calculate the mean and the median number of occupied seats.

3. *Sports*   The times, in seconds, for a 100-meter dash at a college track meet were 10.45, 10.23, 10.57, 11.01, 10.26, 10.90, 10.74, 10.64, 10.52, and 10.78. Calculate the mean and median times for the 100-meter dash.

4. *Consumerism*   A consumer research group purchased identical items in 8 grocery stores. The costs for the purchased items were $45.89, $52.12, $41.43, $40.67, $48.73, $42.45, $47.81, and $45.82. Calculate the mean and the median cost of the purchased items.

5. *Education*   Your scores on six history tests were 78, 92, 95, 77, 94, and 88. If an "average score" of 90 receives an A for the course, which average, the mean or the median, would you prefer the instructor to use?

6. *Computers*   One measure of a computer's hard drive speed, *access time*, is measured in milliseconds (thousandths of a second). Find the mean and median for 11 hard drives whose access times were 18, 17, 16, 17, 18, 19, 20, 19, 14, 17, and 17.

7. *Health Insurance*   Eight health maintenance organizations (HMOs) presented group health insurance plans to a company. The monthly rates per employee were $423, $390, $405, $396, $426, $355, $404, and $430. Calculate the mean and the median monthly rates for these eight companies.

8. *Sports*   The number of yards gained by a college running back for 6 games was recorded. The numbers were 98, 105, 120, 90, 111, and 104. How many yards must this running back gain in the next game so that the average for the seven games is 100 yd?

9. *Sports*   The number of unforced errors a tennis player made in four sets of tennis was recorded. The numbers were 15, 22, 24, and 18. How many unforced errors did this player make in the fifth set so that the mean number of unforced errors for the 5 sets was 20?

Solve.

10. *Sports* The last five golf scores for a player were 78, 82, 75, 77, and 79. What score on the next round of golf will give the player a mean score of 78 for all six rounds?

11. *Business* A survey by an ice cream store asked people to name their favorite ice cream from five flavors. The responses were mint chocolate chip, 34; pralines and cream, 27; German chocolate cake, 44; chocolate raspberry swirl, 34; rocky road, 42. What was the modal response?

12. *Physical Characteristics* The eye colors of 100 students were recorded. The results were blue, 35; brown, 38; hazel, 14; green, 3; grey, 9. What was the modal eye color?

13. *Politics* A newspaper survey asked people to rate the performance of the city's mayor. The responses were very unsatisfactory, 230; unsatisfactory, 403; satisfactory, 1,237; very satisfactory, 403. What was the modal response for this survey?

14. *Business* The patrons of a restaurant were asked to rate the quality of food. The responses were bad, 8; good, 21; very good, 43; excellent, 21. What was the modal response for this survey?

OBJECTIVE B

Solve.

15. *Compensation* The hourly wage for an entry-level position at various firms was recorded by a labor research firm. The results were as follows.

Starting Hourly Wages for 16 Companies

7.09	10.50	6.46	6.70	8.85	8.03	10.40	8.01
9.35	6.45	6.35	7.64	8.02	7.12	9.05	7.94

Find the first quartile and the third quartile, and draw a box-and-whiskers plot of the data.

16. *Health* The cholesterol levels for 14 adults are recorded in the table below. Find the first quartile and the third quartile, and draw a box-and-whiskers plot of the data.

Cholesterol Levels for 14 Adults

375	185	254	221	183	251	258
292	214	172	233	208	198	211

Solve.

17.   *Automobiles*   The gasoline consumption of 19 cars was tested and the results recorded in the following table. Find the first quartile and the third quartile, and draw a box-and-whiskers plot of the data.

Miles Per Gallon for 19 Cars

33	21	30	32	20	31	25	20	16	24
22	31	30	28	26	19	21	17	26	

18.   *Education*   The ages of the accountants who passed the certified public accountant (CPA) exam at one test center are recorded in the table below. Find the first quartile and the third quartile, and draw a box-and-whiskers plot of the data.

Ages of Accountants Passing the CPA Exam

24	42	35	26	24	37	27	26	28
34	43	46	29	34	25	30	28	

19.   *Manufacturing*   The times for new employees to learn how to assemble a toy are recorded in the table below. Find the first quartile and the third quartile, and draw a box-and-whiskers plot of the data.

Time to Train Employees (in hours)

4.3	3.1	5.3	8.0	2.6	3.5	4.9	4.3
6.2	6.8	5.4	6.0	5.1	4.8	5.3	6.7

20.   *Manufacturing*   A manufacturer of light bulbs tested the life of 20 light bulbs. The results are recorded in the table below. Find the first quartile and the third quartile, and draw a box-and-whiskers plot of the data.

Life of 20 Light Bulbs (in hours)

1010	1235	1200	998	1400	789	986	905	1050	1100
1180	1020	1381	992	1106	1298	1268	1309	1390	890

OBJECTIVE C

Solve.

21.   *The Airline Industry*   An airline recorded the times for a ground crew to unload the baggage from an airplane. The recorded times, in minutes, were 12, 18, 20, 14, and 16. Find the standard deviation of these times.

Solve.

22. *Health* The weight in ounces of newborn infants was recorded by a hospital. The weights were 96, 105, 84, 90, 102, and 99. Find the standard deviation of the weights.

23. *Business* The numbers of rooms occupied in a hotel on six consecutive days were 234, 321, 222, 246, 312, and 396. Find the standard deviation for the number of rooms occupied.

24. *Coin Tosses* Seven coins were tossed 100 times. The numbers of heads recorded were 56, 63, 49, 50, 48, 53, and 52. Find the standard deviation of the number of heads.

25. *Consumerism* The prices for unleaded gasoline at seven service stations were $1.41, $1.34, $1.28, $1.31, $1.29, $1.34, and $1.27. Find the standard deviation of the prices of gasoline.

26. *Weather* The temperatures for eleven consecutive days at a desert resort were 95°, 98°, 98°, 104°, 97°, 100°, 96°, 97°, 108°, 93°, and 104°. For the same days, the temperatures in Antarctica were 27°, 28°, 28°, 30°, 28°, 27°, 30°, 25°, 24°, 26°, and 21°. Which location has the greater standard deviation of temperatures?

27. *Sports* The scores for five college basketball games were 56, 68, 60, 72, and 64. The scores for five professional basketball games were 106, 118, 110, 122, and 114. Which scores have the larger standard deviation?

CRITICAL THINKING

28. One student received scores of 85, 92, 86, and 89. A second student received scores of 90, 97, 91, and 94 (exactly 5 points more on each test). Are the means of the two students the same? If not, what is the relationship between the means of the two students? Are the standard deviations of the scores of the two students the same? If not, what is the relationship between the standard deviations of the two students?

29. Grade point average (GPA) is a *weighted* mean. It is called a weighted mean because a grade in a 5-unit course has more influence on your GPA than a grade in a 2-unit course. GPA is calculated by multiplying the numerical equivalent of each grade by the number of units, adding those products, and then dividing by the number of units. Calculate your GPA for the last quarter or semester.

30. A company is negotiating with its employees for a raise in salary. One proposal would add $500 a year to each employee's salary. The second proposal would give each employee a 4% raise. Explain how these proposals would affect the current mean and standard deviation of salaries for the company.

SECTION **10.3**   Introduction to Probability

OBJECTIVE **A**

The probability of simple events

A weather forecaster estimates that there is a 75% chance of rain. A state lottery director claims that there is a $\frac{1}{9}$ chance of winning a prize offered by the lottery. Each of these statements involves uncertainty to some extent. The degree of uncertainty is called **probability**. For the statements above, the probability of rain is 75% and the probability of winning a prize in the lottery is $\frac{1}{9}$.

A probability is determined from an **experiment**, which is any activity that has an observable outcome. Examples of experiments are:

> Tossing a coin and observing whether it lands heads or tails
> Interviewing voters to determine their preference for a political candidate
> Recording the percent change in the price of a stock

All the possible outcomes of an experiment are called the **sample space** of the experiment. The outcomes of an experiment are listed between braces and frequently designated by S.

> For each experiment, list all the possible outcomes.
>
> **a.** A number cube, which has the numbers from 1 to 6 written on its sides, is rolled once.
> **b.** A fair coin is tossed once.
> **c.** The spinner at the right is spun once.
>
> **a.** Any of the numbers from 1 to 6 could show on the top of the cube. $S = \{1, 2, 3, 4, 5, 6\}$.
> **b.** A fair coin is one for which heads and tails have an equal chance of being tossed. $S = \{H, T\}$, where H represents heads and T represents tails.
> **c.** Assuming that the spinner does not come to rest on a line, the arrow could come to rest in any one of the four sectors. $S = \{1, 2, 3, 4\}$.

An **event** is one or more outcomes of an experiment. Events are denoted by capital letters. Consider the experiment of rolling the number cube given above. Some possible events are:

> The number is even. $E = \{2, 4, 6\}$
> The number is a prime number. $P = \{2, 3, 5\}$
> The number is less than 10. $T = \{1, 2, 3, 4, 5, 6\}$. Note that in this case, the event is the entire sample space.
> The number is greater than 20. This event is impossible for the given sample space. The impossible event is symbolized by $\varnothing$.

When discussing experiments and events, it is convenient to refer to the *favorable outcomes* of an experiment. These are the outcomes of an experiment that satisfy the requirements of the particular event. For instance, consider the experiment of rolling a fair die once. The sample space is $\{1, 2, 3, 4, 5, 6\}$, and one possible event E would be rolling a number that is divisible by 3. The outcomes of the experiment that are favorable to E are 3 and 6, and $E = \{3, 6\}$.

POINT OF INTEREST

It was dice-playing that led Antoine Gombaud, Chevalier de Mere, to ask Blaise Pascal, a French mathematician, to figure out the probability of throwing two sixes. Pascal and Pierre Fermat solved the problem, and their explorations led to the birth of probability theory.

Probability Formula

The probability of an event E, written $P(E)$, is the ratio of the number of favorable outcomes of an experiment to the total number of possible outcomes of the experiment.

$$P(E) = \frac{\text{number of favorable outcomes}}{\text{number of possible outcomes}}$$

The outcomes of the experiment of tossing a fair coin are *equally likely*. Any one of the outcomes is just as likely as another. If a fair coin is tossed once, the probability of a head or a tail is $\frac{1}{2}$. Each event, heads or tails, is equally likely.

The probability formula applies to experiments for which the outcomes are equally likely.

Not all experiments have equally likely outcomes. Consider an exhibition baseball game between a professional team and a college team. Although either team *could* win the game, the probability that the professional team will win is greater than that of the college team. The outcomes are not equally likely. For the experiments in this section, assume that the outcomes of an experiment are equally likely.

There are five choices, *a* through *e*, for each question on a multiple-choice test. By just guessing, what is the probability of choosing the correct answer for a certain question?

It is possible to select any of the letters a, b, c, d, or e.	There are 5 possible outcomes of the experiment.
The event E is the correct answer.	There is 1 favorable outcome, guessing the correct answer.
Use the probability formula.	$P(E) = \dfrac{\text{number of favorable outcomes}}{\text{number of possible outcomes}} = \dfrac{1}{5}$

The probability of guessing the correct answer is $\frac{1}{5}$.

Each of the letters of the word *Tennessee* is written on a card, and the cards are placed in a hat. If one card is drawn at random from the hat, what is the probability that the card has the letter *e* on it?

The phrase "at random" means that each card has an equal chance of being drawn.	There are 9 letters in *Tennessee*. Therefore, there are 9 possible outcomes of the experiment.
There are 4 cards with an *e* on them.	There are 4 favorable outcomes of the experiment, the 4 *e*'s.
Use the probability formula.	$P(E) = \dfrac{\text{number of favorable outcomes}}{\text{number of possible outcomes}} = \dfrac{4}{9}$

The probability is $\frac{4}{9}$.

Calculating the probability of an event requires counting the number of possible outcomes of an experiment and the number of outcomes that are favorable to the event. One way to do this is to list the outcomes of the experiment in some systematic way. Using a table is often very helpful.

Q_1	Q_2	Q_3
T	T	T
T	T	F
T	F	T
T	F	F
F	T	T
F	T	F
F	F	T
F	F	F

A professor writes three true/false questions for a test. If the professor randomly chooses which questions will have a true answer and which will have a false answer, what is the probability that the test will have 2 true questions and 1 false question?

The experiment S consists of choosing T or F for each of the 3 questions. The possible outcomes of the experiment are shown in the table at the right.

$S = \{TTT, TTF, TFT, TFF,$
$\qquad FTT, FTF, FFT, FFF\}$

There are 8 outcomes for S.

The event E consists of 2 true questions and 1 false question.

$E = \{TTF, TFT, FTT\}$

There are 3 outcomes for E.

Use the probability formula.

$P(E) = \dfrac{3}{8}$

The probability that there are 2 true questions and 1 false question is $\dfrac{3}{8}$.

The probabilities that have been calculated so far are referred to as *mathematical* or *theoretical* probabilities. The calculations are based on theory—for example, that either side of a coin is equally likely or that each of the six sides of a fair die is equally likely to be face up. Not all probabilities arise from such assumptions.

Empirical probabilities are based on observations of certain events. For instance, a weather forecast of a 75% chance of rain is an empirical probability. From historical records kept by the weather bureau, when a similar weather pattern existed, rain occurred 75% of the time. It is theoretically impossible to predict the weather, and only observations of past weather patterns can be used to predict future weather conditions.

Empirical Probability Formula

The empirical probability of an event E is the ratio of the number of observations of E to the total number of observations.

$$P(E) = \frac{\text{number of observations of } E}{\text{total number of observations}}$$

Records of an insurance company show that of 2,549 claims for theft filed by policy holders, 927 were claims for more than $5,000. What is the empirical probability that the next claim for theft this company receives will be a claim for more than $5,000?

The empirical probability of E is the ratio of the number of claims for over $5,000 to the total number of claims.

$P(E) = \dfrac{927}{2,549} \approx 0.36$

The probability is approximately 0.36.

Empirical probabilities can be used to test theoretical probabilities. For example, suppose you want to determine whether a coin is fair. After tossing the coin 1,000 times, you note that heads occurred 527 times. The empirical probability of heads is $\frac{527}{1,000} = 0.527$. This differs from the theoretical value of 0.5. Is this difference enough to suggest that the coin is not fair? That is a more difficult question, but one that can be answered using statistics.

If two dice are rolled, the sample space for the experiment can be recorded systematically in a table. Use this table for Example 1 and You Try It 1.

POINT OF INTEREST

Romans called a die that was marked on four faces a *talus*, which meant "anklebone." The anklebone was considered an ideal die because it is roughly a rectangular solid and it has no marrow, so loose ones from sheep were more likely to be lying around after the wolves had left their prey.

Possible Outcomes from Rolling Two Dice

1,1	2,1	3,1	4,1	5,1	6,1
1,2	2,2	3,2	4,2	5,2	6,2
1,3	2,3	3,3	4,3	5,3	6,3
1,4	2,4	3,4	4,4	5,4	6,4
1,5	2,5	3,5	4,5	5,5	6,5
1,6	2,6	3,6	4,6	5,6	6,6

Example 1

Two dice are rolled once. Calculate the probability that the sum of the numbers on the two dice is 7.

Strategy

To find the probability:

▶ Count the number of possible outcomes of the experiment.

▶ Count the outcomes of the experiment that are favorable to the event the sum is 7.

▶ Use the probability formula.

Solution

There are 36 possible outcomes.

There are 6 outcomes favorable for *E*: (1, 6), (2, 5), (3, 4), (4, 3), (5, 2), and (6, 1).

$P(E) = \dfrac{6}{36} = \dfrac{1}{6}$

The probability that the sum is 7 is $\dfrac{1}{6}$.

You Try It 1

Two dice are rolled once. Calculate the probability that the two numbers on the dice are equal.

Your Strategy

Your Solution

Solution on p. A30

Example 2

A large box contains 25 red, 35 blue, and 40 white balls. If one ball is randomly selected from the box, what is the probability that it is blue? Write the answer as a percent.

You Try It 2

There are 8 covered circles on a "scratcher card" that is given to each customer at a fast-food restaurant. Under one of the circles is a symbol for a free soft drink. If the customer scratches off one circle, what is the probability that the soft drink symbol will be uncovered?

Strategy

To find the probability:
▶ Count the number of outcomes of the experiment.
▶ Count the number of outcomes of the experiment favorable to the event E that the ball is blue.
▶ Use the probability formula.

Your Strategy

Solution

There are 100 (25 + 35 + 40) balls in the box.

35 balls of the 100 are blue.

$$P(E) = \frac{35}{100} = 0.35 = 35\%$$

There is a 35% chance of selecting a blue ball.

Your Solution

Solution on p. A30

Objective B

The odds of an event

Sometimes the chances of an event occurring are given in terms of *odds*. This concept is closely related to probability.

Odds in Favor of an Event

The **odds in favor** of an event is the ratio of the number of favorable outcomes of an experiment to the number of unfavorable outcomes

$$\text{Odds in favor} = \frac{\text{number of favorable outcomes}}{\text{number of unfavorable outcomes}}$$

Odds Against an Event

The **odds against** an event is the ratio of the number of unfavorable outcomes of an experiment to the number of favorable outcomes.

$$\text{Odds against} = \frac{\text{number of unfavorable outcomes}}{\text{number of favorable outcomes}}$$

To find the odds in favor of a 4 when a die is rolled once, list the favorable outcomes and the unfavorable outcomes.

favorable outcomes: 4   unfavorable outcomes, 1, 2, 3, 5, 6

$$\text{Odds in favor of a 4} = \frac{\text{number of favorable outcomes}}{\text{number of unfavorable outcomes}} = \frac{1}{5}$$

Frequently the odds of an event are expressed as a ratio using the word TO. For the last problem, the odds in favor of a 4 are 1 TO 5.

It is possible to compute the probability of an event from the odds in favor fraction. The probability of an event is the ratio of the numerator to the sum of the numerator and denominator.

The odds in favor of winning a prize in a charity drawing are 1 TO 19. What is the probability of winning a prize?

Write the ratio 1 TO 19 as a fraction. $1 \text{ TO } 19 = \frac{1}{19}$

The probability of winning a prize is the ratio of the numerator to the sum of the numerator and denominator. $\text{Probability} = \frac{1}{1 + 19} = \frac{1}{20}$

The probability of winning a prize is $\frac{1}{20}$.

Example 3

In a horse race, the odds against a horse winning the race are posted as 9 TO 2. What is the probability of the horse's winning the race?

Strategy

To calculate the probability of winning:

▶ Restate the odds against as odds in favor.
▶ Using the odds in favor fraction, the probability of winning is the ratio of the numerator to the sum of the numerator and denominator.

Solution

The odds against winning are 9 TO 2. Therefore, the odds in favor of winning are 2 TO 9.

$$\text{Probability of winning} = \frac{2}{2 + 9} = \frac{2}{11}$$

The probability of the horse's winning the race is $\frac{2}{11}$.

You Try It 3

The odds in favor of contracting the flu during a flu epidemic are 2 TO 13. Calculate the probability of getting the flu.

Your Strategy

Your Solution

Solution on p. A30

10.3 EXERCISES

OBJECTIVE A

Solve.

1. A coin is tossed 4 times. List all the possible outcomes of the experiment as a sample space. The table on page 571 shows a systematic way of recording results.

2. Three cards—one red, one green, and one blue—are to be arranged in a stack. Using R for red, G for green, and B for blue, make a list of all the different stacks that can be formed. (Some computer monitors are called RGB monitors for the colors red, green, and blue.)

3. A tetrahedral die is one with four triangular sides. If two tetrahedral dice are rolled, list all the possible outcomes of the experiment as a sample space. (See the Table of Two Dice, page 572, for a systematic method of listing the outcomes.)

4. A coin is tossed and then a die is rolled. List all the possible outcomes of the experiment as a sample space. (To get you started, (H, 1) is one of the possible outcomes.)

5. Some people who cheat at gambling use dice that are loaded so that 7 occurs more frequently than expected. If these dice are used, are the probabilities of the outcomes equal? Why or why not?

6. If the spinner at the right is spun once, is each of the numbers 1 through 5 equally likely? Why or why not?

7. A coin is tossed four times. What is the probability that the outcomes of the tosses are exactly in the order HHTT? (See Exercise 1.)

8. A coin is tossed four times. What is the probability that the outcomes of the tosses are exactly in the order HTTH? (See Exercise 1.)

9. A coin is tossed four times. What is the probability that the outcomes of the tosses consist of two heads and two tails? (See Exercise 1.)

Solve.

10. A coin is tossed four times. What is the probability that the outcomes of the tosses consist of one head and three tails? (See Exercise 1.)

11. If two dice are rolled, what is the probability that the sum of the dots on the upward faces is 5?

12. If two dice are rolled, what is the probability that the sum of the dots on the upward faces is 9?

13. If two dice are rolled, what is the probability that the sum of the dots on the upward faces is 15?

14. If two dice are rolled, what is the probability that the sum of the dots on the upward faces is less than 15?

15. If two dice are rolled, what is the probability that the sum of the dots on the upward faces is 2?

16. If two dice are rolled, what is the probability that the sum of the dots on the upward faces is 12?

17. A dodecahedral die has 12 sides. If the die is rolled once, what is the probability that the upward face shows an 11?

18. A dodecahedral die has 12 sides. If the die is rolled once, what is the probability that the upward face shows 5?

19. If two tetrahedral dice are rolled (see Exercise 3), what is the probability that the sum on the upward faces is 4?

20. If two tetrahedral dice are rolled (see Exercise 3), what is the probability that the sum on the upward faces is 6?

21. A dodecahedral die has 12 sides. If the die is rolled once, what is the probability that the upward face shows a number divisible by 4?

Solve.

22. A dodecahedral die has 12 sides. If the die is rolled once, what is the probability that the upward face shows a number that is a multiple of 3?

23. A survey of 95 people showed that 37 preferred a cash discount of 2% if an item was purchased using cash or a check. Based on this survey, what is the empirical probability that a person prefers a cash discount?

24. A survey of 725 people showed that 587 had a group health insurance plan where they worked. Based on this survey, what is the empirical probability that an employee has a group health insurance plan?

25. A signal light is green for 3 min, yellow for 15 s, and red for 2 min. If you drive up to this light, what is the probability that it will be green when you reach the intersection?

26. In a history class, a professor gave 4 A's, 8 B's, 22 C's, 10 D's, and 3 F's. If a single student's paper is chosen from this class, what is the probability that it received a B?

27. A television cable company surveyed some of its customers and asked them to rate the cable service as excellent, satisfactory, average, unsatisfactory, or poor. The results are recorded in the table at the right. What is the probability that a customer who was surveyed rated the service as satisfactory or excellent?

Quality of Service	Number Who Voted
Excellent	98
Satisfactory	87
Average	129
Unsatisfactory	42
Poor	21

28. Using the television cable survey in Exercise 27, what is the probability that a customer who was surveyed rated the service as unsatisfactory or poor?

OBJECTIVE B

Solve.

29. A fair coin is tossed once. What are the odds of its showing heads?

30. A fair coin is tossed twice. What are the odds of its showing tails both times?

31. At the beginning of the professional football season, one team was given 40 TO 1 odds against winning the Super Bowl. What is the probability of this team winning the Super Bowl?

Solve.

32. At the beginning of the professional baseball season, one team was given 25 TO 1 odds against winning the World Series. What is the probability of this team winning the World Series?

33. Two fair dice are rolled. What are the odds in favor of rolling a seven?

34. Two fair dice are rolled. What are the odds in favor of rolling a twelve?

35. A single card is selected from a regular deck of playing cards. What are the odds against its being an ace?

36. A single card is selected from a regular deck of playing cards. What are the odds against its being a heart?

37. The odds in favor of a candidate winning an election are 3 TO 2. What is the probability of the candidate winning the election?

38. On a board game, the odds in favor of winning $5,000 are 3 TO 7. What is the probability of winning the $5,000?

39. A stock market analyst estimates that the odds in favor of a stock going up in value are 2 TO 1. What is the probability of the stock's not going up in value?

40. The odds in favor of the occurrence of an event A are given as 5 TO 2, and the odds against a second event B are given as 1 TO 7. Which event, A or B, has the greater probability of occurring?

CRITICAL THINKING

41. A box contains only white, blue, and red balls. You are told that the probability of choosing a white ball is $\frac{1}{2}$, that of choosing a blue ball is $\frac{1}{3}$, and that of choosing a red ball is $\frac{1}{9}$. What is wrong with that statement?

42. Three line segments are randomly chosen from line segments whose lengths are 1 cm, 2 cm, 3 cm, 4 cm, and 5 cm. What is the probability that a triangle can be formed from the line segments?

43. ✏ The probability of tossing a fair coin and having it land heads is $\frac{1}{2}$.

    Does this mean that if that coin is tossed 100 times, it will land heads 50 times? Explain your answer.

44. ✏ Suppose a surgeon tells you that an operation that you need has a 90% success rate. Explain what that means.

PROJECTS IN MATHEMATICS

Random Samples

When a survey is taken to determine, for example, who is the most popular choice for a political office such as a governor or president, it would not be appropriate to survey only one ethnic group, or survey only one religious group, or survey only one political group. A survey done in that way would reflect only the views of that particular group of people. Instead, a *random sample* of people must be chosen. This sample would include people from different ethnic, religious, political, and income groups. The purpose of the random sample is to identify the popular choice of all the people by interviewing only a few people, the people in the random sample.

One way of choosing a random sample is to use a table of random numbers. An example of a portion of such a table is shown below.

Random Digits

40784	38916	12949
29798	57707	57392
42228	94940	10668
02218	89355	76117
15736	08506	29759
42658	32502	99698
98670	57794	64795
38266	30138	61250
68249	32459	41627
36910	85225	78541

In a random number table, each digit should occur with approximately the same frequency as any other digit.

1. Make a histogram for the table above. For the horizontal axis, use the digits 0, 1, 2, . . . , 8, 9. The vertical axis is the frequency of each digit. Do the digits occur with approximately the same frequency?

As an example of how to use a random number table, suppose there are 33 students in your class and you want to randomly select 6 students. Using the numbers 01, 02, 03, . . . , 31, 32, 33, assign each student a two-digit number. Starting with the first column of random numbers, move down the column looking at only the first two digits. If the two digits are one of the numbers 01 through 33, write them down. If not, move to the next number. Continue in this way until six numbers have been selected.

For instance, the first two digits of 40784 are 40, which is not between 01 and 33. Therefore, move to the next number. The first two digits of 29798 are 29. Since 29 is between 01 and 33, write it down. Continuing in this way, the random sample would be the students with numbers 29, 02, 15, 08, 32, and 30.

2. Once a random sample has been selected, it is possible to use this sample to estimate characteristics of the entire class. For example, find the mean height of the students in the random sample and compare it to the mean height of the entire class.

Random numbers also are used to *simulate* random events. In a simulation, it is not the actual activity that is performed, but instead one that is very similar.

Random Digits

40784	38916	12949
29798	57707	57392
42228	94940	10668
02218	89355	76117
15736	08506	29759
42658	32502	99698
98670	57794	64795
38266	30138	61250
68249	32459	41627
36910	85225	78541

For instance, suppose a coin is tossed repeatedly, and the result, head or tail, is recorded. The actual activity is tossing the coin. A simulation of tossing a coin uses a random number table. Starting with the first column of random numbers, move down the column looking at the first digit. Associate an even digit with heads and an odd digit with tails. Using the first column of the table at the left, a simulation of the first ten tosses of the coin would be H, H, H, H, T, H, T, T, H, and T.

Probabilities can be approximated by simulating an event. Consider the event of tossing two heads when a fair coin is tossed twice. We can simulate the event by associating a two-digit number with two even digits with two heads, one with two odd digits with two tails, and any other pair of digits with a head and tail or tail and head.

3. Using the 30 numbers in the table above, show that the probability of tossing two heads would be $\frac{7}{30} \approx 0.233$. Explain why the actual probability is 0.25.

4. Go to the library and find a table of random numbers. Use this table to simulate rolling a pair of dice. Using the first two digits of a column, a valid roll consists of two digits that are both between 1 and 6, inclusive. Simulate the empirical probability that the sum of the dice is 7 by using 100 rolls of the dice.

The Game of Nim

Sometimes the solution to a problem can be found by *working backward*. This technique can be used to find a winning strategy for a game called Nim.

There are many variations of this game. For our game, there are two players, Player A and Player B, who alternately place 1, 2, or 3 matchsticks in a pile. The object of the game is to place the 32nd matchstick in the pile. Is there a strategy that Player A can use to guarantee winning the game?

Working backward, if there are 29, 30, or 31 matchsticks in the pile when it is A's turn to play, A can win by placing 3 (29 + 3 = 32), 2 (30 + 2 = 32), or 1 (31 + 1 = 32) matchsticks on the pile. If there are to be 29, 30, or 31 matchsticks in the pile when it is A's turn, there must be 28 matchsticks in the pile when it is B's turn.

Working backward from 28, if there are to be 28 matches in the pile at B's turn, there must be 25, 26, or 27 at A's turn. Player A can then add 3, 2 or 1 matchsticks to the pile to bring the number to 28. For there to be 25, 26, or 27 matchsticks in the pile at A's turn, there must be 24 matchsticks in the pile at B's turn.

Now working backward from 24, if there are to be 24 matches in the pile at B's turn, there must be 21, 22, or 23 at A's turn. Player A can then add 3, 2, or 1 matchsticks to the pile to bring the number to 24. For there to be 21, 22, or 23 matchsticks in the pile at A's turn, there must be 20 matchsticks in the pile at B's turn.

So far, we have found that for Player A to win, there must be 28, 24, or 20 matchsticks in the pile when it is B's turn to play. Note that each time, the number is decreasing by 4. Continuing this pattern, Player A will win if there are 16, 12, 8, or 4 matchsticks in the pile when it is B's turn.

Player A can guarantee winning by making sure that the number of matchsticks in the pile is a multiple of 4. To ensure this, Player A allows Player B to go first and then adds exactly enough matchsticks to the pile to bring the total to a multiple of 4.

For example, suppose B places 3 matchsticks in the pile; then A places 1 matchstick (3 + 1 = 4) in the pile. Now B places 2 matchsticks in the pile. The total is now 6 matchsticks. Player A then places 2 matchsticks in the pile to bring the total to 8, a multiple of 4. If play continues in this way, Player A will win.

Here are some variations of Nim. See whether you can develop a winning strategy for Player A. *Hint:* It may not be possible.

1. Suppose the goal is to place the last matchstick in a pile of 30 matches.

2. Suppose the players make two piles of matchsticks, with the maximum number of matchsticks in each pile to be 20.

3. In this variation of Nim, there are 40 matchsticks in a pile. Each player alternately selects 1, 2, or 3 matches from the pile. The player who selects the last match wins.

CHAPTER SUMMARY

Key Words

Statistics is the study of collecting, organizing, and interpreting data.

A *population* is the set of all observed outcomes of an experiment.

A *frequency distribution* is one method of organizing data.

A *class* in a frequency distribution is a portion of all the data.

The *range* of a set of numerical data is the difference between the largest and smallest values.

A *histogram* is a bar graph of the data in a frequency distribution.

A *class midpoint* is the center of a class.

A *frequency polygon* is a broken-line graph of the data in a frequency distribution.

The *mean, median,* and *mode* are three types of averages used in statistics. The *median* of data is the number that separates the data into two equal parts when the data have been arranged from smallest to largest (or largest to smallest). The *mode* is the most frequently occurring data value.

The *first quartile* Q_1 is the number below which one-fourth of the data lie.

The *third quartile* Q_3 is a number above which one-fourth of the data lie.

A *box-and-whiskers plot* is a graph that shows five numbers: the smallest value, the first quartile, the median, the third quartile, and the largest value. The box is placed around the values between the first quartile and the third quartile.

Standard deviation is a measure of the clustering of data near the mean.

An *experiment* is an activity with an observable outcome.

All the possible outcomes of an experiment are called the *sample space* of the experiment.

An *event* is one or more outcomes of an experiment.

Essential Rules	**Mean of a set of data**	$\bar{x} = \dfrac{\text{sum of all data values}}{\text{number of data values}}$
	Interquartile range	Interquartile range $= Q_3 - Q_1$
	Standard deviation	1. Sum the squares of the differences between each value of data and the mean. 2. Divide the result in Step 1 by the number of items in the set of data. 3. Take the square root of the result in Step 2.
	Probability formula	$P(E) = \dfrac{\text{number of favorable outcomes}}{\text{number of possible outcomes}}$
	Empirical probability formula	$P(E) = \dfrac{\text{number of observations of } E}{\text{total number of observations}}$
	Odds in favor of an event	Odds in favor $= \dfrac{\text{number of favorable outcomes}}{\text{number of unfavorable outcomes}}$
	Odds against an event	Odds against $= \dfrac{\text{number of unfavorable outcomes}}{\text{number of favorable outcomes}}$

CHAPTER REVIEW EXERCISES

Education   Use the data in the table below for Exercises 1–5.

Number of Students in 40 Mathematics Classes

30	45	54	24	48	12	38	31
15	36	37	27	40	35	55	32
42	14	21	18	29	25	16	42
44	41	28	32	27	24	30	24
21	35	27	32	39	41	35	48

1.   Make a frequency distribution for these data using 6 classes.

2.   Which class has the greatest frequency?

3.   How many math classes have 35 or fewer students?

4.   What percent of the math classes have 44 or more students?

5.   What percent of the math classes have 27 or fewer students?

Temperature   The high temperatures at a ski resort during a 125-day ski season are recorded in the figure at the right. Use this histogram for Exercises 6 and 7.

6.   Find the number of days the high temperature was 45° or above.

7.   How many days had a high temperature of 25° or below?

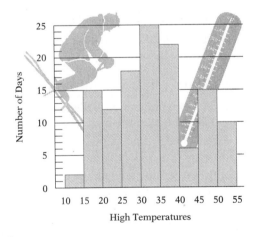

Solve.

8.   *Health*   A health clinic administered a test for cholesterol to 11 people. The results were 180, 220, 160, 230, 280, 200, 210, 250, 190, 230, and 210. Find the mean and median of these data.

9.   *Health*   The weights, in pounds, of 10 babies born at a hospital were recorded as 6.3, 5.9, 8.1, 6.5, 7.2, 5.6, 8.9, 9.1, 6.9, and 7.2. Find the mean and median of these data.

10.   *The Arts*   People leaving a new movie were asked to rate the movie as bad, good, very good, or excellent. The responses were: bad, 28; good, 65; very good, 49; excellent, 28. What was the modal response for this survey?

Investments The frequency polygon in the figure at the right shows the number of shares of stock that were sold on a stock exchange. Use the frequency polygon for Exercises 11–13.

11. How many shares of stock were sold between 7 A.M. and 10 A.M.?

12. Between which hours were less than 15 million shares sold?

13. What is the ratio of the number of shares of stock sold between 9 A.M. and 10 A.M. to the number that were sold between 11 A.M. and 12 P.M.?

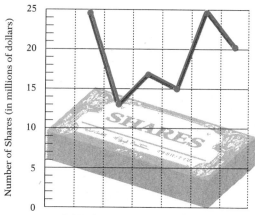

Shares of Stock Sold between 7 A.M. and 1 P.M.

Solve.

14. *Sports* The numbers of points scored by a basketball team for fifteen games were 89, 102, 134, 110, 121, 124, 111, 116, 99, 120, 105, 109, 110, 124, and 131. Find the first quartile, median, and third quartile. Draw a box-and-whiskers plot.

15. *Automotive Industry* A consumer research group tested the average miles per gallon for six cars. The results were 24, 28, 22, 25, 26, and 27. Find the standard deviation of the gasoline mileage ratings.

16. A charity raffle sells 2,500 raffle tickets for a big-screen television set. If you purchase 5 tickets, what is the probability that you will win the television?

17. A box contains 50 balls, of which 15 are red. If one ball is randomly selected from the box, what are the odds in favor of the ball's being red?

18. In professional jai alai, a gambler can wager on who will win the event. The odds against one of the players winning are given as 5 TO 2. What is the probability of that player's winning?

19. A dodecahedral die has 12 sides numbered from 1 to 12. If this die is rolled once, what is the probability that a number divisible by 6 will be on the upward face?

20. One student is randomly selected from 3 first-year students, 4 sophomores, 5 juniors, and 2 seniors. What is the probability that the student is a junior?

CUMULATIVE REVIEW EXERCISES

1. Simplify: $\sqrt{200}$

2. Solve: $7p - 2(3p - 1) = 5p + 6$

3. Evaluate $3a^2b - 4ab^2$ when $a = -1$ and $b = 2$.

4. Simplify: $-2[2 - 4(3x - 1) + 2(3x - 1)]$

5. Solve: $-\frac{2}{3}y - 5 = 7$

6. Simplify: $-\frac{4}{5}\left[\frac{3}{4} - \frac{7}{8} - \left(\frac{2}{3}\right)^2\right]$

7. Graph $y = \frac{4}{3}x - 3$.

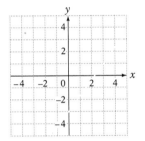

8. Graph $y = \frac{1}{3}x$.

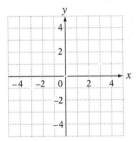

9. Subtract: $(7y^2 + 5y - 8) - (4y^2 - 3y + 1)$

10. Simplify: $(4a^2b)^3$

11. $16\frac{2}{3}\%$ of what number is 24?

12. Solve: $\frac{9}{8} = \frac{3}{n}$

13. Write 87,600,000,000 in scientific notation.

14. A landscape architect designed the cement patio shown below. Determine the area of the patio.

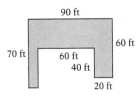

15. Multiply: $(5c^2d^4)(-3cd^6)$

16. Convert 40 km/h to meters per second.

17. What is the measure of $\angle n$ in the figure below?

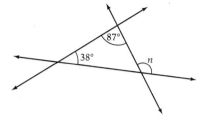

18. Find the area of the parallelogram shown below.

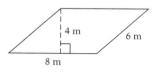

Solve.

19. Find the simple interest on a 3-month loan of $25,000 at an annual interest rate of 7.5%.

20. A box contains 12 white, 15 blue, and 9 red balls. If one ball is randomly chosen from the box, what is the probability that the ball is not white?

21. *Education* The scores for six students on an achievement test were 24, 38, 22, 34, 37, and 31. What were the mean and median scores for these students?

22. *Elections* The results of a recent city election showed that 55,000 people voted, out of a possible 230,000 registered voters. What percent of the registered voters did not vote in the election? Round to the nearest tenth of a percent.

23. *Measurement* A Greek mathematician, Eratosthenes (circa 300 B.C.) calculated that an angle of 7.5° at Earth's center cuts an arc of 1,600 km on Earth's surface. Using this information, what is the approximate circumference of Earth?

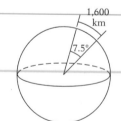

24. *Meteorology* The annual rainfall totals, in inches, for a certain region for the last five years were 12, 16, 20, 18, and 14. Find the standard deviation of these rainfall totals.

25. *Compensation* A chef's helper received a 10% hourly wage increase to $16.50 per hour. What was the chef's helper's hourly wage before the increase?

FINAL EXAMINATION

1. Estimate the sum of 672, 843, 509, and 417.

2. Simplify: $18 + 3(6 - 4)^2 \div 2$

3. Simplify: $-8 - (-13) - 10 + 7$

4. Evaluate $|a - b| - 3bc^3$ when $a = -2$, $b = 4$, and $c = -1$.

5. What is $5\frac{3}{8}$ minus $2\frac{11}{16}$?

6. Find the quotient of $\frac{7}{9}$ and $\frac{5}{6}$.

7. Simplify: $\dfrac{\frac{3}{4} - \frac{1}{2}}{\frac{5}{8} + \frac{1}{2}}$

8. Place the correct symbol, $<$ or $>$, between the two numbers.

   $\dfrac{5}{16}$ 0.313

9. Evaluate $-10qr$ when $q = -8.1$ and $r = -9.5$.

10. Divide and round to the nearest hundredth: $-15.32 \div 4.67$

11. Is -0.5 a solution of the equation $-90y = 45$?

12. Simplify: $\sqrt{162}$

13. Graph $x \geq -4$.

14. Identify the property that justifies the statement.
    $8 + (y + 4) = (y + 4) + 8$

15. Simplify: $-\dfrac{5}{6}(-12t)$

16. Simplify: $2(x - 3y) - 4(x + 2y)$

17. Subtract: $(5z^3 + 2z^2 - 1) - (4z^3 + 6z - 8)$

18. Multiply: $(4x^2)(2x^5y)$

19. Multiply: $2a^2b^2(5a^2 - 3ab + 4b^2)$

20. Multiply: $(3x - 2)(5x + 3)$

21. Simplify: $(3x^2y)^4$

22. Evaluate 4^{-3}.

23. Simplify: $\dfrac{m^5n^8}{m^3n^4}$

24. Solve: $2 - \dfrac{4}{3}y = 10$

25. Solve: $6z + 8 = 5 - 3z$

26. Solve: $8 + 2(6c - 7) = 4$

27. Convert 2.6 mi to feet.

28. Solve: $\dfrac{n + 2}{8} = \dfrac{5}{12}$

29. Given that $\ell_1 \parallel \ell_2$, find the measures of angles a and b.

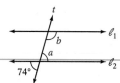

30. Find the unknown side of the triangle. Round to the nearest tenth.

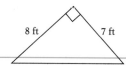

31. Find the perimeter of the composite figure. Round to the nearest hundredth.

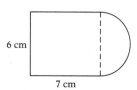

32. Find the volume of the composite figure. Round to the nearest hundredth.

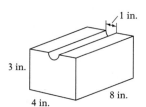

33. Graph $y = -2x + 3$.

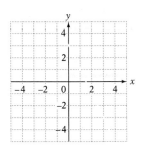

34. Graph $y = \dfrac{3}{5}x - 4$.

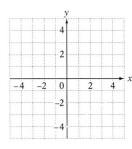

Solve.

35. *Physics*   Find the ground speed of an airplane traveling into a 22-mph wind with an air speed of 386 mph. Use the formula $g = a - h$, where g is the ground speed, a is the air speed, and h is the speed of the head wind.

36. *Business*   A factory worker can inspect a product in $1\dfrac{1}{2}$ min. How many products can the worker inspect during an 8-hour day?

37. *Chemistry*   The boiling point of bromine is 58.78°C. The melting point of bromine is −7.2°C. Find the difference between the boiling point and the melting point of bromine.

38. *Physics*   One light-year, which is the distance that light travels through empty space in one year, is approximately 5,880,000,000,000 mi. Write this number in scientific notation.

39. *Physics*   Two children are sitting on a seesaw that is 10 ft long. One child weighs 50 lb, and the second child weighs 75 lb. How far from the 50-pound child should the fulcrum be placed so that the seesaw balances? Use the formula $F_1 x = F_2(d - x)$.

40. *Consumerism*   The fee charged by a ticketing agency for a concert is $5.50 plus $22.50 for each ticket purchased. If your total charge for tickets is $140.50, how many tickets are you purchasing?

41. *Taxes*   The property tax on a $125,000 house is $3,750. At this rate, what is the property tax on a home appraised at $157,000?

Solve.

42. 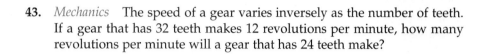 *Geography* The figure at the right represents the land area of the states in the United States. What percent of the states have a land area of 75,000 mi² or more?

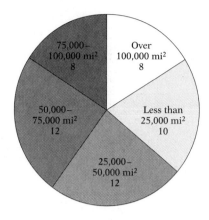

43. *Mechanics* The speed of a gear varies inversely as the number of teeth. If a gear that has 32 teeth makes 12 revolutions per minute, how many revolutions per minute will a gear that has 24 teeth make?

Land Area of the States in the United States

44. *Consumerism* A customer purchased a car for $12,500 and paid a sales tax of 5.5% of the cost. Find the total cost of the car including sales tax.

45. *Economics* Due to a recession, the number of housing starts in a community decreased from 124 to 96. What percent decrease does this represent? Round to the nearest tenth of a percent.

46. *Business* A necklace with a regular price of $245 is on sale for 35% off the regular price. Find the sale price.

47. *Finances* Find the simple interest on a 9-month loan of $25,000 at an annual interest rate of 8.6%.

48. 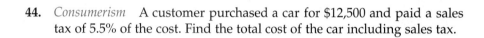 *Labor Force* The number of hours per week that 80 twelfth grade students spend at paid jobs are given in the figure at the right. What percent of the students work more than 15 hours per week?

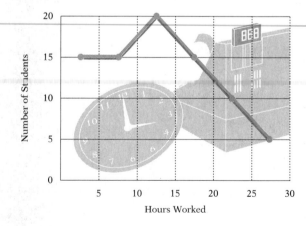

49. *Insurance* You requested rates for term life insurance from five different insurance companies. The annual premiums were $297, $425, $362, $281, and $309. Calculate the mean and median annual premiums for these five insurance companies.

50. *Probability* If two dice are tossed, what is the probability that the sum of the dots on the upward faces is divisible by 3?

The Metric System of Measurement

In 1789, an attempt was made to standardize units of measurement internationally in order to simplify trade between nations. A commission in France developed a system of measurement known as the **metric system.**

The basic unit of *length,* or distance, in the metric system is the **meter** (m). One meter is approximately the distance from a doorknob to the floor. All units of length in the metric system are derived from the meter. Prefixes to the basic unit denote the length of each unit. For example, the prefix "centi-" means one-hundredth; therefore, one centimeter is 1 one-hundredth of a meter.

kilo-	= 1,000	1 kilometer (km)	= 1,000 meters (m)
hecto-	= 100	1 hectometer (hm)	= 100 m
deca-	= 10	1 decameter (dam)	= 10 m
		1 meter (m)	= 1 m
deci-	= 0.1	1 decimeter (dm)	= 0.1 m
centi-	= 0.01	1 centimeter (cm)	= 0.01 m
milli-	= 0.001	1 millimeter (mm)	= 0.001 m

Mass and weight are closely related. *Weight* is a measure of how strongly gravity is pulling on an object. Therefore, an object's weight is less in space than on the earth's surface. However, the amount of material in the object, its *mass,* remains the same. On the surface of the earth, the terms *mass* and *weight* can be used interchangeably.

The basic unit of mass in the metric system is the **gram** (g). If a box that is 1 centimeter long on each side is filled with water, the mass of that water is 1 gram.

1 gram = the mass of water in a box that is
1 centimeter long on each side

The units of mass in the metric system have the same prefixes as the units of length.

1 kilogram (kg)	= 1,000 grams (g)
1 hectogram (hg)	= 100 g
1 decagram (dag)	= 10 g
1 gram (g)	= 1 g
1 decigram (dg)	= 0.1 g
1 centigram (cg)	= 0.01 g
1 milligram (mg)	= 0.001 g

The gram is a very small unit of mass. A paperclip weighs about one gram. In applications, the kilogram (1,000 grams) is a more useful unit of mass. This textbook weighs about 1 kilogram.

Liquid substances are measured in units of *capacity*.

The basic unit of capacity in the metric system is the **liter** (L). One liter is defined as the capacity of a box that is 10 centimeters long on each side.

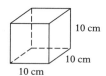

1 liter = the capacity of a box that is 10 centimeters long on each side

The units of capacity in the metric system have the same prefixes as the units of length.

1 kiloliter (kl)	=	1,000 liters (L)
1 hectoliter (hl)	=	100 L
1 decaliter (dal)	=	10 L
1 liter (L)	=	1 L
1 deciliter (dl)	=	0.1 L
1 centiliter (cl)	=	0.01 L
1 milliliter (ml)	=	0.001 L

Converting between units in the metric system involves moving the decimal point to the right or to the left. Listing the units in order from largest to smallest will indicate how many places to move the decimal point and in which direction.

To convert 3,800 cm to meters, write the units of length in order from largest to smallest.

km   hm   dam   m   dm   cm   mm
2 positions

Converting from cm to m requires moving 2 places to the left.

3,800 cm = 38.00 m
2 places

Move the decimal point the same number of places and in the same direction.

Convert 627 g to kilograms.

kg   hg   dag   g   dg   cg   mg
3 positions

Write the units of mass in order from largest to smallest. Converting g to kg requires moving 3 positions to the left.

627 g = 0.627 kg
3 places

Move the decimal point the same number of places and in the same direction.

Example 1   Convert 4.08 m to centimeters.

   Solution   4.08 m = 408 cm

You Try It 1   Convert 1,295 m to kilometers.

   Your Solution

Example 2   Convert 5.93 g to milligrams.

   Solution   5.93 g = 5,930 mg

You Try It 2   Convert 7,543 g to kilograms.

   Your Solution

Example 3   Convert 9 kl to liters.

   Solution   9 kl = 9,000 L

You Try It 3   Convert 6.3 L to milliliters.

   Your Solution

Example 4

The thickness of a single sheet of paper is 0.07 mm. Find the height in centimeters of a ream of paper. A ream is 500 sheets of paper.

Strategy

To find the height:

▶ Multiply the thickness per sheet (0.07) by the number of sheets in a ream (500) to find the height in millimeters.
▶ Convert millimeters to centimeters.

Solution

(0.07)(500) = 35

35 mm = 3.5 cm

The height of a ream of paper is 3.5 cm.

You Try It 4

One egg contains 274 mg of cholesterol. How many grams of cholesterol are in one dozen eggs?

Your Strategy

Your Solution

Solutions on p. A30

Exercises

Convert.

1. 42 cm = _____ mm

2. 91 cm = _____ mm

3. 360 g = _____ kg

4. 1,856 g = _____ kg

5. 5,194 ml = _____ L

6. 7,285 ml = _____ L

7. 2 m = _____ mm

8. 8 m = _____ mm

9. 217 mg = _____ g

10. 34 mg = _____ g

11. 4.52 L = _____ ml

12. 0.0297 L = _____ ml

Convert.

13. 8,406 m = _____ km

14. 7,530 m = _____ km

15. 2.4 kg = _____ g

16. 9.2 kg = _____ g

17. 6.18 kl = _____ L

18. 0.036 kl = _____ L

19. 9.612 km = _____ m

20. 2.35 km = _____ m

21. 0.24 g = _____ mg

22. 0.083 g = _____ mg

23. 298 cm = _____ m

24. 71.6 cm = _____ m

25. 2,431 L = _____ kl

26. 6,302 L = _____ kl

27. 0.66 m = _____ cm

28. 4.58 m = _____ cm

29. 243 mm = _____ cm

30. 92 mm = _____ cm

Solve.

31. One of the events in the summer Olympics is the 50,000-meter walk. How many kilometers do the entrants in this event walk?

32. The height of Mt. Everest is 8,882 m. Find Mt. Everest's height in kilometers.

33. The length of the Delaware River is 628 km. Find the length of the Delaware River in meters.

34. One of the events in the winter Olympic Games is the 10,000-meter speed skating. How many kilometers do the entrants in this event skate?

35. A nickel is 2 mm thick. Find the length in centimeters of the 40 coins in a roll of nickels.

36. One glass of milk contains 33 mg of cholesterol. How many grams of cholesterol are in four glasses of milk?

37. One milliliter of salt contains 400 mg of sodium. Find the number of grams of sodium in 5 ml of salt.

38. A carat is a unit of weight equal to 200 mg. Find the weight in grams of a ten-carat precious stone.

39. At a fabric shop, 800 cm of ribbon are cut from a 15-meter spool. How many meters of ribbon remain on the spool?

40. How many pieces of material, each 75 cm long, can be cut from a bolt of fabric that is 6 m long?

41. How many 240-milliliter servings are in a 2-liter bottle of cola? Round to the nearest whole number.

42. Each of the four shelves in a bookcase measures 175 cm. Find the cost of the shelves when the price of lumber is $12.75 per meter.

SOLUTIONS TO CHAPTER 1   You Try Its

SECTION 1.1 (*pages 3–12*)

You Try It 1

0 1 2 3 4 5 6 7 8 9 10 11 12

You Try It 2

4

0 1 2 3 4 5 6 7 8 9 10 11 12
7 is 4 units to the left of 11.

You Try It 3   **a.** $47 > 19$ **b.** $26 > 0$

You Try It 4   0, 3, 17, 52, 68, 94

You Try It 5   forty-six million thirty-two thousand seven hundred fifteen

You Try It 6   920,008

You Try It 7 $70,000 + 6,000 + 200 + 40 + 5$

You Try It 8   ⌐――――― Given place value
529,374
   ∟――― $9 > 5$

529,374 rounded to the nearest ten-thousand is 530,000.

You Try It 9   ⌐――――― Given place value
7,985
   ∟――― $8 > 5$

7,985 rounded to the nearest hundred is 8,000.

You Try It 10
Strategy   To find the sport named by the greatest number of people, find the largest number given in the circle graph.

Solution   The largest number given in the graph is 80.

The sport named by the greatest number of people was football.

You Try It 11
Strategy   To find the shorter distance, compare the numbers 347 and 387.

Solution $347 < 387$

The shorter distance is between Los Angeles and San Jose.

You Try It 12
Strategy   To determine in which city there were fewer letter carriers bitten, compare the numbers 101 and 75.

Solution $75 < 101$

There were fewer letter carriers bitten in San Jose.

You Try It 13
Strategy   To find the land area to the nearest thousand square miles, round 3,851,809 to the nearest thousand.

Solution   3,851,809 rounded to the nearest thousand is 3,852,000.

To the nearest thousand, the land area of Canada is 3,852,000 mi².

SECTION 1.2 (*pages 19–32*)

You Try It 1   The net revenue from the Defense Systems segment is $1,740,000,000. The net revenue from the Components segment is $9,480,000,000.

   1,740,000,000
 + 9,480,000,000
―――――――――――
  11,220,000,000

The sum of the net revenues from these two segments is $11,220,000,000.

You Try It 2   6,285 ⟶   6,000
3,972 ⟶   4,000
5,140 ⟶ + 5,000
―――――――――
         15,000

You Try It 3 $x + y + z$
$1,692 + 4,783 + 5,046$

    ¹ ²¹
   1,692
   4,783
 + 5,046
―――――――
  11,521

You Try It 4   The Addition Property of Zero

You Try It 5 $\dfrac{13 = b + 6}{13 \mid 7 + 6}$
$13 = 13$
Yes, 7 is a solution of the equation.

You Try It 6   The number of stores in 1995 was 634.
The number of stores in 1993 was 605.

   634
 − 605
―――――
    29

The difference is 29 stores.

You Try It 7
    ⁸ ⁹ ⁹ ¹²
   4̶9̶,0̶0̶2̶   Check: 3 1 , 8 6 5
 − 3 1 , 8 6 5   + 1 7 , 1 3 7
―――――――――   ――――――――――
   1 7 , 1 3 7   4 9 , 0 0 2

You Try It 8

$$8{,}544 \longrightarrow 9{,}000 \qquad 8{,}544$$
$$3{,}621 \longrightarrow \underline{-4{,}000} \qquad \underline{-3{,}621}$$
$$5{,}000 \qquad 4{,}923$$

You Try It 9

$x - y$
$7{,}061 - 3{,}229$

$$\overset{6 \ \ 10 \ 5 \ 11}{7{,}0\not{6}\not{1}}$$
$$\underline{-3{,}229}$$
$$3{,}832$$

You Try It 10

$$\frac{46 = 58 - p}{46 \ \big| \ 58 - 11}$$
$$46 \neq 47$$

No, 11 is not a solution of the equation.

You Try It 11

Strategy

To find the total number of screens owned by the three chains that own the most screens:

▶ Find the number of screens owned by each of the three chains that own the most screens.
▶ Add the three numbers.

Solution

The three chains that own the most screens:

United Artists: 2,295
Carmike Cinemas: 2,037
AMC Entertainment: 1,632

$$2{,}295$$
$$2{,}037$$
$$\underline{+1{,}632}$$
$$5{,}964$$

The three chains that own the most screens own a total of 5,964 screens.

You Try It 12

Strategy

To find how much taller the World Trade Center is, subtract the height of the Empire State Building (1,250) from the height of the World Trade Center (1,368).

Solution

$$1{,}368$$
$$\underline{-1{,}250}$$
$$118$$

The World Trade Center is 118 ft taller than the Empire State Building.

You Try It 13

Strategy

To find the price, replace C by 148 and M by 74 in the given formula and solve for P.

Solution

$P = C + M$
$P = 148 + 74$
$P = 222$

The price of the leather jacket is $222.

SECTION 1.3 (pages 41–58)

You Try It 1

The cost of travel in Chicago is $229 per day.

$$229$$
$$\underline{\times 14}$$
$$916$$
$$\underline{2\ 29}$$
$$3{,}206$$

The cost would be $3,206.

You Try It 2

$$8{,}704 \longrightarrow 9{,}000$$
$$93 \longrightarrow 90$$
$$9{,}000 \cdot 90 = 810{,}000$$

You Try It 3

$5xy$
$5(20)(60) = 100(60)$
$= 6{,}000$

You Try It 4

$90(7{,}000) = 630{,}000$

You Try It 5

$0 \cdot 10 = 0$

You Try It 6

$$\frac{7a = 77}{7 \cdot 11 \ \big| \ 77}$$
$$77 = 77$$

Yes, 11 is a solution of the equation.

You Try It 7

$2 \cdot 2 \cdot 2 \cdot 3 \cdot 3 \cdot 3 \cdot 3 = 2^3 \cdot 3^4$

You Try It 8

$6^4 = 6 \cdot 6 \cdot 6 \cdot 6 = 36 \cdot 6 \cdot 6$
$= 216 \cdot 6 = 1{,}296$

You Try It 9

$10^8 = 100{,}000{,}000$

You Try It 10

$2^4 \cdot 3^2 = (2 \cdot 2 \cdot 2 \cdot 2) \cdot (3 \cdot 3)$
$= 16 \cdot 9 = 144$

You Try It 11

$5^3 = 5 \cdot 5 \cdot 5 = 25 \cdot 5 = 125$

You Try It 12

$x^4 y^2$
$1^4 \cdot 3^2 = (1 \cdot 1 \cdot 1 \cdot 1) \cdot (3 \cdot 3)$
$= 1 \cdot 9$
$= 9$

You Try It 13

$$\begin{array}{r} 320 \ \text{r}14 \\ 24\overline{)7{,}694} \\ \underline{-7\ 2} \\ 49 \\ \underline{-48} \\ 14 \\ \underline{-\ 0} \\ 14 \end{array}$$

Check: $(320 \cdot 24) + 14 = 7{,}680 + 14$
$= 7{,}694$

You Try It 14 The annual expense for food is $7,200.

$$7{,}200 \div 12 = 600$$

The monthly expense for food is $600.

You Try It 15 $216{,}936 \longrightarrow 200{,}000$
$207 \longrightarrow 200$
$200{,}000 \div 200 = 1{,}000$

You Try It 16 $\dfrac{x}{y}$

$$\dfrac{672}{8} = 84$$

You Try It 17 $\dfrac{60}{y} = 2$

$$\dfrac{60}{12}\;\bigg|\;2$$

$$5 \neq 2$$

No, 12 is not a solution of the equation.

You Try It 18 $30 \div 1 = 30$
$30 \div 2 = 15$
$30 \div 3 = 10$
$30 \div 4 \qquad$ Does not divide evenly.
$30 \div 5 = 6$
$30 \div 6 = 5 \qquad$ The factors are repeating.

The factors of 30 are 1, 2, 3, 5, 6, 10, 15, and 30.

You Try It 19 $\begin{array}{r} 11 \\ 2\overline{)22} \\ 2\overline{)44} \\ 2\overline{)88} \end{array}$

$$88 = 2 \cdot 2 \cdot 2 \cdot 11 = 2^3 \cdot 11$$

You Try It 20 $\begin{array}{r} 59 \\ 5\overline{)295} \end{array}$

$$295 = 5 \cdot 59$$

You Try It 21

Strategy To find how many times more expensive a stamp was, divide the cost in 1997 (32) by the cost in 1960 (4).

Solution $32 \div 4 = 8$

A stamp was 8 times more expensive in 1997.

You Try It 22

Strategy To find the number of calories burned, multiply the number of calories burned per hour (480) by the number of hours (4) of tennis played.

Solution $\begin{array}{r} 480 \\ \times\ 4 \\ \hline 1{,}920 \end{array}$

The number of calories burned is 1,920.

You Try It 23

Strategy To find the speed, replace d by 486 and t by 9 in the given formula and solve for r.

Solution $r = \dfrac{d}{t}$

$$r = \dfrac{486}{9}$$

$$r = 54$$

You would need to travel at a speed of 54 mph.

SECTION 1.4 (*pages 67–70*)

You Try It 1 $37 = a + 12$
$37 - 12 = a + 12 - 12$
$25 = a + 0$
$25 = a$

Check: $\dfrac{37 = a + 12}{37\;\big|\;25 + 12}$

$$37 = 37$$

The solution is 25.

You Try It 2 $3z = 36$
$\dfrac{3z}{3} = \dfrac{36}{3}$
$1z = 12$
$z = 12$

Check: $\dfrac{3z = 36}{3(12)\;\big|\;36}$

$$36 = 36$$

The solution is 12.

You Try It 3 The unknown number: n

A number increased by four	is	seventeen

$$n + 4 = 17$$
$$n + 4 - 4 = 17 - 4$$
$$n = 13$$

The number is 13.

You Try It 4

Strategy To find the price of the Mercury Villager, write and solve an equation using x to represent the price of the Mercury Villager.

Solution

The price of a used Ford Windstar	is	$900 more than the price of a used Mercury Villager

$$16{,}495 = x + 900$$
$$16{,}495 - 900 = x + 900 - 900$$
$$15{,}595 = x$$

The price of the Mercury Villager is $15,595.

You Try It 5

Strategy

To find the interest earned, replace A by 21,060 and P by 18,000 in the given formula and solve for I.

Solution

$$A = P + I$$
$$21,060 = 18,000 + I$$
$$21,060 - 18,000 = 18,000 - 18,000 + I$$
$$3,060 = I$$

The interest earned on the investment is $3,060.

SECTION 1.5 (*pages 73–74*)

You Try It 1

$$4 \cdot (8 - 3) \div 5 - 2 = 4 \cdot 5 \div 5 - 2$$
$$= 20 \div 5 - 2$$
$$= 4 - 2$$
$$= 2$$

You Try It 2

$$16 + 3(6 - 1)^2 \div 5 = 16 + 3(5)^2 \div 5$$
$$= 16 + 3(25) \div 5$$
$$= 16 + 75 \div 5$$
$$= 16 + 15$$
$$= 31$$

You Try It 3

$$(a - b)^2 + 5c$$
$$(7 - 2)^2 + 5(4) = 5^2 + 5(4)$$
$$= 25 + 5(4)$$
$$= 25 + 20$$
$$= 45$$

SOLUTIONS TO CHAPTER 2   You Try Its

SECTION 2.1 (*pages 85–90*)

You Try It 1

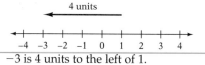

4 units

-3 is 4 units to the left of 1.

You Try It 2

A is -5, and C is -3.

You Try It 3

a. 2 is to the right of -5 on the number line.

$$2 > -5$$

b. -4 is to the left of 3 on the number line.

$$-4 < 3$$

You Try It 4 $-7, -1, 0, 4, 8$

You Try It 5   a. -24 b. 13   c. b

You Try It 6   a. negative three minus twelve

b. eight plus negative five

You Try It 7   a. $-(-59) = 59$ b. $-(y) = -y$

You Try It 8   a. $|-8| = 8$ b. $|12| = 12$

You Try It 9   a. $|0| = 0$ b. $-|35| = -35$

You Try It 10 $|-y| = |-2| = 2$

You Try It 11 $|6| = 6, |-2| = 2, -(-1) = 1, -|-8| = -8$

$-8, -4, 1, 2, 6$

$-|-8|, -4, -(-1), |-2|, |6|$

You Try It 12

Strategy

To find the player that came in third, find the player with the third lowest number for a score.

Solution $-16 < -13 < -11 < -10 < -9$

The third lowest number among the scores is -11.

Robbins came in third in the tournament.

You Try It 13

Strategy

To determine which is closer to blastoff, find the absolute value of each number. The number with the smaller absolute value is closer to zero and, therefore, closer to blastoff.

Solution $|-9| = 9, |-7| = 7$

$7 < 9$

-7 s and counting is closer to blastoff than -9 s and counting.

SECTION 2.2 (*pages 97–104*)

You Try It 1 $-36 + 17 + (-21) = -19 + (-21)$
$$= -40$$

You Try It 2 $-154 + (-37) = -191$

You Try It 3 $-x + y$
$$(-9) + (-10) = 9 + (-10)$$
$$= -7$$

You Try It 4 $\dfrac{2 = 11 + a}{2 \mid 11 + (-9)}$

$$2 = 2$$

Yes, -9 is a solution of the equation.

You Try It 5   The boiling point of xenon is -107. The melting point of xenon is -112.

$$-107 - (-112) = -107 + 112$$
$$= 5$$

The difference is 5°C.

You Try It 6 $-8 - 14 = -8 + (-14)$
$$= -22$$

You Try It 7
$$-4 - (-3) + 12 - (-7) - 20$$
$$= -4 + 3 + 12 + 7 + (-20)$$
$$= -1 + 12 + 7 + (-20)$$
$$= 11 + 7 + (-20)$$
$$= 18 + (-20)$$
$$= -2$$

You Try It 8
$$x - y$$
$$-9 - 7 = -9 + (-7)$$
$$= -16$$

You Try It 9
$$\underline{a - 5 = -8}$$
$$\dfrac{-3 - 5}{} \Big| -8$$
$$-3 + (-5) \Big| -8$$
$$-8 = -8$$

Yes, -3 is a solution of the equation.

You Try It 10

Strategy To find the difference, subtract the lowest melting point shown (-259) from the highest melting point shown (181).

Solution
$$181 - (-259) = 181 + 259$$
$$= 440$$
The difference is 440°C.

You Try It 11

Strategy To find the temperature, add the increase (10) to the previous temperature (-3).

Solution
$$-3 + 10 = 7$$
The temperature is 7°C.

You Try It 12

Strategy To find the difference, subtract the lower temperature (-70) from the higher temperature (57).

Solution
$$57 - (-70) = 57 + 70$$
$$= 127$$
The difference between the average temperatures is 127°F.

You Try It 13

Strategy To find d, replace a by -6 and b by 5 in the given formula and solve for d.

Solution
$$d = |a - b|$$
$$d = |-6 - 5|$$
$$d = |-11|$$
$$d = 11$$
The distance between the two points is 11 units.

SECTION 2.3 (*pages 111–116*)

You Try It 1 $-38(51) = -1,938$

You Try It 2
$$-7(-8)(9)(-2) = 56(9)(-2)$$
$$= 504(-2)$$
$$= -1,008$$

You Try It 3
$$-9y$$
$$-9(20) = -180$$

You Try It 4
$$12 = -4a$$
$$12 \Big| -4(-3)$$
$$12 = 12$$
Yes, -3 is a solution of the equation.

You Try It 5 $0 \div (-17) = 0$

You Try It 6 $\dfrac{84}{-6} = -14$

You Try It 7   Any number divided by one is the number.
$$x \div 1 = x$$

You Try It 8
$$\dfrac{a}{-b}$$
$$\dfrac{-14}{-(-7)} = \dfrac{-14}{7} = -2$$

You Try It 9
$$\dfrac{-6}{y} = -2$$
$$\dfrac{-6}{-3} \Big| -2$$
$$2 \neq -2$$
No, -3 is not a solution of the equation.

You Try It 10

Strategy To find the average daily high temperature:
▶ Add the seven temperature readings.
▶ Divide by 7.

Solution
$$-7 + (-8) + 0 + (-1) + (-6) + (-11) + (-2) = -35$$
$$-35 \div 7 = -5$$
The average daily high temperature was $-5°$.

SECTION 2.4 (*pages 123–126*)

You Try It 1
$$-12 = x + 12$$
$$-12 - 12 = x + 12 - 12$$
$$-24 = x$$
The solution is -24.

You Try It 2
$$14a = -28$$
$$\dfrac{14a}{14} = \dfrac{-28}{14}$$
$$a = -2$$
The solution is -2.

You Try It 3

Strategy To find the number of people employed by Ryder in 1995, write and solve an equation using x to represent the number of people employed by Ryder in 1995.

Solution

The number of people employed by Ryder in 1994	was	1,408 less than the number the company employed in 1995

$$43{,}095 = x - 1{,}408$$
$$43{,}095 + 1{,}408 = x - 1{,}408 + 1{,}408$$
$$44{,}503 = x$$

Ryder employed 44,503 people in 1995.

You Try It 4

Strategy   To find the airspeed, replace g by 250 and h by 50 in the given formula and solve for a.

Solution

$$g = a - h$$
$$250 = a - 50$$
$$250 + 50 = a - 50 + 50$$
$$300 = a$$

The airspeed of the plane is 300 mph.

SECTION 2.5 (*pages 129–130*)

You Try It 1 $(-5)^2 = (-5)(-5) = 25$
$-5^2 = -(5 \cdot 5) = -25$

You Try It 2 $8 \div 4 \cdot 4 - (-2)^2 = 8 \div 4 \cdot 4 - 4$
$= 2 \cdot 4 - 4$
$= 8 - 4$
$= 4$

You Try It 3 $(-2)^2(3 - 7)^2 - (-16) \div (-4)$
$= (-2)^2(-4)^2 - (-16) \div (-4)$
$= (4)(16) - (-16) \div (-4)$
$= 64 - (-16) \div (-4)$
$= 64 - 4$
$= 60$

You Try It 4 $3a - 4b$
$3(-2) - 4(5) = -6 - 4(5)$
$= -6 - 20$
$= -6 + (-20)$
$= -26$

SOLUTIONS TO CHAPTER 3   You Try Its

SECTION 3.1 (*pages 143–146*)

You Try It 1 $16 = \boxed{2^4}$
$24 = 2^3 \cdot \boxed{3}$
$28 = 2^2 \cdot \boxed{7}$

The LCM $= 2^4 \cdot 3 \cdot 7 = 16 \cdot 3 \cdot 7 = 336.$

You Try It 2 $25 = 5^2$
$52 = 2^2 \cdot 13$

No prime factor occurs in both factorizations. The GCF is 1.

You Try It 3 $32 = 2^5$
$40 = \boxed{2^3} \cdot 5$
$56 = 2^3 \cdot 7$

The GCF $= 2^3 = 8.$

You Try It 4

Strategy   To find the number of diskettes to be packaged together, find the GCF of 20, 50, and 100.

Solution $20 = 2^2 \cdot \boxed{5}$
$50 = \boxed{2} \cdot 5^2$
$100 = 2^2 \cdot 5^2$

The GCF $= 2 \cdot 5 = 10.$

Each package should contain 10 diskettes.

You Try It 5

Strategy   To find how long it will be before both of you are at the starting point again, find the LCM of 3 and 4.

Solution $3 = \boxed{3}$
$4 = \boxed{2^2}$

The LCM $= 3 \cdot 2^2 = 12.$

In 12 min both of you will be at the starting point again.

You will not have passed each other at some other point on the track prior to that time. It is as if it takes the faster runner 4 laps to "catch up to" the slower runner.

SECTION 3.2 (*pages 149–156*)

You Try It 1 $\dfrac{19}{6}; 3\dfrac{1}{6}$

You Try It 2
$$\begin{array}{r} 8 \\ 3\overline{)26} \\ -24 \\ \hline 2 \end{array} \qquad \dfrac{26}{3} = 8\dfrac{2}{3}$$

You Try It 3
$$\begin{array}{r} 9 \\ 4\overline{)36} \\ -36 \\ \hline 0 \end{array} \qquad \dfrac{36}{4} = 9$$

You Try It 4 $9\dfrac{4}{7} = \dfrac{(7 \cdot 9) + 4}{7} = \dfrac{63 + 4}{7} = \dfrac{67}{7}$

You Try It 5 $3 = \dfrac{3}{1}$

You Try It 6 $48 \div 8 = 6$
$$\dfrac{5}{8} = \dfrac{5 \cdot 6}{8 \cdot 6} = \dfrac{30}{48}$$
$\dfrac{30}{48}$ is equivalent to $\dfrac{5}{8}$.

You Try It 7 $8 = \dfrac{8}{1}$ $12 \div 1 = 12$

$8 = \dfrac{8}{1} = \dfrac{8 \cdot 12}{1 \cdot 12} = \dfrac{96}{12}$

$\dfrac{96}{12}$ is equivalent to 8.

You Try It 8 $\dfrac{21}{84} = \dfrac{3 \cdot 7}{2 \cdot 2 \cdot 3 \cdot 7} = \dfrac{1}{4}$

You Try It 9 $\dfrac{32}{12} = \dfrac{2 \cdot 2 \cdot 2 \cdot 2 \cdot 2}{2 \cdot 2 \cdot 3} = \dfrac{8}{3}$

You Try It 10 $\dfrac{11t}{11} = \dfrac{11 \cdot t}{11} = t$

You Try It 11 $\dfrac{4}{9} = \dfrac{28}{63}$ $\dfrac{8}{21} = \dfrac{24}{63}$

$\dfrac{28}{63} > \dfrac{24}{63}$

$\dfrac{4}{9} > \dfrac{8}{21}$

You Try It 12   The LCM of 24 and 21 is 168.

$\dfrac{17}{24} = \dfrac{119}{168}$ $\dfrac{8}{21} = \dfrac{64}{168}$

$\dfrac{119}{168} > \dfrac{64}{168}$

$\dfrac{17}{24} > \dfrac{8}{21}$

You Try It 13

Strategy   To find the fraction:

▶ Add to find the total of the first four companies listed (895 + 599 + 411 + 199).
▶ Write a fraction with Quaker Oats' sales in the numerator and the total sales in the denominator.

Solution $895 + 599 + 411 + 199 = 2{,}104$

$\dfrac{199}{2{,}104}$

Quaker Oats' fraction of the sales was

$\dfrac{199}{2{,}104}$.

You Try It 14

Strategy   To find the fraction:

▶ Convert one dollar to cents.
▶ Write a fraction with the number of cents earned in profit in the numerator and the number of cents in one dollar in the denominator.
▶ Write the fraction in simplest form.

Solution   1 dollar = 100 cents

$\dfrac{30}{100} = \dfrac{3}{10}$

For product sold outside the United States, $\dfrac{3}{10}$ of every dollar's worth of product is profit.

SECTION 3.3 (*pages 163–172*)

You Try It 1 $\dfrac{7}{12} + \dfrac{3}{8} = \dfrac{14}{24} + \dfrac{9}{24} = \dfrac{23}{24}$

You Try It 2 $\dfrac{3}{5} + \dfrac{2}{3} + \dfrac{5}{6} = \dfrac{18}{30} + \dfrac{20}{30} + \dfrac{25}{30} = \dfrac{63}{30}$

$= 2\dfrac{3}{30} = 2\dfrac{1}{10}$

You Try It 3 $16 + 8\dfrac{5}{9} = 24\dfrac{5}{9}$

You Try It 4 $-\dfrac{5}{12} + \dfrac{5}{8} + \left(-\dfrac{1}{6}\right) = \dfrac{-5}{12} + \dfrac{5}{8} + \dfrac{-1}{6}$

$= \dfrac{-10}{24} + \dfrac{15}{24} + \dfrac{-4}{24}$

$= \dfrac{-10 + 15 + (-4)}{24}$

$= \dfrac{1}{24}$

You Try It 5 $x + y + z$

$3\dfrac{5}{6} + 2\dfrac{1}{9} + 5\dfrac{5}{12} = 3\dfrac{30}{36} + 2\dfrac{4}{36} + 5\dfrac{15}{36}$

$= 10\dfrac{49}{36}$

$= 11\dfrac{13}{36}$

You Try It 6 $-\dfrac{5}{6} - \dfrac{7}{9} = \dfrac{-5}{6} - \dfrac{7}{9}$

$= \dfrac{-15}{18} - \dfrac{14}{18}$

$= \dfrac{-15 - 14}{18}$

$= \dfrac{-29}{18}$

$= -\dfrac{29}{18} = -1\dfrac{11}{18}$

You Try It 7 $9\dfrac{7}{8} - 5\dfrac{2}{3} = 9\dfrac{21}{24} - 5\dfrac{16}{24} = 4\dfrac{5}{24}$

You Try It 8 $6 - 4\dfrac{2}{11} = 5\dfrac{11}{11} - 4\dfrac{2}{11} = 1\dfrac{9}{11}$

You Try It 9

$$\frac{2}{3} - v = \frac{11}{12}$$

$$\frac{\dfrac{2}{3} - \left(-\dfrac{1}{4}\right) \ \Big| \ \dfrac{11}{12}}{}$$

$$\frac{\dfrac{2}{3} + \dfrac{1}{4} \ \Big| \ \dfrac{11}{12}}{}$$

$$\frac{\dfrac{8}{12} + \dfrac{3}{12} \ \Big| \ \dfrac{11}{12}}{}$$

$$\frac{11}{12} = \frac{11}{12}$$

Yes, $-\dfrac{1}{4}$ is a solution of the equation.

You Try It 10

Strategy　To find the fraction that did not vote for any of these three candidates:

▶ Add the three fractions to find the fraction of the voters who voted for one of the three candidates.
▶ Subtract the fraction that did vote for one of the three candidates from 1, the entire voting population.

Solution　$\dfrac{49}{100} + \dfrac{41}{100} + \dfrac{2}{25}$

$$= \frac{49}{100} + \frac{41}{100} + \frac{8}{100}$$

$$= \frac{98}{100} = \frac{49}{50}$$

$$1 - \frac{49}{50} = \frac{50}{50} - \frac{49}{50} = \frac{1}{50}$$

$\dfrac{1}{50}$ of the voters did not vote for any of these three candidates.

You Try It 11

Strategy　To find the size of the penny nail needed:

▶ Find the thickness of one board by subtracting $\dfrac{1}{4}$ in. from the given thickness (1 in.).
▶ Multiply the thickness of one board by 3 to find the thickness of 3 boards.
▶ To the thickness of 3 boards, add $\dfrac{1}{2}$ in., as we want the nail to extend $\dfrac{1}{2}$ in. into the fourth board. This is the length of the penny nail needed.
▶ To calculate the size penny nail needed, use the facts that the length of a nail increases by $\dfrac{1}{4}$ in. for each 1 penny increase in size and that a 4-penny nail is $1\dfrac{1}{2}$ in. long.

Solution　Given thickness of one board

minus $\dfrac{1}{4}$ in. $= 1 - \dfrac{1}{4} = \dfrac{3}{4}$

Thickness of 3 boards $= 3 \cdot \dfrac{3}{4} = \dfrac{9}{4} = 2\dfrac{1}{4}$

Length of penny nail needed $= 2\dfrac{1}{4} + \dfrac{1}{2}$

$$= 2\frac{1}{4} + \frac{2}{4} = 2\frac{3}{4}$$

A 4-penny nail is $1\dfrac{1}{2}$ in. long.

A 5-penny nail is $1\dfrac{1}{2} + \dfrac{1}{4} = 1\dfrac{3}{4}$ in. long.

A 6-penny nail is $1\dfrac{3}{4} + \dfrac{1}{4} = 2$ in. long.

A 7-penny nail is $2 + \dfrac{1}{4} = 2\dfrac{1}{4}$ in. long.

An 8-penny nail is $2\dfrac{1}{4} + \dfrac{1}{4} = 2\dfrac{1}{2}$ in. long.

A 9-penny nail is $2\dfrac{1}{2} + \dfrac{1}{4} = 2\dfrac{3}{4}$ in. long.

A 9-penny nail is needed.

SECTION 3.4 (*pages 181–190*)

You Try It 1　$\dfrac{5}{12} \cdot \dfrac{9}{35} \cdot \dfrac{7}{8} = \dfrac{5 \cdot 9 \cdot 7}{12 \cdot 35 \cdot 8}$

$$= \frac{5 \cdot 3 \cdot 3 \cdot 7}{2 \cdot 2 \cdot 3 \cdot 5 \cdot 7 \cdot 2 \cdot 2 \cdot 2}$$

$$= \frac{3}{32}$$

You Try It 2　$\dfrac{y}{10} \cdot \dfrac{z}{7} = \dfrac{y \cdot z}{10 \cdot 7} = \dfrac{yz}{70}$

You Try It 3　$-\dfrac{1}{3}\left(-\dfrac{5}{12}\right)\left(\dfrac{8}{15}\right) = \dfrac{1}{3} \cdot \dfrac{5}{12} \cdot \dfrac{8}{15}$

$$= \frac{1 \cdot 5 \cdot 8}{3 \cdot 12 \cdot 15}$$

$$= \frac{1 \cdot 5 \cdot 2 \cdot 2 \cdot 2}{3 \cdot 2 \cdot 2 \cdot 3 \cdot 3 \cdot 5}$$

$$= \frac{2}{27}$$

You Try It 4　$\dfrac{8}{9} \cdot 6 = \dfrac{8}{9} \cdot \dfrac{6}{1} = \dfrac{8 \cdot 6}{9 \cdot 1}$

$$= \frac{2 \cdot 2 \cdot 2 \cdot 2 \cdot 3}{3 \cdot 3 \cdot 1} = \frac{16}{3} = 5\frac{1}{3}$$

You Try It 5　$3\dfrac{6}{7} \cdot 2\dfrac{4}{9} = \dfrac{27}{7} \cdot \dfrac{22}{9} = \dfrac{27 \cdot 22}{7 \cdot 9}$

$$= \frac{3 \cdot 3 \cdot 3 \cdot 2 \cdot 11}{7 \cdot 3 \cdot 3} = \frac{66}{7} = 9\frac{3}{7}$$

You Try It 6 xy

$$5\frac{1}{8} \cdot \frac{2}{3} = \frac{41}{8} \cdot \frac{2}{3}$$

$$= \frac{41 \cdot 2}{8 \cdot 3}$$

$$= \frac{41 \cdot 2}{2 \cdot 2 \cdot 2 \cdot 3}$$

$$= \frac{41}{12} = 3\frac{5}{12}$$

You Try It 7

$$\frac{5}{6} \div \frac{10}{27} = \frac{5}{6} \cdot \frac{27}{10} = \frac{5 \cdot 27}{6 \cdot 10}$$

$$= \frac{5 \cdot 3 \cdot 3 \cdot 3}{2 \cdot 3 \cdot 2 \cdot 5} = \frac{9}{4} = 2\frac{1}{4}$$

You Try It 8

$$\frac{x}{8} \div \frac{y}{6} = \frac{x}{8} \cdot \frac{6}{y}$$

$$= \frac{x \cdot 6}{8 \cdot y} = \frac{x \cdot 2 \cdot 3}{2 \cdot 2 \cdot 2 \cdot y} = \frac{3x}{4y}$$

You Try It 9

$$4 \div \left(-\frac{6}{7}\right) = -\left(\frac{4}{1} \div \frac{6}{7}\right)$$

$$= -\left(\frac{4}{1} \cdot \frac{7}{6}\right)$$

$$= -\frac{4 \cdot 7}{1 \cdot 6}$$

$$= -\frac{2 \cdot 2 \cdot 7}{1 \cdot 2 \cdot 3} = -\frac{14}{3} = -4\frac{2}{3}$$

You Try It 10

$$4\frac{3}{8} \div 3\frac{1}{2} = \frac{35}{8} \div \frac{7}{2} = \frac{35}{8} \cdot \frac{2}{7} = \frac{35 \cdot 2}{8 \cdot 7}$$

$$= \frac{5 \cdot 7 \cdot 2}{2 \cdot 2 \cdot 2 \cdot 7} = \frac{5}{4} = 1\frac{1}{4}$$

You Try It 11 $x \div y$

$$2\frac{1}{4} \div 9 = \frac{9}{4} \div \frac{9}{1} = \frac{9}{4} \cdot \frac{1}{9} = \frac{9 \cdot 1}{4 \cdot 9}$$

$$= \frac{3 \cdot 3 \cdot 1}{2 \cdot 2 \cdot 3 \cdot 3} = \frac{1}{4}$$

You Try It 12

Strategy To find the total cost:
▶ Multiply the amount of material per sash $\left(1\frac{3}{8}\right)$ by the number of sashes (22) to find the total number of yards of material needed.
▶ Multiply the total number of yards of material needed by the cost per yard (8).

Solution

$$1\frac{3}{8} \cdot 22 = \frac{11}{8} \cdot \frac{22}{1} = \frac{11 \cdot 22}{8 \cdot 1} = \frac{11 \cdot 2 \cdot 11}{2 \cdot 2 \cdot 2 \cdot 1}$$

$$= \frac{121}{4} = 30\frac{1}{4}$$

$$30\frac{1}{4} \cdot 8 = \frac{121}{4} \cdot \frac{8}{1} = \frac{121 \cdot 8}{4 \cdot 1}$$

$$= \frac{11 \cdot 11 \cdot 2 \cdot 2 \cdot 2}{2 \cdot 2 \cdot 1} = 242$$

The total cost of the material is $242.

You Try It 13

Strategy To find the Celsius temperature, replace F by 68 in the given formula and solve for C.

Solution

$$C = \frac{5}{9}(F - 32)$$

$$C = \frac{5}{9}(68 - 32) = \frac{5}{9}(36) = \frac{5}{9} \cdot \frac{36}{1} = \frac{5 \cdot 36}{9 \cdot 1} = 20$$

The Celsius temperature is $20°$.

SECTION 3.5 (pages 199–202)

You Try It 1

$$-\frac{1}{5} = z - \frac{5}{6}$$

$$-\frac{1}{5} + \frac{5}{6} = z - \frac{5}{6} + \frac{5}{6}$$

$$-\frac{6}{30} + \frac{25}{30} = z$$

$$\frac{-6 + 25}{30} = z$$

$$\frac{19}{30} = z \quad \text{The solution is } \frac{19}{30}.$$

You Try It 2

$$26 = 4x$$

$$\frac{26}{4} = \frac{4x}{4}$$

$$\frac{13}{2} = x$$

$$6\frac{1}{2} = x \quad \text{The solution is } 6\frac{1}{2}.$$

You Try It 3 The unknown number: x

Negative five-sixths	is equal to	ten-thirds of a number

$$-\frac{5}{6} = \frac{10}{3}x$$

$$\frac{3}{10}\left(-\frac{5}{6}\right) = \frac{3}{10} \cdot \frac{10}{3}x$$

$$-\frac{15}{60} = x$$

$$-\frac{1}{4} = x$$

The number is $-\frac{1}{4}$.

You Try It 4

Strategy To find the total number of software products sold in January, write and solve an equation using s to represent the number of software products sold in January.

Solution

The number of computer software games sold in January	was	three-fifths of all the software products sold

$$450 = \frac{3}{5}s$$

$$\frac{5}{3} \cdot 450 = \frac{5}{3} \cdot \frac{3}{5}s$$

$$750 = s$$

BAL Software sold a total of 750 software products in January.

You Try It 5

Strategy To find the total number of points scored, replace A by 73 and N by 5 in the given formula and solve for T.

Solution

$$A = \frac{T}{N}$$

$$73 = \frac{T}{5}$$

$$5 \cdot 73 = 5 \cdot \frac{T}{5}$$

$$365 = T$$

The total number of points scored was 365.

SECTION 3.6 (*pages 205–210*)

You Try It 1 $\left(\frac{2}{9}\right)^2 \cdot (-3)^4$

$$= \frac{2}{9} \cdot \frac{2}{9} \cdot (-3)(-3)(-3)(-3)$$

$$= \frac{2}{9} \cdot \frac{2}{9} \cdot 3 \cdot 3 \cdot 3 \cdot 3$$

$$= \frac{2}{9} \cdot \frac{2}{9} \cdot \frac{3}{1} \cdot \frac{3}{1} \cdot \frac{3}{1} \cdot \frac{3}{1}$$

$$= \frac{2 \cdot 2 \cdot 3 \cdot 3 \cdot 3 \cdot 3}{9 \cdot 9 \cdot 1 \cdot 1 \cdot 1 \cdot 1} = 4$$

You Try It 2 $x^4 y^3$

$$\left(2\frac{1}{3}\right)^4 \cdot \left(\frac{3}{7}\right)^3 = \left(\frac{7}{3}\right)^4 \cdot \left(\frac{3}{7}\right)^3$$

$$= \frac{7}{3} \cdot \frac{7}{3} \cdot \frac{7}{3} \cdot \frac{7}{3} \cdot \frac{3}{7} \cdot \frac{3}{7} \cdot \frac{3}{7}$$

$$= \frac{7 \cdot 7 \cdot 7 \cdot 7 \cdot 3 \cdot 3 \cdot 3}{3 \cdot 3 \cdot 3 \cdot 3 \cdot 7 \cdot 7 \cdot 7} = \frac{7}{3} = 2\frac{1}{3}$$

You Try It 3 $\dfrac{2y - 3}{y} = -2$

$\dfrac{2\left(-\frac{1}{2}\right) - 3}{-\frac{1}{2}}$	-2
$\dfrac{-1 - 3}{-\frac{1}{2}}$	-2
$\dfrac{-4}{-\frac{1}{2}}$	-2
$-4(-2)$	-2
$8 \neq -2$	

No, $-\dfrac{1}{2}$ is not a solution of the equation.

You Try It 4 $\dfrac{x}{y - z}$

$$\frac{2\frac{4}{9}}{3 - 1\frac{1}{3}} = \frac{\frac{22}{9}}{\frac{5}{3}} = \frac{22}{9} \div \frac{5}{3} = \frac{22}{9} \cdot \frac{3}{5}$$

$$= \frac{22}{15} = 1\frac{7}{15}$$

You Try It 5 $\left(-\dfrac{1}{2}\right)^3 \cdot \dfrac{7 - 3}{4 - 9} + \dfrac{4}{5}$

$$= \left(-\frac{1}{2}\right)^3 \cdot \frac{4}{-5} + \frac{4}{5}$$

$$= -\frac{1}{8} \cdot \frac{4}{-5} + \frac{4}{5}$$

$$= \frac{1}{10} + \frac{4}{5} = \frac{9}{10}$$

SOLUTIONS TO CHAPTER 4   You Try Its

SECTION 4.1 (*pages 225–230*)

You Try It 1 The digit 4 is in the thousandths' place.

You Try It 2 $\dfrac{501}{1,000} = 0.501$

[five hundred one thousandths]

You Try It 3 $0.67 = \dfrac{67}{100}$ [sixty-seven hundredths]

You Try It 4   fifty-five and six thousand eighty-three ten-thousandths

You Try It 5   806.00491

You Try It 6 $0.065 = 0.0650$

$0.0650 < 0.0802$

$0.065 < 0.0802$

You Try It 7   3.03, 0.33, 0.30, 3.30, 0.03

0.03, 0.30, 0.33, 3.03, 3.30

0.03, 0.3, 0.33, 3.03, 3.3

You Try It 8

┌────── Given place value
3.675849
└────── $4 < 5$

3.675849 rounded to the nearest ten-thousandth is 3.6758.

You Try It 9

┌────── Given place value
48.907
└────── $0 < 5$

48.907 rounded to the nearest tenth is 48.9.

You Try It 10

┌────── Given place value
31.8652
└────── $8 > 5$

31.8652 rounded to the nearest whole number is 32.

You Try It 11

Strategy   To determine who had more home runs for every 100 times at bat, compare the numbers 7.03 and 7.09.

Solution $7.09 > 7.03$

Ralph Kiner had more home runs for every 100 times at bat.

You Try It 12

Strategy   To determine how fast Johnson ran, to the nearest second, round the number 19.32 to the nearest whole number.

Solution   19.32 rounded to the nearest whole number is 19.

To the nearest second, Johnson ran the 200 m race in 19 s.

SECTION 4.2 (*pages 235–250*)

You Try It 1

$$
\begin{array}{r}
{\scriptstyle 1\ 1} \\
8.64 \\
52.7 \\
+\ 0.39105 \\
\hline
61.73105
\end{array}
$$

You Try It 2 $4.002 - 9.378 = 4.002 + (-9.378)$

$= -5.376$

You Try It 3

$$
\begin{array}{r}
{\scriptstyle 4\ 9\ 10} \\
2\cancel{5}.\cancel{0}\cancel{0} \\
-\ 4.91 \\
\hline
20.09
\end{array}
\qquad
\text{Check:}
\begin{array}{r}
4.91 \\
+20.09 \\
\hline
25.00
\end{array}
$$

You Try It 4

$$
\begin{array}{r}
6.514 \longrightarrow\ \ \ 7 \\
8.903 \longrightarrow\ \ \ 9 \\
2.275 \longrightarrow +\ 2 \\
\hline
18
\end{array}
$$

You Try It 5 $x + y + z$

$-7.84 + (-3.05) + 2.19$

$= -10.89 + 2.19$

$= -8.7$

You Try It 6

$$
\begin{array}{r|l}
\multicolumn{2}{c}{-m + 16.9 = 40.7} \\
\hline
-(-23.8) + 16.9 & 40.7 \\
23.8 + 16.9 & 40.7 \\
40.7 & = 40.7
\end{array}
$$

Yes, -23.8 is a solution of the equation.

You Try It 7

$$
\begin{array}{r}
0.000081 \\
\times\ \ \ \ 0.025 \\
\hline
405 \\
162 \\
\hline
0.000002025
\end{array}
$$

You Try It 8

$$
\begin{array}{r}
6.407 \longrightarrow\ \ 6 \\
0.959 \longrightarrow \times 1 \\
\hline
6
\end{array}
$$

You Try It 9 $1.756 \cdot 10^4 = 17{,}560$

You Try It 10 $(-0.7)(-5.8) = 4.06$

You Try It 11 $25xy$

$25(-0.8)(0.6) = -20(0.6) = -12$

You Try It 12

$$
\begin{array}{r}
48.2 \\
6.53.\overline{)314.74.6} \\
-\,261\,2 \\
\hline
53\,54 \\
-\,52\,24 \\
\hline
1\,30\,6 \\
-\,1\,30\,6 \\
\hline
0
\end{array}
$$

You Try It 13

$$62.7 \longrightarrow 60$$
$$3.45 \longrightarrow 3$$
$$60 \div 3 = 20$$

You Try It 14

$$
\begin{array}{r}
6.0391 \approx 6.039 \\
86\overline{)519.3700} \\
-516 \\
\hline
33 \\
-0 \\
\hline
3\,37 \\
-2\,58 \\
\hline
790 \\
-774 \\
\hline
160 \\
-86 \\
\hline
74
\end{array}
$$

You Try It 15 $63.7 \div 100 = 0.637$

You Try It 16   The quotient is negative.

$$-25.7 \div 0.31 \approx -82.9$$

You Try It 17 $\dfrac{x}{y}$

$$\dfrac{-40.6}{-0.7} = -40.6 \div (-0.7) = 58$$

You Try It 18 $-2 = \dfrac{d}{-0.6}$

$$
\begin{array}{c|c}
-2 & \dfrac{-1.2}{-0.6} \\
\end{array}
$$

$$-2 \neq 2$$

No, -1.2 is not a solution of the equation.

You Try It 19 $5\overline{)4.0}^{\,0.8}$ $\dfrac{4}{5} = 0.8$

You Try It 20 $6\overline{)5.0000}^{\,0.8333}$ $1\dfrac{5}{6} = 1.8\overline{3}$

You Try It 21 $6.2 = 6\dfrac{2}{10} = 6\dfrac{1}{5}$

You Try It 22 $\dfrac{7}{12} \approx 0.5833$

$$0.5880 > 0.5833$$

$$0.588 > \dfrac{7}{12}$$

You Try It 23

Strategy   To find the change you receive:

▶ Multiply the number of stamps (12) by the cost of each stamp (32¢) to find the total cost of the stamps.

▶ Convert the total cost of the stamps to dollars and cents.

▶ Subtract the total cost of the stamps from $5.

Solution

$$12(32) = 384 \qquad \text{The stamps cost } 384¢.$$
$$384¢ = \$3.84 \qquad \text{The stamps cost } \$3.84.$$
$$5.00 - 3.84 = 1.16$$

You receive $1.16 in change.

You Try It 24

Strategy

To find the increase from 1990 to 1995:

▶ Determine which credit card had the greatest volume.

▶ For the credit card with the greatest volume, find the amount of the annual purchases in 1990 and 1995.

▶ Subtract the amount of the annual purchases in 1990 from those in 1995.

To determine whether the amount is less than or greater than the annual purchases in 1990, compare the difference in purchases from 1990 to 1995 with the amount of annual purchases in 1990.

Solution

The card with the greatest volume is Visa.

Visa's amount of annual purchases in 1990: $136.92 billion

Visa's amount of annual purchases in 1995: $288.86 billion.

288.86 billion − 136.92 billion = 151.94 billion.

The increase was $151.94 billion.

151.94 billion > 136.92 billion

The amount of increase was greater than the annual purchases made in 1990.

You Try It 25

Strategy

To find the profit:

▶ Divide the number of pounds per 100-pound container (100) by the number of pounds packaged in each bag (2) to find the number of bags sold.

▶ Multiply the number of bags sold by the selling price per bag (8.50) to find the income from selling the nuts.

▶ Multiply the number of bags sold by the cost for each bag (.04) to find the total cost of the bags.

▶ Subtract the cost of the bags and the cost of the nuts (325) from the income.

Solution

$$100 \div 2 = 50 \qquad \text{Each container makes 50 bags of nuts.}$$
$$50(8.50) = 425 \qquad \text{The income from the 50 bags is \$425.}$$
$$50(.04) = 2 \qquad \text{The total cost of the bags is \$2.}$$
$$425 - 2 - 325 = 98$$

The profit is $98.

You Try It 26

Strategy To find the insurance premium due, replace B by 276.25 and F by 1.8 in the given formula and solve for P.

Solution
$$P = BF$$
$$P = 276.25(1.8)$$
$$P = 497.25$$

The insurance premium due is $497.25.

SECTION 4.3 (*pages 263–264*)

You Try It 1
$$a - 1.23 = -6$$
$$a - 1.23 + 1.23 = -6 + 1.23$$
$$a = -4.77$$

The solution is -4.77.

You Try It 2
$$-2.13 = -0.71c$$
$$\frac{-2.13}{-0.71} = \frac{-0.71c}{-0.71}$$
$$3 = c$$

The solution is 3.

You Try It 3

Strategy To find the assets, replace N by 24.3 and L by 17.9 in the given formula and solve for A.

Solution
$$N = A - L$$
$$24.3 = A - 17.9$$
$$24.3 + 17.9 = A - 17.9 + 17.9$$
$$42.2 = A$$

The assets of the business are $42.2 billion.

You Try It 4

Strategy

To find the markup, write and solve an equation using M to represent the amount of the markup.

Solution

The selling price	is	the sum of the amount paid by the store and the amount of the markup

$$295.50 = 223.75 + M$$
$$295.50 - 223.75 = 223.75 - 223.75 + M$$
$$71.75 = M$$

The markup is $71.75.

SECTION 4.4 (*pages 267–272*)

You Try It 1 Since $12^2 = 144$, $-\sqrt{144} = -12$.

You Try It 2 Since $\left(\frac{9}{10}\right)^2 = \frac{81}{100}$, $\sqrt{\frac{81}{100}} = \frac{9}{10}$.

You Try It 3
$$4\sqrt{16} - \sqrt{9} = 4 \cdot 4 - 3$$
$$= 16 - 3 = 13$$

You Try It 4 $5\sqrt{a + b}$
$$5\sqrt{17 + 19} = 5\sqrt{36}$$
$$= 5 \cdot 6$$
$$= 30$$

You Try It 5 $5\sqrt{23} \approx 23.9792$

You Try It 6 57 is between the perfect squares 49 and 64.
$$\sqrt{49} = 7 \quad \text{and} \quad \sqrt{64} = 8.$$
$$7 < \sqrt{57} < 8$$

You Try It 7 $9^2 = 81$; 81 is too big.
$8^2 = 64$; 64 is not a factor of 80.
$7^2 = 49$; 49 is not a factor of 80.
$6^2 = 36$; 36 is not a factor of 80.
$5^2 = 25$; 25 is not a factor of 80.
$4^2 = 16$; 16 is a factor of 80. $(80 = 16 \cdot 5)$
$$\sqrt{80} = \sqrt{16 \cdot 5} = \sqrt{16} \cdot \sqrt{5}$$
$$= 4 \cdot \sqrt{5} = 4\sqrt{5}$$

You Try It 8

Strategy To find the range, replace h by 6 in the given formula and solve for R.

Solution
$$R = 1.4\sqrt{h}$$
$$R = 1.4\sqrt{6}$$
$$R \approx 3.43$$

The range of the periscope is 3.43 mi.

SECTION 4.5 (*pages 277–282*)

You Try It 1

You Try It 2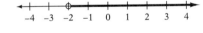

You Try It 3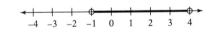

You Try It 4
a. $x \geq 4$
$-1 \geq 4$ False
b. $x \geq 4$
$0 \geq 4$ False
c. $x \geq 4$
$4 \geq 4$ True
d. $x \geq 4$
$\sqrt{26} \geq 4$ True

The numbers 4 and $\sqrt{26}$ make the inequality true.

You Try It 5 All real numbers greater than -7 make the inequality $x > -7$ true.

You Try It 6

You Try It 7

Strategy
▶ To write the inequality, let s represent the speeds at which a motorist is ticketed. Motorists are ticketed at speeds greater than 55.
▶ To determine whether a motorist traveling at 58 mph will be ticketed, replace s in the inequality by 58. If the inequality is true, the motorist will be ticketed. If the inequality is false, the motorist will not be ticketed.

Solution
$s > 55$

$58 > 55$ True

Yes, a motorist traveling at 58 mph will be ticketed.

SOLUTIONS TO CHAPTER 5   You Try Its

SECTION 5.1 (*pages 297–302*)

You Try It 1 $-6(-3p) = [-6(-3)]p = 18p$

You Try It 2
$(-2m)(-8n) = [(-2)(-8)](m \cdot n)$
$= 16mn$

You Try It 3
$(-12)(-d) = (-12)(-1d)$
$= [(-12)(-1)]d$
$= 12d$

You Try It 4
$6n + 9 + (-6n) = 6n + (-6n) + 9$
$= [6n + (-6n)] + 9$
$= 0 + 9$
$= 9$

You Try It 5
$-7(2k - 5) = -7(2k) - (-7)(5)$
$= -14k + 35$

You Try It 6
$-4(x - 2y) = -4(x) - (-4)(2y)$
$= -4x + 8y$

You Try It 7
$3(-2v + 3w - 7) = 3(-2v) + 3(3w) - 3(7)$
$= -6v + 9w - 21$

You Try It 8
$-4(2x - 7y - z) = -4(2x) - (-4)(7y) - (-4)(z)$
$= -8x + 28y + 4z$

You Try It 9 $-(c - 9d + 1) = -c + 9d - 1$

SECTION 5.2 (*pages 307–310*)

You Try It 1
$12a^2 - 8a + 3 - 16a^2 + 8a = 12a^2 - 16a^2 - 8a + 8a + 3$
$= -4a^2 + 0a + 3$
$= -4a^2 + 3$

You Try It 2
$-7x^2 + 4xy + 8x^2 - 12xy = -7x^2 + 8x^2 + 4xy - 12xy$
$= x^2 - 8xy$

You Try It 3
$-2r + 7s - 12 - 8r + s + 8$
$= -2r - 8r + 7s + s - 12 + 8$
$= -10r + 8s - 4$

You Try It 4
$8x^2y - 15xy^2 + 12xy^2 - 7x^2y$
$= 8x^2y - 7x^2y - 15xy^2 + 12xy^2$
$= x^2y - 3xy^2$

You Try It 5
$6 - 4(2x - y) + 3(x - 4y)$
$= 6 - 8x + 4y + 3x - 12y$
$= -5x - 8y + 6$

You Try It 6
$8c - 4(3c - 8) - 5(c + 4)$
$= 8c - 12c + 32 - 5c - 20$
$= -9c + 12$

You Try It 7
$6p + 5[3(2 - 3p) - 2(5 - 4p)]$
$= 6p + 5[6 - 9p - 10 + 8p]$
$= 6p + 5[-p - 4]$
$= 6p - 5p - 20$
$= p - 20$

SECTION 5.3 (*pages 315–316*)

You Try It 1
$(-4x^3 + 2x^2 - 8) + (4x^3 + 6x^2 - 7x + 5)$
$= (-4x^3 + 4x^3) + (2x^2 + 6x^2) - 7x + (-8 + 5)$
$= 8x^2 - 7x - 3$

You Try It 2
$$\begin{array}{r} 6x^3 \qquad\quad + 2x + 8 \\ -9x^3 + 2x^2 - 12x - 8 \\ \hline -3x^3 + 2x^2 - 10x \end{array}$$

You Try It 3
$$\begin{array}{r} 13y^3 \qquad - 6y - 7 \\ - 4y^2 + 6y + 9 \\ \hline 13y^3 - 4y^2 \qquad + 2 \end{array}$$

SECTION 5.4 (*pages 319–322*)

You Try It 1
$(-7a^4)(4a^2) = [-7(4)](a^4 \cdot a^2)$
$= -28a^{4+2}$
$= -28a^6$

You Try It 2
$(8m^3n)(-3n^5) = [8(-3)](m^3)(n \cdot n^5)$
$= -24m^3n^{1+5}$
$= -24m^3n^6$

You Try It 3
$(12p^4q^3)(-3p^5q^2) = [12(-3)](p^4 \cdot p^5)(q^3 \cdot q^2)$
$= -36p^{4+5}q^{3+2}$
$= -36p^9q^5$

You Try It 4
$(-y^4)^5 = [(-1)y^4]^5$
$= (-1)^{1 \cdot 5}y^{4 \cdot 5}$
$= (-1)^5y^{20}$
$= -1y^{20}$
$= -y^{20}$

You Try It 5 $(-3a^4bc^2)^3 = (-3)^{1\cdot3}a^{4\cdot3}b^{1\cdot3}c^{2\cdot3}$
$= (-3)^3a^{12}b^3c^6 = -27a^{12}b^3c^6$

SECTION 5.5 (*pages 325–326*)

You Try It 1 $-3a(-6a + 5b) = (-3a)(-6a) + (-3a)(5b)$
$= 18a^2 - 15ab$

You Try It 2 $3mn^2(2m^2 - 3mn - 1)$
$= (3mn^2)(2m^2) - (3mn^2)(3mn) - (3mn^2)1$
$= 6m^3n^2 - 9m^2n^3 - 3mn^2$

You Try It 3 $(3c + 7)(3c - 7)$
$= (3c)(3c) + (3c)(-7) + 7(3c) + (7)(-7)$
$= 9c^2 - 21c + 21c - 49$
$= 9c^2 - 49$

SECTION 5.6 (*pages 329–332*)

You Try It 1 $\dfrac{1}{d^{-6}} = d^6$

You Try It 2 $4^{-2} = \dfrac{1}{4^2} = \dfrac{1}{16}$

You Try It 3 $\dfrac{n^6}{n^{11}} = n^{6-11} = n^{-5} = \dfrac{1}{n^5}$

You Try It 4 $0.000000961 = 9.61 \times 10^{-7}$

You Try It 5 $7.329 \times 10^6 = 7,329,000$

SECTION 5.7 (*pages 335–338*)

You Try It 1   twice x divided by the <u>difference</u> between x and 7

$\dfrac{2x}{x - 7}$

You Try It 2   the <u>product</u> of negative three and the <u>square</u> of d
$-3d^2$

You Try It 3   Let the smaller number be x.
The larger number is $16 - x$.

the <u>difference</u> between the larger number and <u>twice</u> the smaller number
$16 - x - 2x$
$16 - 3x$

You Try It 4   the <u>difference</u> between fourteen and the <u>sum</u> of a number and seven
Let the unknown number be x.
$14 - (x + 7)$
$14 - x - 7$
$7 - x$

You Try It 5   the pounds of carmel: c
the pounds of milk chocolate: $c + 3$

SOLUTIONS TO CHAPTER 6   You Try Its

SECTION 6.1 (*pages 353–358*)

You Try It 1
$7 + y = 12$
$7 - 7 + y = 12 - 7$
$y = 5$
The solution is 5.

You Try It 2
$19 = b - 23$
$19 + 23 = b - 23 + 23$
$42 = b$
The solution is 42.

You Try It 3
$-5r + 3 + 6r = 1$
$r + 3 = 1$ Combine like terms.
$r + 3 - 3 = 1 - 3$
$r = -2$
The solution is -2.

You Try It 4
$-60 = 5d$
$\dfrac{-60}{5} = \dfrac{5d}{5}$
$-12 = d$
The solution is -12.

You Try It 5
$10 = \dfrac{-2x}{5}$
$\left(-\dfrac{5}{2}\right)10 = \left(-\dfrac{5}{2}\right)\left(-\dfrac{2}{5}x\right)$ ▶ $-\dfrac{2x}{5} = -\dfrac{2}{5}x$
$-25 = x$
The solution is -25.

You Try It 6
$\dfrac{1}{3}x - \dfrac{5}{6}x = 4$

$\dfrac{2}{6}x - \dfrac{5}{6}x = 4$

$-\dfrac{1}{2}x = 4$ ▶ $-\dfrac{3}{6} = -\dfrac{1}{2}$

$-2\left(-\dfrac{1}{2}x\right) = -2(4)$

$x = -8$

Check:
$$\dfrac{1}{3}x - \dfrac{5}{6}x = 4$$

$\dfrac{1}{3}(-8) - \dfrac{5}{6}(-8)$	4
$-\dfrac{8}{3} - \left(-\dfrac{20}{3}\right)$	4
$-\dfrac{8}{3} + \dfrac{20}{3}$	4
$4 = 4$	

-8 checks as the solution.
The solution is -8.

SECTION 6.2 (*pages 361–362*)

You Try It 1

$$-5 - 4t = 7$$
$$-5 + 5 - 4t = 7 + 5$$
$$-4 = 12$$
$$\frac{-4t}{-4} = \frac{12}{-4}$$
$$t = -3$$

The solution is −3.

You Try It 2

$$5v + 3 - 9v = 9$$
$$-4v + 3 = 9 \qquad \text{Combine like terms.}$$
$$-4v + 3 - 3 = 9 - 3$$
$$-4v = 6$$
$$\frac{-4v}{-4} = \frac{6}{-4}$$
$$v = -\frac{3}{2}$$

The solution is $-\frac{3}{2}$.

You Try It 3

Strategy To find the pressure, replace P by its value and solve for D. $P = 45$.

Solution

$$P = 15 + \frac{1}{2}D$$
$$45 = 15 + \frac{1}{2}D$$
$$45 - 15 = 15 - 15 + \frac{1}{2}D$$
$$30 = \frac{1}{2}D$$
$$2(30) = 2\left(\frac{1}{2}D\right)$$
$$60 = D$$

When the pressure is 45 pounds per square inch, the depth is 60 ft.

You Try It 3

$$6 - 5(3y + 2) = 26$$
$$6 - 15y - 10 = 26$$
$$-15y - 4 = 26$$
$$-15y - 4 + 4 = 26 + 4$$
$$-15y = 30$$
$$\frac{-15y}{-15} = \frac{30}{-15}$$
$$y = -2$$

The solution is −2.

You Try It 4

$$2w - 7(3w + 1) = 5(5 - 3w)$$
$$2w - 21w - 7 = 25 - 15w$$
$$-19w - 7 = 25 - 15w$$
$$-19w + 15w - 7 = 25 - 15w + 15w$$
$$-4w - 7 = 25$$
$$-4w - 7 + 7 = 25 + 7$$
$$-4w = 32$$
$$\frac{-4w}{-4} = \frac{32}{-4}$$
$$w = -8$$

The solution is −8.

You Try It 5

Strategy To find the force applied on the lip of the can, replace the variables, F_1, d, and x by the given values and solve for F_2.

Solution

$$F_1 x = F_2(d - x)$$
$$F_1(0.15) = 30(9 - 0.15)$$
$$0.15F_1 = 30(8.85)$$
$$0.15F_1 = 265.5$$
$$\frac{0.15F_1}{0.15} = \frac{265.5}{0.15}$$
$$F_1 = 1{,}770$$

A force of 1,770 lb is applied to the lip of the can.

SECTION 6.3 (*pages 367–370*)

You Try It 1

$$r - 7 = 5 - 3r$$
$$r + 3r - 7 = 5 - 3r + 3r$$
$$4r - 7 = 5$$
$$4r - 7 + 7 = 5 + 7$$
$$4r = 12$$
$$\frac{4r}{4} = \frac{12}{4}$$
$$r = 3 \qquad \text{The solution is 3.}$$

You Try It 2

$$4a - 2 + 5a = 2a - 2 + 3a$$
$$9a - 2 = 5a - 2$$
$$9a - 5a - 2 = 5a - 5a - 2$$
$$4a - 2 = -2$$
$$4a - 2 + 2 = -2 + 2$$
$$4a = 0$$
$$\frac{4a}{4} = \frac{0}{4}$$
$$a = 0 \qquad \text{The solution is 0.}$$

SECTION 6.4 (*pages 375–378*)

You Try It 1

The unknown number: x

Six more than one-half a number	is	the total of the number and nine

$$\frac{1}{2}x + 6 = x + 9$$
$$\frac{1}{2}x - x + 6 = x - x + 9$$
$$-\frac{1}{2}x + 6 = 9$$
$$-\frac{1}{2}x + 6 - 6 = 9 - 6$$
$$-\frac{1}{2}x = 3$$
$$(-2)\left(-\frac{1}{2}x\right) = (-2)3$$
$$x = -6$$

−6 checks as the solution. The solution is −6.

You Try It 2

The unknown number: x

Seven less than a number	is equal to	five more than three times the number

$$x - 7 = 3x + 5$$
$$x - 3x - 7 = 3x - 3x + 5$$
$$-2x - 7 = 5$$
$$-2x - 7 + 7 = 5 + 7$$
$$-2x = 12$$
$$\frac{-2x}{-2} = \frac{12}{-2}$$
$$x = -6$$

-6 checks as the solution.

The solution is -6.

You Try It 3

The smaller number: n
The larger number: $14 - n$

One more than three times the smaller number	equals	the sum of the larger number and three

$$3n + 1 = (14 - n) + 3$$
$$3n + 1 = 17 - n$$
$$3n + n + 1 = 17 - n + n$$
$$4n + 1 = 17$$
$$4n + 1 - 1 = 17 - 1$$
$$4n = 16$$
$$\frac{4n}{4} = \frac{16}{4}$$
$$n = 4$$

$$14 - n = 14 - 4 = 10$$

These numbers check as solutions.

The smaller number is 4.
The larger number is 10.

You Try It 4

Strategy To find the price of the large-screen television today, write and solve an equation using t to represent the price today.

Solution

4,200	is	$1,700 more than the price today

$$4,200 = t + 1,700$$
$$4,200 - 1,700 = t + 1,700 - 1,700$$
$$2,500 = t$$

The price today is $2,500.

You Try It 5

Strategy To find the number of tickets that you are purchasing, write and solve an equation using x to represent the number of tickets purchased.

Solution

$3.50 plus $17.50 for each ticket	equals	$161

$$3.50 + 17.50x = 161$$
$$3.50 - 3.50 + 17.50x = 161 - 3.50$$
$$17.50x = 157.50$$
$$\frac{17.50x}{17.50} = \frac{157.50}{17.50}$$
$$x = 9$$

You are purchasing 9 tickets.

You Try It 6

Strategy To find the number of minutes of access time, write and solve an equation using m to represent the number of minutes.

Solution

$9.95 plus $.13 per minute	equals	$24.77

$$9.95 + 0.13m = 24.77$$
$$9.95 - 9.95 + 0.13m = 24.77 - 9.95$$
$$0.13m = 14.82$$
$$\frac{0.13m}{0.13} = \frac{14.82}{0.13}$$
$$m = 114$$

The customer used 114 min of access time.

You Try It 7

Strategy To find the length of each piece, write and solve an equation using x to represent the length of the shorter piece and $18 - x$ to represent the length of the longer piece.

Solution

1 ft more than twice the shorter piece	is	2 ft less than the longer piece

$$2x + 1 = (18 - x) - 2$$
$$2x + 1 = 16 - x$$
$$2x + x + 1 = 16 - x + x$$
$$3x + 1 = 16$$
$$3x + 1 - 1 = 16 - 1$$
$$3x = 15$$
$$\frac{3x}{3} = \frac{15}{3}$$
$$x = 5$$

$$18 - x = 18 - 5 = 13$$

The length of the shorter piece is 5 ft, and the length of the longer piece is 13 ft.

SECTION 6.5 (*pages 383–388*)

You Try It 1

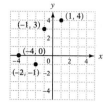

You Try It 2
$A(4, 2)$
$B(-3, 4)$
$C(-3, 0)$
$D(0, 0)$

You Try It 3

Strategy Graph the ordered pairs on a rectangular coordinate system where the horizontal axis represents the total yards gained and the vertical axis represents the number of points scored.

Solution

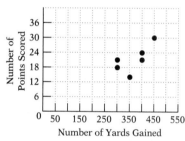

You Try It 4 Substitute the values of 2 for x and -4 for y into the equation and simplify.

$$y = -\frac{1}{2}x - 3$$

-4	$\left(-\frac{1}{2}\right)(2) - 3$
-4	$-1 - 3$
$-4 = -4$	

Yes, $(2, -4)$ is a solution of the equation $y = -\frac{1}{2}x - 3$.

You Try It 5 $y = 3x + 1$

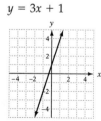

SOLUTIONS TO CHAPTER 7 You Try Its

SECTION 7.1 (*pages 403–408*)

You Try It 1 $12:20 = 3:5$
12 TO 20 = 3 TO 5

You Try It 2 $\dfrac{20 \text{ bags}}{12 \text{ acres}} = \dfrac{5 \text{ bags}}{3 \text{ acres}}$

You Try It 3 $\dfrac{\$4.48}{3.5 \text{ lb}} = \$1.28/\text{lb}$

You Try It 4

Strategy To convert square miles to acres, use the conversion factor $\dfrac{640 \text{ acres}}{1 \text{ mi}^2}$.

Solution $1.5 \text{ mi}^2 = 1.5 \text{ mi}^2 \cdot \dfrac{640 \text{ acres}}{1 \text{ mi}^2}$
$= 1.5 \cdot 640 \text{ acres} = 960 \text{ acres}$

You Try It 5

Strategy To convert liters to gallons, use the conversion factor $\dfrac{1 \text{ gal}}{3.79 \text{ L}}$.

Solution $10 \text{ L} = \dfrac{10 \text{ L}}{1} \cdot \dfrac{1 \text{ gal}}{3.79 \text{ L}} = \dfrac{10 \text{ gal}}{3.79} \approx 2.6 \text{ gal}$

You Try It 6

Strategy To convert kilometers per hour to meters per second, use the conversion factors $\dfrac{1,000 \text{ m}}{1 \text{ km}}$ and $\dfrac{1 \text{ h}}{3,600 \text{ s}}$.

Solution $\dfrac{90 \text{ km}}{\text{h}} = \dfrac{90 \text{ km}}{\text{h}} \cdot \dfrac{1,000 \text{ m}}{1 \text{ km}} \cdot \dfrac{1 \text{ h}}{3,600 \text{ s}}$
$= \dfrac{90 \cdot 1,000 \text{ m}}{3,600 \text{ s}} = 25 \text{ m/s}$

You Try It 7

Strategy

To find the number of gallons:

▶ Use the formula $V = LWH$ to find the volume in cubic inches.

▶ Use the conversion factor $\dfrac{1 \text{ gal}}{231 \text{ in}^3}$ to convert cubic inches to gallons.

Solution

$V = LWH = 36 \text{ in.} \cdot 24 \text{ in.} \cdot 16 \text{ in.} = 13,824 \text{ in}^3$

$13,824 \text{ in}^3 = 13,824 \text{ in}^3 \cdot \dfrac{1 \text{ gal}}{231 \text{ in}^3} \approx 59.8 \text{ gal}$

The fishtank holds 59.8 gal of water.

SECTION 7.2 (*pages 413–416*)

You Try It 1 $\dfrac{50}{3} \diagdown\diagup \dfrac{250}{12}$ $\longrightarrow 3 \cdot 250 = 750$
$\longrightarrow 50 \cdot 12 = 600$

$750 \neq 600$

The proportion is not true.

You Try It 2 $\dfrac{7}{12} = \dfrac{42}{x}$
$12 \cdot 42 = 7 \cdot x$
$504 = 7x$
$72 = x$

You Try It 3

$$\frac{6}{n} = \frac{3}{321}$$
$$n \cdot 3 = 6 \cdot 321$$
$$3n = 1,926$$
$$n = 642$$

You Try It 4

$$\frac{4}{5} = \frac{3}{x - 3}$$
$$5 \cdot 3 = 4(x - 3)$$
$$15 = 4x - 12$$
$$27 = 4x$$
$$\frac{27}{4} = x$$

You Try It 5

Strategy To find the number of gallons, write and solve a proportion using n to represent the number of gallons needed to travel 832 mi.

Solution
$$\frac{396 \text{ mi}}{11 \text{ gal}} = \frac{832 \text{ mi}}{n \text{ gal}}$$
$$11 \cdot 832 = 396 \cdot n$$
$$9,152 = 396n$$
$$23.1 \approx n$$

To travel 832 mi, approximately 23.1 gal of gas are needed.

You Try It 6

Strategy

To find the number of defective transmissions, write and solve a proportion using n to represent the number of defective transmissions in 120,000 cars.

Solution
$$\frac{15 \text{ defective transmissions}}{1,200 \text{ cars}} = \frac{n \text{ defective transmissions}}{120,000 \text{ cars}}$$
$$1,200 \cdot n = 15 \cdot 120,000$$
$$1,200n = 1,800,000$$
$$n = 1,500$$

1,500 defective transmissions would be found in 120,000 cars.

SECTION 7.3 (*pages 421–424*)

You Try It 1

Strategy To find the constant of variation, substitute 120 for y and 8 for x in the direct variation equation $y = kx$ and solve for k.

Solution
$$y = kx$$
$$120 = k \cdot 8$$
$$15 = k$$

The constant of variation is 15.

You Try It 2

Strategy To find S when $R = 200$:

▶ Write the basic direct variation equation, replace the variables by the given values, and solve for k.
▶ Write the direct variation equation, replacing k by its value. Substitute 200 for R and solve for S.

Solution
$$S = kR$$
$$8 = k \cdot 30$$
$$\frac{8}{30} = k$$
$$\frac{4}{15} = k$$
$$S = \frac{4}{15}R = \frac{4}{15}(200) = \frac{160}{3} \approx 53.3$$

The value of S is approximately 53.3 when $R = 200$.

You Try It 3

Strategy To find the distance:

▶ Write the basic direct variation equation, replace the variables by the given values, and solve for k.
▶ Write the direct variation equation, replacing k by its value. Substitute 9 for the time and solve for the distance.

Solution
$$d = kt^2$$
$$64 = k \cdot 2^2$$
$$64 = k \cdot 4$$
$$16 = k$$
$$d = 16t^2 = 16 \cdot 9^2 = 16 \cdot 81 = 1,296$$

The object will fall 1,296 ft.

You Try It 4

Strategy To find the resistance:

▶ Write the basic inverse variation equation, replace the variables by the given values, and solve for k.
▶ Write the inverse variation equation, replacing k by its value. Substitute 0.02 for the diameter and solve for the resistance.

Solution
$$R = \frac{k}{d^2}$$
$$0.5 = \frac{k}{(0.01)^2}$$
$$0.5 = \frac{k}{0.0001}$$
$$0.00005 = k$$
$$R = \frac{0.00005}{d^2} = \frac{0.00005}{(0.02)^2} = \frac{0.00005}{0.0004} = 0.125$$

The resistance is 0.125 ohms.

SOLUTIONS TO CHAPTER 8   You Try Its

SECTION 8.1 (*pages 439–440*)

You Try It 1
$$110\% = 110\left(\frac{1}{100}\right) = \left(\frac{110}{100}\right) = 1\frac{1}{10}$$
$$110\% = 110(0.01) = 1.10$$

You Try It 2 $16\dfrac{3}{8}\% = 16\dfrac{3}{8}\left(\dfrac{1}{100}\right) = \dfrac{131}{8}\left(\dfrac{1}{100}\right)$

$$= \dfrac{131}{800}$$

You Try It 3 $0.8\% = 0.8(0.01) = 0.008$

You Try It 4 $0.038 = 0.038(100\%) = 3.8\%$

You Try It 5 $\dfrac{9}{7} = \dfrac{9}{7}(100\%) = \dfrac{900}{7}\% = 128\dfrac{4}{7}\%$

You Try It 6 $1\dfrac{5}{9} = \dfrac{14}{9} = \dfrac{14}{9}(100\%) = \dfrac{1400}{9}\% \approx 155.6\%$

Section 8.2 (*pages 443–448*)

You Try It 1

Strategy   To find the amount, solve the basic percent equation. Percent $= 33\dfrac{1}{3}\% = \dfrac{1}{3}$, base $= 45$, amount $= n$

Solution   Percent · base = amount

$$\dfrac{1}{3}(45) = n$$

$$15 = n$$

15 is $33\dfrac{1}{3}\%$ of 45.

You Try It 2

Strategy   To find the percent, solve the basic percent equation. Percent $= n$, base $= 40$, amount $= 25$

Solution   Percent · base = amount

$$n \cdot 40 = 25$$

$$n = \dfrac{25}{40} = 0.625$$

$$n = 62.5\%$$

25 is 62.5% of 40.

You Try It 3

Strategy   To find the base, solve the basic percent equation. Percent $= 16\dfrac{2}{3}\% = \dfrac{1}{6}$, base $= n$, amount $= 15$

Solution   Percent · base = amount

$$\dfrac{1}{6} \cdot n = 15$$

$$n = 15 \cdot 6$$

$$n = 90$$

$16\dfrac{2}{3}\%$ of 90 is 15.

You Try It 4   Percent $= n$, base $= 182$, amount $= 56$

$$\dfrac{n}{100} = \dfrac{56}{182}$$

$$n \cdot 182 = 100 \cdot 56$$

$$182n = 5{,}600$$

$$n = \dfrac{5{,}600}{182}$$

$$n \approx 30.77$$

30.77% of 182 is approximately 56.

You Try It 5   Percent $= 0.74$, base $= 1{,}200$, amount $= n$

$$\dfrac{0.74}{100} = \dfrac{n}{1{,}200}$$

$$100 \cdot n = 0.74 \cdot 1{,}200$$

$$100n = 888$$

$$n = \dfrac{888}{100}$$

$$n = 8.88$$

0.74% of 1,200 is 8.88.

You Try It 6   Percent $= 25$, base $= n$, amount $= 8$

$$\dfrac{25}{100} = \dfrac{8}{n}$$

$$25 \cdot n = 100 \cdot 8$$

$$25n = 800$$

$$n = \dfrac{800}{25}$$

$$n = 32$$

8 is 25% of 32.

You Try It 7

Strategy   To find the percent, use the basic percent equation. Percent $= n$, base $= 2{,}165$, amount $= 324.75$

Solution   Percent · base = amount

$$n \cdot 2{,}165 = 324.75$$

$$n = \dfrac{324.75}{2{,}165}$$

$$n = 0.15$$

15% of the instructor's salary is deducted for income tax.

You Try It 8

Strategy   To find the number polled, solve the basic percent equation. Percent $= 70\% = 0.70$, base $= n$, amount $= 210$

Solution   Percent · base = amount

$$0.70 \cdot n = 210$$

$$n = \dfrac{210}{0.70}$$

$$n = 300$$

300 people were polled.

You Try It 9

Strategy   To find the increase in the hourly wage:

▶ Find last year's wage. Solve the basic percent equation.
Percent $= 115\% = 1.15$, base $= n$, amount $= 20.01$

▶ Subtract last year's wage from this year's wage.

Solution Percent · base = amount

$$1.15 \cdot n = 20.01$$

$$n = \frac{20.01}{1.15}$$

$$n = 17.40$$

$$20.01 - 17.40 = 2.61$$

The increase in the hourly wage was $2.61.

SECTION 8.3 (*pages 453–454*)

You Try It 1

Strategy To find the percent increase in mileage:

▶ Find the amount of increase in mileage.
▶ Solve the basic percent equation. Percent = n, base = 17.5, amount = amount of increase

Solution $18.2 - 17.5 = 0.7$

Percent · base = amount

$$n \cdot 17.5 = 0.7$$

$$n = \frac{0.7}{17.5}$$

$$n = 0.04$$

The percent increase in mileage is 4%.

You Try It 2

Strategy To find the value of the car:

▶ Solve the basic percent equation to find the amount of decrease in value. Percent = 24% = 0.24, base = 47,000, amount = n
▶ Subtract the amount of decrease from the cost.

Solution Percent · base = amount

$$0.24 \cdot 47{,}000 = n$$

$$11{,}280 = n$$

$$47{,}000 - 11{,}280 = 35{,}720$$

The value of the car is $35,720.

SECTION 8.4 (*pages 457–462*)

You Try It 1

Strategy To find the markup rate:

▶ Solve the formula $M = S - C$ for M. $S = 1{,}450, C = 950$
▶ Solve the formula $M = r \cdot C$ for r.

Solution $M = S - C$

$$M = 1{,}450 - 950$$

$$M = 500$$

$$M = r \cdot C$$

$$500 = r \cdot 950$$

$$\frac{500}{950} = r$$

$$0.526 \approx r$$

The markup rate on the cost is 52.6%.

You Try It 2

Strategy To find the markup rate:

▶ Solve the formula $M = S - C$ for M. $S = 940, C = 517$
▶ Solve the formula $M = r \cdot S$ for r.

Solution $M = S - C$

$$M = 940 - 517$$

$$M = 423$$

$$M = r \cdot S$$

$$423 = r \cdot 940$$

$$\frac{423}{940} = r$$

$$0.45 = r$$

The markup rate on the selling price is 45%.

You Try It 3

Strategy To find the sale price, solve the formula $S = (1 - r)R$ for S. $r = 25\% = 0.25$, $R = 212$

Solution $S = (1 - r)R$

$$S = (1 - 0.25) \cdot 212$$

$$S = 0.75 \cdot 212$$

$$S = 159$$

The sale price is $159.

You Try It 4

Strategy To find the discount rate:

▶ Solve the formula $M = R - S$ for M. $R = 2{,}300, S = 1{,}495$
▶ Solve the formula $M = r \cdot R$ for r.

Solution $M = R - S$

$$M = 2{,}300 - 1{,}495$$

$$M = 805$$

$$M = r \cdot R$$

$$805 = r \cdot 2{,}300$$

$$\frac{805}{2{,}300} = r$$

$$0.35 = r$$

The discount rate is 35%.

You Try It 5

Strategy To find the simple interest, solve the formula $I = Prt$ for I.

$P = 4{,}000, r = 8.6\% = 0.086, t = \dfrac{60}{360}$

Solution $I = Prt$

$$I = 4{,}000(0.086)\left(\frac{60}{360}\right)$$

$$I \approx 57.33$$

The interest on the loan is $57.33.

You Try It 6

Strategy To find the simple interest, solve the formula $I = Prt$ for r.

$I = 225, P = 5{,}000, t = \dfrac{180}{360}$

Solution $I = Prt$

$225 = 5{,}000(r)\dfrac{180}{360}$

$225 = 2{,}500r$

$\dfrac{225}{2{,}500} = r$

$0.09 = r$

The simple interest rate is 9%.

SOLUTIONS TO CHAPTER 9   You Try Its

SECTION 9.1 (pages 477–486)

You Try It 1 $QR + RS + ST = QT$
$24 + RS + 17 = 62$
$41 + RS = 62$
$RS = 21$

$RS = 21$ cm

You Try It 2 $AC = AB + BC$

$AC = \dfrac{1}{4}(BC) + BC$

$AC = \dfrac{1}{4}(16) + 16$

$AC = 4 + 16$
$AC = 20$

$AC = 20$ ft

You Try It 3

Strategy Supplementary angles are two angles whose sum is 180°. To find the supplement, let x represent the supplement of a 129° angle. Write an equation and solve for x.

Solution $x + 129° = 180°$
$x = 51°$

The supplement of a 129° angle is a 51° angle.

You Try It 4

Strategy To find the measure of $\angle a$, write an equation using the fact that the sum of the measure of $\angle a$ and 68° is 118°. Solve for $\angle a$.

Solution $\angle a + 68° = 118°$
$\angle a = 50°$

The measure of $\angle a$ is 50°.

You Try It 5

Strategy The angles labeled are adjacent angles of intersecting lines and are therefore supplementary angles. To find x, write an equation and solve for x.

Solution $(x + 16°) + 3x = 180°$
$4x + 16° = 180°$
$4x = 164°$
$x = 41°$

You Try It 6

Strategy $3x = y$ because corresponding angles have the same measure. $y + (x + 40°) = 180°$ because adjacent angles of intersecting lines are supplementary angles. Substitute $3x$ for y and solve for x.

Solution $3x + (x + 40°) = 180°$
$4x + 40° = 180°$
$4x = 140°$
$x = 35°$

You Try It 7

Strategy
- To find the measure of angle b, use the fact that $\angle b$ and $\angle x$ are supplementary angles.
- To find the measure of angle c, use the fact that the sum of the interior angles of a triangle is 180°.
- To find the measure of angle y, use the fact that $\angle c$ and $\angle y$ are vertical angles.

Solution $\angle b + \angle x = 180°$
$\angle b + 100° = 180°$
$\angle b = 80°$

$\angle a + \angle b + \angle c = 180°$
$45° + 80° + \angle c = 180°$
$125° + \angle c = 180°$
$\angle c = 55°$

$\angle y = \angle c = 55°$

You Try It 8

Strategy To find the measure of the third angle, use the facts that the measure of a right angle is 90° and the sum of the measures of the interior angles of a triangle is 180°. Write an equation using x to represent the measure of the third angle. Solve the equation for x.

Solution $x + 90° + 34° = 180°$
$x + 124° = 180°$
$x = 56°$

The measure of the third angle is 56°.

SECTION 9.2 (pages 493–502)

You Try It 1

Strategy To find the perimeter, use the formula for the perimeter of a square. Substitute 60 for s and solve for P.

Solution $P = 4s$
$P = 4(60)$
$P = 240$

The perimeter of the infield is 240 ft.

You Try It 2

Strategy To find the perimeter, use the formula for the perimeter of a rectangle. Substitute 11 for L and $8\dfrac{1}{2}$ for W and solve for P.

Solution

$$P = 2L + 2W$$
$$P = 2(11) + 2\left(8\frac{1}{2}\right)$$
$$P = 2(11) + 2\left(\frac{17}{2}\right)$$
$$P = 22 + 17$$
$$P = 39$$

The perimeter of a standard piece of typing paper is 39 in.

You Try It 3

Strategy   To find the circumference, use the circumference formula that involves the diameter. Leave the answer in terms of π.

Solution $C = \pi d$
$C = \pi(9)$
$C = 9\pi$

The circumference is 9π in.

You Try It 4

Strategy

To find the number of rolls of wallpaper to be purchased:

► Use the formula for the area of a rectangle to find the area of one wall.
► Multiply the area of one wall by the number of walls to be covered (2).
► Divide the area of wall to be covered by the area one roll of wallpaper will cover (30).

Solution

$A = LW$
$A = 12 \cdot 8 = 96$ The area of one wall is 96 ft².

$2(96) = 192$ The area of the two walls is 192 ft².

$192 \div 30 = 6.4$

Because a portion of a seventh roll is needed, 7 rolls of wallpaper should be purchased.

You Try It 5

Strategy   To find the area, use the formula for the area of a circle. An approximation is asked for; use the π key on a calculator. $r = 11$.

Solution $A = \pi r^2$
$A = \pi(11)^2$
$A = 121\pi$
$A \approx 380.13$

The area is approximately 380.13 cm².

SECTION 9.3 (*pages 511–516*)

You Try It 1

Strategy   To find the measure of the other leg, use the Pythagorean Theorem. $a = 2, c = 6$

Solution $a^2 + b^2 = c^2$
$2^2 + b^2 = 6^2$
$4 + b^2 = 36$
$b^2 = 32$

$b = \sqrt{32}$
$b \approx 5.66$

The measure of the other leg is approximately 5.66 m.

You Try It 2

Strategy   To find *FG*, write a proportion using the fact that, in similar triangles, the ratio of corresponding sides equals the ratio of corresponding heights. Solve the proportion for *FG*.

Solution $\dfrac{AC}{DF} = \dfrac{CH}{FG}$

$\dfrac{10}{15} = \dfrac{7}{FG}$

$10(FG) = 15(7)$
$10(FG) = 105$
$FG = 10.5$

The height *FG* of triangle *DEF* is 10.5 m.

You Try It 3

Strategy   To determine whether the triangles are congruent, determine whether one of the rules for congruence is satisfied.

Solution $PR = MN, QR = MO,$ and $\angle QRP = \angle OMN$.
Two sides and the included angle of one triangle equal two sides and the included angle of the other triangle.

The triangles are congruent by the SAS rule.

SECTION 9.4 (*pages 521–526*)

You Try It 1

Strategy   To find the volume, use the formula for the volume of a cube. $s = 2.5$.

Solution $V = s^3$
$V = (2.5)^3 = 15.625$

The volume of the cube is 15.625 m³.

You Try It 2

Strategy   To find the volume:

► Find the radius of the base of the cylinder. $d = 8$.
► Use the formula for the volume of a cylinder. Leave the answer in terms of π.

Solution $r = \dfrac{1}{2}d = \dfrac{1}{2}(8) = 4$

$V = \pi r^2 h = \pi(4)^2(22) = \pi(16)(22) = 352\pi$

The volume of the cylinder is 352π ft³.

You Try It 3

Strategy   To find the surface area of the cylinder:

► Find the radius of the base of the cylinder. $d = 6$.

▶ Use the formula for the surface area of a cylinder. An approximation is asked for; use the π key on a calculator.

Solution

$r = \dfrac{1}{2}d = \dfrac{1}{2}(6) = 3$

$SA = 2\pi r^2 + 2\pi rh$
$SA = 2\pi(3)^2 + 2\pi(3)(8)$
$\quad = 2\pi(9) + 2\pi(3)(8)$
$\quad = 18\pi + 48\pi$
$\quad = 66\pi$
$\quad \approx 207.35$

The surface area of the cylinder is approximately 207.35 ft².

You Try It 4

Strategy To find which solid has the larger surface area:

▶ Use the formula for the surface area of a cube to find the surface area of the cube. $s = 10$.

▶ Find the radius of the sphere. $d = 8$.

▶ Use the formula for the surface area of a sphere to find the surface area of the sphere. Since this number is to be compared to another number, use the π key on a calculator to approximate the surface area.

▶ Compare the two numbers.

Solution

$SA = 6s^2$
$SA = 6(10)^2 = 6(100) = 600$

The surface area of the cube is 600 cm².

$r = \dfrac{1}{2}d = \dfrac{1}{2}(8) = 4$

$SA = 4\pi r^2$
$SA = 4\pi(4)^2 = 4\pi(16) = 64\pi \approx 201.06$

The surface area of the sphere is 201.06 cm².

$600 > 201.06$

The cube has a larger surface area than the sphere.

SECTION 9.5 (*pages 531–534*)

You Try It 1

Strategy To find the perimeter of square $ABCD$:

▶ Use the circumference formula that involves the diameter to find a diameter of the circle. A diameter of the circle is equal to the length of a side of the square.

▶ Use the formula for the perimeter of a square.

Solution

$C = \pi d$
$6\pi = \pi d$
$6 = d$ The diameter of the circle is 6 cm.

$P = 4s$
$P = 4(6) = 24$

The perimeter of square $ABCD$ is 24 cm.

You Try It 2

Strategy The area is equal to the area of the rectangle minus the area of the triangle. The base of the triangle is equal to the width of the rectangle.

Solution

$A = LW - \dfrac{1}{2}bh$

$A = 10(6) - \dfrac{1}{2}(6)(4) = 60 - 12 = 48$

The area of the composite figure is 48 in².

You Try It 3

Strategy The volume is equal to the volume of the rectangular solid plus the volume of the cylinder. Leave the answer in terms of π.

Solution

$V = LWH + \pi r^2 h$
$V = (8)(8)(2) + \pi(1)^2(2)$
$\quad = (8)(8)(2) + \pi(1)(2) = 128 + 2\pi$

The volume of the solid is $(128 + 2\pi)$ cm³.

You Try It 4

Strategy

The total surface area equals the surface area of the rectangular solid, minus the top of the rectangular solid, plus one half the surface area of the cylinder. The radius of the base of the cylinder is one half the width of the rectangular solid. The height of the cylinder is equal to the length of the rectangular solid. An approximation is asked for; use the π key on a calculator.

Solution

$r = \dfrac{1}{2}(W) = \dfrac{1}{2}(6) = 3$

$SA = LW + 2LH + 2WH + \dfrac{1}{2}(2\pi r^2 + 2\pi rh)$

$SA = (24)(6) + 2(24)(4) + 2(6)(4) + \dfrac{1}{2}[2\pi(3)^2 + 2\pi(3)(24)]$

$\quad = 144 + 192 + 48 + \dfrac{1}{2}(18\pi + 144\pi)$

$\quad = 144 + 192 + 48 + \dfrac{1}{2}(162\pi)$

$\quad = 384 + 81\pi$
$\quad \approx 638.47$

The surface area of the solid is approximately 638.47 in².

SOLUTIONS TO CHAPTER 10   You Try Its

SECTION 10.1 (*pages 551–554*)

You Try It 1

Strategy To make the frequency distribution:

▶ Find the range.

▶ Divide the range by 8, the number of classes. The quotient is the class width.

▶ Tabulate the data for each class.

Solution

Range = 998 − 118 = 880

Class width = $\dfrac{880}{8}$ = 110

Dollar Amount of Insurance Claims

Classes	Tally	Frequency
118–228	///	3
229–339	////	4
340–450	/////	5
451–561	////////	8
562–672	014	6
673–783	///////	7
784–894	//////////	10
895–1,005	///////	7

You Try It 2

Strategy To find the number:

▸ Read the histogram to find the number of households using between 1,100 and 1,150 kWh and the number using between 1,150 and 1,200 kWh.
▸ Add the two numbers.

Solution Number between 1,100 and 1,150 kWh: 14
Number between 1,150 and 1,200 kWh: 10

14 + 10 = 24

24 households used 1,100 kWh of electricity or more during the month.

You Try It 3

Strategy To find the ratio:

▸ Read the frequency polygon to find the number of states with a per capita income between $15,000 and $18,000 and between $24,000 and $27,000.
▸ Write the ratio of the number of states with a per capita income between $15,000 and $18,000 to the number of states with a per capita income between $24,000 and $27,000.

Solution Number of states with a per capita income between $15,000 and $18,000: 10

Number of states with a per capita income between $24,000 and $27,000: 5

$\dfrac{\text{income between \$15,000 and \$18,000}}{\text{income between \$24,000 and \$27,000}}$

$= \dfrac{10}{5} = \dfrac{2}{1}$

The ratio is $\dfrac{2}{1}$ or 2 TO 1.

SECTION 10.2 (*pages 559–564*)

You Try It 1

Strategy To calculate the mean amount spent:

▸ Calculate the sum of the amounts spent by the customers.
▸ Divide the sum by the number of customers.

To calculate the median amount spent by the customers:

▸ Arrange the numbers from smallest to largest.
▸ Because there is an even number of values, the median is the sum of the two middle numbers divided by 2.

Solution The sum of the numbers is 31.30.

$\bar{x} = \dfrac{31.30}{10} = 3.13$

The mean amount spent by the customers was $3.13.

Arrange the numbers from smallest to largest.

1.59   2.32   2.45   2.75   3.05
3.45   3.58   3.90   4.00   4.21

median = $\dfrac{3.05 + 3.45}{2}$ = 3.25

The median is $3.25.

You Try It 2

Strategy To find the score, use the formula for the mean, letting n be the score on the fifth exam.

Solution $84 = \dfrac{82 + 91 + 79 + 83 + n}{5}$

$84 = \dfrac{335 + n}{5}$

$5 \cdot 84 = 5\left(\dfrac{335 + n}{5}\right)$

$420 = 335 + n$

$85 = n$

The score on the fifth test must be 85.

You Try It 3

Strategy To draw the box-and-whiskers plot:

▸ Use the value of the first quartile, the median, and the third quartile from Example 3.
▸ Determine the smallest and largest data value.
▸ Draw the box-and-whiskers plot.

Solution $Q_1 = 38$, median = 44, $Q_3 = 53$

Smallest value: 24
Largest value: 64

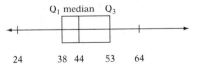

You Try It 4

Strategy To calculate the standard deviation:

▸ Find the mean of the number of miles run.
▸ Use the procedure for calculating the standard deviation.

Solution

$$\bar{x} = \frac{5 + 7 + 3 + 6 + 9 + 6}{6} = 6$$

Step 1

x	$(x - \bar{x})^2$
5	$(5 - 6)^2 = 1$
7	$(7 - 6)^2 = 1$
3	$(3 - 6)^2 = 9$
6	$(6 - 6)^2 = 0$
9	$(9 - 6)^2 = 9$
6	$(6 - 6)^2 = 0$
	Total $= 20$

Step 2 $\dfrac{20}{6} = \dfrac{10}{3}$

Step 3 $\sigma = \sqrt{\dfrac{10}{3}} \approx 1.826$

The standard deviation of the number of miles run is 1.826.

SECTION 10.3 (*pages 569–574*)

You Try It 1

Strategy

To calculate the probability:

▶ Count the number of possible outcomes of the experiment.
▶ Count the outcomes of the experiment that are favorable to the event the two numbers are the same.
▶ Use the probability formula.

Solution

There are 36 possible outcomes.
There are 6 favorable outcomes: (1, 1), (2, 2), (3, 3), (4, 4), (5, 5), and (6, 6).

$$P(E) = \frac{6}{36} = \frac{1}{6}$$

The probability that the two numbers are equal is $\dfrac{1}{6}$.

You Try It 2

Strategy

To calculate the probability:

▶ Count the number of possible outcomes of the experiment.
▶ Count the number of outcomes of the experiment that are favorable to the event E that the soft drink symbol is uncovered.
▶ Use the probability formula.

Solution

There are 8 possible outcomes of the experiment.
There is 1 favorable outcome for E, uncovering the soft drink symbol.

$$P(E) = \frac{1}{8}$$

The probability is $\dfrac{1}{8}$ that the soft drink symbol will be uncovered.

You Try It 3

Strategy

To calculate the probability:

▶ Write the odds in favor of contracting the flu as a fraction.
▶ The probability of contracting the flu is the numerator of the odds in favor fraction over the sum of the numerator and the denominator.

Solution

Odds in favor of contracting the flu $= \dfrac{2}{13}$

Probability of contracting the flu $= \dfrac{2}{2 + 13} = \dfrac{2}{15}$

The probability of contracting the flu is $\dfrac{2}{15}$.

SOLUTIONS TO THE METRIC SYSTEM OF MEASUREMENT APPENDIX

(*pages A1–A3*)

You Try It 1 1,295 m = 1.295 km

You Try It 2 7,543 g = 7.543 kg

You Try It 3 6.3 L = 6,300 ml

You Try It 4

Strategy To find the number of grams:

▶ Multiply the number of milligrams of cholesterol in one egg (274) by the number of eggs in one dozen (12).
▶ Convert milligrams to grams.

Solution 274(12) = 3,288

3,288 mg = 3.288 g

There are 3.288 g of cholesterol in one dozen eggs.

ANSWERS TO CHAPTER 1 Exercises

1.1 EXERCISES (*pages 13–18*)

1. 0 1 2 3 4 5 6 7 8 9 10 11 12   **3.** 0 1 2 3 4 5 6 7 8 9 10 11 12
5. 0 1 2 3 4 5 6 7 8 9 10 11 12   **7.** 5   **9.** 5   **11.** 0   **13.** $27 < 39$ **15.** $0 < 52$ **17.** $273 > 194$
19. $2,761 < 3,857$ **21.** $4,610 > 4,061$ **23.** $8,005 < 8,050$ **25.** 11, 14, 16, 21, 32   **27.** 13, 48, 72, 84, 93   **29.** 26, 49, 77, 90, 106   **31.** 204, 399, 662, 736, 981   **33.** 307, 370, 377, 3,077, 3,700   **35.** five hundred eight   **37.** six hundred thirty-five   **39.** four thousand seven hundred ninety   **41.** fifty-three thousand six hundred fourteen   **43.** two hundred forty-six thousand fifty-three   **45.** three million eight hundred forty-two thousand nine hundred five   **47.** 496
49. 53,340   **51.** 502,140   **53.** 9,706   **55.** 5,012,907   **57.** 8,005,010   **59.** $7,000 + 200 + 40 + 5$
61. $500,000 + 30,000 + 2,000 + 700 + 90 + 1$ **63.** $5,000 + 60 + 4$ **65.** $20,000 + 300 + 90 + 7$
67. $400,000 + 2,000 + 700 + 8$ **69.** $8,000,000 + 300 + 10 + 6$ **71.** 7,110   **73.** 5,000   **75.** 28,600   **77.** 7,000
79. 94,000   **81.** 630,000   **83.** 350,000   **85.** 72,000,000   **87.** Billy Hamilton   **89.** *Fiddler on the Roof*
91. peanut butter   **93.** St. Louis to San Diego   **95.** Neptune   **97a.** 481 fatalities   **b.** Thanksgiving
99. 160,000 acres   **101a.** 1993–94   **b.** 1993–94   **103.** 300,000 km/s   **105.** 999; 10,000

1.2 EXERCISES (*pages 33–40*)

1. 1,383,659   **3.** 6,043   **5.** 112,152   **7.** 12,548   **9.** 199,556   **11.** 327,473   **13.** 168,574   **15.** 7,947
17. 99,637   **19.** 1,872 students   **21.** 15,000; 15,040   **23.** 1,400,000; 1,388,917   **25.** 2,000; 1,998   **27.** 307,000; 329,801   **29.** 1,272   **31.** 12,150   **33.** 89,900   **35.** 1,572   **37.** 14,591   **39.** 97,413   **41.** Commutative Property of Addition   **43.** Associative Property of Addition   **45.** Addition Property of Zero   **47.** 28   **49.** 4
51. 15   **53.** Yes   **55.** No   **57.** Yes   **59.** 416   **61.** 188   **63.** 464   **65.** 208   **67.** 3,557   **69.** 2,836
71. 1,437   **73.** 20,148   **75.** 1,618   **77.** 7,378   **79.** 17,548   **81.** 15 ft   **83.** 2,000; 2,136   **85.** 40,000; 38,283
87. 35,000; 31,195   **89.** 100,000; 125,665   **91.** 13   **93.** 643   **95.** 355   **97.** 5,211   **99.** 766   **101.** 18,231
103. Yes   **105.** No   **107.** Yes   **109.** 210   **111.** 901   **113.** 370 calories   **115a.** 324 wins   **b.** 165 wins
117. less than   **119.** $1,924   **121.** $272   **123.** 280,000 mi^2 **125.** March to April of 1996; 26 cars   **127.** 1996
129. $9,284   **131.** $94,400   **133.** 410 mph   **135.** No   **137.** at or below the posted speed limit   **139.** 90; 900

1.3 EXERCISES (*pages 59–66*)

1. 1,143   **3.** 46,963   **5.** 470,152   **7.** 48,493   **9.** 324,438   **11.** 3,206,160   **13.** 1,500   **15.** 2,000   **17.** 0
19. qrs **21.** 1,200,000; 1,244,653   **23.** 1,200,000; 1,138,134   **25.** 42,000; 46,935   **27.** 6,300,000; 6,491,166
29. 14,880   **31.** 3,255   **33.** 1,800   **35.** 3,082   **37.** Multiplication Property of One   **39.** Commutative Property of Multiplication   **41.** 30   **43.** 0   **45.** Yes   **47.** No   **49.** Yes   **51.** $2^3 \cdot 7^5$ **53.** $2^2 \cdot 3^3 \cdot 5^4$ **55.** c^2
57. $x^3 y^3$ **59.** 32   **61.** 1,000,000   **63.** 200   **65.** 9,000   **67.** 0   **69.** 540   **71.** 144   **73.** 512   **75.** a^4
77. 24   **79.** 320   **81.** 225   **83.** 307   **85.** 309 r4   **87.** 2,550   **89.** 21 r9   **91.** 147 r38   **93.** 200 r8
95. 404 r34   **97.** 16 r97   **99.** 907   **101.** 881 r1   **103.** $\dfrac{c}{d}$ **105.** 800; 776   **107.** 5,000; 5,129
109. 500; 493 r37   **111.** 1,500; 1,516   **113.** 48   **115.** undefined   **117.** 9,800   **119.** Yes   **121.** No
123. 1, 2, 5, 10   **125.** 1, 2, 3, 4, 6, 12   **127.** 1, 2, 4, 8   **129.** 1, 13   **131.** 1, 2, 3, 6, 9, 18   **133.** 1, 5, 25
135. 1, 2, 4, 7, 8, 14, 28, 56   **137.** 1, 2, 4, 7, 14, 28   **139.** 1, 2, 3, 4, 6, 8, 12, 16, 24, 48   **141.** 1, 2, 3, 6, 9, 18, 27, 54
143. 2^4 **145.** $2^2 \cdot 3$ **147.** $3 \cdot 5$ **149.** $2^3 \cdot 5$ **151.** prime   **153.** $5 \cdot 13$ **155.** $2^2 \cdot 7$ **157.** $2 \cdot 3 \cdot 7$
159. $3 \cdot 17$ **161.** $2 \cdot 23$ **163.** 460 calories   **165.** 368,640 bytes   **167a.** $396,408   **b.** $480,000   **c.** $290,520
169. $32   **171a.** $17,000   **b.** $8,000   **173.** $6,732   **175.** 8 h   **177.** $18   **179.** 87,312   **181a.** sometimes true   **b.** sometimes true

1.4 EXERCISES (*pages 71–72*)

1. 14   **3.** 25   **5.** 5   **7.** 13   **9.** 7   **11.** 9   **13.** 8   **15.** 1   **17.** 0   **19.** 76   **21.** 24   **23.** 6
25. 12   **27.** 58   **29.** $760   **31.** 190 mi   **33.** 24 payments   **35.** 8 h   **37.** Answers will vary. For example: **a.** $5x = 0$, and **b.** $8x = 8$.

1.5 EXERCISES (*pages 75–76*)

1. 4   **3.** 29   **5.** 13   **7.** 19   **9.** 11   **11.** 6   **13.** 61   **15.** 54   **17.** 19   **19.** 24   **21.** 186
23. 39   **25.** 18   **27.** 14   **29.** 14   **31.** 2   **33.** 57   **35.** 8   **37.** 68   **39.** 16   **41.** 97

CHAPTER REVIEW EXERCISES (*pages 81–82*)

1. (Objective 1.1A)   **2.** 10,000 (Objective 1.3B)   **3.** 2,583 (Objective 1.2B)
4. $3^2 \cdot 5^4$ (Objective 1.3B)   **5.** 1,389 (Objective 1.2A)   **6.** 38,700 (Objective 1.1C)   **7.** 247 > 163 (Objective 1.1A)
8. 32,509 (Objective 1.1B)   **9.** 700 (Objective 1.3A)   **10.** 2,607 (Objective 1.3C)   **11.** 4,048 (Objective 1.2B)
12. 1,500 (Objective 1.2A)   **13.** 1, 2, 5, 10, 25, 50 (Objective 1.3D)   **14.** Yes (Objective 1.2B)   **15.** 18 (Objective 1.5A)
16. Commutative Property of Addition (Objective 1.2A)   **17.** four million nine hundred twenty-seven thousand thirty-six
(Objective 1.1B)   **18.** 675 (Objective 1.3B)   **19.** 3 times (Objective 1.3E)   **20.** 67 r70 (Objective 1.3C)   **21.** 2,636
(Objective 1.3A)   **22.** 137 (Objective 1.2B)   **23.** $2 \cdot 3^2 \cdot 5$ (Objective 1.3D)   **24.** 80 (Objective 1.3C)   **25.** 1
(Objective 1.5A)   **26.** 9 (Objective 1.4A)   **27.** 932 (Objective 1.2A)   **28.** 432 (Objective 1.3A)   **29.** 56
(Objective 1.5A)   **30.** Karem Abdul-Jabbar (Objective 1.2C)   **31.** $182,000 (Objective 1.3E)   **32.** December 1995 to
January 1996; 306,000 people (Objective 1.2C)   **33.** 42 mi (Objective 1.3E)   **34.** $449 (Objective 1.2C)

ANSWERS TO CHAPTER 2   Exercises

2.1 EXERCISES (*pages 91–96*)

1.   **3.**

5.   **7.**   **9.** 1   **11.** −1   **13.** 3

15. A is −4. C is −2.   **17.** A is −7. D is −4   **19.** −2 > −5   **21.** 3 > −7   **23.** −42 < 27   **25.** 53 > −46
27. −51 < −20   **29.** −131 < 101   **31.** −7, −2, 0, 3   **33.** −5, −3, 1, 4   **35.** −4, 0, 5, 9   **37.** −10, −7, −5, 4, 12
39. −11, −7, −2, 5, 10   **41.** −45   **43.** 88   **45.** −n **47.** d **49.** the opposite of negative thirteen   **51.** the
opposite of negative p **53.** five plus negative ten   **55.** negative fourteen minus negative three   **57.** negative thirteen
minus eight   **59.** m plus the opposite of n **61.** 7   **63.** 61   **65.** −46   **67.** 73   **69.** z **71.** −p **73.** 4
75. 9   **77.** 11   **79.** 12   **81.** 23   **83.** −27   **85.** 25   **87.** −41   **89.** −93   **91.** 10   **93.** 8   **95.** 6
97. $|-12| > |8|$ **99.** $|6| < |13|$ **101.** $|-1| < |-17|$ **103.** $|x| = |-x|$ **105.** −$|6|$, −(4), $|-7|$, −(−9)
107. −9, −$|-7|$, −(5), $|4|$ **109.** −$|10|$, −$|-8|$, −(−2), −(−3), $|5|$ **111.** 11, −11   **113.** −6, −5, −4, −3, −2, −1, 0, 1, 2,
3, 4, 5, 6   **115.** −24°F   **117.** −40°F   **119.** −20°F with a 10 mph wind   **121a.** −27¢   **b.** −40¢   **123.** Yes;
1994   **125.** Stock B   **127.** the third quarter   **129a.** −2 and 6   **b.** −2 and 8   **133.** −9, −8, −7, 7, 8, 9

2.2 EXERCISES (*pages 105–110*)

1. −2   **3.** −9   **5.** 1   **7.** −15   **9.** 0   **11.** −21   **13.** −14   **15.** 19   **17.** −5   **19.** −30   **21.** 9
23. −12   **25.** −28   **27.** −13   **29.** −18   **31.** 11   **33.** 1   **35.** $x + (-7)$ **37a.** −95,392,900,000
b. −26,482,700,000   **c.** −78,403,300,000   **39.** 5   **41.** −2   **43.** −11   **45.** −17   **47.** Addition Property of
Zero   **49.** Associative Property of Addition   **51.** 0   **53.** 18   **55.** No   **57.** Yes   **59.** No   **61.** −3
63. −13   **65.** 7   **67.** 0   **69.** −17   **71.** −3   **73.** 12   **75.** 27   **77.** −106   **79.** −67   **81.** −6
83. −15   **85.** −$t - r$ **87.** 82°C   **89.** −9   **91.** 11   **93.** 0   **95.** −138   **97.** 26   **99.** 13   **101.** −8
103. 5   **105.** 2   **107.** −6   **109.** 12   **111.** −3   **113.** 18   **115.** Yes   **117.** No   **119.** Yes
121a. 7,046 m   **b.** 6,028 m   **123.** Europe   **125a.** $71,000,000   **b.** $29,333,000   **127.** Kmart   **129.** −3
131. 19   **133a.** sometimes true   **b.** always true

2.3 EXERCISES (*pages 117–122*)

1. −24   **3.** 6   **5.** 18   **7.** −20   **9.** −16   **11.** 25   **13.** 0   **15.** 42   **17.** −128   **19.** 208
21. −243   **23.** −115   **25.** 238   **27.** −96   **29.** −210   **31.** −224   **33.** −40   **35.** 180   **37.** −qr
39a. −231,200,000   **b.** −46,872,000   **c.** −264,372,000   **41.** Multiplication Property of One   **43.** Associative
Property of Multiplication   **45.** −6   **47.** 1   **49.** −24   **51.** −60   **53.** 357   **55.** −56   **57.** −1,600
59. No   **61.** No   **63.** Yes   **65.** −6   **67.** 8   **69.** −49   **71.** 8   **73.** −11   **75.** 14   **77.** 13   **79.** 1

81. 26 **83.** 23 **85.** −110 **87.** 111 **89.** $\dfrac{-9}{x}$ **91.** −1,000,000 **93.** −9 **95.** 9 **97.** −6 **99.** 6
101. Yes **103.** No **105.** Yes **107.** −3 **109.** −62°F **111.** −4° **113.** −60°F **115.** −16, 32, −64
117. −125, −625, −3,125 **119a.** 81 **b.** −17 **121.** 3, −2, −1

2.4 EXERCISES (*pages 127–128*)

1. 15 **3.** 11 **5.** −7 **7.** −16 **9.** −8 **11.** 0 **13.** −5 **15.** −2 **17.** 5 **19.** 10 **21.** −20
23. 0 **25.** 25 **27.** −15 **29.** −8 **31.** 0 **33.** −$115,568 million **35.** 3°C **37.** $13,525
39. $15 million **41a.** False; for example, 0 is the solution of $3x = 0$. **b.** False; for example, −2 is the solution of
$-3x = 6$. **c.** False; for example, 5 is the solution of $-2x = -10$.

2.5 EXERCISES (*pages 131–132*)

1. −3 **3.** −6 **5.** −5 **7.** −12 **9.** −3 **11.** 19 **13.** 2 **15.** 1 **17.** 14 **19.** 42 **21.** −13
23. −12 **25.** 32 **27.** 30 **29.** −27 **31.** 27 **33.** 2 **35.** 8 **37.** 1 **39.** 15 **41.** 32 **43.** 1
45. 1 **47.** 5 **49.** 28 **51.** −4 **53a.** No **b.** Yes

CHAPTER REVIEW EXERCISES (*pages 137–138*)

1. eight minus negative one (Objective 2.1B) **2.** −36 (Objective 2.1C) **3.** 200 (Objective 2.3A) **4.** −9
(Objective 2.3B) **5.** −14 (Objective 2.2A) **6.** 13 (Objective 2.1B) **7.** ◄─┼─┼─┼─┼─◆─┼─┼─┼─┼─┼─┼─►
                                  −6 −5 −4 −3 −2 −1 0 1 2 3 4 5 6
(Objective 2.1A) **8.** 4 (Objective 2.4A) **9.** 17 (Objective 2.3B) **10.** −210 (Objective 2.3B) **11.** −2
(Objective 2.2B) **12.** −18 (Objective 2.3A) **13.** −1 (Objective 2.2A) **14.** −72 (Objective 2.3A) **15.** −4
(Objective 2.5A) **16.** −2 (Objective 2.2B) **17.** 14 strokes (Objective 2.2B) **18.** 13 (Objective 2.2B)
19. Commutative Property of Multiplication (Objective 2.3A) **20.** Yes (Objective 2.2B) **21.** 14 (Objective 2.2B)
22. 0 (Objective 2.3B) **23.** −60 (Objective 2.3A) **24.** −12 (Objective 2.2A) **25.** 5 (Objective 2.5A)
26. −8 > −10 (Objective 2.1A) **27.** 21 (Objective 2.2A) **28.** 27 (Objective 2.1C) **29.** −8 (Objective 2.4B)
30. −12°C (Objective 2.1D) **31.** −245°C (Objective 2.3C) **32.** −3°C (Objective 2.2C) **33.** 12 (Objective 2.2C)

CUMULATIVE REVIEW EXERCISES (*pages 139–140*)

1. 5 (Objective 2.2B) **2.** 12,000 (Objective 1.3A) **3.** 3,209 (Objective 1.3C) **4.** 2 (Objective 1.5A) **5.** −82
(Objective 2.1C) **6.** 309,480 (Objective 1.1B) **7.** 2,400 (Objective 1.3A) **8.** 21 (Objective 2.3B) **9.** −11
(Objective 2.2B) **10.** −40 (Objective 2.2A) **11.** 1, 2, 4, 11, 22, 44 (Objective 1.3D) **12.** 1,936 (Objective 1.3B)
13. 630,000 (Objective 1.1C) **14.** 1,300 (Objective 1.2A) **15.** 9 (Objective 2.2B) **16.** −2,500 (Objective 2.3A)
17. $3 \cdot 23$ (Objective 1.3D) **18.** −16 (Objective 2.4A) **19.** −32 (Objective 2.5A) **20.** −4 (Objective 2.3B)
21. $435 (Objective 1.3A) **22.** −3 (Objective 2.3B) **23.** −62 < 26 (Objective 2.1A) **24.** 126 (Objective 2.3A)
25. −9 (Objective 2.4A) **26.** $2^5 \cdot 7^2$ (Objective 1.3B) **27.** 47 (Objective 1.5A) **28.** 10,062 (Objective 1.2A)
29. −26 (Objective 2.2B) **30.** 5,000 (Objective 1.2B) **31.** 2,025 (Objective 1.3B) **32.** 1,722,685 mi² (Objective 1.2C)
33. 76 years old (Objective 1.2B) **34.** $14,200 (Objective 1.2B) **35.** $92,250 (Objective 1.3E) **36.** −5°C (Objective 2.2C)
37. $24,900 (Objective 1.2C) **38.** −8 (Objective 2.2C)

ANSWERS TO CHAPTER 3 Exercises

3.1 EXERCISES (*pages 147–148*)

1. 8 **3.** 14 **5.** 30 **7.** 45 **9.** 48 **11.** 20 **13.** 42 **15.** 72 **17.** 120 **19.** 30 **21.** 24
23. 180 **25.** 90 **27.** 78 **29.** 3 **31.** 6 **33.** 14 **35.** 16 **37.** 1 **39.** 4 **41.** 4 **43.** 6
45. 12 **47.** 15 **49.** 2 **51.** 3 **53.** 21 **55.** 12 **57.** every 6 min **59.** 25 copies **61.** 12:20 P.M.;
12:20 P.M. **63.** $2x; x$

3.2 Exercises (pages 157–162)

1. $\frac{4}{5}$ 3. $\frac{1}{4}$ 5. $\frac{4}{3}$; $1\frac{1}{3}$ 7. $\frac{13}{5}$; $2\frac{3}{5}$ 9. $3\frac{1}{4}$ 11. 4 13. $2\frac{7}{10}$ 15. 7 17. $1\frac{8}{9}$ 19. $2\frac{2}{5}$ 21. 18

23. $2\frac{2}{15}$ 25. 1 27. $9\frac{1}{3}$ 29. $\frac{9}{4}$ 31. $\frac{11}{2}$ 33. $\frac{14}{5}$ 35. $\frac{47}{6}$ 37. $\frac{7}{1}$ 39. $\frac{33}{4}$ 41. $\frac{31}{3}$ 43. $\frac{55}{12}$

45. $\frac{8}{1}$ 47. $\frac{64}{5}$ 49. $\frac{6}{12}$ 51. $\frac{9}{24}$ 53. $\frac{6}{51}$ 55. $\frac{24}{32}$ 57. $\frac{108}{18}$ 59. $\frac{30}{90}$ 61. $\frac{14}{21}$ 63. $\frac{42}{49}$ 65. $\frac{8}{18}$

67. $\frac{28}{4}$ 69. $\frac{1}{4}$ 71. $\frac{3}{4}$ 73. $\frac{1}{6}$ 75. $\frac{8}{33}$ 77. 0 79. $\frac{7}{6}$ 81. 1 83. $\frac{3}{5}$ 85. $\frac{4}{15}$ 87. $\frac{3}{5}$

89. $\frac{2m}{3}$ 91. $\frac{y}{2}$ 93. $\frac{2a}{3}$ 95. c 97. $6k$ 99. $\frac{3}{8} < \frac{2}{5}$ 101. $\frac{3}{4} < \frac{7}{9}$ 103. $\frac{2}{3} > \frac{7}{11}$ 105. $\frac{17}{24} > \frac{11}{16}$

107. $\frac{7}{15} > \frac{5}{12}$ 109. $\frac{5}{9} > \frac{11}{21}$ 111. $\frac{7}{12} < \frac{13}{18}$ 113. $\frac{4}{5} > \frac{7}{9}$ 115. $\frac{9}{16} > \frac{5}{9}$ 117. $\frac{5}{8} < \frac{13}{20}$ 119. $\frac{1}{8}$ 121. $\frac{5}{6}$

123. $\frac{3}{4}$ 125. $\frac{1}{4}$ 127. more 129. $\frac{3}{7}$ 131a. 2-door sedan b. station wagon 133. $\frac{6}{13}$ 135. less

137a. $\frac{3}{40}$ b. $\frac{7}{80}$ 139. $m + (n - 1)$ 141a. 1996 b. 1997

3.3 Exercises (pages 173–180)

1. $\frac{9}{11}$ 3. 1 5. $1\frac{2}{3}$ 7. $1\frac{1}{6}$ 9. $\frac{16}{b}$ 11. $\frac{9}{c}$ 13. $\frac{11}{x}$ 15. $\frac{11}{12}$ 17. $\frac{11}{12}$ 19. $1\frac{7}{12}$ 21. $2\frac{2}{15}$

23. $-\frac{1}{12}$ 25. $-\frac{1}{3}$ 27. $\frac{11}{24}$ 29. $\frac{1}{12}$ 31. $15\frac{2}{3}$ 33. $5\frac{2}{3}$ 35. $15\frac{1}{20}$ 37. $10\frac{7}{36}$ 39. $7\frac{5}{12}$ 41. $-\frac{7}{18}$

43. $\frac{3}{4}$ 45. $-1\frac{1}{2}$ 47. $6\frac{5}{24}$ 49. $2\frac{5}{24}$ 51. $1\frac{2}{5}$ 53. $-\frac{1}{12}$ 55. $1\frac{13}{18}$ 57. $-\frac{19}{24}$ 59. $1\frac{5}{24}$ 61. $11\frac{2}{3}$

63. $14\frac{3}{4}$ 65. Yes 67. Yes 69. $\frac{13}{20}$ 71. $\frac{1}{6}$ 73. $\frac{1}{6}$ 75. $\frac{5}{d}$ 77. $-\frac{5}{n}$ 79. $\frac{1}{14}$ 81. $\frac{1}{2}$ 83. $\frac{1}{4}$

85. $-\frac{7}{8}$ 87. $-1\frac{1}{10}$ 89. $\frac{1}{4}$ 91. $\frac{13}{36}$ 93. $2\frac{1}{3}$ 95. $6\frac{3}{4}$ 97. $1\frac{1}{12}$ 99. $3\frac{3}{8}$ 101. $5\frac{1}{9}$ 103. $2\frac{3}{4}$

105. $1\frac{17}{24}$ 107. $4\frac{19}{24}$ 109. $1\frac{7}{10}$ 111. $-1\frac{13}{36}$ 113. $-\frac{5}{24}$ 115. $6\frac{5}{12}$ 117. $\frac{1}{3}$ 119. $-1\frac{1}{3}$ 121. $\frac{1}{12}$

123. $\frac{1}{6}$ 125. $1\frac{1}{9}$ 127. $4\frac{2}{5}$ 129. $2\frac{2}{9}$ 131. $4\frac{11}{12}$ 133. No 135. Yes 137. $\frac{1}{12}$ 139. $1\frac{3}{4}$ acres

141. $3\frac{1}{4}$ ft 143. $\frac{1}{16}$ mi; $\frac{5}{16}$ mi 145. $6\frac{3}{4}$ lb 147. $91 149. $\frac{7}{20}$; yes 151a. $11\frac{1}{2}$ b. 21 153. $3\frac{7}{8}$

155. $2\frac{3}{8}$ 157. No, because the parts are not equal in size.

3.4 Exercises (pages 191–198)

1. $\frac{3}{5}$ 3. $-\frac{11}{14}$ 5. $\frac{4}{5}$ 7. 0 9. $\frac{1}{10}$ 11. $-\frac{3}{8}$ 13. $\frac{63}{xy}$ 15. $-\frac{yz}{30}$ 17. $\frac{1}{9}$ 19. $-\frac{7}{30}$ 21. $\frac{3}{16}$

23. 1 25. 6 27. $-7\frac{1}{2}$ 29. $-3\frac{11}{15}$ 31. 0 33. $\frac{1}{2}$ 35. 19 37. $-2\frac{1}{3}$ 39. 1 41. $7\frac{7}{9}$

43. -30 45. 42 47. $5\frac{1}{2}$ 49. $\frac{7}{10}$ 51. $-\frac{1}{12}$ 53. $-\frac{1}{21}$ 55. $1\frac{4}{5}$ 57. $4\frac{1}{2}$ 59a. 70 more

b. 105 more 61. $-\frac{7}{48}$ 63. $3\frac{1}{2}$ 65. $-17\frac{1}{2}$ 67. -8 69. $\frac{1}{5}$ 71. $\frac{1}{6}$ 73. $-3\frac{2}{3}$ 75. Yes 77. No

79. No 81. $1\frac{11}{14}$ 83. -1 85. 0 87. $-\frac{2}{3}$ 89. $\frac{5}{6}$ 91. 0 93. 8 95. $-\frac{1}{8}$ 97. undefined

99. $-\frac{8}{9}$ 101. $\frac{1}{6}$ 103. $-\frac{32}{xy}$ 105. $\frac{bd}{30}$ 107. $5\frac{1}{3}$ 109. -8 111. $-\frac{6}{7}$ 113. $\frac{1}{2}$ 115. $5\frac{2}{7}$

117. -12 119. $1\frac{29}{31}$ 121. $1\frac{1}{5}$ 123. $-1\frac{1}{24}$ 125. $\frac{7}{26}$ 127. $-\frac{10}{11}$ 129. $\frac{7}{10}$ 131. $\frac{1}{12}$ 133. -48

135. undefined   **137.** $2\frac{1}{7}$ **139.** $\frac{4}{29}$ **141.** $-1\frac{3}{5}$ **143.** 2   **145.** 30 min   **147.** 930 min   **149.** $\frac{7}{8}$ c

151. 234 h   **153.** 30 houses   **155.** 14 in. by 7 in. by $1\frac{3}{4}$ in.   **157.** $3,318\frac{3}{4}$ **159.** $318   **161.** $21\frac{1}{4}$ lb/in^2

163. $2\frac{1}{2}$ mph   **165.** 1,250 mi   **167a.** sometimes true   **b.** sometimes true

3.5 EXERCISES (*pages 203–204*)

1. 36   **3.** -12 **5.** 25   **7.** -12 **9.** $\frac{1}{2}$ **11.** $\frac{7}{12}$ **13.** $\frac{3}{4}$ **15.** $-\frac{4}{5}$ **17.** $\frac{2}{3}$ **19.** $1\frac{1}{2}$ **21.** $-1\frac{1}{3}$

23. $-\frac{4}{9}$ **25.** $\frac{5}{6}$ **27.** $1\frac{1}{2}$ **29.** -3 **31.** $-\frac{2}{9}$ **33.** 1,600 square feet   **35.** 25 quarts   **37.** $1,500

39. 532 mi   **41.** a

3.6 EXERCISES (*pages 211–214*)

1. $\frac{9}{16}$ **3.** $-\frac{1}{216}$ **5.** $5\frac{1}{16}$ **7.** $\frac{5}{128}$ **9.** $\frac{4}{45}$ **11.** $-\frac{1}{10}$ **13.** $1\frac{1}{7}$ **15.** $-\frac{27}{49}$ **17.** $\frac{16}{81}$ **19.** $\frac{25}{144}$

21. $\frac{2}{3}$ **23.** $\frac{3}{4}$ **25.** $-\frac{8}{9}$ **27.** $\frac{1}{6}$ **29.** 6   **31.** $\frac{18}{35}$ **33.** $-\frac{1}{2}$ **35.** $1\frac{7}{25}$ **37.** $3\frac{3}{11}$ **39.** 17   **41.** $-\frac{4}{5}$

43. 1   **45.** No   **47.** $1\frac{1}{5}$ **49.** $\frac{5}{36}$ **51.** $\frac{11}{32}$ **53.** 1   **55.** 4   **57.** 0   **59.** $1\frac{3}{10}$ **61.** $1\frac{1}{9}$ **63.** $1\frac{15}{16}$

65. $\frac{1}{2}$ **67.** 1   **69.** No   **71.** 3 min   **73a.** x **b.** y

CHAPTER REVIEW EXERCISES (*pages 219–220*)

1. $9\frac{1}{2}$ (Objective 3.2A)   **2.** $2\frac{5}{6}$ (Objective 3.3B)   **3.** $1\frac{1}{2}$ (Objective 3.4B)   **4.** -1 (Objective 3.4A)   **5.** 2

(Objective 3.4B)   **6.** $2\frac{2}{3}$ (Objective 3.4A)   **7.** $2\frac{11}{12}$ (Objective 3.6B)   **8.** $\frac{3}{5} > \frac{7}{15}$ (Objective 3.2C)   **9.** 150

(Objective 3.1A)   **10.** $11\frac{13}{30}$ (Objective 3.3A)   **11.** $3\frac{1}{3}$ (Objective 3.4A)   **12.** $\frac{10}{7}$; $1\frac{3}{7}$ (Objective 3.2A)   **13.** $\frac{7}{8} > \frac{17}{20}$

(Objective 3.2C)   **14.** $\frac{3}{5}$ (Objective 3.6B)   **15.** $\frac{32}{72}$ (Objective 3.2B)   **16.** $-\frac{1}{3}$ (Objective 3.6A)   **17.** $\frac{2}{7}$ (Objective 3.6C)

18. 21 (Objective 3.1B)   **19.** $\frac{33}{14}$ (Objective 3.2A)   **20.** $\frac{3}{8}$ (Objective 3.3A)   **21.** $-\frac{5}{6}$ (Objective 3.4B)   **22.** $1\frac{3}{40}$

(Objective 3.6C)   **23.** -14 (Objective 3.4A)   **24.** $\frac{1}{18}$ (Objective 3.3B)   **25.** $1\frac{17}{24}$ (Objective 3.3B)   **26.** $2\frac{1}{4}$

(Objective 3.6A)   **27.** $9\frac{1}{12}$ (Objective 3.3A)   **28.** $\frac{2}{7}$ (Objective 3.2B)   **29.** $4\frac{7}{10}$ (Objective 3.3B)   **30.** $-\frac{13}{18}$

(Objective 3.5A)   **31.** $\frac{2}{3}$ (Objective 3.2D)   **32.** 40 copies (Objective 3.1C)   **33.** $6\frac{1}{4}$ lb (Objective 3.3C)

34. 192 units (Objective 3.4C)   **35.** $150 (Objective 3.4C)   **36.** 496 ft/s (Objective 3.4C)

CUMULATIVE REVIEW EXERCISES (*pages 221–222*)

1. 39 (Objective 1.5A)   **2.** $3\frac{1}{2}$ (Objective 3.4A)   **3.** $8\frac{11}{18}$ (Objective 3.3A)   **4.** -15 (Objective 2.2B)   **5.** 36

(Objective 3.1B)   **6.** 16 (Objective 3.4A)   **7.** $-1\frac{1}{9}$ (Objective 3.4B)   **8.** $-\frac{4}{15}$ (Objective 3.3B)   **9.** 9 (Objective 3.6B)

10. $\frac{7}{11} < \frac{4}{5}$ (Objective 3.2C)   **11.** $-1\frac{22}{27}$ (Objective 3.4B)   **12.** $\frac{1}{15}$ (Objective 3.4A)   **13.** 2 (Objective 3.4A)

14. $7\dfrac{1}{28}$ (Objective 3.3B)   **15.** $\dfrac{23}{24}$ (Objective 3.3B)   **16.** $1\dfrac{7}{12}$ (Objective 3.6C)   **17.** $1\dfrac{5}{8}$ (Objective 3.3B)   **18.** $6\dfrac{3}{16}$

(Objective 3.3A)   **19.** -4 (Objective 2.4A)   **20.** $4\dfrac{5}{9}$ (Objective 3.2A)   **21.** $\dfrac{1}{7}$ (Objective 3.3B)   **22.** $\dfrac{3}{28}$

(Objective 3.6A)   **23.** -21 (Objective 2.5A)   **24.** 11,272 (Objective 1.2A)   **25.** 48 (Objective 1.5A)   **26.** $-\dfrac{11}{20}$

(Objective 3.5A)   **27.** 20,000 (Objective 1.2B)   **28.** -13 (Objective 2.2B)   **29.** $\dfrac{31}{4}$ (Objective 3.2A)   **30.** $2^2 \cdot 5 \cdot 7$

(Objective 1.3D)   **31.** 40 calories (Objective 1.3E)   **32.** 330,000 people (Objective 1.2C)   **33.** $\$10\dfrac{7}{8}$ (Objective 3.3C)

34. $4\dfrac{1}{8}$ mi (Objective 3.4C)   **35.** 45 mph (Objective 1.4B)   **36.** $22\dfrac{3}{8}$ lb/in² (Objective 3.4C)

ANSWERS TO CHAPTER 4 Exercises

4.1 EXERCISES (*pages 231–234*)

1. thousandths   **3.** ten-thousandths   **5.** hundredths   **7.** 0.3   **9.** 0.21   **11.** 0.461   **13.** 0.093
15. $\dfrac{1}{10}$ **17.** $\dfrac{47}{100}$ **19.** $\dfrac{289}{1,000}$ **21.** $\dfrac{9}{100}$ **23.** thirty-seven hundredths   **25.** nine and four tenths
27. fifty-three ten-thousands   **29.** forty-five thousandths   **31.** twenty-six and four hundredths   **33.** 3.0806
35. 407.03   **37.** 246.024   **39.** 73.02684   **41.** $0.7 > 0.56$ **43.** $3.605 > 3.065$ **45.** $9.004 < 9.04$
47. $9.31 > 9.031$ **49.** $4.6 < 40.6$ **51.** $0.07046 > 0.07036$ **53.** 0.609, 0.66, 0.696, 0.699   **55.** 1.237, 1.327,
1.372, 1.732   **57.** 21.78, 21.805, 21.87, 21.875   **59.** 5.4   **61.** 30.0   **63.** 413.60   **65.** 6.062   **67.** 97
69. 5,440   **71.** 0.0236   **73.** 0.18 oz   **75.** 26.2 mi   **77.** Italy   **79.** $6,824   **81a.** $2.40   **b.** $3.60
c. $6.00   **d.** $7.00   **e.** $4.70   **f.** $2.40   **g.** $2.40   **83.** For example: **a.** 0.15   **b.** 1.05   **c.** 0.001

4.2 EXERCISES (*pages 251–262*)

1. 65.9421   **3.** 190.857   **5.** 21.26   **7.** 21.26   **9.** 2.768   **11.** -50.7 **13.** -3.312 **15.** -5.905
17. -16.35 **19.** -9.55 **21.** -19.189 **23.** 56.361   **25.** 53.67   **27.** -98.38 **29.** -649.36 **31.** 31.09
33. 12; 12.325   **35.** 40; 33.63   **37.** 0.3; 0.303   **39.** 40; 38.618   **41a.** $3.51 billion   **b.** $8.67 billion
43. -1.159 **45.** -25.665 **47.** 13.535   **49.** 28.3925   **51.** 10.737   **53.** -27.553 **55.** -1.412 **57.** Yes
59. Yes   **61.** 1.70   **63.** 0.03316   **65.** 15.12   **67.** -5.46 **69.** -0.00786 **71.** -473 **73.** 4,250
75. 67,100   **77.** 0.036   **79.** 8.0; 7.5537   **81.** 70; 68.5936   **83.** 30; 32.1485   **85.** 9,651.3 pounds   **87.** 50.16
89. -48 **91.** -0.08338 **93.** 23.0867   **95.** Yes   **97.** No   **99.** 32.3   **101.** -67.7 **103.** 4.14
105. -6.1 **107.** 6.3   **109.** 5.8   **111.** 0.81   **113.** -0.08 **115.** 5.278   **117.** 0.4805   **119.** -25.4
121. -0.5 **123.** 10; 11.17   **125.** 1; 1.16   **127.** 50; 58.90   **129.** 6; 7.20   **131.** 2.3 times greater   **133.** 5.06
135. 0.24   **137.** 2.06   **139.** 6.1   **141.** Yes   **143.** No   **145.** 0.375   **147.** $0.\overline{72}$ **149.** $0.58\overline{3}$

151. 1.75   **153.** 1.5   **155.** $4.1\overline{6}$ **157.** 2.25   **159.** $3.\overline{8}$ **161.** $\dfrac{1}{5}$ **163.** $\dfrac{3}{4}$ **165.** $\dfrac{1}{8}$ **167.** $2\dfrac{1}{2}$

169. $4\dfrac{11}{20}$ **171.** $1\dfrac{18}{25}$ **173.** $\dfrac{9}{200}$ **175.** $\dfrac{9}{10} > 0.89$ **177.** $\dfrac{4}{5} < 0.803$ **179.** $0.444 < \dfrac{4}{9}$ **181.** $0.13 > \dfrac{3}{25}$

183. $\dfrac{5}{16} > 0.312$ **185.** $\dfrac{10}{11} > 0.909$ **187.** $3,468.25   **189.** 32.22°C   **191.** $.28   **193.** $3.42   **195.** $24
197. $840.06   **199.** $473.72   **201.** $416.50   **203.** $50   **205a.** Yes   **b.** Females; 6.8 years   **c.** 1970
207. 5,256 lb   **209.** $578.62   **211.** $562.20   **213.** $.48   **215.** -41.65 newtons   **217.** $57,146.75   **219.** $.54
221. 0.098   **223a.** $(1.1)^3 > 1.31$ **b.** $(0.9)^3 < 1^5$ **c.** $(1.2)^3 > (0.8)^3$ **225.** 31¢

4.3 EXERCISES (*pages 265–266*)

1. 4.49   **3.** 5.7   **5.** -8.03 **7.** -1.2 **9.** 0.144   **11.** -0.3 **13.** -0.01 **15.** -1.86 **17.** -0.21
19. -2.5 **21.** -2.005 **23.** 16.38   **25.** $8.73   **27.** 100.8 ft/s   **29.** $.00125   **31.** $256.45   **33.** 0.2 lb
35. Answers will vary. For example: **a.** $x - 0.04 = -1$ **b.** $-3x = -6.3$

4.4 EXERCISES (*pages 273–276*)

1. 6 **3.** −3 **5.** 13 **7.** 15 **9.** −5 **11.** −10 **13.** 5 **15.** 10 **17.** 9 **19.** 27 **21.** −14

23. 16 **25.** 13 **27.** −6 **29.** 32 **31.** $\frac{1}{10}$ **33.** $\frac{3}{4}$ **35.** $\frac{5}{8}$ **37.** −24 **39.** 40 **41.** 23 **43.** 5

45. 5 **47.** 8 **49.** 1 **51.** −36 **53.** 1.7321 **55.** 3.1623 **57.** 4.8990 **59.** 11.2250 **61.** −5.6569
63. −43.8178 **65.** 4 and 5 **67.** 5 and 6 **69.** 7 and 8 **71.** 11 and 12 **73.** $2\sqrt{2}$ **75.** $3\sqrt{5}$ **77.** $2\sqrt{5}$
79. $3\sqrt{3}$ **81.** $4\sqrt{3}$ **83.** $5\sqrt{3}$ **85.** $3\sqrt{7}$ **87.** $7\sqrt{2}$ **89.** $4\sqrt{7}$ **91.** $5\sqrt{7}$ **93.** 30 ft/s **95.** 3 s

97. 4,000 mi **99.** 3, 4, 5, 6, 7, 8, 9 **101.** $\sqrt{\frac{1}{5}+\frac{1}{6}}, \sqrt{\frac{1}{4}+\frac{1}{8}}, \sqrt{\frac{1}{3}+\frac{1}{9}}$ **103a.** 36 **b.** 4 and 14

4.5 EXERCISES (*pages 283–286*)

1. **3.** **5.**

7. **9.** **11.** **13.**

15. **17.** **19.** **21.**

23. **25.** $\sqrt{101}$ **27.** −2, 0.4, $\sqrt{17}$ **29.** all real numbers less than 3 **31.** all real

numbers greater than or equal to −1 **33.** **35.**

37. **39.** **41.** $s \geq 50{,}000$; no **43.** $h \leq 9$; yes **45.** $b \leq 1{,}200$; yes

47. $T > 50$; no **49a.** integer, negative integer, rational number, real number **b.** whole number, integer, positive integer, rational number, real number **c.** rational number, real number **d.** rational number, real number **e.** rational number, real number **f.** irrational number, real number **51a.** −2.5, 0 **b.** −6.3, −3, 0, 6.7 **c.** 4, 13.6 **d.** −4.9, 0, 2.1, 5 **53a.** always true **b.** always true **c.** sometimes true

CHAPTER REVIEW EXERCISES (*pages 291–292*)

1. 20.5670 (Objective 4.4B) **2.** 91,800 (Objective 4.2B) **3.** −11 (Objective 4.4A) **4.** −8.301 (Objective 4.2A)
5. 89.243 (Objective 4.2A) **6.** 5.034 (Objective 4.1A) **7.** −4 (Objective 4.4A) **8.** 0.0142 (Objective 4.2C)
9. −0.34 (Objective 4.3A) **10.** 8.039 < 8.31 (Objective 4.1B) **11.** 0.11 (Objective 4.2C) **12.** 2.4622 (Objective 4.2B)

13. −1, −0.5, $\sqrt{10}$ (Objective 4.5B) **14.** $\frac{3}{7} < 0.429$ (Objective 4.2D) **15.** $\frac{7}{25}$ (Objective 4.2D) **16.** −0.1

(Objective 4.2C) **17.** 7.3 h (Objective 4.2A) **18.** (Objective 4.5A) **19.**

(Objective 4.5B) **20.** −441.2 (Objective 4.2A) **21.** 6.143 (Objective 4.2C) **22.** 50.743 (Objective 4.2A) **23.** $3\sqrt{10}$
(Objective 4.4B) **24.** −1,110 (Objective 4.2B) **25.** 440 (Objective 4.2A) **26.** $G \geq 3.5$; No (Objective 4.5C)
27. 395.45°C (Objective 4.2E) **28a.** $2.72 trillion **b.** 1.5 times (Objective 4.2E) **29.** $.84 (Objective 4.2E)
30. $207.87 (Objective 4.3B) **31.** $499.49 (Objective 4.2E) **32.** 40 ft/s (Objective 4.4C)

CUMULATIVE REVIEW EXERCISES (*pages 293–294*)

1. 0.03879 (Objective 4.2C) **2.** 11 (Objective 2.5A) **3.** 20 (Objective 4.3A) **4.** 8,072,092 (Objective 1.1B)

5. (Objective 4.5A) **6.** (Objective 4.5B) **7.** −4 (Objective 2.2B)

8. 1,900 (Objective 1.2A) **9.** $8\sqrt{3}$ (Objective 4.4B) **10.** $1\frac{1}{2}$ (Objective 3.4B) **11.** −18.42 (Objective 4.2A) **12.** $\frac{1}{7}$

(Objective 3.4A) **13.** 1,600 (Objective 1.3B) **14.** $2^2 \cdot 5 \cdot 13$ (Objective 1.3D) **15.** 0.76 (Objective 4.2D)
16. 95.3939 (Objective 4.4B) **17a.** Finland **b.** 3 times (Objective 1.1D/1.3E) **18.** undefined (Objective 2.3B)
19. $-\frac{11}{21}$ (Objective 3.3A) **20.** 11 (Objective 4.4A) **21.** 30 (Objective 4.2B) **22.** 17 (Objective 2.5A)

23. $\dfrac{3}{10}$ (Objective 3.6B)   **24.** $\dfrac{1}{24}$ (Objective 3.3B)   **25.** 2.8 (Objective 4.2C)   **26.** \$50.24 (Objective 4.2E)
27. 46.62°C (Objective 4.2E)   **28.** \$1.56 (Objective 4.2E)   **29a.** 46.5 h   **b.** face-to-face selling (Objective 4.2E)
30. 30 mph (Objective 4.4C)

ANSWERS TO CHAPTER 5 Exercises

5.1 EXERCISES (*pages 303–306*)

1. Associative Property of Multiplication   **3.** Commutative Property of Addition   **5.** Inverse Property of Addition
7. Inverse Property of Multiplication   **9.** Commutative Property of Multiplication   **11a.** Associative Property of
Multiplication   **b.** Inverse Property of Multiplication   **c.** Multiplication Property of One   **13.** $(x + 4) + y$ **15.** $\dfrac{1}{5}$
17. 0   **19.** $7y$ **21.** $-\dfrac{3}{2}$ **23.** $12x$ **25.** $-15x$ **27.** $21t$ **29.** $-21p$ **31.** $12q$ **33.** $2x$ **35.** $-15w$
37. x **39.** $6x^2$ **41.** $-27x^2$ **43.** x^2 **45.** x **47.** c **49.** a **51.** $12w$ **53.** $16vw$ **55.** $-28bc$
57. 0   **59.** 0   **61.** 9   **63.** 7   **65.** -15 **67.** $-5y$ **69.** $13b$ **71.** $10z + 4$ **73.** $12y + 30z$
75. $21x - 27$ **77.** $-2x + 7$ **79.** $4x + 9$ **81.** $-5y - 15$ **83.** $-12x + 18$ **85.** $-20n + 40$ **87.** $48z - 24$
89. $24p + 42$ **91.** $10a + 15b + 5$ **93.** $12x - 4y - 4$ **95.** $36m - 9n + 18$ **97.** $12v - 18w - 42$ **99.** $20x + 4$
101. $20a - 25b + 5c$ **103.** $-18p + 12r + 54$ **105.** $-5a + 9b - 7$ **107.** $-11p + 2q + r$ **109.** No. Zero
111. No. Zero

5.2 EXERCISES (*pages 311–314*)

1. $3x^2, 4x, \underline{-9}$ **3.** $b, \underline{5}$ **5.** $9a^2, -12a, 4b^2$ **7.** $3x^2$ **9.** $1, -6$ **11.** $12, 4$ **13.** $16a$ **15.** $27x$ **17.** $3z$
19. $8x$ **21.** $-7z$ **23.** $-6w$ **25.** $\underline{0}$ **27.** s **29.** $6x - 3y$ **31.** $2r + 13p$ **33.** $-3w + 2v$
35. $-9p + 11$ **37.** $2p$ **39.** 6   **41.** $13y^2 + 1$ **43.** $12w^2 - 16$ **45.** $-14w$ **47.** $5a^2b + 8ab^2$ **49.** 5
51. $11x^2 - 2x$ **53.** $8b^2 - 2b$ **55.** $7x + 2$ **57.** $3n + 3$ **59.** $4a + 4$ **61.** $4a + 1$ **63.** $8x + 42$
65. $-12x + 28$ **67.** $-18m - 52$ **69.** $20c + 23$ **71.** $8a + 5b$ **73.** $15z - 12$ **75.** -19 **77.** $-13x - 2y$
79. $-2v + 13$ **81.** $-5c - 6$ **83.** $2a + 21$ **85.** $11n - 26$ **87.** $-9x + 6$ **89.** $111v - 246$ **91.** $-3r - 24$
93. $27z^2 - 24z - 90$ **95.** $6 \cdot 527 = 6(500 + 20 + 7) = 6 \cdot 500 + 6 \cdot 20 + 6 \cdot 7 = 3{,}000 + 120 + 42 = 3{,}162$

5.3 EXERCISES (*pages 317–318*)

1. $11y^2 - 4y + 2$ **3.** $3w^3 + 13w^2 - 8w - 5$ **5.** $-2a^3 - 9a^2 - 8a + 1$ **7.** $11k^2 + 2k - 18$
9. $17x^3 - 9x^2 + 9x - 5$ **11.** $16b^3 + 14b^2 + 1$ **13.** $13t^2 - 35$ **15.** $2x^2 - 9x - 2$ **17.** $11b^3 - 8b^2 + 12b + 14$
19. $-2z^3 - 8z^2 + 21$ **21.** $-12r^3 + 9r^2 + 16r$ **23.** $2a^2 + 12a - 2$ **25.** $3z^3 + 4z^2 + 8z + 10$ **27.** $3n^3 - 8n$
29. $2b^2 - 12b + 18$ **31a.** 8   **b.** 9   **c.** 4

5.4 EXERCISES (*pages 323–324*)

1. a^9 **3.** z^8 **5.** a^8b^3 **7.** $-m^9n^3$ **9.** $10x^7$ **11.** $8x^3y^6$ **13.** $-12m^7$ **15.** $-14v^3w$ **17.** $-2ab^5c^5$
19. $24a^3b^5c^2$ **21.** $40r^3t^7v$ **23.** $-27m^2n^4p^3$ **25.** $24x^7$ **27.** $6a^6b^5$ **29.** $-15x^3y^8$ **31.** $48r^4t^8v^4$
33. $-60a^3b^5c^4$ **35.** $-8a^6b^{10}$ **37.** b^8 **39.** p^{28} **41.** $8y^3$ **43.** $m^{12}n^6$ **45.** $a^{10}b^5$ **47.** z^6 **49.** $27n^9$
51. $9b^6$ **53.** $64a^{12}b^{15}$ **55.** $16x^2y^6z^4$ **57.** $(2^3)^2 = 2^6 = 64,\ 2^{(3^2)} = 2^9 = 512,\ 2^{(3^2)} > (2^3)^2$

5.5 EXERCISES (*pages 327–328*)

1. $x^3 - 3x^2 - 4x$ **3.** $8a^3 + 12a^2 - 24a$ **5.** $-6a^3 - 18a^2 + 14a$ **7.** $4m^4 - 9m^3$ **9.** $10x^5 - 12x^4y + 4x^3y^2$
11. $-6r^7 + 12r^6 + 36r^5$ **13.** $12a^4 + 24a^3 - 28a^2$ **15.** $-6n^2 + 8n^5 + 10n^7$ **17.** $3a^3b^2 - 4a^2b^3 + ab^4$
19. $-4x^7y^5 + 5x^5y^4 + 7x^3y^3$ **21.** $6r^2t^3 - 6r^3t^4 - 6r^5t^6$ **23.** $36q^2 - 28q$ **25.** $y^2 + 12y + 27$ **27.** $x^2 + 11x + 30$
29. $a^2 - 11a + 24$ **31.** $10z^2 + 9z + 2$ **33.** $40c^2 - 11c - 21$ **35.** $10v^2 - 11v + 3$ **37.** $35t^2 + 18t - 8$
39. $24x^2 - x - 10$ **41.** $25r^2 - 4$ **43.** $21y^2 - 41y - 40$ **45.** 0   **47.** $x^3 + 7x^2 + 17x + 20$

5.6 EXERCISES (*pages 333–334*)

1. 1 **3.** −1 **5.** $\frac{1}{9}$ **7.** $\frac{1}{8}$ **9.** $\frac{1}{x^5}$ **11.** $\frac{1}{w^8}$ **13.** a^5 **15.** b^3 **17.** a^6 **19.** q^4 **21.** mn^2

23. t^2u^3 **25.** $\frac{1}{x^5}$ **27.** $\frac{1}{b^4}$ **29.** 2.37×10^6 **31.** 4.5×10^{-4} **33.** 3.09×10^5 **35.** 6.01×10^{-7}

37. 5.7×10^{10} **39.** 1.7×10^{-8} **41.** 710,000 **43.** 0.000043 **45.** 671,000,000 **47.** 0.00000713
49. 5,000,000,000,000 **51.** 0.00801 **53.** 1.6×10^{10} **55.** 1.6×10^{-19} **57.** 1×10^{-9} s **59a.** > **b.** >
c. > **d.** <

5.7 EXERCISES (*pages 339–342*)

1. $t + 3$ **3.** $6m - 5$ **5.** $3b - 7$ **7.** $7n$ **9.** $2(3 + w)$ **11.** $4(2r - 5)$ **13.** $\frac{v}{v - 4}$ **15.** $4t^2$

17. $m^2 + m^3$ **19.** $(31 - s) + 5$ **21.** -12 **23.** $\frac{7}{24}x$ **25.** $14x + 12$ **27.** $14x$ **29.** $9x + 63$ **31.** $x + 12$

33. $7x - 28$ **35.** $7x$ **37.** $3x - 8$ **39.** $7x - 98$ **41.** $8x + 80$ **43.** $8x - 8$ **45.** $2x + 35$ **47.** $1 - x$

49. $16x$ **51.** $5x + 6$ **53.** $45 - 5y$ **55.** $42 - 3m$ **57.** $390d$ **59.** $H + 4,430$ **61.** $3A$ **63.** $\frac{3}{4}c$

65. $3 - L$ **67.** $12 - L$ **69.** $2x$ **71.** $\frac{4}{7}n$

CHAPTER REVIEW EXERCISES (*pages 347–348*)

1. $2x - 16$ (Objective 5.2A) **2.** $-18z - 2$ (Objective 5.1B) **3.** $10z^2 - z - 15$ (Objective 5.3A) **4.** $-8m^5n^2$
(Objective 5.4A) **5.** $\frac{1}{243}$ (Objective 5.6A) **6.** $-\frac{3}{7}$ (Objective 5.1A) **7.** x (Objective 5.1A) **8.** $-4s + 43t$
(Objective 5.2B) **9.** $15x^3y^7$ (Objective 5.4A) **10.** $21a^2 - 10a - 24$ (Objective 5.5B) **11.** $-3b^3 + 4b - 18$
(Objective 5.3B) **12.** $32z^{20}$ (Objective 5.4B) **13.** $6w$ (Objective 5.1A) **14.** $-15x^3yz^3 + 30xy^2z^4 - 5x^4y^5z^2$
(Objective 5.5A) **15.** $-\frac{4}{9}$ (Objective 5.1A) **16.** $-12c + 32$ (Objective 5.1B) **17.** $-2m + 16$ (Objective 5.2A)

18. $-12a^5b^{15}$ (Objective 5.4A) **19.** The Distributive Property (Objective 5.1B) **20.** p^6q^9 (Objective 5.4B) **21.** $\frac{1}{a^7}$
(Objective 5.6A) **22.** 3.97×10^{-5} (Objective 5.6B) **23.** The Commutative Property of Addition (Objective 5.1A)
24. $3y^3 + 8y^2 + 8y - 19$ (Objective 5.3A) **25.** $12c - 4d$ (Objective 5.2B) **26.** $14m - 42$ (Objective 5.1B) **27.** x^2y^4
(Objective 5.6A) **28.** $-5a^2 + 3a + 9$ (Objective 5.2A) **29.** $12p^2 - 15p - 63$ (Objective 5.5B)

30. $-8a^5b + 10a^3b^3 - 6a^2b^5$ (Objective 5.5A) **31.** $3x - 4y$ (Objective 5.2A) **32.** $-a + 13b$ (Objective 5.2B) **33.** $\frac{1}{c^5}$
(Objective 5.6A) **34.** $6x^3 - 10x$ (Objective 5.3B) **35.** 240,000 (Objective 5.6B) **36.** $6z^2 - 6z$ (Objective 5.2A)

37. $\frac{4x}{7} - 9$ (Objective 5.7A) **38.** $5x - 14$ (Objective 5.7B) **39.** 6.023×10^{23} (Objective 5.6B) **40.** $30 - p$
(Objective 5.7C)

CUMULATIVE REVIEW EXERCISES (*pages 349–350*)

1. -12.4 (Objective 4.2C) **2.** $14v - 2$ (Objective 5.2A) **3.** $6x^2 + 2x - 20$ (Objective 5.5B) **4.** $\frac{3}{8}$ (Objective 3.3B)

5. 24 (Objective 4.4A) **6.** (Objective 4.5A) **7.** x^7 (Objective 5.6A) **8.** -9 (Objective 2.4A)

9. 8.4×10^{-7} (Objective 5.6B) **10.** $9x^2 - 2x - 4$ (Objective 5.3A) **11.** -35 (Objective 4.4A) **12.** $\frac{11}{20}$ (Objective 3.6B)

13. $-12a^7b^9$ (Objective 5.4A) **14.** $\frac{1}{x^2}$ (Objective 5.6A) **15.** $\frac{2}{5}$ (Objective 3.6A) **16.** $-48p$ (Objective 5.1A)

17. 200 (Objective 4.2A) **18.** $-12a^3b^3 - 15a^2b^3 + 6a^2b^4$ (Objective 5.5A) **19.** $18x$ (Objective 5.2B)

20. −7 (Objective 2.3B)   **21.** $\dfrac{9}{16}$ (Objective 4.2D)   **22.** 4 (Objective 2.5A)   **23.** $10\sqrt{3}$ (Objective 4.4B)

24. $5y^2 - 2y - 5$ (Objective 5.3B)   **25.** 1 (Objective 3.4A)   **26.** −1 (Objective 5.6A)   **27.** $32a^{20}b^{15}$ (Objective 5.4B)

28. 90 (Objective 2.5A)   **29.** $2\dfrac{2}{5}$ (Objective 3.4A)   **30.** 0.0000623 (Objective 5.6B)   **31.** $\dfrac{10}{x-9}$ (Objective 5.7A)

32. $2x + 6$ (Objective 5.7B)   **33.** 30.78 in. (Objective 4.2E)   **34.** 1.1 million units (Objective 4.2E)   **35.** $30d$ (Objective 5.7C)   **36.** $3,075 (Objective 3.4C)

ANSWERS TO CHAPTER 6   Exercises

6.1 EXERCISES (*pages 359–360*)

1. 6   **3.** 9   **5.** 17   **7.** 1   **9.** −7   **11.** −9   **13.** 5   **15.** 9   **17.** 0   **19.** −8   **21.** −6   **23.** 12
25. 6   **27.** −12   **29.** $\dfrac{2}{7}$ **31.** $\dfrac{1}{2}$ **33.** $\dfrac{1}{12}$ **35.** $\dfrac{5}{8}$ **37.** 3   **39.** −3   **41.** −8   **43.** 7   **45.** 0
47. 6   **49.** −4   **51.** 4   **53.** $\dfrac{5}{2}$ **55.** $-\dfrac{7}{2}$ **57.** $-\dfrac{7}{3}$ **59.** $\dfrac{26}{9}$ **61.** 6   **63.** −36   **65.** −28   **67.** 12
69. 10   **71.** $\dfrac{7}{10}$ **73.** −5   **75.** −7   **77a.** $x = b - a$; Yes   **b.** $x = \dfrac{b}{a}$; No, $a \neq 0$

6.2 EXERCISES (*pages 363–366*)

1. 2   **3.** 10   **5.** 2   **7.** −1   **9.** −2   **11.** −4   **13.** 3   **15.** 9   **17.** 4   **19.** $\dfrac{3}{5}$ **21.** $\dfrac{7}{2}$ **23.** $\dfrac{3}{4}$
25. $\dfrac{3}{2}$ **27.** $\dfrac{7}{4}$ **29.** $-\dfrac{5}{2}$ **31.** $\dfrac{9}{2}$ **33.** 0   **35.** $-\dfrac{7}{6}$ **37.** $\dfrac{1}{3}$ **39.** 1   **41.** 10   **43.** 28   **45.** −45
47. −12   **49.** 10   **51.** $\dfrac{24}{7}$ **53.** $\dfrac{5}{6}$ **55.** $\dfrac{7}{5}$ **57.** $\dfrac{9}{4}$ **59.** 1.1   **61.** 5.8   **63.** 4   **65.** 1   **67.** $\dfrac{11}{3}$
69. 5   **71.** 2.8 years   **73.** $15,655.11   **75.** 1987   **77.** 136 ft   **79.** For example, $2x + 5 = -1$.   **81.** No; it is not an equation.

6.3 EXERCISES (*pages 371–374*)

1. 3   **3.** 3   **5.** −1   **7.** −4   **9.** 5   **11.** 2   **13.** 1   **15.** $\dfrac{7}{6}$ **17.** 3   **19.** $\dfrac{8}{5}$ **21.** $\dfrac{7}{2}$ **23.** $\dfrac{3}{2}$
25. $\dfrac{7}{2}$ **27.** 3   **29.** −1   **31.** 0   **33.** $\dfrac{5}{6}$ **35.** −3   **37.** 3   **39.** 2   **41.** 2   **43.** 2   **45.** 3   **47.** $\dfrac{1}{2}$
49. 8   **51.** −2   **53.** $-\dfrac{6}{5}$ **55.** $-\dfrac{1}{3}$ **57.** 6 ft   **59.** 34.6 lb   **61.** 325 units   **63.** 400 units   **65.** 108

6.4 EXERCISES (*pages 379–382*)

1. $x + 12 = 20$; 8   **3.** $\dfrac{3}{5}x = -30$; −50   **5.** $3x + 4 = 13$; 3   **7.** $9x - 6 = 12$; 2   **9.** $11x - 8 = -19$; −1
11. $4x - 15 = 6(x - 11)$; $\dfrac{51}{2}$ **13.** $2(3x + 8) + 6 = -2$; −4   **15.** $2x = 21 - x + 3$; 8 and 13   **17.** $23 - x = 2x + 5$; 6 and 17   **19.** $16,875   **21.** 350,000 bytes   **23.** $3,600   **25.** $30 million   **27.** 37 h   **29.** 3 ft and 9 ft
31. $5,000   **33.** 2 lb of Columbian; 3 lb of French Roast; 5 lb of Java   **35.** Identity   **37.** Identity
39. Contradiction

6.5 EXERCISES (*pages 389–394*)

1.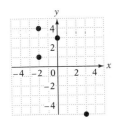

3.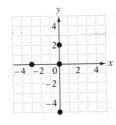

5.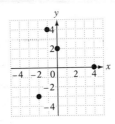

7. $A(2, 3)$; $B(4, 0)$; $C(-4, 1)$; $D(-2, -2)$

9. $A(-2, 5)$; $B(3, 4)$; $C(0, 0)$; $D(-3, -2)$

11a. Abscissa of A: 2; abscissa of C: -4 **b.** Ordinate of B: 1; ordinate of D: -3

17. Yes **19.** No **21.** No **23.** Yes

13.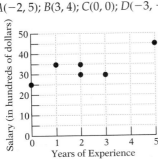

15.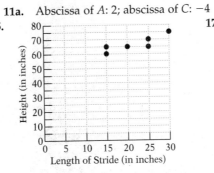

25. No **27.** $(3, 7)$ **29.** $(6, 3)$ **31.** $(0, 1)$ **33.** $(-5, 0)$

35.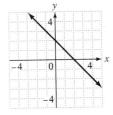

37.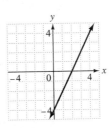

39.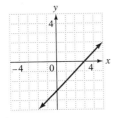

41.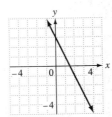

43.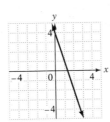

45.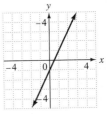

47.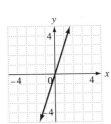

49.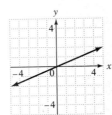

51.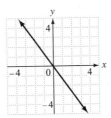

53.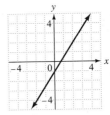

55.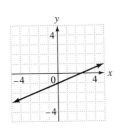

57.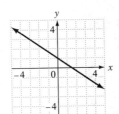

59.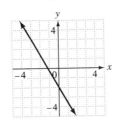

61.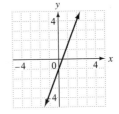

63. 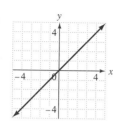 **65.**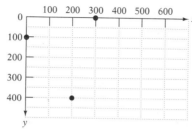

67a. *y* increases by 3 **b.** *y* decreases by 2

CHAPTER REVIEW EXERCISES (*pages 397–398*)

1. −3 (Objective 6.1A) **2.** 3 (Objective 6.1B) **3.** $\frac{3}{2}$ (Objective 6.2A) **4.** −7 (Objective 6.1A) **5.** −24

(Objective 6.1B) **6.** $-\frac{15}{32}$ (Objective 6.1B) **7.** 2 (Objective 6.2A) **8.** $\frac{17}{3}$ (Objective 6.3B) **9.** $\frac{1}{2}$ (Objective 6.3A)

10. −2 (Objective 6.3A) **11.** 3 (Objective 6.3B) **12.** −4 (Objective 6.1B) **13.** $-\frac{10}{9}$ (Objective 6.3B) **14.** 4

(Objective 6.3B) **15.** −4 (Objective 6.2A) **16.** Yes (Objective 6.5C) **17.** (Objective 6.5A)

18. 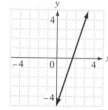 (Objective 6.5D) **19.** 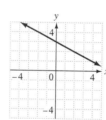 (Objective 6.5D) **20.** (2, −1) (Objective 6.5C)

21. $7 − 5x = 37$; −6 (Objective 6.4A) **22.** 16 in. (Objective 6.4B) **23.** 7 h (Objective 6.4B) **24.** 55 m (Objective 6.4B)
25. 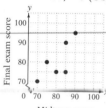 (Objective 6.5B) **26.** 12.5 lb (Objective 6.3C) **27.** 147 amplifiers (Objective 6.2B)

CUMULATIVE REVIEW EXERCISES (*pages 399–400*)

1. 18 (Objective 2.3A) **2.** $−12p + 21$ (Objective 5.1B) **3.** 0 (Objective 3.6C) **4.** −18 (Objective 6.1B) **5.** 8
(Objective 2.5A) **6.** −60 (Objective 2.5A) **7.** 11 (Objective 4.4A) **8.** $4\sqrt{3}$ (Objective 4.4B) **9.** $2v + 7$
(Objective 5.2B) **10.** $12m$ (Objective 5.1A) **11.** No (Objective 2.3A) **12.** 1 (Objective 6.3A) **13.** $−10z + 12$
(Objective 5.2B) **14.** $\frac{5}{4}$ (Objective 3.6C) **15.** $\frac{3}{2}$ (Objective 6.2A) **16.** $32m^{10}n^{25}$ (Objective 5.4B)

17. $−6a^5 − 9a^4b + 12a^3b^2$ (Objective 5.5A) **18.** $6x^2 − 7x − 3$ (Objective 5.5B) **19.** $\frac{1}{16}$ (Objective 5.6A)

20. x^6 (Objective 5.6A) **21.** $15x^8y^3$ (Objective 5.4A) **22.** $\frac{31}{8}$ (Objective 6.3B)

23. 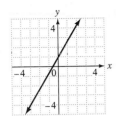 (Objective 6.5D)   **24.** 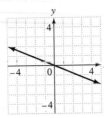 (Objective 6.5D)   **25.** 0.000000035 (Objective 5.6B)

26. $5(n + 2)$; $5n + 10$ (Objective 5.7B)   **27.** 1.5 s (Objective 6.2B)   **28.** $\frac{1}{2}x$ (Objective 5.7C)   **29.** $43.76 million

(Objective 4.2A)   **30.** $576.50 (Objective 6.4B)   **31.** 1,929 stories (Objective 4.2E)   **32.** $4,000 and $8,000 (Objective 6.4B)

ANSWERS TO CHAPTER 7 Exercises

7.1 EXERCISES (*pages 409–412*)

1. $\frac{2}{3}$, 2:3, 2 TO 3   **3.** $\frac{3}{8}$, 3:8, 3 TO 8   **5.** $\frac{9}{2}$, 9:2, 9 TO 2   **7.** $\frac{1}{2}$, 1:2, 1 TO 2   **9.** $\frac{2}{3}$ **11.** $\frac{9}{13}$ **13.** $\frac{25 \text{ mi}}{1 \text{ h}}$

15. $\frac{\$1.64}{3 \text{ bars}}$ **17.** $\frac{9 \text{ children}}{4 \text{ families}}$ **19.** $3,225/month   **21.** 38.4 mi/gal   **23.** $.224/oz   **25.** 21.4 mph   **27.** 0.332

29. 7.63 lb   **31.** 2 gal   **33.** 16,896 ft   **35.** 10,890 ft²   **37.** 215.6 lb   **39.** 1.4 in.   **41.** 96.8 ft/s
43. 104.7 km/h   **45.** $.39/L   **47.** 49.7 mph   **49.** 25 gal   **51.** $5,520   **53.** $3   **55a.** 25,465.8 mi
b. 149,730,000 km   **57.** 23.4 kg   **59.** 182.3 mph   **61.** 0.46

7.2 EXERCISES (*pages 417–420*)

1. Not true   **3.** True   **5.** True   **7.** True   **9.** True   **11.** True   **13.** 10   **15.** 2.4   **17.** 4.5   **19.** 17.14
21. 25.6   **23.** 20.83   **25.** 4.35   **27.** 10.97   **29.** 1.15   **31.** 38.73   **33.** 0.5   **35.** 10   **37.** 0.43
39. −1.6   **41.** 6.25   **43.** 32   **45.** 6.2   **47.** 5.8   **49.** 0.0052 in.   **51.** 24 robes   **53.** $6,720   **55.** 406 mi
57. $1.82   **59.** 438 lights   **61.** 60 weeks   **63.** 2,750 defects   **65.** 198,000 mi   **67.** $1,500   **69.** 9 in.
71. $63,000   **73.** Yes   **75.** No

7.3 EXERCISES (*pages 425–428*)

1. $\frac{15}{2}$ **3.** 16   **5.** 24   **7.** 10   **9.** 0.625   **11.** 12.56   **13.** $307.50   **15.** 5.4 lb/in²   **17.** 287.3 ft

19. 3 amps   **21.** 50   **23.** 100   **25.** 200   **27.** 6.4   **29.** 10 ft   **31.** 1.67 ohms   **33.** 2,160 computers

35. 51.2 lumens   **37.** 55.6 decibels   **39a.** True   **b.** False   **c.** False   **41a.** 8 times   **b.** $\frac{1}{8}$

CHAPTER REVIEW EXERCISES (*pages 433–434*)

1. $\frac{1}{1}$ (Objective 7.1A)   **2.** $\frac{2 \text{ roof supports}}{3 \text{ ft}}$ (Objective 7.1A)   **3.** 40.25 km (Objective 7.1B)   **4.** 1.6 (Objective 7.2A)

5. $\frac{8}{15}$ (Objective 7.1A)   **6.** $6.85/h (Objective 7.1A)   **7.** 12.8 (Objective 7.2A)   **8.** 1.1 acres (Objective 7.1B)

9. 57 mph (Objective 7.1A)   **10.** 35.2 (Objective 7.2A)   **11.** 2.0 mi (Objective 7.1B)   **12.** 6.86 (Objective 7.2A)

13. $\frac{1}{8}$ (Objective 7.1A)   **14.** $\frac{5 \text{ lb}}{4 \text{ trees}}$ (Objective 7.1A)   **15.** 3.2 (Objective 7.2A)   **16.** 7.4 yd³ (Objective 7.1B)   **17.** $\frac{4}{1}$

(Objective 7.1A)   **18.** 374.4 (Objective 7.2A)   **19.** $\frac{1}{3}$ (Objective 7.3A)   **20.** 20 (Objective 7.3B)   **21.** 75

(Objective 7.3A)   **22.** 28,800 (Objective 7.3A)   **23.** $\frac{1}{25}$ (Objective 7.3B)   **24.** 80 (Objective 7.3B)   **25.** $\frac{2}{5}$

(Objective 7.1A)   **26.** $12,000 (Objective 7.2B)   **27.** $928 (Objective 7.2B)   **28.** 2.75 lb (Objective 7.2B)   **29.** 243,750
voters (Objective 7.2B)   **30.** 11.2 in. (Objective 7.3A)   **31.** 127.6 ft/s (Objective 7.1C)   **32.** 1.25 ft³ (Objective 7.3B)
33. 61.2 ft (Objective 7.3A)

CUMULATIVE REVIEW EXERCISES (*pages 435–436*)

1. 57 (Objective 3.6C) **2.** 4.8 qt (Objective 7.1B) **3.** $3\frac{31}{36}$ (Objective 3.3B) **4.** $1\frac{2}{3}$ (Objective 3.6C)

5. −114 (Objective 2.3B) **6.** 22 (Objective 2.5A) **7.** 4 (Objective 6.2A) **8.** 3 (Objective 6.3B)

9. (Objective 4.5A) **10.** (Objective 4.5B) **11.** 21 (Objective 2.5A)

12. $-\frac{3}{8}$ (Objective 3.6A) **13.** 13 (Objective 4.4A) **14.** $8a - 3$ (Objective 5.2B) **15.** $-20a^5b^4$ (Objective 5.4A)

16. $-4y^2 - 3y$ (Objective 5.2A) **17.** $(-1, -5)$ (Objective 6.5C) **18.** $\frac{3}{10}$ (Objective 7.1A) **19.** \$1,885/month

(Objective 7.1A) **20.** 66 ft/s (Objective 7.1B) **21.** 32 (Objective 7.2A) **22.** $\frac{10}{11}$ (Objective 3.6B) **23.** −10

(Objective 4.4A) **24.** −2 (Objective 6.3B) **25.** \$3,916.67 (Objective 4.2C) **26.** 12 (Objective 6.4A)

27. $4x - 3(x + 2); x - 6$ (Objective 5.7B) **28.** 64 mi (Objective 1.2C) **29.** \$265.48 (Objective 4.2E) **30.** $\frac{4}{15}$

(Objective 3.3C) **31.** 20,854 votes (Objective 7.2B) **32.** 35 mi (Objective 7.1C) **33.** 3,750 rpm (Objective 6.4B)

ANSWERS TO CHAPTER 8 Exercises

8.1 EXERCISES (*pages 441–442*)

1. $\frac{1}{20}$, 0.05 **3.** $\frac{3}{10}$, 0.30 **5.** $\frac{5}{2}$, 2.50 **7.** $\frac{7}{25}$, 0.28 **9.** $\frac{7}{20}$, 0.35 **11.** $\frac{3}{50}$, 0.06 **13.** $\frac{61}{50}$, 1.22 **15.** $\frac{29}{100}$, 0.29

17. $\frac{1}{9}$ **19.** $\frac{3}{8}$ **21.** $\frac{1}{32}$ **23.** $\frac{5}{11}$ **25.** $\frac{3}{800}$ **27.** $\frac{1}{200}$ **29.** $\frac{5}{6}$ **31.** $\frac{7}{8}$ **33.** 0.073 **35.** 0.158

37. 0.003 **39.** 1.212 **41.** 0.6214 **43.** 0.0825 **45.** $\frac{6}{25}$ **47.** 37% **49.** 2% **51.** 12.5% **53.** 136%

55. 96% **57.** 7% **59.** 83% **61.** 33.3% **63.** 45.5% **65.** 87.5% **67.** 166.7% **69.** 128.6% **71.** 40%

73. 16.7% **75.** 34% **77.** $37\frac{1}{2}$% **79.** $35\frac{5}{7}$% **81.** $57\frac{1}{7}$% **83.** $262\frac{1}{2}$% **85.** $283\frac{1}{3}$% **87.** $23\frac{1}{3}$%

89. $22\frac{2}{9}$%

8.2 EXERCISES (*pages 449–452*)

1. 8 **3.** 0.075 **5.** $16\frac{2}{3}$% **7.** 37.5% **9.** 100 **11.** 1,200 **13.** 51.895 **15.** 13 **17.** 2.7%

19. 400% **21.** 7.5 **23.** 65 **25.** 25% **27.** 75 **29.** 12.5% **31.** 400 **33.** 19.5 **35.** 14.8%
37. 62.62 **39.** 5 **41.** 45 **43.** 300% **45.** 50,000 mi **47.** \$403.20 **49.** 12% **51.** 5.0%
53. 584 million pounds **55.** \$12,750 **57.** 0.015 g **59.** 10,875% **61.** better **63.** 7,944 computer boards
65. No **67.** less than

8.3 EXERCISES (*pages 455–456*)

1. 8.7% **3.** 200% **5.** 2,400% **7.** \$100 **9.** 11.1% **11.** 6.6% **13.** 4.1% **15.** \$7,331.25 **17.** \$71.76
19. No

8.4 EXERCISES (*pages 463–466*)

1. 60% **3.** 28% **5.** \$60.50 **7.** \$2,187.50 **9.** \$19.50 **11.** \$25.60 **13.** 44.4% **15.** 64.1% **17.** \$110
19. 23.2% **21.** \$1,396.50 **23.** \$25.20 **25.** \$300 **27.** \$123.08 **29.** 42% **31.** 12.9% **33.** \$277.50
35. \$306 **37.** \$1,440 **39.** 7.7% **41.** 9% **43.** \$8,200 **45.** \$1,997.78

CHAPTER REVIEW EXERCISES (*pages 471–472*)

1. $\frac{8}{25}$ (Objective 8.1A)   2. 0.22 (Objective 8.1A)   3. $\frac{1}{4}$, 0.25 (Objective 8.1A)   4. $\frac{17}{500}$ (Objective 8.1A)   5. 17.5% (Objective 8.1B)   6. 128.6% (Objective 8.1B)   7. 280% (Objective 8.1B)   8. 21 (Objective 8.2A/8.2B)   9. 500% (Objective 8.2A/8.2B)   10. 66.7% (Objective 8.2A/8.2B)   11. 30 (Objective 8.2A/8.2B)   12. 15.3 (Objective 8.2A/8.2B)   13. 562.5 (Objective 8.2A/8.2B)   14. 5.625% (Objective 8.2A/8.2B)   15. 0.014732 (Objective 8.2A/8.2B)   16. 9.6 (Objective 8.2A/8.2B)   17. 37.7% (Objective 8.2C)   18. $8,400 (Objective 8.2C)   19. 3,952 telephones (Objective 8.2C)   20. 6.1% (Objective 8.2C)   21. $165,000 (Objective 8.2C)   22. 1,620 seats (Objective 8.2C)   23. 196 tickets (Objective 8.2C)   24. 22.7% (Objective 8.2C)   25. $11.34 (Objective 8.3A)   26. 25% (Objective 8.3B)   27. $15,370 (Objective 8.4A)   28. $209.30 (Objective 8.4A)   29. $56 (Objective 8.4B)   30. $390 (Objective 8.4B)   31. $32.25 (Objective 8.4C)   32. $234 (Objective 8.4C)   33. $2,822.22 (Objective 8.4C)

CUMULATIVE REVIEW EXERCISES (*pages 473–474*)

1. 24.954 (Objective 4.2A)   2. 625 (Objective 1.3B)   3. 14.04269 (Objective 4.2B)   4. $4x^2 - 16x + 15$ (Objective 5.5B)   5. $1\frac{25}{62}$ (Objective 3.4B)   6. $6a^3b^3 - 8a^4b^4 + 2a^3b^4$ (Objective 5.5A)   7. 42 (Objective 8.2A/8.2B)   8. -3 (Objective 6.1A)   9. 100,500 (Objective 4.2B)   10. $\frac{23}{24}$ (Objective 3.3B)   11. $1\frac{23}{28}$ (Objective 3.6B)   12. $-12a^7b^5$ (Objective 5.4A)   13. (Objective 6.5D)   14. (Objective 6.5D)

15. $2\frac{4}{5}$ (Objective 3.4B)   16. 4 (Objective 2.2B)   17. -12 (Objective 6.1B)   18. 5 (Objective 6.3A)   19. 64.48 mph (Objective 7.1A)   20. 44.8 (Objective 7.2A)   21. 8.3% (Objective 8.2A/8.2B)   22. 67.2 (Objective 8.2A/8.2B)   23. 58 (Objective 2.5A)   24. 5 (Objective 6.3B)   25. $\frac{1}{10}$ (Objective 8.1A)   26. $129.60 (Objective 8.4B)   27. $42 (Objective 8.4A)   28. 117 games (Objective 7.2B)   29. $2\frac{1}{4}$ lb (Objective 3.3C)   30. 72 ft/s (Objective 4.4C)   31. 33.7 km/h (Objective 7.1C)   32. 36 h (Objective 6.4B)   33. 5 ohms (Objective 7.3B)

ANSWERS TO CHAPTER 9   Exercises

9.1 EXERCISES (*pages 487–492*)

1. 40°; acute   3. 115°; obtuse   5. 90°; right   7. 28°   9. 18°   11. 14 cm   13. 28 ft   15. 30 m   17. 86°   19. 71°   21. 30°   23. 36°   25. 127°   27. 116°   29. 20°   31. 20°   33. 20°   35. 141°   37. 106°   39. 11°   41. $\angle a = 38°, \angle b = 142°$ 43. $\angle a = 47°, \angle b = 133°$ 45. 20°   47. 47°   49. $\angle x = 155°, \angle y = 70°$ 51. $\angle a = 45°, \angle b = 135°$ 53. $90° - x$ 55. 60°   57. 35°   59. 102°   61. The three angles form a straight angle. The sum of the measures of the three angles of a triangle is 180°.   63a. always true   b. always true   c. sometimes true

9.2 EXERCISES (*pages 503–510*)

1. hexagon   3. pentagon   5. scalene   7. equilateral   9. obtuse   11. acute   13. 56 in.   15. 14 ft   17. 47 mi   19. 8π cm or approximately 25.13 cm   21. 11π mi or approximately 34.56 mi   23. 17π ft or approximately 53.41 ft   25. 17.4 cm   27. 8 cm   29. 24 m   31. 48.8 cm   33. 17.5 in.   35. 1.5π in.   37. 226.19 cm   39. 60 ft   41. 44 ft   43. 120 ft   45. 10 in.   47. 12 in.   49. 2.55 cm   51. 13.19 ft

53. 50.27 ft **55.** 39,935.93 km **57.** 60 ft² **59.** 20.25 in² **61.** 546 ft² **63.** 16π cm² or approximately 50.27 cm²
65. 30.25π mi² or approximately 95.03 mi² **67.** 72.25π ft² or approximately 226.98 ft² **69.** 156.25 cm² **71.** 570 in²
73. 192 in² **75.** 13.5 ft² **77.** 330 cm² **79.** 25π in² **81.** $10,000\pi$ in² **83.** 126 ft² **85.** 7,500 yd²
87. 10 in. **89.** 20 m **91.** 2 qt **93.** $74 **95.** 113.10 in² **97.** $638 **99.** 216 m² **101.** 4:9
105a. sometimes true **b.** sometimes true **c.** always true **d.** always true **e.** always true **f.** always true

9.3 EXERCISES (*pages 517–520*)

1. 5 in. **3.** 8.6 cm **5.** 11.2 ft **7.** 4.5 cm **9.** 12.7 yd **11.** 8.5 cm **13.** 24.3 cm **15.** $\frac{1}{2}$ **17.** $\frac{3}{4}$
19. 7.2 cm **21.** 3.3 m **23.** 12 m **25.** 12 in. **27.** 56.3 cm² **29.** 18 ft **31.** 16 m **33.** The triangles are
congruent by the SAS Rule. **35.** The triangles are congruent by the SSS Rule. **37.** The triangles are congruent by the
ASA Rule. **39.** Yes, the triangles are congruent by the SAS Rule. **41.** No **43.** No **45a.** always true
b. sometimes true **c.** always true

9.4 EXERCISES (*pages 527–530*)

1. 840 in³ **3.** 15 ft³ **5.** 4.5π cm³ or approximately 14.14 cm³ **7.** 34 m³ **9.** 15.625 in³ **11.** 36π ft³
13. 8,143.01 cm³ **15.** 75π in³ **17.** 120 in³ **19.** 2.5 ft **21.** 4.00 in. **23.** length: 5 in.; width: 5 in.
25. 75.40 m³ **27.** 94 m² **29.** 56 m² **31.** 96π in² or approximately 301.59 in² **33.** 184 ft² **35.** 69.36 m²
37. 225π cm² **39.** 402.12 in² **41.** 6π ft² **43.** 297 in² **45.** 3 cm **47.** 11 cans of paint **49.** 456 in²
51. 22.53 cm² **53a.** always true **b.** never true **c.** sometimes true

9.5 EXERCISES (*pages 535–540*)

1. 121 cm **3.** $(4 + 2\pi)$ ft or approximately 10.28 ft **5.** $(38 + 4\pi)$ m or approximately 50.57 m **7.** 176 ft
9. $(4 + \pi)$ ft or approximately 7.14 ft **11.** 24 m **13.** 20.71 ft **15.** $9,360 **17.** 26 cm² **19.** $(36 + 4.5\pi)$ in² or
approximately 50.14 in² **21.** 48π in² or approximately 150.80 in² **23.** 30 in² **25.** 16 in² **27.** $(40 + 12.5\pi)$ in² or
approximately 79.27 in² **29.** $640.10 **31.** 19,026.55 ft² **33.** $(1.92 - 0.08\pi)$ m³ or approximately 1.67 m³
35. 126π ft³ or approximately 395.84 ft³ **37.** 272 ft³ **39.** 8.25π in³ or approximately 25.92 in³ **41.** 36π in³ or
approximately 113.10 in³ **43.** $1,458\pi$ cm³ or approximately 4,580.44 cm³ **45.** 212.60 in³ **47.** $3,515.00 **49.** 19 m²
51. 93π cm² or approximately 292.17 cm² **53.** $(120 + 160\pi)$ m² or approximately 622.65 m² **55.** 56π cm² or
approximately 175.93 cm² **57.** 324 ft² **59.** $(126 + 15\pi)$ in² or approximately 173.12 in² **61.** 158 cans of paint
63. $498.75 **65.** $12\frac{3}{8}$ in. $\times\ 8\frac{5}{16}$ in. $\times\ 8\frac{1}{4}$ in.

CHAPTER REVIEW EXERCISES (*pages 545–546*)

1. $\angle x = 22°, \angle y = 158°$ (Objective 9.1C) **2.** 24 in. (Objective 9.3B) **3.** 240 in³ (Objective 9.5C) **4.** 68°
(Objective 9.1B) **5.** Yes, the triangles are congruent by the SAS Rule. (Objective 9.3C) **6.** 138.23 m² (Objective 9.5D)
7. 44 cm (Objective 9.1A) **8.** 19° (Objective 9.1A) **9.** 57.13 in² (Objective 9.5B) **10.** 96 cm³ (Objective 9.4A)
11. 47.71 in. (Objective 9.5A) **12.** $\angle a = 138°, \angle b = 42°$ (Objective 9.1B) **13.** 220 ft² (Objective 9.4B) **14.** 9.75 ft
(Objective 9.3A) **15.** 148° (Objective 9.1A) **16.** 39 ft³ (Objective 9.4A) **17.** 95° (Objective 9.1C) **18.** 8 cm
(Objective 9.2B) **19.** 288π mm³ (Objective 9.4A) **20.** 21.5 cm (Objective 9.2A) **21.** 4 cans of paint (Objective 9.4B)
22. 208 yd (Objective 9.2A) **23.** 90.25 m² (Objective 9.2B) **24.** 276 m² (Objective 9.2B)

CUMULATIVE REVIEW EXERCISES (*pages 547–548*)

1. 204 (Objective 8.2A/8.2B) **2.** 1, 2, 3, 6, 13, 26, 39, 78 (Objective 1.3D) **3.** $\frac{5}{6}$ (Objective 3.4B) **4.** $7x^2 + 4x + 5$
(Objective 5.3A) **5.** 12.8 (Objective 4.2C) **6.** 2.9×10^{-5} (Objective 5.6B) **7.** 131° (Objective 9.1B) **8.** 26 cm
(Objective 9.3A) **9.** 50 in² (Objective 9.5B) **10.** 1,407.43 cm³ (Objective 9.5C) **11.** $-12x^5y^3$ (Objective 5.4A)
12. $\frac{1}{2}$ (Objective 6.3B) **13.** 11.14 cm (Objective 9.5A) **14.** ←——⊕——|——→ (Objective 4.5B) **15.** $7x + 18$
    $\qquad\qquad\qquad -3\quad 0$

(Objective 5.2B) **16.** −2 (Objective 2.5A) **17.** 5 (Objective 6.3A) **18.** 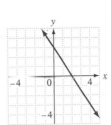 (Objective 6.5D)

19. $\frac{4}{5}$ (Objective 3.6C) **20.** 23 gal (Objective 7.1C) **21.** 87 min (Objective 6.4B) **22.** $4.50 (Objective 7.2B)

23. 27.3% (Objective 8.3B) **24.** 3 ft (Objective 9.4A) **25.** 40 ft (Objective 6.2B) **26.** 4,000 mi (Objective 4.4C)

ANSWERS TO CHAPTER 10   Exercises

10.1 EXERCISES (*pages 555–558*)

1. 64 **3.** 76–84 **5.** 5 universities **7.** 10% **9.** 52.5% **11.**

Classes	Tally	Frequency
37–47	/////	5
48–58	//////////	10
59–69	///////	7
70–80	///////////	11
81–91	//////////	10
92–102	//////	6
103–113	/	1

13. 5 hotels **15.** 33 hotels **17.** 22% **19.** 30% **21.** 13 account balances **23.** 62% **25.** $\frac{1}{2}$ **27.** 60%

29. 12.5% **31.** 55% **33.** 22 nurses **35.** 66% **37.** $\frac{6}{1}$ **39.** 53.3% **41.** Answers will vary.

10.2 EXERCISES (*pages 565–568*)

1. mean: 19; median: 19.5 **3.** mean: 10.61; median: 10.605 **5.** median **7.** mean: 403.625; median: 404.50
9. 21 unforced errors **11.** German chocolate cake **13.** satisfactory
15. $Q_1 = 6.895$, $Q_3 = 8.95$

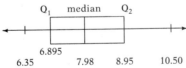

6.35   6.895   7.98   8.95   10.50

17. $Q_1 = 20$, $Q_3 = 30$

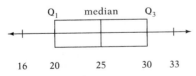

16   20   25   30   33

19. $Q_1 = 4.3$, $Q_3 = 6.1$

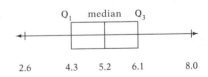

2.6   4.3   5.2   6.1   8.0

21. 2.828 **23.** 61.05 **25.** 0.045 **27.** The standard deviations are the same.

10.3 EXERCISES (*pages 575–578*)

1. HHHH, HHHT, HHTH, HTHH, THHH, HHTT, HTTH, TTHH, HTHT, THTH, THHT, HTTT, TTTH, TTHT, THTT, TTTT
3. (1, 1), (1, 2), (1, 3), (1, 4), (2, 1), (2, 2), (2, 3), (2, 4), (3, 1), (3, 2), (3, 3), (3, 4), (4, 1), (4, 2), (4, 3), (4, 4) **5.** No. Because the dice are weighted so that some numbers occur more often than other numbers. **7.** $\frac{1}{16}$ **9.** $\frac{3}{8}$ **11.** $\frac{1}{9}$ **13.** 0

15. $\frac{1}{36}$ **17.** $\frac{1}{12}$ **19.** $\frac{3}{16}$ **21.** $\frac{1}{4}$ **23.** $\frac{37}{95}$ **25.** $\frac{4}{7}$ **27.** $\frac{185}{377}$ **29.** 1 TO 1 **31.** $\frac{1}{41}$ **33.** $\frac{1}{5}$ **35.** $\frac{12}{1}$

37. $\frac{3}{5}$ **39.** $\frac{1}{3}$ **41.** The sum of the probabilities is not 1.

CHAPTER REVIEW EXERCISES (*pages 583–584*)

1.

Classes	Tally	Frequency (Objective 10.1A)
12–19	/////	5
20–27	/////////	9
28–35	///////////	11
36–43	/////////	9
44–51	////	4
52–59	//	2

2. 28–35 (Objective 10.1A)   **3.** 25 classes (Objective 10.1A)

4. 15% (Objective 10.1A)   **5.** 35% (Objective 10.1A)   **6.** 25 days (Objective 10.1B)   **7.** 29 days (Objective 10.1B)
8. mean: $214.\overline{54}$; median: 210 (Objective 10.2A)   **9.** mean: 7.17; median: 7.05 (Objective 10.2A)   **10.** good
(Objective 10.2A)   **11.** 55 million shares (Objective 10.1C)   **12.** 8 A.M.–9 A.M. (Objective 10.1C)   **13.** $\dfrac{17}{25}$
(Objective 10.1C)   **14.** $Q_1 = 105$, median = 111, $Q_3 = 124$ (Objective 10.2B)

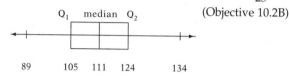

15. 1.97 (Objective 10.2C)   **16.** $\dfrac{1}{500}$ (Objective 10.3A)   **17.** $\dfrac{3}{7}$ (Objective 10.3B)   **18.** $\dfrac{2}{7}$ (Objective 10.3B)   **19.** $\dfrac{1}{6}$
(Objective 10.3A)   **20.** $\dfrac{5}{14}$ (Objective 10.3A)

CUMULATIVE REVIEW EXERCISES (*pages 585–586*)

1. $10\sqrt{2}$ (Objective 4.4B)   **2.** -1 (Objective 6.3B)   **3.** 22 (Objective 2.5A)   **4.** $12x - 8$ (Objective 5.2B)   **5.** -18
(Objective 6.2A)   **6.** $\dfrac{41}{90}$ (Objective 3.6C)   **7.**

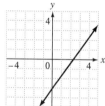

(Objective 6.5D)

8. 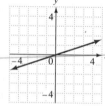 (Objective 6.5D)   **9.** $3y^2 + 8y - 9$ (Objective 5.3B)   **10.** $64a^6b^3$ (Objective 5.4B)

11. 144 (Objective 8.2A)   **12.** $\dfrac{8}{3}$ (Objective 7.2A)   **13.** 8.76×10^{10} (Objective 5.6B)   **14.** 3,100 ft² (Objective 9.5B)
15. $-15c^3d^{10}$ (Objective 5.4A)   **16.** $11.\overline{1}$ m/s (Objective 7.1B)   **17.** 125° (Objective 9.1C)   **18.** 32 m² (Objective 9.2B)
19. $468.75 (Objective 8.4C)   **20.** $\dfrac{2}{3}$ (Objective 10.3A)   **21.** mean: 31; median: 32.5 (Objective 10.2A)   **22.** 76.1%
(Objective 8.2C)   **23.** 76,800 km (Objective 7.2B)   **24.** 2.83 (Objective 10.2C)   **25.** $15.00 (Objective 8.3A)

ANSWERS TO FINAL EXAMINATION (*pages 587–590*)

1. 2,400 (Objective 1.2A)   **2.** 24 (Objective 1.5A)   **3.** 2 (Objective 2.2B)   **4.** 18 (Objective 2.5A)   **5.** $2\dfrac{11}{16}$
(Objective 3.3B)   **6.** $\dfrac{14}{15}$ (Objective 3.4B)   **7.** $\dfrac{2}{9}$ (Objective 3.6B)   **8.** $\dfrac{5}{16} < 0.313$ (Objective 4.2D)   **9.** -769.5
(Objective 4.2B)   **10.** -3.28 (Objective 4.2C)   **11.** Yes (Objective 4.2B)   **12.** $9\sqrt{2}$ (Objective 4.4B)

13. 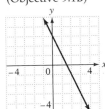 (Objective 4.5B)   **14.** The Commutative Property of Addition (Objective 5.1A)

15. $10t$ (Objective 5.1A)   **16.** $-2x - 14y$ (Objective 5.2B)   **17.** $z^3 + 2z^2 - 6z + 7$ (Objective 5.3B)   **18.** $8x^7y$

(Objective 5.4A)   **19.** $10a^4b^2 - 6a^3b^3 + 8a^2b^4$ (Objective 5.5A)   **20.** $15x^2 - x - 6$ (Objective 5.5B)   **21.** $81x^8y^4$

(Objective 5.4B)   **22.** $\frac{1}{64}$ (Objective 5.6A)   **23.** m^2n^4 (Objective 5.6A)   **24.** -6 (Objective 6.2A)   **25.** $-\frac{1}{3}$

(Objective 6.3A)   **26.** $\frac{5}{6}$ (Objective 6.3B)   **27.** 13,728 ft (Objective 7.1B)   **28.** $\frac{4}{3}$ (Objective 7.2A)

29. $\angle a = 74°$; $\angle b = 106°$ (Objective 9.1B)   **30.** 10.6 ft (Objective 9.3A)   **31.** 29.42 cm (Objective 9.5A)   **32.** 92.86 in³

(Objective 9.5C)   **33.**   (Objective 6.5D)   **34.**   (Objective 6.5D)

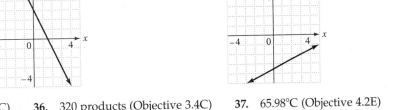

35. 364 mph (Objective 1.2C)   **36.** 320 products (Objective 3.4C)   **37.** 65.98°C (Objective 4.2E)   **38.** 5.88×10^{12}
(Objective 5.6B)   **39.** 6 ft (Objective 6.3C)   **40.** 6 tickets (Objective 6.4B)   **41.** $4,710 (Objective 7.2B)   **42.** 32%
(Objective 8.2C)   **43.** 16 revolutions/min (Objective 7.3B)   **44.** $13,187.50 (Objective 8.2C)   **45.** 22.6% (Objective 8.3B)
46. $159.25 (Objective 8.4B)   **47.** $1,612.50 (Objective 8.4C)   **48.** 37.5% (Objective 10.1C)   **49.** mean: $334.80;

median: $309 (Objective 10.2A)   **50.** $\frac{1}{3}$ (Objective 10.3A)

ANSWERS TO APPENDIX: THE METRIC SYSTEM OF MEASUREMENT

EXERCISES (*pages A3–A4*)

1. 420 mm   **3.** 0.36 kg   **5.** 5.194 L   **7.** 2,000 mm   **9.** 0.217 g   **11.** 4,520 ml   **13.** 8.406 km
15. 2,400 g   **17.** 6,180 L   **19.** 9,612 m   **21.** 240 mg   **23.** 2.98 m   **25.** 2.431 kl   **27.** 66 cm   **29.** 24.3 cm
31. 50 km   **33.** 628,000 m   **35.** 8 cm   **37.** 2 g   **39.** 7 m   **41.** 8 servings

Index